Lecture Notes in Chemistry

P. Pyykkö

Relativistic Theory of Atoms and Molecules II

A Bibliography 1986-1992

Springer-Verlag Berlin Heidelberg GmbH

Author

Pekka Pyykkö
University of Helsinki
Department of Chemistry
Et. Hesperiankatu 4
00014 Helsinki, Finland

ISBN 978-3-540-57219-0 ISBN 978-3-642-51488-3 (eBook)
DOI 10.1007/978-3-642-51488-3

Library of Congress Cataloging-in-Publication Data
Pyykkö, Pekka, 1941- . Relativistic theory of atoms and molecules II : a bibliography, 1986-1992 / P. Pyykkö. p. cm. – (Lectures notes in chemistry; 60)
ISBN 978-3-540-57219-0
1. Atomic theory–Bibliography. 2. Molecular theory–Bibliography. I. Title. II. Series. Z5524.A6P99 1993 [QD461] 016.5412–dc20 93-31082 CIP

Originally published by Springer-Verlag Berlin Heidelberg New York in 1993

Typesetting: Camera ready by author
52/3140 - 5 4 3 2 1 0 - Printed on acid-free paper

PREFACE

Seven years have elapsed since "RTAM I" was published. It contained 3119 references from the years 1916 to 1985. The present volume covers the years 1986 - 1992 and also contains a number of references previously missed or now purposely inserted. The pre-1986 references are marked with a "*". The running numbers in the present Bibliography begin from 3120 and end at 6577.

A source book like this, is inevitably very superficial and could be characterized by the Chinese expression "to watch flowers riding a horse" [走马观花]. It's virtue, if any, lies in being reasonably complete.

A further problem may be in recovering the information on a particular subject. To find the calculations on a particular molecule, the Quantum Chemistry Literature Data Base (QCLDB) is recommended as a companion. To facilitate finding the references by a given author or with a given title, the combined reference files of RTAM I-II will be made available on a diskette through Springer-Verlag.

We apologise for any inadvertent omissions or characterisations that the author would not find fitting. No legal responsibility is accepted in such cases.

Thanks are due to colleagues who have contributed material. Mr. Chen Jing gave most helpful practical assistance, the text part was typed by Monika Holmberg, Susanne Lundberg and Johan Sandberg. The project received support from Finska Vetenskaps-Societeten.

I also thank the Chemistry Editor of Springer-Verlag, Dr. Rainer Stumpe, for constructive cooperation during the past years.

Helsinki, 4th June, 1993

Pekka Pyykkö

CONTENTS

1. Introduction 1

Table 1.1. Monographs and other general references 2

2. One-particle problems 4

Table 2.1. The Dirac equation: interpretative studies, symmetry properties and non-relativistic limits 4
Table 2.2. The Dirac equation: further transformations 11
Table 2.3. The Dirac equation: solutions for hydrogen-like systems 15
Table 2.4. The Dirac equations: solutions for various non-coulomb fields 22
Table 2.5. Relativistic virial theorems 34

3. Quantum electrodynamical effects 35

Table 3.1. Higher-order corrections: methods 35
Table 3.2. Higher-order corrections: hyperfine interactions 41
Table 3.3. Higher-order corrections: energy levels 42
Table 3.4. Higher-order corrections: interatomic and -molecular interactions 45

4. Multielectron atoms: methods 46

Table 4.1. General methods and basic theory for multielectron atoms 46
Table 4.2. Published programs for atoms 51
Table 4.3. Numerical, non-statistical four-component methods 52
Table 4.4. Four-component LCAO approaches for many-electron atoms 53
Table 4.5. Various four-component local-density methods 56
Table 4.6. Thomas-Fermi calculations 57
Table 4.7. Independent-particle models 58
Table 4.8. Definitions, reviews and background for effective potential calculations 59
Table 4.9. Effective-potential methods 60
Table 4.10. Available relativistic effective potentials 62
Table 4.11. One-component and perturbation calculations 63
Table 4.12. (1/Z)- and other similar expansions for many-electron atoms 63
Table 4.13. Relativistic nuclear calculations 64

5. Multielectron atoms: results 66

Table 5.1. Tabulations of atomic ground-state properties 66
Table 5.2. Data on atomic energy levels 67
Table 5.3. Auger and autoionization processes 79
Table 5.4. Ionization potentials and electron affinities 82
Table 5.5. Supercritical (Z > 137) systems 84
Table 5.6. Electromagnetic transition probabilities 86
Table 5.7. Polarisabilities and screening constants 92
Table 5.8. Electric and magnetic hyperfine properties 93
Table 5.9. Average radii $<r^n>$ and magnetic g-factors 96
Table 5.10. Compton profiles, momentum distributions and spin densities 97
Table 5.11. X-ray scattering factors 98
Table 5.12. Electron and positron scattering 99
Table 5.13. Particle-atom collisions 105
Table 5.14. Photon scattering and photoionization 108
Table 5.15. Atom-atom collisions and interatomic potentials 112
Table 5.16. Nuclear processes involving electronic wave functions 114
Table 5.17. Parity-violation effects in atoms and molecules 115

6. Symmetry 117

Table 6.1. Group theory and symmetry aspects 117

7. Molecular calculations 118

Table 7.1. One-electron systems 118
Table 7.2. LCAO-DF calculations on molecules 120
Table 7.3. Molecular all-electron calculations using "no-pair" or other transformed Dirac Hamiltonians 122
Table 7.4. Molecules treated by the DS-DVM method 124
Table 7.5. Molecules treated by the DS-MS Xα method 126
Table 7.6. Molecules treated by the quasirelativistic DS-MS Xα approach 127
Table 7.7. Molecules treated by pseudopotential methods 131
Table 7.8. Molecules treated by the Perturbative Hartree-Fock-Slater (P-HFS) method 154
Table 7.9. First-order perturbation theory on molecules 156
Table 7.10. Density functional calculations 160
Table 7.11. Semiempirical methods 161
Table 7.12. Relativistic crystal field theory 166
Table 7.13. Relativistic theories of molecular properties 168

8. Solid-state theory 170

Table 8.1. Band-structure calculations 170
Table 8.2. Relativistic theories of solid- and liquid-state phenomena 180

9. Relativistic effects and heavy-element chemistry 186

Table 9.1. "Relativity and the periodic system". Periodic trends, reviews and pedagogical papers 187
Table 9.2. Bond lengths 191
Table 9.3. Magnetic resonance parameters 193
Table 9.4. Electric dipole moments and molecular charge distributions 195
Table 9.5. Molecular energy levels and energy transfer 196
Table 9.6. Molecular ionization potentials and electron affinities 196

Corrections to Volume I 197

Some comments on notations and terminology 198

List of acronyms and symbols 199

Bibliography 203

1. INTRODUCTION

The area of the relativistic theory of atoms and molecules, with its contacts "upstream" to relativistic quantum mechanics and QED, and "downstream" to atomic and molecular physics and all branches of heavy-element chemistry, has now become so vast that a detailed exposition of the entire domain by the same author is hardly thinkable. The purpose of the present compilation is to make available the comprehensive bibliography, assembled by the author over the years, on the art of solving the Dirac equation, or approximations thereof, for atoms and molecules.

Most of the material is given in tabular form. After the general references in ch. 1., the ch. 2 covers the relativistic single-particle wave functions. Ch. 3 gives some references on QED aspects. The methods for multielectron atoms are discussed in ch. 4 and the available results in ch. 5. Symmetry aspects are covered in ch. 6. Molecular calculations are reviewed in ch. 7, with a slightly more detailed analysis than in other areas. Some solid-state problems and band-structure calculations are mentioned in ch. 8. Chapter 9 summarises the subject of "relativity and the periodic system", and the various molecular properties.

The fringe areas where no complete coverage was attempted, include elementary particle models (included in Table 2.4.), Breit-Pauli perturbation calculations on atoms and molecules (Tables 4.11. and 7.9., respectively), the large area of relativistic nuclear calculations (Table 4.13.), band-structure calculations (Tables 8.1-2.), or otherwise excellent review articles with a very slight 'relativistic' content. In each of these cases it was thought useful to give an idea of current acitivity, however.

The 3458 new references are given in the Bibliography. No "distinctions" are awarded in it. Thus, in the present, tabular form of organizing the material, central contributions and small curiosities receive the same amount of attention. We apologize for any inadvertent omissions, or summaries that the author would not find fitting.

Table 1.1. gives a list of treatises and other general references.

Table 1.1. Monographs and other general references.

Reference	Area
Kramers (1957)	Foundations of quantum theory. Translation of Kramers (1938). Chapter on relativistic theories.
Green (1965)	Book on QM, including Dirac theory.
Hestenes (1966)	Space-time algebra.
Aitchison (1972)	Relativistic QM.
Flügge (1974)	Properties of the Dirac equation.
Pilkuhn (1979)	Relativistic particle physics.
Altmann (1986)	Quaternions and double groups.
Dyall (1986)	Review on relativistic atomic and molecular calculations.
Grant (1986b)	Review on relativistic atomic and molecular calculations.
Pyykkö (1986a)	Part I of the present bibliography.
Safronova (1986b)	Relativistic and correlation effects in atoms and ions. Symposium proceedings.
Sokolov and Ternov (1986)	Radiation from relativistic electrons.
Sucher (1986, 1987, 1989ab)	Reviews on QED in atomic physics.
Greiner (1987)	Physics of strong fields.
Karazija (1987)	Theory of x-ray and electronic spectra of free atoms. An introduction.
Kutzelnigg (1987)	The relativistic many-body problem in molecular theory: a review.
Lindgren (1987a)	Relativistic many-body problems: Proceedings of Trieste, 1986.
Schwarz (1987, 1990)	Reviews on relativistic molecular calculations.
Englert (1988)	Semiclassical theory of atoms.
Ermler et al. (1988a)	Review on SO coupling and other relativistic effects in atoms and molecules.
Grant (1988a)	High-Z atoms.
Grant (1988b)	Review on atoms.
Grant and Quiney (1988)	Foundations of relativistic atomic and molecular theory.
L. N. Ivanov et al. (1988)	Multicharged ions.
Johnson (1988)	Review on RRPA.
Khriplovich (1988, 1991)	Book on PNC in atoms.
Levitan and Sargsyan (1988, 1991)	Book on Sturm-Liouville and Dirac operators.
Malli (1988)	Review on relativistic quantum chemistry.
Pyykkö (1988a)	Relativistic effects in structural chemistry.
Quiney (1988)	Relativistic MBPT.
Teitelboim (1988)	QM of fundamental systems.
Wilson (1988)	Edited volume on atomic and molecular calculations. Reviews by Grant, Gropen, Pyykkö, Quiney and Wilson.

Table 1.1. (continued).

Johnson et al. (1989b)	Relativistic, QED and PNC effects in atoms (Santa Barbara 1988).
Kiselev and Lyaptsev (1989)	Diagrammatic PT, including QR corrections.
Malli (1989)	Review on 5d and 5f elements.
Mårtensson-Pendrill (1989ab)	Reviews on relativistic pair-correlation methods for atoms.
Bagrov and Gitman (1990)	Exact solutions of relativistic wave equations.
Greiner (1990)	Relativistic QM : wave equations.
Y.-K. Kim (1990)	Relativistic and QED effects in highly-charged ions.
Kinoshita (ed.)(1990)	Precision tests on QED.
Landau (1990)	Book on QM. §13-18: Relativistic QM.
Dreizler and Gross (1990)	Density functional theory. Book.
Migdalek (1990)	Book on model potentials for atoms.
Moiseiwitsch (1990)	Electron capture at relativistic energies.
Quiney (1990ab, 1991)	Relativistic finite-basis calculations.
Koga and Morokuma (1991)	Homogeneous catalysis by transition-metal complexes.
Schmutzer (1991)	Textbook on relativistic QM.
Wilson et al. (1991)	Relativistic effects in atoms, molecules and solids. Proceedings of the Abingdon 1990 conference.
Bransden and McDowell (1992)	Charge exchange in ion-atom collisions.
Connerade (1992)	Review on cooperative effects in atomic physics, especially "giant resonances".
Grant (1992)	Review on atomic theory.
Greiner and Reinhardt (1992)	Book on QED.
Jones (1992)	Lecture notes on R QM, QED and QCD.
Szász (1992)	Book on HF-level theory of atoms.
Thaller (1992)	Book on the Dirac equation.
Wilson (1992ab)	Relativistic MBPT for atoms and molecules.

2. ONE-PARTICLE PROBLEMS

This chapter gives the references to the one-particle Dirac equation. Occasional references to other relativistic equations are included.

For detailed solutions of the Dirac equation, or for further details on its mathematical properties, we recommend the recent monographs by Bagrov and Gitman (1990) and by Thaller (1992).

Table 2.1. The Dirac equation: interpretative studies, symmetry properties and non-relativistic limits. For literature on the Klein paradox, see Table 2.4.

Reference	Comments
Kudar (1926)	A further original discovery of the "Klein-Gordon" equation, besides Fock, Gordon, Klein and Schrödinger.
Breit (1931c)	Interpretation of Dirac's **α**-matrices.
Hönl and Papapetrou (1940)	Interpretation of the Dirac equation.
Tani (1949, 1951)	Physical interpretation of the Dirac equation.
Foldy (1956)	Synthesis of covariant particle equations (Dirac, KG, Proca) from the Lorentz group.
Gürsey (1957)	Relativistic kinematics of a classical point particle in spinor form.
Kramers (1957)	Translation of Kramers (1938). Chapter on relativistic theory.
Moses (1959)	Maxwell equations in a spinor notation.
Hilgevoord and Wouthuysen (1963)	Covariant splitting of **j** into **l** and **s** for a spin-½ particle.
Barthélémy (1967)	Levinson theorem for Dirac.
Forman (1968)	History of the relativistic doublet problem around 1924.
Dittrich (1970)	Equivalence of the Dirac equation to a subclass of Feynman diagrams.
Lévy-Leblond (1974)	Group theory in QM. Spin obtained already in NR theory. Criticises several textbooks.
Perkins (1976)	Direction of Zitterbewegung.
Ni (1979)	Levinson theorem for Dirac and KG.
Berthier and Georgescu (1980)	Unique continuation for a Dirac operator.
Klaus (1980a)	Number of discrete Dirac eigenvalues for a potential τV, $V \le 0$, is proportional to τ^3.
Barut and Bracken (1981)	Zitterbewegung and the internal geometry of the electron.

Table 2.1. (continued).

Hostler (1982)	An SL(2,C)-invariant representation of the Dirac equation.
Bessis et al. (1984)	Atomic fine structure in curved space.
Biedenharn and Horwitz (1984)	Chiral two-component spinors and the factorization of Kramers' equation.
Biedenharn et al. (1984)	Spinorial relativistic rotator.
Grosse (1984, 1986)	Solitons connected to the 1D Dirac equation.
Blockley and Stedman (1985)	Supersymmetry: the 2D Pauli equation and a Dirac particle in constant **B**.
Gesztesy et al. (1985)	Mathematical properties of the Pauli Hamiltonian.
Hostler (1985c)	Scalar formalism for QED.
Shirokov (1985)	Irreducible representations of the inhomogeneous Lorentz group as basis of relativistic QM.
Belloni and Reina (1986)	Sommerfeld's (1931) way to the Thomas precession.
Bismut and Freed (1986)	Eta invariants and the holonomy theorem for Dirac operators.
Chang et al. (1986)	Two-component Pauli-like effective Hamiltonians in Dirac theory. H-like tests, Z = 1-100. Rare-gas SCF tests.
Chatterjee (1986)	Large-D solution of the KG equation in spherical potentials , including Coulomb.
Cohen and Mustafa (1986)	Classical (non-quantum) special relativity can give g-factors > 1.
Fanchi (1986)	Time parameters in relativistic QM: Galilean time, Minkowski time and historical time.
Gough (1986)	Analysis of spin and spin-orbit effects using quaternionic algebra.
Gross (1986)	Relativistic equations for nucleon systems.
Hostler (1986)	Scalar formalism for non-Abelian gauge theory.
Keller (1986)	Generalized Dirac equation admitting isospin and color symmetries.
Z.-Q. Ma (1986)	Levinson's theorem for potentials with attractive or repulsive Coulomb tails.
Morita (1986)	Role of quaternions in Dirac theory.
Qian et al. (1986)	Double wave description for a single relativistic free electron.
Rudra (1986)	Maximal symmetry groups of quantum relativistic equations.
Sallhofer (1986)	Elementary derivation of the Dirac equation.
Takahashi and Ropchan (1986)	First-order Schrödinger equation. Includes SO.
Szwed (1986)	'Square root' of the Dirac equation within supersymmetry.

Table 2.1. (continued)

Talman (1986)	Minimax principle for the Dirac equation.
Haar and Curtis (1987)	The Thomas procession gives g_e-1, not $g_e/2$.
Hall (1987)	"Dirac spinor orbits". L^2 orbits with eigenvalue return to origin.
Hokkyo (1987)	Zitterbewegung and mass formulas
Ichinose and Tamura (1987)	The Zitterbewegung of a Dirac particle in a 1+1 D space-time.
Lipkin (1987)	Criticizes C.-Y. Wong (1986).
Malvetti (1987)	Recoil corrections to M1 hfs, including the anomalous magnetic moment of the proton.
Marshakov and Fainberg (1987)	Representation of a Dirac particle as a sum over paths.
Moore et al. (1987, 1988)	Relativistic, many-particle Lagrangian for EM interactions.
Polyzou (1987)	Numerical methods for the discrete states of the squareroot operator $h = (-\frac{1}{2}\nabla^2 + m^2)^{\frac{1}{2}} + V$.
Arodz and Ruijgrok (1988)	Classical limit of the Dirac eq. with an external EM field.
Bessis and Bessis (1988)	Space-curvature effects in atomic and molecular calculations: a review.
Cannata and Ferrari (1988)	"Hausdorff dimension of trajectory" for a spin-1/2 particle in a FW representation.
Cooper et al. (1988)	Supersymmetry as guide for exact solutions for the Dirac equation. Chiral fields. 2D, 4D fields.
Hinton et al. (1988)	Levinson's theorem and Titchmarsh-Weyl theory.
Ionescu et al. (1988)	Nonlinear extensions of the Dirac equation.
Katz (1988)	Semiclassical (h->0) limit of the Dirac equation and the g-2 experiment.
Levitan and Sargsyan (1988)	Book on Sturm-Liouville and Dirac operators.
Ley-Koo and Wang (1988)	Dirac equation in orthogonal curvilinear coordinates.
Lieb and Yau (1988)	Discuss $Z\alpha$-related instabilities for point-nucleus atoms, described by "relativistic Schrödinger" equations. For the many-electron case, a bound on α only is obtained.
Z.-Q. Ma and A.-Y. Dai (1988)	Integral equations with symmetrical kernel applied to a system with a Dirac-type spectrum.
Moore and Scott (1988, 1989)	A class of physically acceptable, classical, relativistic many-particle Lagrangians.
Rastall (1988)	New form of the Dirac equation: wave function a 2 x 2 matrix.

Table 2.1. (continued).

Sherry (1988)	Zitterbewegung. See Epstein (1989) for comments.
Teitelboim (1988)	Quaternionic quantum field theory.
Anghel (1989)	L^2-index theorems for perturbed Dirac operators.
Batenburg and Schakel (1989)	A geometrical interpretation of spin.
Belov and Maslov (1989)	Semiclassical path-coherent states of a Dirac operator with an anomalous Pauli interaction.
Berezin et al. (1989)	Quaternions in relativistic physics. Book, 155 references.
Epstein (1989)	Points out an error in Sherry's (1988) discussion of Zitterbewegung.
Good (1989)	Relativistic particles in EM fields.
Gough (1989)	Probability density of a relativistic free electron.
Grigore et al. (1989)	NR limit of h_D.
Gulmaro Corona (1989)	A one-parameter family of approximations to the initial Cauchy problem for the Dirac equation, see Suárez (1984).
Hillion (1989)	"Front form" of electromagnetism.
Kutzelnigg (1989ab)	NR limit: a Lévy-Leblond (1967, 1970) equation. nth-order PT derived.
Kutzelnigg (1989c)	Relativistic version of Kato's cusp condition.
Okun (1989ab)	Discusses the concept of mass in relativistic theory. Criticizes some interpretations of $E=mc^2$.
Omote et al. (1989)	Galilean covariance leads to the Schrödinger equation.
Pan and Zhao (1989)	The usual mass-velocity term h_m is OK in 1st-order PT, contrary to claims of Farazdel and Smith (1986).
Percoco and Villalba (1989a)	Aharonov-Bohm effect [of potentials in regions where they vanish] on Dirac electrons: solution along a line of magnetic flux.
Polyzou (1989)	Relativistic two-body models.
Sherry (1989)	Pseudo-classical phase space description of the relativistic eletron.
Sokalski (1989)	2D Dirac equation has no magnetic moment. The $+\frac{1}{2}$ and $-\frac{1}{2}$ solutions are separated by 2 mc^2.
Ungar (1989)	Relativistic velocity composition paradox and the Thomas rotation.
White (1989)	NR limit of the Dirac operator.
Ambjørn et al. (1990)	Random walk representation of the Dirac propagator.
Baerends et al. (1990)	Suggest the use of different radii, r and r_{FW} (Foldy-Wouthuysen) for discussing relativistic orbital contraction.

Table 2.1. (continued).

Benítez et al. (1990)	Solution and hidden supersymmetry of a Dirac oscillator.
Chatterjee (1990)	Review on large-D expansions for Dirac and KG equations.
Cho et al. (1990)	A Clifford algebra quantization of Dirac's electron-positron field. Free particle.
Clemence (1990)	Levinson's theorem for a Dirac system.
Davies (1990)	Quaternionic Dirac equation.
Dieks and Nienhuis (1990)	Relativistic aspects of NR QM.
Dreitlein and Sauer (1990)	Manipulation of spinors and the group SL(2,C) using te language C++.
Ferrarese et al. (1990)	Interpretation of Thomas precession.
Ferrari and Russo (1990)	"Zitterbewegung" and localization for a NR two-band system in solids.
Greiner (1990)	Book on relativistic wave equations.
Hestenes (1990)	Zitterbewegung attributed, not necessarily to E>0 and E<0 interference, but to a complex phase factor in the Dirac equation.
Keller and Rodriguez-Romo (1990)	A "multivectorial Dirac equation".
Ketov and Prager (1990ab)	Square root of the Dirac equation within extended supersymmetry.
Klaus (1990)	Levinson theorem for the Dirac equation. Connects the number of bound states to phase shifts of continuum functions.
Kutzelnigg (1990)	A PT-based analysis and classification of relativistic approaches.
Rumrich et al. (1990)	Basis-set truncation violates gauge invariance in the coupled-channel representation of the Dirac equation.
Vonsovskii et al. (1990)	Zitterbewegung of a band electron.
Al-Jaber et al. (1991)	NR derivation of SO coupling.
Bordovitsyn and Ternov (1991)	Zitterbewegung with definite parity operators.
Crawford (1991)	Clifford algebra.
Domingos and Caldeira (1991)	Localisation of relativistic quantum particles is not in conflict with causality.
Evans (1991)	Klein's paradox for a modified, "4-space" Dirac equation.
Gilbert and Murray (1991)	Clifford algebras and Dirac operators in harmonic analysis.
Hamacher and Hinze (1991)	"Finite-volume variational method" for the Dirac equation. No applications reported.
Hartley and Tucker (1991)	Classical relativistic motion of a charged particle with spin in an external EM field.
Hestenes (1991)	The electron, new theory and experiment.

Table 2.1. (continued).

Hinton et al. (1991)	Inverse scattering in 1D.
Jackiw and Nair (1991)	Relativistic wave equation for anyons in 2+1 D.
Jaroszewicz and Kurzepa (1991)	Dirac propagator in nD represented as a random walk.
Kato and Yajima (1991)	A time-dependent Dirac equation for moving nuclei.
Kijowski (1991)	Lattice approximation of the Dirac equation.
Kostin (1991)	Linearized relativistic wave equations: Dirac, Feshbach-Villars and others.
Martínez y Romero et al. (1991)	Supersymmetric properties and stability of the Dirac sea.
Mathur (1991)	Thomas precession, SO interaction and Berry's phase.
Moshinsky et al. (1991)	A "many-body Dirac oscillator" basis for the Poincaré group.
Mustafa and Sever (1991)	Shifted 1/D expansion for the KG equation with vector and scalar potentials (D = number of dimensions).
Parisse (1991, 1992)	'Resonances of the Dirac operator' (Kramers degeneracy etc.)
Pavsic (1991)	Interpretation of R QM with invariant evolution parameter.
Sandin (1991)	Interpretation of $E = mc^2$
Simulik (1991)	Connection between the symmetry properties of teh Dirac and Maxwell equations. Conservation laws.
Sonego (1991)	Quasiprobabilities and explicitly covariant QT.
Todorov (1991)	Position-velocity coexistence in a 'minimal' statistical interpretation of the Dirac electron states.
Ungar (1991)	Thomas precession and the associated grouplike structure.
Yamasaki (1991)	Relativistic composition of velocities and the Thomas precession.
Arik and Mungan (1992)	q-oscillators and relativistic position operators.
Baylis (1992)	Classical (non-quantum) eigenspinors and the Dirac equation (cp. Gürsey (1957).
Baylis et al. (1992)	Pauli algebra and $i = (-1)^{1/2}$ in physics.
Beckers et al. (1992a)	Extended symmetries and hidden supersymmetry of the free Dirac equation.
Beckers et al. (1992bc)	Lie extended symmetries for spin 0, 1/2 or 1 particles with non-zero rest mass.
Buenker (1992a)	"Photons with zero kinetic energy".
Bulla et al. (1992)	Holomorphy of Dirac scattering matrix with respect to c^{-2}.
de la Jara and Tabensky (1992)	Spinor operators.

Table 2.1. (continued).

Domínguez-Adame (1992)	A relativistic interaction without Klein paradox.
Gudder (1992)	Discretized Schrödinger and Dirac quantum random walks.
Hegerfeldt et al. (1992)	Quantization may break down the relativistic invariance of wave equations.
Hillion (1992)	Nondispersive solutions of the KG equation.
Holladay (1992)	Derivation of relativistic energy from Lorentz gamma.
Holland (1992)	Dirac equation in the de Broglie-Bohm theory of motion.
Keller and Rodriguez (1992)	Geometric superalgebra and the Dirac equation.
Littlejohn and Flynn (1992)	Semiclassical theory of SO coupling.
Lozada (1992)	Equations of motion for an arbitrary multipole point particle in an external EM field.
McKeon and Ord (1992)	1+1 D Dirac equation obtained from a stochastic process.
Menon (1992)	Although $[h_D, \mathbf{s}] \neq 0$ in Dirac theory, the expectation value $\langle \mathbf{s}(t) \rangle$ is constant.
Muller (1992)	Thomas precession: where is the torque?
Plyushchay (1992)	Free relativistic particle with fractional spin.
Rawnsley (1992)	Exact symplectic structures and a classical model for the Dirac electron.
Schulman (1992)	Optimum two-time localization for massive, free relativistic particles.
Stepanov and Tutik (1992)	1/D expansion of the Dirac equation via an $\hbar$ expansion.
Thaller (1992)	The Dirac equation. Book. 506 references.
Ungar (1992)	The abstract Lorentz transformation group.

Table 2.2. The Dirac equation: further transformations

Reference	Comments
Tani (1949, 1951)	FW-like transformations discovered independently.
Eriksen (1958)	FW transformations for two-particle problems.
Eriksen and Kolsrud (1960)	Canonical transformation of Dirac's equation to even forms. Expansion in terms of the external fields.
McVoy and van Hove (1962)	Formulae for electron-scattering from "FW" nucleons.
Brodsky and Primack (1968)	Electromagnetic interactions of composite systems: not equal to sum of the individual FW transformations.
Chiu (1968)	Dirac and Pauli-level operators for radiative transitions of molecules.
Tsai (1973)	Uses a FW transformation to diagonalise a spin-($\frac{1}{2}$) particle with an anomalous magnetic moment in **B**.
Moss and Okninski (1976)	Diagonal forms of the free-particle Dirac Hamiltonian.
Saroja et al. (1978)	The Cini-Touschek transformation of a free Dirac particle expressed as product of two FW transformations.
Gesztesy et al. (1985b)	Selfadjointness of Dirac operators with anomalous magnetic moment.
Baretty et al. (1986)	A "no pair" equation in momentum space. Tested on H-like 1s states.
Ch. Chang et al. (1986)	A variational 2-component approximation without projection operators.
Durand (1986b)	Transformation to a Pauli Hamiltonian.
Farazdel and Smith (1986)	Claims that the mass-velocity term h_m would be incorrect. See Pan and Zhao (1989).
Hess (1986)	A 1-component "no-pair" Hamiltonian for variational calculations. Good up to order $(\alpha Z)^3$ for H-like atoms.
Heully et al. (1986a)	Transformed 1- and 2-electron Dirac Hamiltonians for MB calculations.
Heully et al. (1986b)	Comment on Hardekopf's and Sucher's (1984) projection operators for 1-electron atoms.
Jarvis and Stedman (1986)	Supersymmetry in 2nd-order Dirac-Coulomb equations.
Jaskolski and Karwowski (1986)	QR approach to low-energy electron scattering from closed-shell atoms.
Karwowski and Kobus (1986)	A new QR equation, tested on a Th model potential. Beats Barthelat et al. (1980).
Kobus (1986)	Argues that the Cowan-Griffin Hamiltonian requires non-orthogonal orbitals for closer parallelism with the exact Dirac solution.

Table 2.2. (continued).

Rutkowski (1986ab, 1988)	A new perturbative approach to relativistic effects. Applied on H using GTOs.
Rutkowski (1986c)	New PT approach for two-electron atoms.
Goldberg and Pratt (1987a); Goldberg et al. (1989)	A 2nd-order Dirac equation derived for screened Coulomb potentials, used in single-channel quantum defect theory.
Hardekopf (1987)	Relativistic momentum-space equations with application to atomic physics. See Heully et al. (1986b) for comments.
Jankowski and Rutkowski (1987)	Double PT for relativity and correlation. He-like ground states, Z=2...30, as example.
Kobus and Jaskólski (1987)	Comparisons between the QR Cowan-Griffin and full Dirac solutions for total and orbital energies and $<r^n>$. Low-energy positron scattering from Hg.
Kobus et al. (1987)	H-like $<r^n>$ as tests between the QR and Dirac approaches.
Kutzelnigg (1987)	Reviews transformed Dirac equations for molecular work.
Mohanty and Sampson (1987, 1988)	H-like E1 TP as tests of various QR approximations.
Rutkowski and Rutkowska (1987)	H_2^+ using a new PT approach. Terms of order c^n, n = 2, 4, 6 obtained.
Thakur (1987)	Semiclassical approximation for massless particles, with a non-local kinetic energy, in relativistic potentials.
Baretty and Garcia (1988)	A modified Darwin term proposed for variational calculations.
Bel and Ruiz (1988)	Relativistic Schrödinger equations.
Delgado-Barrio et al. (1988)	QR HF model for Ne solved at SZ level.
Karwowski et al. (1988)	Improve the QR method of Karwowski and Kobus (1985) for transition probabilities. See also Mohanty and Sampson (1987, 1988).
Katsuki and Huzinaga (1988)	Variational use of h_p <u>or</u> h_{CG} is possible if the innermost GTOs are contracted to act as a boundary condition (H-like 1s, 2p states, Z= 1-100.
Kissel-Phillip and Schwarz (1988)	Relativistically corrected Schrödinger equation: A relativistic kinetic energy $(m^2c^4+p^2c^2)^{\frac{1}{2}}$ combined with a simulated Zitterbewegung potential and a V_{so}.
Rustgi et al. (1988a)	Limitations of semirelativistic approaches in the Thomas-Reiche-Kuhn sum rule.

Table 2.2. (continued).

Thaller (1988)	Unitary transformations of the Dirac operator into an "even" form. Applications to scattering theory.
Baylis and Jones (1989ab)	A Pauli-algebra simplification of the Clifford algebra.
Cortona (1989)	Transforms the relativistic spin-polarized atomic equations of Doniach and Sommers (1981) to two uncoupled Dirac equations.
Hostler (1989)	FW-type reduction of the 2nd-order Dirac equation
Jankowski and Rutkowski (1989)	Double-PT for relativity and correlation.
Jansen and Hess (1989a)	Revision of the transformation of Douglas and Kroll (1974) to the form of Hess (1986).
Jaskolski (1989)	QR approach to electron-scattering from atoms.
Kutzelnigg (1989ab)	PT of the Dirac equation, starting from a Lévy-Leblond equation.
Beckers and Debergh (1990)	FW transformations and supersymmetry applied on oscillators.
Goldman (1990)	Shows that the squared Dirac Hamiltonian of Volkov (1989) gives no advantages.
Moreno et al. (1990)	Supersymmetry, FW transformation and stability of the Dirac sea.
Ternov (1990)	Spin of a relativistic electron in the Heisenberg represenation.
Karwowski and Martin (1991); Martin and Karwowski (1991)	QR formulation of the quantum-defect-orbital method.
Kosaka and Yonei (1991ab)	WKB method for an electron in screened Coulomb potentials.
Kosik (1991)	Approximate Dirac-Coulomb radial integrals.
Layton and Chu (1991)	A variational, Fourier-grid approach to the Dirac equation.
Weigel et al. (1991)	Semiclassical expansions in nuclear DF theory.
Adler and Martin (1992)	The electron g = 2 derived for a Pauli equation.
Arredondo Ruiz (1992)	A KG equation with indefinite form.
Buenker (1992b)	'Exponentially damped BP hamiltonian' and positronium decay.
Cardoso (1992)	Bessel function expressions for Dirac fields.
Clerk and McKellar (1992)	Scalar symmetry of the massless Dirac equation.
DeAngelis and Serva (1992)	Probabilistic, Brownian path integral approach to the FW transformation.
Lucha et al. (1992); Lucha and Schöberl (1992)	Breit Hamiltonian for quark-antiquark pairs. A new semirelativistic equation.

Table 2.2. (continued).

Malik et al. (1992)	Relativistic terms in the reduction of the Bethe-Salpeter equation.
Mourad and Vashakidze (1992)	A relativistic Schrödinger equation with SO interaction: $h = \boldsymbol{\alpha}\cdot\mathbf{p} + \beta m + V(1+\beta)/2$.
D.-D. Wu (1992)	Ambiguity of perturbative Dirac theory.

Table 2.3. The Dirac equation: solutions for hydrogen-like system.

Reference	Comments
Sommerfeld (1915)	FS of H-like atoms. Precursor of Sommerfeld (1916).
Rozhanskii and Frenkel' (1938)	Connections between the Sommerfeld and Dirac theories. Points out the equivalence with a NR potential $V= -Z/r-a/r^2$.
Bethe (1948)	Virial theorems and momentum space eigenfunctions for the Dirac-Coulomb case.
Infeld and Hull (1951)	Dirac-Coulomb radial functions from a factorization method. Matrix elements of V(r) for bound and continuum states.
Salpeter (1958)	Dirac-Coulomb problem for an electron with an electric dipole moment.
Epstein and Epstein (1962)	Hypervirial theorems involving α, β and r^n.
Armstrong (1963)	Anomalous solutions with j=0 to the Dirac-Coulomb problem.
Rosen (1963)	Relativistic Coulomb scattering.
Pratt et al. (1964)	K-shell photoionization cross sections.
Green (1965)	Applies the factorisation technique to the 2nd-order radial Dirac-Coulomb problem.
Dollard and Velo (1966)	Asymptotic ($r \to \infty$) behaviour of a Dirac particle scattered by a Coulomb field.
Bacry and Richard (1967)	Group theory of the KG Coulomb problem.
Kiefer and Fradkin (1969)	Noninvariance group for the Dirac-Coulomb problem.
Dittrich (1970, 1974)	Feynman diagrams of the Dirac-Coulomb problem.
Hojman (1971)	Algebraic method for the Dirac-Coulomb problem.
Johnson (1972)	2s lifetime of H-like atoms, Z = 1-92.
Lin and Feinberg (1974)	M1 decays of 2s in H-like ions.
Narnhofer (1974)	The KG and Dirac-Coulomb problems. The former leads to r^{-2} Schrödinger terms. Analytical properties as function of Z.
Cizek and Paldus (1977)	D-dimensional KG and Dirac hydrogen.
Friar (1979b)	Analytical solution on finite-nucleus contributions to order $(\alpha Z)^6$.
Nieto (1979)	KG hydrogen solved in D dimensions. The unphysical V=-1/r used for all D.
Mil'shtein and Strakhovenko (1982)	O(2,1) algebra and the Dirac-Coulomb Green function.
Durand and Durand (1983)	Analytic s-state solution for the "relativistic Schrödinger equation", $T = (p^2 + m^2)^{\frac{1}{2}} - m$.
Castorina et al. 1984)	Partial-wave analysis for potentials $\lim_{r \to 0} V(r)=-\alpha/r$. WKB fits.

Table 2.3. (continued).

Friar and Tomusiak (1984)	Alternative analytic s-state solution for the "relativistic Schrödinger equation".
Miramontes and Pajares (1984)	D-dimensional KG and Dirac H-like atoms.
Borisoglebskii and Yakimovich (1985)	Nuclear-quadrupole-moment-induced $\Delta L = \pm 3$ TP for hydrogen-like atoms.
Thaler (1985)	Relativistic effects in protonium ($p\bar{p}$). (Strong-interaction effects with relativistic wave functions.)
Wong and Yeh (1985)	Dirac-Coulomb Green function and its application to Raleigh scattering.
Aerts and Nieuwpoort (1986)	GTO solutions for 1s-3d* with $Z \leq 92$.
Ayres (1986)	Fermion-antifermion bound states.
Bruce (1986)	2D Dirac-Coulomb problem.
Chang et al. (1986)	Two-component Pauli-like effective Hamiltonian in Dirac theory.
Chatterjee (1986)	Large-D solution of the KG equation.
Durand (1986b)	Transformation from Dirac to Pauli. Variational.
Dutta et al. (1986)	Compton profiles of Kr-Rn using screened hydrogenic functions.
Emss and Thaller (1986)	Dirac-Coulomb scattering asymptotics.
Garcia (1986)	Classical (non-quantum) limit of the Dirac-Coulomb problem.
Hall (1986)	Eigenvalues of screened Coulomb potentials related to the screening function.
Hegarty (1986); Hegarty and Aerts (1987)	Discusses conditions for finite-basis solution of the Dirac equation. Example: U^{91+}.
Hess (1986)	"No-pair" formalism tested on the 1s state.
Heully et al. (1986b)	Explain the deviation of the H-like results of Hardekopf and Sucher (1984) from Dirac-Coulomb ones by negative-energy-state contamination.
Indelicato (1986)	Binding energies, Z=100...170 for finite-nucleus 1s levels.
Jarvis and Stedman (1986)	Supersymmetry in 2nd-order Dirac-Coulomb equations. 1D and 3D. Also KG 1D and 3D.
Krause (1986)	He-like systems treated variationally using a screened H-like basis.
Z.-Q. Ma (1986)	Levinson's theorem for potentials with (attractive or repulsive) Coulomb tails.
Moore and Scott (1986)	An α expansion for general V(r), tested on H.
Mukoyama (1986)	Excitation of H-like atoms by heavy charged particles.
Papp (1986a)	Ground-state energy of relativistic two-body Hamiltonians with a Coulomb potential.

Table 2.3. (continued).

Rutkowski (1986ab)	A new PT approach tested using GTOs.
Talman (1986)	The H-like $2p_{\frac{1}{2}}$ energy for Z = 92 as test of a new "minimax" method.
Thaller and Enss (1986)	Asymptotic observables and Coulomb scattering of the Dirac equation.
M.K.F. Wong (1986)	Infeld-Hull factorisation and the simplified solution of the Dirac-Coulomb equation.
Bhattacharya et al. (1987)	Uses screened Dirac-Coulomb functions to discuss M-shell vacancy effects on L-shell x-ray spectra.
Eichler (1987)	Relativistic charge exchange with Coulomb boundary conditions.
Grosse (1987)	Perturbed Dirac-Coulomb problem: order of split levels related to sign of $\nabla^2 V_{pert}$.
Hall (1987)	"Dirac spinor orbits" for the large versus the small component.
Hostler (1987a)	Self-energy operator for an electron in an external Coulomb potential.
Hostler (1987b)	Calculates the required matrix elements using the Green function of a 2nd-order Dirac equation.
Ishikawa and Quiney (1987)	Finite-nucleus GTO solution for 1s and $2p_{1/2}$ of Hg^{79+}.
Ishikawa et al. (1987)	O^{7+} using momentum-space GTOs.
Kobus et al. (1987)	$\langle r^n \rangle$ as test between QR and Dirac wave functions.
Malvetti (1987)	M1 hfs of H atoms, including recoil corrections and proton anomalous magnetic moment.
Manakov et al. (1987)	Sturm expansion of the Coulomb Green function in QED. Photon scattering by a bound electron.
Moss (1987)	D-dimensional hydrogen, V=-Z/r. KG and Dirac.
Mrówczynski (1987)	Interaction of relativistic, Coulomb bound states (positronium, pionium etc.) with matter. Part II: Denisenko and Mrówczynski (1987).
Pilkuhn (1987)	See Yelkhovsky (1987).
Stepanovskij (1987)	H-like atom in external **E** or **B**.
Syros (1987)	Non-perturbative evolution operator for the relativistic H atom.
Visser et al. (1987)	H-like atoms with Gaussian nuclear charge distributions solved using GTO:s.
S. Wilson (1987)	Basis sets tested on Dirac H atoms. Also PT using "wrong nuclear charge".
M.K.F. Wong (1987)	Values of the Dirac-Coulomb wave function at origin. (for Lamb shift or positronium decay). The 1s correction factor 1.000 118, 2s only 1.000 004 1.

Table 2.3. (continued).

J.-S. Wu et al. (1987)	Group-theory for scattering S matrix.
Yelkhovsky (1987)	The nuclear magnetic moment-induced atomic quadrupole moment for the 1s, F=3/2 state of hydrogen by Pilkuhn (1984) is correct, after all. See Pilkuhn (1987).
Yin and Pratt (1987b)	Point-Coulomb relativistic Cooper minima in photoionization.
Yoshida (1987, 1988)	Rotating Laplace-Runge-Lenz vector for the relativistic Coulomb problem.
Zimmermann (1987)	Uses Dirac-Coulomb wavefunctions to discuss ns-state hfs.
Zygelman and Dalgarno (1987)	Electron and proton impact excitation of the H-like n = 2 FS levels.
Barut and Salamin (1988)	Spontaneous 2p-1s and 2s-1s emission in hydrogen and muonium.
Carson (1988)	Dirac-Coulomb bound-bound multipole transitions.
Cheon (1988)	$2S_{1/2}$-$2P_{1/2}$ energy shift influenced by conducting boundaries, $\Delta\nu$= 0.894 MHz for 1μm separation between plates.
Galić (1988)	1+1D hydrogen, V=|x|, has no bound states. Example on Klein's paradox.
Katsuki and Huzinaga (1988)	1s and 2p energies for Z=1-100 using h_D or h_{CG} variationally with contracted inner GTOs.
Mohanty and Sampson (1988)	H-like E1 TP as tests of various QR approximations. See also Karwowski et al. (1988).
Mukoyama and Sarkadi (1988)	Dirac-Coulomb continuum functions at zero kinetic energy in ion-atom collisions.
Nagiev (1988, 1989)	Dynamical symmetry group of the relativistic Coulomb problem for spinless particles.
Panja and Dutt (1988)	1/N theory for spin-0 and spin-½ particles in a Coulomb potential or in arbitary V(r).
Stahlhofen and Biedenharn (1988)	Supersymmetry and scattering in the Dirac-Coulomb problem.
Struensee and Cohen (1988)	n = 2 FS transitions in H + p collisions.
Arbuzov et al. (1989, 1990ab)	Bound states immersed in the continuum.
Atag (1989)	Large-D solution.
Boschi Filho and Vaidya (1989)	Green function for the KG Coulomb problem.
Bottcher et al. (1989)	A "damped relaxation" method tested.
Darewych and Horbatsch (1989, 1990)	Two-body Coulomb system with spins 0, 1/2. KG and Dirac α^4 limits obeyed, respectively.
de Lange (1989a)	Operator analysis of KG and Dirac Coulomb problems.
de Lange (1989b)	A simple solution for the Dirac Coulomb problem.

Table 2.3. (continued).

Goldman (1989a)	Accurate dipole polarizabilities for Z=1...115 from a gauge-invariance method.
Goldman (1989b)	Laguerre basis functions give superior numerical stability to STO:s. Two-photon decay rates of 2s states calculated for Z=1...100.
Kuten' and Rapoport (1989)	M1-hfs-induced atomic M3 moments in F=3/2 j=1/2 states (1s and $2p_{1/2}$).
Lieber (1989)	The Dirac-Coulomb Green's function.
Matsuoka and Okomura (1989)	Accurately energy-optimized GTO basis for H.
Mrówczynski (1989)	Review on relativistic elementary atoms.
Mukoyama (1989b)	K-shell ionization by protons: screened Coulomb and DS functions compared.
Pekeris and Frankowski (1989)	Hyperfine splitting of muonium, positronium and hydrogen from solutions of the Dirac equation in Kerr-Newman geometry. Curved space introduces a trace of imaginary energy.
S. Rashid (1989)	An H-like atom in an intense laser field.
Ritchie (1989)	Scattering of a continuum electron in a Dirac-Coulomb field of another electron.
T. Roy (1989)	Eigenvalues of a Dirac hydrogen atom.
Talwar and Wright (1989)	Dirac-Coulomb functions used to evaluate the "radiation tail" accompanying electron scattering from an extended nucleus (Z, A).
Volkov (1989)	H-like atom using Laguerre-type basis. See Goldman (1990) for comments.
Yoshida (1989)	KG eigenvalues for a Coulomb potential from a rotating Laplace-Runge-Lenz vector.
Boston and Sandars (1990)	The P-violating, T-invariant L=1, "anapole moment" of hydrogenic atoms.
Dominguez-Adame (1990b)	1D H atom in momentum representation.
Dyall and Faegri (1990)	LCAO solution for $2p_{1/2}$ of Hg^{79+}. Kinetic balance and intruder states discussed.
Florescu et al. (1990)	Low-frequency Rayleigh scattering including spin flip.
Kargabaev et al. (1990)	Dirac-Coulomb Green function for finite nuclei.
Krstic and Mittleman (1990)	Bound states of atomic hydrogen in a laser field.
Matsuoka (1990a)	H-like H, Ag, Hg with STO basis and a new finite nuclear model with constant dV/dr.
Parera López (1990)	Scaled hydrogenic solutions as DF basis for He-like systems.
Reiss (1990)	Dirac hydrogen atom in dominant plane-wave field.

Table 2.3. (continued).

Sallhofer (1990ab, 1992)	The Dirac-Coulomb problem using new operators.
Seke (1990)	Deviations from exponential decay for the $2p_{\frac{1}{2}}$ - $1S_{\frac{1}{2}}$ transition.
Shinada (1990)	H atom in static, uniform **B**.
Tong et al. (1990)	s-s, d-s two-photon decay rates.
M.K.F. Wong (1990)	nD Dirac-Coulomb problem solved.
Aissing (1991)	Monte Carlo variance minimization with STOs.
Bawin and Cugnon (1991)	"Partially screened solutions" for the Dirac equation in $Z>Z_c$ Coulomb fields.
Belinfante (1991)	Continuum functions at larger r by summing divergent terms.
Bosanac (1991)	Scattering of a KG wave packet from a Coulomb potential.
Z-H. Chen and Goldman (1991ab)	H-like atoms in strong **B**, up to 10^{12}G. STO basis.
Eberlein (1991)	Level shifts for a hydrogen atom between two conducting plates. Dirac-Coulomb basis.
Goldman and Chen (1991)	H-like atoms in magnetic fields.
Goodson et al. (1991)	D-dimensional KG or Dirac hydrogenic atoms. Singularity analysis.
Grosse et al. (1991)	Order of KG-Coulomb levels, split by a perturbation, related to the sign of its Laplacian.
S.H. Guo et al. (1991)	2D hydrogen (V=-1/r), with and without Chern-Simons potential.
Hasegawa et al. (1991)	Finite-nucleus solutions for various states of Z=1...92.
Hersbach and Ruijgrok (1991)	A quasipotential theory for a spin-1/2 electron bound to a spinless nucleus of finite mass and charge +Ze. Agrees with Dirac-Coulomb up to order α^4.
Kosik (1991)	Approximate Dirac-Coulomb radial integrals.
Layton and Chu (1991)	Solve the $S_{\frac{1}{2}}$ Dirac-Coulomb eigenstates for Z=2 using a variational Fourier-grid approach. With a 41-point radial grid, 2-3 lowest bound states have accurate E_{rel}.
Legrand et al. (1991)	Dirac-Coulomb basis used to describe L-shell ionization in p + Sm collisions.
Matsuoka et al. (1991)	H-like atoms, Z=1...92, using GTO basis and a uniformly charged, finite nucleus.
Nandi and Shastry (1991)	2D and 3D KG-Coulomb problems. Classical (small-h) limits discussed.
Pyykkö and Zhao (1991d)	Notice that for V =-Z/r, the 2D Dirac and 3D KG eigenvalue expressions are similar.
Schlageter and Pilkuhn (1991)	Relativistic polarizabilities of H-like systems for "electrons" with arbitrary g.

Table 2.3. (continued)

Shabaev (1991a)	Virial relations for M1 and E2 hfs integrals and other integrals.
Smit (1991, 1992)	Light-ion induced K-shell ionization in the adiabatic region. H-like wave function.
Spence and Vary (1991a)	Electron-positron scattering resonances from relativistic two-body wave equations.
Spence and Vary (1991b)	Electron-proton resonances at low energy.
Swainson and Drake (1991a-c)	Unified treatment of R and NR H atom. Green functions.
Visscher et al. (1991ab)	Kinetic balance and basis-set contraction tested on H-like atoms.
Yudin (1991)	Excitation and ionization of H-like atoms by fast, multiply charged ions.
Zapryagaev and Nefedov (1991)	Photoionization of multiply charged H-like ions.
Z.-H. Chen and Goldman (1992)	H atom in **B**. R/NR.
Cheung and Li (1992)	A QR equation for a system consisting of a $S=\frac{1}{2}$ fermion and a S=0 boson.
Datta and Datta (1992)	'Minimax' techniques for LCAO methods tested on the Dirac-Coulomb problem.
de Lange and Welter (1992ab)	'Shape invariance' of the Dirac and KG Coulomb problems.
Goldman (1992)	Fundamentals of a basis-set approach tested on the Dirac-Coulomb case.
Grozdanov et al. (1992)	Microwave excitation of H-like Rydberg states.
Kao et al. (1992)	Electron-impact ionization of H-like atoms, Z = 1...47. R/NR.
Kuchiev et al. (1992)	Atomic quadrupole moment of $ns_{\frac{1}{2}}(F = 1)$ states.
Papp (1992)	'Fixed point structure' of the Dirac-Coulomb problem. 1/D approach.
Seke (1992ac)	Non-exponential $2p_{\frac{1}{2}}$-1s decay of Dirac hydrogen.
Shen et al. (1992c)	2D Dirac-Coulomb problem.
Shin et al. (1992)	Wigner formulation for the density matrix of the Dirac-Coulomb problem.
Tutik (1992)	Exact solution of the nD Dirac-Coulomb problem.

Table 2.4. The Dirac equation: solutions for various non-Coulomb fields.

Reference	Comments
Volkov (1937)	Dirac electron in a plane, non-polarised EM wave.
Pomeranchuk and Smorodinsky (1945)	Dirac-Coulomb problem for a charged-shell nuclear model.
Voronkov and Kolesnikov (1960)	ns states for finite nuclei with radius r_0, $V=-Z/r_0$, $r<r_0$.
Roos and Sangren (1962)	Spectral theory of the radial Dirac equation.
Jannussis (1966)	**B** using "Schraubenfunktionen".
Barthélémy (1967)	Central-potential scattering.
Canuto and Chinderi (1969)	**E** ⊥ **B**.
Lam (1970ab)	**E** ⊥ **B**.
Lipas (1970)	3D KG harmonic oscillator solved to order α^2.
Steslicka and Davison (1970)	Boundary conditions for the relativistic Kronig-Penney model.
Steslicka et al. (1971a)	1D Kronig-Penney model for impurity states of solids.
Steslicka et al. (1971b)	Kronig-Penney model for heavy diatomic crystals.
Bhagwat and Subramanian (1972)	Impurity states in a relativistic Saxton-Hunter model (chain of δ functions).
Fedorov and Kazakov (1973)	Quantized plane wave, directed along a constant **B**.
Hautot (1973)	Coulomb field + **B**.
Tsai (1973)	Spin-(½) and spin-0 particles with anomalous magnetic moments in **B**.
Wüst (1973, 1975, 1977)	Conditions for selfadjoint Dirac operators with a cut-off potential.
Chodos et al. (1974)	The "bag" model for hadrons. Spurious solutions avoided by the boundary conditions r^x, $x>0$, at $r \to 0$, see Johnson and Sapirstein (1986).
Flügge (1974)	Properties of the Dirac equation.
Sen Gupta (1974)	Monoatomic and diatomic 1D lattices of δ-function potentials.
Gyulassy (1975)	Radial Green functions for finite-nucleus (muonic) atoms.
Bongaarts and Ruijsenaars (1976, 1977)	The Klein paradox analysed.
Moss and Okninski (1976)	Dirac particle with anomalous magnetic moment in **B**.
Friar (1979b)	Analytical evaluation of finite-nucleus effects in hydrogenic atoms through order $(Z\alpha)^6$.
Kaburagi et al. (1980)	$V= -\ln r$ as model for meson spectra. Variational method.
Klaus (1980a)	Asymptotic properties of a Dirac particle in a potential V.

Table 2.4. (continued).

Klaus (1980b)	General properties of Dirac operators with several Coulomb singularities.
Eisenberg (1981)	A relativistic nucleon in scalar and vector potentials.
Shen (1981a)	A semi-infinite 1D Kronig-Penney model with a step-function surface.
Shen (1981b)	Surface states of a semi-infinite, 1D Kronig-Penney model in a constant electric field.
Borie and Rinker (1982)	Muonic atoms.
Rabin (1982)	Solves the lattice fermion doubling problem. 1+1 D.
Silagadze and Khelashvili (1984)	A two-fermion bound state with $V(r) = gr + b$, without a Klein paradox.
Stacey (1982)	Elimination of fermion doubling.
Bergou and Ehlotzky (1983)	Exact solutions for KG or Dirac particles in a plane-wave radiation field, directed along a constant **B**.
Jena (1983)	A power-law, Ar^x, model for meson mass spectra; $x = 0.1$.
Atiyah and Singer (1984)	Dirac eq. coupled to vector potentials.
Fleischer and Soff (1984)	Bound states of KG particles for strong potentials. (Coulomb, Hulthén, Yukawa, exponential, finite nucleus, square well...).
Miramontes and Pajares (1984)	nD KG and Dirac hydrogen.
Nickish et al. (1984)	Numerical solutions for a two-particle problem with $h_i = (-\nabla^2 + m^2)^{\frac{1}{2}}$ and $h_{12} = V(r_{12})$.
Behncke (1985)	Spectral properties of the Dirac equation with anomalous magnetic moment.
Bruce (1986)	2D Dirac-Coulomb problem.
Fröhlich et al. (1986)	H-like atom in **B**.
Hall (1986)	Eigenvalues of screened Coulomb potentials related to the screening function.
Moore and Scott (1986)	A perturbation expansion in α for arbitrary, spherically symmetrical potential $V(r)$. Wave functions to order α^2, energies to α^4 given. Tests on H.
Ni (1986)	Exact solution for a Dirac electron in constant **E**.
Nogami and van Dijk (1986)	Discuss bound and scattering solutions of the Dirac equation in a central field.
Papp (1986a)	Ground-state energies for two-body Hamiltonians with Coulomb and Yukawa potentials.
Papp (1986b)	Quasiclassical ground-state energies for quarkonia potentials: $\ln r$, $r-\alpha/r$, C_n/r^n with $C_n n<0$ etc.
Rosenberg (1986)	Relativistic scattering in a slowly time-dependent **E**.

Table 2.4. (continued).

Rosenberg and Spruch (1986)	Extremum principles for relativistic bound-state energies.
C. L. Roy (1986)	1D disordered system, modelled by rectangular wells. DOS for an approximate Dirac equation.
Roy and Pandey (1986a)	Relativistic Kronig-Penney equation and surface states of a distorted surface.
Ray and Pandey (1986b)	Relativistic surface states of a finite crystal.
Sokolov and Ternov (1986)	Radiation from relativistic electrons.
Su and Ma (1986)	(3+1)D and (1+1)D Dirac equations with both scalar ($\beta\phi$) and vector (V) like potentials. No confinement for $\phi=\pm V$.
Thaller (1986)	Scattering theory for long-range Lorentz scalar potentials $\beta V(\mathbf{x})$, such as $\beta/\lvert\mathbf{x}\rvert$.
Ulehla and Horejsi (1986)	Generalized Prüfer transformation and the eigenvalue problem of radial Dirac equations. $V(r)= -5\ e^{-r}$.
Abe and Fujita (1987)	Approximate analytic solution for a scalar linear potential (see ref. 688): $h_D=\boldsymbol{\alpha}\cdot\mathbf{p} + \beta\mu r$.
Abounadi and Darkhosh (1987)	Dirac wave equations in external fields.
Barut and Pavsic (1987)	A classical Dirac electron in curved space.
Bottcher and Strayer (1987)	Time-dependent fermion problems solved on a collocation lattice using splines. Fermion doubling avoided by different discretisation for G and F (forward and backward, respectively).
Calkin et al. (1987)	Delta-function potential in a 1D Dirac eqn.
Coutinho and Nogami (1987a)	A 1D "DF" model for a 2-electron "atom" with δ-function interactions solved to order α^2 (=NR result).
Coutinho and Nogami (1987b); Coutinho et al. (1988)	Conditions for existence of bound states for 1D Dirac problems.
Dominguez-Adame (1987, 1989)	Relativistic Kronig-Penney model.
Falkensteiner and Grosse (1987)	Quantization of fermions interacting with point-like external fields.
Gesztesy and Seba (1987)	New analytically solvable models of relativistic point interactions.
Hall (1987)	"Dirac spinor orbits" for the large versus the small component.
Karpenko (1987)	Relativistic fermion in a square-well potential.
McKellar and Stephenson (1987)	Klein paradox and the Dirac-Kronig-Penney model.
Misra et al. (1987)	Relativistic effects on meson spectra.
Moss (1987)	D-dimentional hydrogen, $V=-Z/r$, solved for KG and Dirac. No bound states for D=1. Cp. Capri and Ferrari (1985).

Table 2.4. (continued).

Mukoyama and Lin (1987b)	Dirac equation in a DS atomic potential using STOs.
Palladino and Leal Ferreira (1987)	Level order for a Dirac particle in mixed scalar and vector potential, $h=\boldsymbol{\alpha}\cdot\mathbf{p} + \beta m + \frac{1}{2}(1+\beta)V(r)$, without Klein paradox.
Polyzou (1987)	The square-root operator solved for Coulomb, Yukawa and 'confining' ($-C_1/r + C_2r$) potentials.
Rosenhouse and Katriel (1987)	Isospectral transformations in R and NR mechanics. The C -> ∞ and h -> 0 limits of 1D Dirac equations.
Salvat et al. (1987a)	Numerical solutions of the Dirac equation for a 5- parameter MP for Z = 1-92.
Saritepe and Swamy (1987)	2D Dirac equation for an electron in a helical wiggler magnetic field.
Solovtsov and Chernichenko (1987)	Inverse scattering problem in the relativistic quasiclassical approximation.
Stepanovskij (1987)	H-like atom in external **E** or **B**.
Arshansky and Horwitz (1988)	Relativistic potential scattering.
Barut and Bozic (1988)	Motion of Dirac particles with anomalous magnetic moments through magnetic fields. A magnetic barrier as example.
Bijtebier (1988, 1989b)	Scattering matrix elements for a relativistic bound state.
Borie and Jödicke (1988)	Antiprotonic atom, with the anomalous magnetic moment and vacuum polarization potential. Solved numerically.
Bulla et al. (1988)	Relativistic energy band corrections for periodic potentials.
Bychkov (1988)	Energy spectra of 2D systems with SO interaction in an inclined magnetic field.
Calkin et al. (1988)	Nonlocal separable potential $(\beta g+h)V(x)V(x')$.
Ebert and Gyorffy (1988)	Scattering solutions of the Dirac equation for nonspherical targets.
Galić (1988)	1+1 D hydrogen, V=1+1, V=|x|, has no bound states. Example on Klein's paradox.
Gjaja and Bhattacharjee (1988)	Dirac theory of the free-electron laser (helical undulator potential). Quantum noise found.
Ixaru and Pantea (1988)	A double square well 1D model for positron creation in superheavy collisions.
Jansen et al. (1988)	Continuum solutions of the KG equation. Point Coulomb, finite nucleus, square well, attractive exponential and 'overcritical' potentials considered. Also homogeneous **B**.
Koller et al. (1988)	Positronium and muonium in strong magnetic fields.

Table 2.4. (continued).

Leal Ferreira (1988)	Two-body Dirac equation with a scalar linear potential.
Mladenov (1988)	Relativistic extension of the Saxon-Hutner theorem: In a 1D Kronig-Penney problem, common energy gaps of A and B are energy gaps of the alloy A+B.
Panja and Dutt (1988)	1/D theory for spin-0 or spin-½ particles in arbitrary V(r). Coulomb case as an example.
C. L. Roy (1988)	A spatially constant solution to the 1D Dirac equation.
Seba (1988)	Absorption of eigenvalues in $E>0$ or $E<0$ continuum of 1D and 3D Dirac equations.
Ternov et al. (1988)	Solutions of the KG and Dirac equations for EM fields characteristic for free-electron lasers. Spinors for wigglers.
Tóth et al. (1988)	Two-centre Dirac equation with separable potentials.
Vrscay and Hamidian (1988)	Eigenvalues for a Dirac-Coulomb problem, perturbed by βr^q or r^q, by hypervirial theorems.
Arshansky and Horwitz (1989ac)	Bound states for spherically symmetric O(3,1) potentials: Coulomb, harmonic and square well.
Arshansky and Horwitz (1989b)	Relativistic potential scattering and phase shift analysis. Levinson theorem.
Arshansky et al. (1989)	Bound states and EM interactions.
Atag (1989)	Large-D solutions for spherical potentials.
Bagrov et al. (1989)	Yang-Mills fields.
Bardsley et al. (1989)	Electrons in intense laser fields.
Bijtebier (1989a)	Two spin-less particles bound in a potential.
Bottcher et al. (1989)	A damped relaxation method for bound states. Coulomb, 1D Gaussian and uniformly charged sphere potentials as test cases.
de Frutos (1989); de Frutos and Sanz-Serna (1989)	Numerical integration of non-linear, 1+1D Dirac equations. Split-step FFT found best.
Detcheva et al. (1989ab)	Effective electron masses in the Kronig-Penney model.
Dittrich et al. (1989, 1990)	Spherically symmetric $\delta(r)$ potentials.
Dominguez-Adame and Maciá (1989a)	Harmonic oscillator + a Lorentz scalar delta function.
Dominguez-Adame and Maciá (1989b)	Point interaction potentials and bound states.
Glass and Mendlowitz (1989)	Spin flips in Dirac scattering by a 1D barrier.
Goldberg et al. (1989)	Phase shifts for screened Coulomb potentials from WKB approximation.
Good (1989)	Particles in EM fields.

Table 2.4. (continued).

D.-S. Guo et al. (1989)	Scattering theory of multiphoton ionization in strong fields. Uses a quantized-field Volkov solution.
Hofer and Stocker (1989)	Establishes a formal analogy between the 3D V(r) and V(z) Dirac problems.
Jaffe and Manohar (1989)	Bound states of the Dirac equation outside a hard sphere in 1, 2 and 3D .
Jagannathan et al. (1989)	Magnetic electron lenses.
Kamath (1989)	Dirac-Coulomb problem in 2+1 D. While V=ln r gives no solution due to Klein's paradox, the Chern-Simons potential does.
Krstic and Milosevic (1989)	Scattering of electrons in an elliptically polarized ultra-strong laser pulse.
Lovas and Molnár (1989)	Dirac particle in a vector meson field, periodic in space, constant in time. See Sen Gupta (1992).
Manoukian (1989)	Propagator for a Dirac δ potential.
U.D. Misra et al. (1989)	Sommerfeld's σ_2 parameter for the screened Dirac-Coulomb problem deduced for Z=10-103 using theoretical energies of Huang et al. (1976). (E_{rel} is proportional to $\alpha^4(Z-\sigma_2)^4/n^4$).
Missana (1989)	The Volkov (1935) solution in astrophysics
Mitchell et al. (1989)	An 1D "relativistic HF" model for the deuteron with δ-function attraction.
Mladenov (1989)	Saxon-Hutter theorem in the relativistic domain: gaps common to A and B are gaps of any AB alloy.
Moreno and Zentella (1989)	Covariance, CPT and the FW transformation for the Dirac oscillator.
Moshinsky and Szczepaniak (1989)	"The Dirac oscillator", $h_D = c\,\boldsymbol{\alpha}\cdot(\mathbf{p} - im\omega\mathbf{r}\beta) + mc^2\beta$. NB: linear in both $\mathbf{p}$ and $\mathbf{r}$. Reduces at NR limit to a harmonic oscillator, $V=m\omega^2r^2/2$, with SO splitting.
Noskov and Shvartsman (1989)	Green's functions for external EM field by the eigenfunction method.
Percoco and Villalba (1989a)	A line of magnetic flux.
Percoco and Villalba (1989b)	1D scalar potentials.
Pimentel (1989)	More solutions for space-times with local rotational symmetry, where the Dirac equation separates.
S. Rashid (1989)	Applies the Volkov solution to a H-like atom in an intense laser field. See Mittleman (1991) for comments.
Raspini (1989)	Relativistic Zeeman effect in positronium, n=2.

Table 2.4. (continued).

C. L. Roy (1989); Roy and Basu (1990)	1D disordered systems modelled by rectangular barriers or wells. R/NR comparison.
Roychoudhury and Varshni (1989)	Relativistic 1/D expansion of the Dirac equation. Solutions for a screened Coulomb potential.
Sahu et al. (1989)	Critical binding ($E \leq 0$) of KG and Dirac particles by δ-function and square-well potentials in 1D...3D.
Seba (1989)	Klein's paradox and the δ-function interaction.
Sen Gupta (1989)	1D Dirac equation in V(x) reduced to a Riccati equation.
Shishkin and Villalba (1989ab)	Dirac equation in external vector fields.
Sitenko (1989)	Zero modes for a 2D Dirac operator on a noncompact singly connected surface in external **B**.
Srivastava (1989)	Solution of the Dirac equation in Kasner's space time.
Thaller (1989)	Dirac particles in magnetic fields.
Volkov (1989)	Confining quark potential $\beta V(r) = 400 r\beta$.
Aly and Barut (1990)	Exact solutions for the E-dependent, 2nd-order Dirac equation.
Anghel (1990b)	2D magnetic field.
Bagrov and Gitman (1990)	Exact solutions: a book.
Beckers and Debergh (1990)	FW transformations and supersymmetry applied on oscillators.
Benítez et al. (1990)	Solution and hidden supersymmetry of a Dirac oscillator, $\boldsymbol{\alpha}\cdot(\mathbf{p}-im\omega\beta\mathbf{r})$.
Bernabeu and Garcia Molina (1990)	Finite-nuclear-size effects on leptonic wave functions.
Chatterjee (1990)	Review on large-D expansions for Dirac and KG problems.
Coutinho and Nogami (1990)	New proof that 2D and 3D δ-function potentials give no solutions for the Dirac equation.
Danilov (1990)	Spectrum of the Dirac operator in R^n with a periodic potential.
Detcheva et al. (1990)	Scattering from an interface model with Kronig-Penney potentials separated by a step.
Dominguez-Adame (1990a)	$V=1/\|x\|$.
Dominguez-Adame (1990b)	1D H atom in momentum representation.
Dominguez-Adame (1990c)	Surface δ interactions.
Dominguez-Adame and Gonzáles (1990)	Solvable linear potetials.
Durand and Gara (1990)	Matrix methods for the numerical solution of relativistic wave equations.
Fassari (1990)	Bound states for relativistic Kronig-Penney hamiltonians with impurities.
Goldberg et al. (1990)	A relativistic Fermi-Segre formula for the wave function near the nucleus from the WKB approximation.

Table 2.4. (continued).

Ivanov et al. (1990)	Dirac Green function for non-singular potentials.
Ivrii (1990)	2D Dirac particle in a strong magnetic field.
Jagannathan (1990)	Theory of magnetic electron lenses, based on the Dirac equation.
Kargabaev et al. (1990)	Dirac-Coulomb Green function for finite nuclei.
Koutroulos (1990)	Gaussian potential model for lambda particles in hypernuclei.
Krstic and Mittleman (1990)	Bound states of atomic hydrogen in a laser field.
Loewe and Sanhueza (1990)	In 1D, no diving to the positron-like continuum by attractive δ-function potentials. In 3D, yes.
Panja et al. (1990)	1/D theory for a spin-½ particle in a screened Coulomb potential.
Pratap (1990)	Dirac electron in axial-wiggler fields.
Quesne and Moshinsky (1990)	Symmetry Lie algebra of the Dirac oscillator.
Reiss (1990)	Hydrogen atom in dominant plane-wave field.
B. Roy and Roychoudhury (1990a)	Shifted 1/D expansion for the Dirac equation for equally mixed vector and scalar potentials.
B. Roy and Roychoudhury (1990b)	Dirac equation with Hulthén potential. Energy levels solved.
Shaarawi et al. (1990)	Nondispersive wave packet solutions to KG and Dirac equations.
Shinada (1990)	H atom in static, uniform **B**. Adiabatic approximation.
Villalba (1990)	Dirac equation in parabolic cylinder and elliptical coordinates.
Anders and Kotlyarov (1991)	Characterisation of the scattering data of Schrödinger and Dirac operators.
Anderson (1991)	Exact solutions for $v=-2k\ \mathrm{sech}^2 kx$, $w^1=-k \tanh kx$.
Bassetto and Griguolo (1991)	Eigenfunctions for a Dirac operator of a vortexlike Abelian potential on a 2D sphere.
Bychkov et al. (1991)	Two-dimensional electrons with SO interaction in a tilted magnetic field.
Castaños et al. (1991)	Soluble extensions of the Dirac oscillator with exact and broken supersymmetry.
Z.-H. Chen and Goldman (1991ab)	H-like atoms in strong **B**. STO-basis.
Clerk and McKellar (1991)	Relativistic band gaps in 1D disordered Kronig-Penney models.
Cordova (1991)	Fractional charge in massive fermion systems in 1+1, 2+1 and 3+1 D.
Coutinho et al. (1991)	Existence of bound states of a Dirac particle in 2D and 3D $U(r)=\beta\ S(r)+V(r)$.
Dominguez-Adame (1991)	A perturbed Dirac oscillator.

Table 2.4. (continued).

Dominguez-Adame and Méndez (1991)	A solvable two-body Dirac equation in 1D.
Dominguez-Adame and Sanchez (1991)	Kronig-Penney models for quasiperiodic lattices.
Franosch and Soff (1991)	Finite-nucleus effects in muonic atoms.
Goldman and Chen (1991)	H-like atoms in magnetic fields. Relativistic correction changes sign around $B=10^{11}$G.
Grochmalicki et al. (1991)	Relativistic effects on ionization of atoms in intense laser fields.
D.S. Guo and Åberg (1991)	Orthogonality and completeness of the Volkov solutions (Wolkow 1935).
S.H. Guo et al. (1991)	2D hydrogen ($V=-Z/r$), with and without Chern-Simons potential.
Hinton et al. (1991)	Inverse scattering on the line for a Dirac system.
Hu and Su (1991)	Hulthén-type potentials.
Iyer and Kamran (1991)	Separability of the Dirac equation in metrics with local rotational symmetry.
Klimenko and Klimenko (1991)	Elastic scattering amplitudes for an electron in a plane-wave EM field.
Kosaka and Yonei (1991b)	Bound states in screened Coulomb potentials. WKB.
Koutroulos (1991)	R/NR lambda particle models with $V(r)=-D[\cosh^2(r/R)]^{-1}$. Results compared with Gaussian and Woods-Saxon models.
Kukulin et al. (1991)	Dirac equation with harmonic oscillator and extra SO potentials.
de Lange (1991ab)	The Dirac oscillator of Moshinsky and Szczepaniak (1989).
J.-F. Lu et al. (1991)	Relativistic electrons in high-power laser light.
Maciá and Dominguez-Adame (1991)	Scattering states of relativistic point-interaction potentials.
Méndez and Dominguez-Adame (1991)	Mathieu potential $V(x)=V_0\cos 2x$ as model for 1D band structures.
Mishchenko and Sitenko (1991)	Ground states of a 2D Dirac operator in a magnetic field with non-local boundary conditions.
Mittleman (1991)	Comments on Rashid (1989).
Molnár and Lovas (1991)	Dirac particle in a periodic field.
Ndimubandi and Ntibashirakandi (1991)	Dirac oscillators through supersymmetry and FW transformations.
Papp (1991ab)	1/D solutions for two-body systems with arbitrary vector and scalar potentials.
Quesne (1991)	Supersymmetry and the Dirac oscillator.
C. L. Roy (1991)	Relativistic surface states.
Roy and Basu (1991)	Band gaps in 1D disordered systems.
Sato et al. (1991)	1D bound-state problems.
Sen Gupta (1991)	Exact solution of the Dirac equation for a helical wiggler with a guide field: $\mathbf{B}=B_{\parallel}\mathbf{k}+B_{\perp}(\mathbf{i}\cos kz+\mathbf{j}\sin kz)$.
Shishkin (1991)	Volkov problem generalised for various tensorial fields.

Table 2.4. (continued).

Shishkin and Ali (1991)	Dirac particle in wave fields (generalized Volkov problem).
Shishkin and Cabos (1991)	Separation of variables for a Dirac equation in Cartesian coordinates.
Shishkin and Yasin (1991ab)	Generalized Volkov problem (external wave fields of different tensor structures).
Thompson and McKellar (1991)	High square barrier.
Toyama et al. (1991)	Relation between the bound-state radius and the scattering length for one and two Dirac particles.
Toyama (1991)	Dito. Solution for the non-local potential of Nogami and van Dijk (1986).
Alan and Barut (1992)	Volkov solution for particles with anomalous magnetic moments.
Arbuzov et al. (1992)	$V(r) = \alpha \cos(Er)/r$ and narrow resonances.
Boudjedaa et al. (1992)	Green functions for KG and Dirac particles in an EM wave field.
Z.-H. Chen and Goldman (1992)	H atom in **B**. R/NR. The correction changes sign around $B=10^{11}$G.
Clerk and Davies (1992b)	Inversion of barrier transmission for a 1D Dirac equation.
de Lange and Raab (1992)	Book on operator methods. Solvable 1D and 3D problems included.
de Lange and Welter (1992a)	'Shape invariance' of the Dirac oscillator.
Domínguez-Adame and González (1992)	Generalized Dirac-Kronig-Penney model with non-local separable potentials. 1D.
Dittrich et al. (1992)	Coulomb potential + spherically symmetric contact interaction.
Domínguez-Adame (1992)	A relativistic interaction without Klein paradox.
González et al. (1992)	A (2+1)D Dirac equation on a sphere used as fullerene model.
Grigis and Mohamed (1992)	Gaps in the Dirac spectrum for periodic V and **B**.
Grosse et al. (1992)	Order of energy levels for a pure Coulomb + vector-like field.
Grotch et al. (1992)	E1 transitions of $b\bar{b}$ bound states.
Grozdanov et al. (1992)	H-like atoms in circularly polarised microwave fields.
Hillion (1992b)	Nondispersive solutions of the Dirac equation.
Hoang et al. (1992)	Exact solution for a Coulomb + Aharonov-Bohm + Dirac monopole field.
Kalnins and Miller (1992)	Dirac equation in Kerr-Newman spacetime. Flat-space applications to bag models.
Krolikowski (1992ab)	Radial equations for the Dirac two-body problem.
Kukulin et al. (1992)	Oscillator potential + SO-coupling.

Table 2.4. (continued).

Lauwers et al. (1992)	Inversion of the Dirac matrix by parallel transported multigrid in SU(2) lattice gauge theory.
Leal Ferreira and Tomio (1992)	Potential bags with smeared surfaces.
Leutwyler and Smilga (1992)	Spectrum of the Dirac operator in vacuum gauge fields, QCD.
Maier and Dreizler (1992)	Poles of the S-matrix for a Dirac square-well potential.
Martínez-y-Romero and Salas-Brito (1992)	Conformal invariance in a Dirac oscillator.
Matsutani (1992)	Berry phase of Dirac particle in thin rod.
Mendel and Trottier (1992)	Model for $Q\bar{q}$ mesons: analytic small-r solution for a $V(r)=-A/(r \ln r)$. Both Dirac and KG solutions finite at origin.
Méndez and Dominguez-Adame (1992)	Level shifts of Dirac particles under point interaction potentials.
Mohamed and Parisse (1992)	Approximations for some singular perturbations.
Moshinsky et al. (1992)	A 1D particle-antiparticle system with a Dirac oscillator interaction.
Mourad and Vashakidze (1992)	$h=h_D + V(1+\beta)/2$.
Nogami and Toyama (1992)	"Transparent potential" (transmission coefficient=1) constructed for the 1D Dirac equation.
Panja et al. (1992)	Large-D expansion for a spin-½ particle in the presence of scalar and vector potentials.
Papp and Micu (1992)	Two-boson systems with $m_1 \neq m_2$ and $V(x) = -g \exp(-ax)$.
Reiss (1992)	Strong EM fields in quantum optics, a review.
Roy and Basu (1992)	Electrical conduction in a 1D disordered system.
Roy and Khan (1992)	Tunneling through a double, rectangular barrier in 1D.
Schramm et al. (1992)	Exact Dirac propagator for the magnetic field of a peripheral collision, $\mathbf{B}= B\,\delta(z)\,\mathbf{e}_x$.
Semay and Silvestre-Brac (1992)	Two-body Dirac and QR equations for a quarkonium model with a Coulomb + confining power law potential. $h=(\boldsymbol{\alpha}_1-\boldsymbol{\alpha}_2)\cdot\mathbf{p}+\beta_2 m_1+\beta_2 m_2+W$, $W= -a/r+br^c(\beta_1+\beta_2)/2 + d$.
Semenov (1992)	A spin-½ neutral particle with an anomalous magnetic moment in a 2D $\mathbf{E}(x,y)$.
Sen Gupta (1992)	Dirac particle in a spatially periodic $\mathbf{B} = B_0 \sin kz$. Comment on Lovas and Molnár (1989).
Shen and Tao (1992)	Exact solution for a 2D Dirac particle with flux ½.

Table 2.4. (continued).

Shin et al. (1992)	Wigner formulation for spin-½ particles in Coulomb and cavity potentials.
Shishkin and Cabos (1992)	Separation of the Dirac equation in curvilinear coordinates for fields of various tensorial characters.
Stepanov and Tutik (1992b)	1/D expansion for the KG equation.
Tamura (1992)	Relativistic Green function for generally shaped and magnetically polarized cell potentials.
van Dijk (1992)	Exact solution of the Dirac equation for a $\beta V_s + V_v$ potential, containing momentum-space form factors $g_0 = 1/(p^2 + \Gamma^2)$, $g_1 = \Gamma p/(p^2 + \Gamma^2)^2$.
Wells et al. (1992)	3+1 D Dirac equation for lepton-pair production in relativistic heavy-ion collisions solved in a Cartesian lattice using splines. Au+Au at 2-100 GeV/nucleon.
Zhou and Rosenberg (1992)	Relativistic scattering of a charged particle from V(r) in presence of a strong, slowly varying multimode EM field.

Table 2.5. Relativistic virial theorems.

Reference	Comments
Bethe (1948)	Rederives $c\langle \boldsymbol{\alpha}\cdot\mathbf{p}\rangle + \langle V\rangle = 0$ (2.11)[a] $E = mc^2\langle \beta\rangle$ (2.12) α^2 limits given.
Epstein and Epstein (1962)	Dirac-Coulomb recursion relations like $s\langle \alpha r^{s-1}\rangle = 0$ $2\langle i\alpha\beta r^s\rangle - s\langle r^{s-1}\rangle - 2k\langle \beta r^{s-1}\rangle = 0$
Albeverio (1972)	VT for Dirac electron in external EM fields.
Kalf (1976)	Derives $\langle \boldsymbol{\alpha}\cdot\mathbf{p}\rangle = \langle r(dV/dr)\rangle$ for a class of potentials, including the Coulomb one, $V(r) = -Z/r$, $Z \le 137$.
Herbst (1977) (ref. 1386)	A KG VT for a Coulomb field: $\langle V\rangle = \langle \mathbf{p}^2/(\mathbf{p}^2 + m_1^2)^{\frac{1}{2}}\rangle$. $E = \langle (\mathbf{p}^2 + m^2)^{-\frac{1}{2}}\rangle$.
Papp (1986c)	Quasiclassical approach to the VT.
Goldman (1987)	Use of the VT $E = mc^2\langle \beta\rangle$, (2.12), in variational energy optimization.
Vrscay and Hamidian (1988)	Hypervirial theorems give PT expansions for central-field Dirac problems.
Lucha and Schöberl (1990a)	Relativistic virial theorem for $T = (p^2+m^2)^{\frac{1}{2}}$: $\langle \mathbf{x}\cdot\nabla V\mathbf{x}\rangle = \sum_{i=1}^{2}\langle \mathbf{p}^2/(\mathbf{p}^2+m^2)^{\frac{1}{2}}\rangle$
Lucha and Schöberl (1990b)	Review.
Pyykkö (1991b)	Comments on relativistic virial theorems. Excellent reply by Ishikawa.
Shabaev (1991a)	Virial relations for off-diagonal radial matrix elements for a Dirac equation in a central field (M1 and E2 hfs integrals etc.) Coulomb field special case.
Luo and Qiu (1992)	VT for KG, Kemmer and spin-less Salpeter equations. KG: $\langle \mathbf{p}^2\rangle = \langle (E-V)(\mathbf{r}\cdot\nabla)V\rangle$.

[a]Equation numbers from Part 1.

3. QUANTUM ELECTRODYNAMICAL EFFECTS

Table 3.1. Higher-order corrections: methods.

Reference	Comments
Todorov (1973)	Recoil corrections and reduced mass in relativistic calculations (for antiprotonic atoms). (Borie and Jödicke 1988).
Gyulassy (1975)	Higher-order vacuum polarisation charge density for finite-radius nuclei.
Barbieri and Sucher (1978)	Radiative corrections to one-photon decay of H-like atoms. Example: 2s -1s.
Dubler (1978)	Calculation method for the vacuum polarisation potential of order $\alpha(Z\alpha)$.
Hossein Partovi and Jabbarian-Lotfabadi (1978)	New fermion-antifermion equation for positronium.
Silagadze and Khelashvili (1984)	Solvable relativistic two-fermion bound-state problem. No Klein paradox.
Hostler (1985b)	Scalar formalism for QED.
Shabaev (1985b)	Two fermions of masses m and M in QED. Mass corrections of order m/M for the limit m<<M without αZ expansion.
Au (1986)	Two-photon exchange in electron-atom scattering. R^{-6} changes to R^{-7} at asymptotic distances ($R>a_0/\alpha$).
Barut (1986)	Reply to comments on Barut and van Heule (1985) by Bialynicki-Birula (1986).
Barut and Strobel (1986)	Center-of-mass motion for a system of relativistic Dirac particles. Exact separation for N particles is possible.
Barut and Ünal (1986)	The two-fermion QED problem solved exactly up to order α^4.
Bialynicki-Birula (1986)	Comments on Barut and van Huele (1985).
Borie (1986)	The role of the fine-structure constant in QED.
Caswell and Lepage (1986)	Effective Lagrangians for bound-state problems in QED.
Dietz (1986)	Vacuum-induced Friedel-type oscillations in relativistic atoms.
Feldman and Fulton (1986, 1987)	Derive the DF, 1/Z and RRPA models from QED.
Grant (1986b)	Review on relativistic atomic theory and QED.
Ni (1986)	Exact solution of the Dirac equation in constant **E** and the electron-positron creation rate in vacuum.
Sazdjian (1986, 1987, 1988)	Two-particle wave functions.
Sucher (1986)	QED and relativity in atomic physics.

Table 3.1. (continued).

Van Alstine and Crater (1986)	Exact parapositronium-like solution to two-body Dirac equations.
Weitsman and Hagelstein (1986)	Self-energy in many-electron atoms.
Yilmazer (1986)	Relativistic two-fermion problem with the most general electric and magnetic potentials.
Holz (1987)	Self-energy of electrons in a Coulomb field: Momentum-space method.
Barut (1987)	Non-perturbative treatment of the Möller potential. Covariant equations for positronium and muonium.
Barut and Dowling (1987)	QED based on self energy, without 2nd quantization: Lamb shift and Casimir-Polder forces.
Barut and Pavsic (1987)	A Dirac electron in curved space.
Dietz (1987b)	A "g-Hartree" approach to QED of many-electron atoms.
Geiger et al. (1987)	A relativistic two-body equation for positronium.
Hostler (1987a)	Self-energy operator for an electron in an external Coulomb potential.
Ladányi (1987)	Scalar Bethe-Salpeter equation and relativistic bound states.
Landro et al. (1987ab)	Two-photon processes for relativistic electrons.
Malvetti and Pilkuhn (1987, 1989, 1990)	Atoms with several spinless particles or one spinless particle among electrons.
Mandelzweig and Wallace (1987)	QED-based two-body Dirac equation.
B. Müller (1987)	The many facets of the Dirac vacuum.
Pacher et al. (1987)	Retardation effects in radiative electron capture.
Sapirstein (1987)	QED of many-electron atoms: a Furry picture based on the many-electron field.
Srivastava and Widom (1987)	QED processes in electrical engineering circuits.
Su and Eberly (1987)	"Essential states" of spontaneous emission in the relativistic theory of electron-photon interactions.
Sucher (1987)	QED and relativistic many-electron Hamiltonians.
Au (1988)	Multipole, nonadiabatic and finite-nuclear-mass corrections for two-Coulomb-photon exchange.
Barut (1988)	Relativistic QM discussed.
Barut et al. (1988)	Self-field QED: g-2.
Barut and Salamin (1988)	Relativistic theory of spontaneus emission.
Bottcher and Strayer (1988)	"DF equations for the vacuum". Pair production in high-energy collision.

Table 3.1. (continued).

Braun (1988)	Decay law of unstable levels in QED. One- and two-photon transitions.
Grant and Quiney (1988)	Foundations of relativistic atomic and molecular theory.
Guo and Åberg (1988)	QED approach to multiphoton ionization in the high-intensity limit.
Hylton (1988)	Nuclear-size correction to the Lamb-shift in hydrogen.
Ionescu et al. (1988)	Nonlinear terms.
Ito (1988)	NR limit of the renormalized Bethe-Salpeter wave function at the origin.
Iwazaki and Kumano (1988)	Oscillations of the polarized vacuum around a large-Z nucleus. Searches neutral modes to explain the GSI positron peaks.
Leal Ferreira (1988)	A Dirac particle and its antiparticle, interacting through a βmr linear scalar potential.
Liaw et al. (1988b)	QED-based theory for atomic energy levels and transitions.
Malenfant (1988, 1989)	The Breit equation for two equal-mass spin-½ particles. Radial equations and small-r behaviour derived.
Nakanishi (1988)	Solutions to the Bethe-Salpeter equation. Contains a 34-page bibliography, until 1987.
Zecca (1988)	Fine structure of equal-mass q-$\bar{q}$ systems.
Au (1989ab)	Transverse-photon exchange in He Rydberg states.
Barut and Dowling (1989ab)	QED without 2nd quantization. No wall effects on g-2 for a confined particle in magnetic field. The cyclotron frequency does change.
Barut and Duru (1989)	QED from classical particle trajectories.
Barut et al. (1989)	Scattering theory of relativistic composite particles.
Dietz and Hess (1989)	Derivation of "g-Hartree" equations from QED.
Kolsrud (1989)	The external field approximation of QED.
Królikowski (1989)	Analytic asymptotic solution for para (S=0) states of two equal-mass s=1/2 particles bound by the Coulomb attraction.
Lindgren (1989c, 1990)	Derives interelectronic potentials from QED. Explains the different results for Feynman and Coulomb gauges in MBPT as missing, irreducible multiphoton exchange.
Lindroth and Mårtensson-Pendrill (1989); Lindroth et al. (1989b)	Analysis of the complete Breit interaction.

Table 3.1. (continued).

Manakov et al. (1989)	Vacuum polarization by a strong Coulomb field. Analytical large-r potential given.
Mittleman (1989)	Theory for three-electron atoms from the Bethe-Salpeter equation. Includes a 3-electron term.
Pindzola et al. (1989)	Electron-impact ionization of highly charged ions in lowest-order QED.
Salomon (1989)	Monte Carlo integration of the Bethe-Salpeter equation in momentum space.
Sazdjian (1989)	N-body bound-state relativistic wave equations for spin 0 and/or ½ particles.
Scott et al. (1989)	Many-particle electrodynamics by symbolic manipulation.
Sucher (1989a)	Bound-state QED.
Sucher (1989b)	Potentials from field theory.
Veltman (1989)	Radiative corrections to polarized Compton scattering.
Barut (1990a)	The electron-positron system at short distances.
Barut (1990b)	Excited states of Zitterbewegung.
Barut and Dowling (1990a)	Self-field-based QED: Thermal radiation seen by an accelerating (a) observer ("Unruh effect"). $T_a = \hbar a/2\pi kc$.
Barut and Dowling (1990b)	Self-field QED on a two-level atom.
Barut et al. (1990)	The classical relativistic two-body problem with spin and self-interactions.
Barut and Ünal (1990)	Analytic evaluation of vacuum polarization in a Coulomb field.
Baur (1990)	Electron-positron pair production in relativistic heavy-ion collisions.
Blunden (1990)	Effect of the Dirac sea on a finite system.
Dykshoorn et al. (1990); Dykshoorn and Koniuk (1990)	Two-particle bound states in QED. Massive and massless cases. Ultra relativistic bound states.
Fainshtein et al. (1990)	Vacuum polarization by a Coulomb field. Analytical approximation of the polarization potential.
Feldman and Fulton (1990)	Log terms for leading radiative corrections to DF.
Leal Ferreira and Galeao (1990)	Two-body Dirac equation with nine different angle- and r-dependent potentials.
Shabaev (1990b)	QED theory for multiply charged ions.
L.-X. Xiao et al. (1990)	Bound-state equations for two spin-0 particles from QED. [KG] Coulomb limit derived.
Barut (1991)	The covariant many-body problem in QED.
Barut and Salamin (1991a)	2s-1s transition probability of hydrogen-like atoms, Z=1-92, from selenergy-based QED. Standard result obtained.

Table 3.1. (continued).

Barut and Salamin (1991b)	Einstein A-coefficients of spontaneous emission.
Bawin and Cugnon (1991)	"Partially screened solutions" for the Dirac equation in $Z>Z_c$ Coulomb fields.
Berseth and Darewych (1991)	Relativistic two fermion one anti-fermion wave equations.
Bialynicki-Birula et al. (1991)	Phase-space structure of the Dirac vacuum.
Blundell and Snyderman (1991)	Basis-set approach to the radiative self-energy in high-Z hydrogenic ions.
De Sanctis and Prosperi (1991)	A two-body equation for nucleons (pion exchange).
Dehnen and Shahin (1991)	Magnetically bound positronium states?
Eby (1991)	Production of low-energy electron-positron pairs by heavy ions.
Elkhovskii and Mil'stein (1991)	Vacuum polarization, induced by nuclear M1 and E2 multipole moments. Dirac-Coulomb Green function used.
Franosch and Soff (1991)	Influence of muonic vacuum polarization of strongly bound electrons.
Graf et al. (1991)	QED in strong fields: a review.
Guo and Åberg (1991)	Orthogonality and completeness of QED Volkov solutions.
Hagelstein (1991)	Partial-wave method for self-energy for non-hydrogenic electrons.
Lucha et al. (1991)	Fermion-antifermion bound states.
Malenfant (1991)	Approximate positronium solutions to the Breit-Coulomb equation through order $O(\alpha^4)$.
Moshinsky and López Laurrabaquio (1991)	Relativistic generalization of the Breit-Wigner two-particle formula, starting from two free KG or Dirac particles.
Pachucki (1991)	Time evolution in QED.
Pindzola et al. (1991)	Lowest-order QED theory using distorted wave states for inelastic electron scattering by highly charged ions.
Shabaev (1991b)	QED theory for the natural spectral line shape of resonance photon scattering on few-electron atoms.
Shvartsman (1991)	Lagrangian of a relativistic spinning particle in QED.
Tesevich (1991)	Vacuum polarisation potential for a finite nucleus near origin.
Thakur (1991)	Spin-0 - spin-($\frac{1}{2}$) and spin($\frac{1}{2}$)-spin($\frac{1}{2}$) two-particle systems.
Thayyullathil (1991)	Fully covariant equation for the bound state problem in field theory. Positronium, compared with Dirac theory.
Zhdanov (1991)	Classical two-body problem including retardation.

Table 3.1. (continued).

Best et al. (1992)	Multiplicity distribution of electron-positron pairs created by strong external fields.
Blundell (1992)	Screened QED calculations on high-Z ions.
Capri and Roy (1992)	A vacuum for a Dirac particle in an external EM field.
Crater et al. (1992)	Nonperturbative solution of two-body Dirac equations in QED. (Applications on meson spectra.)
Darewych and Berseth (1992)	Two- and three-fermion systems in the Hamiltonian QED.
Elkhovskii et al. (1992)	Logarithmic corrections in the two-body problem in QED.
Engel et al. (1992)	Gradient expansion of Green functions of noninteraction particles.
Feinberg and Sucher (1992)	Spin-dependent two-photon exchange forces between a spin-0 particle and a charged spin-½ particle.
Feldman et al. (1992)	Leading radiative corrections to the DF approximation.
Galeao and Leal Ferreira (1992)	General method for reducing the two-body Dirac equation. Unequal masses.
Gorshkov et al. (1992)	QED theory of U^{90+} line shapes.
Greiner and Reinhardt (1992)	Book on QED.
Indelicato and Mohr (1992)	Coordinate-space approach to bound-electron self-energy in V(r), not necessarily Coulomb.
Kinoshita (ed.) (1992)	Book on precision tests on QED.
Klein and Dreizler (1992)	Relativistic two-body equations derived from QED. Equal-mass and Dirac particle + heavy mass limits taken.
Kluger et al. (1992)	Fermion pair production in strong, homogeneous, time-dependent electric field, E(t).
Li and Zhang (1992)	Spinor structures of the Bethe-Salpeter irreducible kernels.
Malik et al. (1992)	Relativistic terms in the reduction of the Bethe-Salpeter equation.
D. Miller et al. (1992)	Finite-element QED in 3+1 D.
Miyazawa and Tanaka (1992)	Relativistic two-body differential equation that reduces to the Dirac equation. Boundary conditions differ from Bethe-Salpeter.
Moore and Scott (1992)	Quantization of a QED Lagrangian via a 1/c series.
Pachucki (1992)	Higher-order binding corrections to the Lamb shift in H-like atoms.
Pilkuhn (1992)	Equal-time relativistic two-body equations for all combinations of spinless and Dirac particles.

Table 3.1. (continued).

Pindzola (1992)	Positron-impact ionization of highly-charged ions in lowest-order QED.
Seke (1992b)	Solution of the renormalization problem in NR QED.
Thakur (1992)	Classical mechanics of directly interacting relativistic particles.
T. Zhang et al. (1992)	Positronium FS and hfs to order α^5 from a new relativistic bound-state formalism for two fermions.

Table 3.2. Higher-order corrections: hyperfine interactions and magnetic moments.

Reference	Comments
Martynenko and Faustov (1987)	Hfs of muonium. Corrections of order $(Z\alpha)^2$ and $(m_1/m_2)(Z\alpha)^2$.
Milstein and Yelkhovsky (1990)	Vacuum polarisation by the nuclear quadrupole moment.
Zhang and Koniuk (1992)	Muonium hfs in a new bound-state formalism.

Table 3.3. Higher-order corrections: energy levels.

Reference	Comments
Hossein Partovi and Jabbarian-Lotfabadi (1978)	Positronium
Borie and Rinker (1982)	Muonic atoms.
Mohr (1983b)	n=1 and 2 states of H-like atoms, Z=10-40.
Borie (1984)	Vacuum polarization and fine structure in antiprotonic atoms.
Ayres (1986)	Fermion-antifermion bound states.
Indelicato (1986)	Various levels of Breit interaction and the 1s binding energy of He-, Ne-, Xe- and Rn-like ions, Z=100...170. SCF Gaunt term tested.
Mohr (1986)	Review on QED effects in many-electron and few-electron atoms.
Indelicato et al. (1987)	Radiative corrections for He-like ions.
W.C. Martin (1987)	Lamb shifts for 1sns and 1snp terms of He I. Various sources combined.
Passante and Power (1987)	Lamb shift in NR QED.
Quiney et al. (1987b)	SCF treatment of the Breit interaction advocated. He- and Be-like systems.
Adkins et al. (1988)	Three-photon contribution to positronium hyperfine interval.
Beier and Soff (1988)	The Källén-Sabry $\alpha^2(Z\alpha)$ contribution to the Lamb shift of H-like atoms, Z=1-110.
Borie and Jödicke (1988)	Energy levels of antiprotonic atoms.
Cheon (1988)	Casimir effect on the $2s_{\frac{1}{2}}$ - $2p_{\frac{1}{2}}$ Lamb shift of a hydrogen atom between two conducting plates.
Drake et al. (1988)	The Lamb shift of He^+.
Gorceix and Indelicato (1988)	Effects of the complete Breit interaction on 1s2s and 1s2p He-like systems, Z=6...92.
Mohr (1988a)	Review on QED calculations on H-like atoms.
Mohr (1988b)	H- and He-like atoms.
Sallhofer (1988, 1989a-c) 1990 ab)	Hydrogen in QED.
Silver (1988)	Precision spectroscopy on 1- and 2-electron ions of medium and high Z.
Soff and Mohr (1988)	Vacuum polarization for finite nuclei. H-like atom K- and L-shells, Z=30...170.
Adkins (1989)	Review on positronium bound states.
Grotch (1989)	Status report on the Lamb shift.
Kim (1989)	Breit correlation in Al-like ions.
Manakov et al. (1989)	Vacuum-polarisation effects on H-like energy levels.
Mohr (1989a)	QED as a perturbation in relativistic MBPT.

Table 3.3. (continued).

Mohr (1989b)	High-Z few-electron atoms.
Petelenz and Smith (1989)	Vacuum polarization in muonic molecules, ppμ, ddμ etc., using the Uehling potential. Earlier values underestimated 2-8 times.
Plunien et al. (1989)	Nuclear polarisation contribution to the Lamb shift in heavy atoms. For $1s_{1/2}$ of $_{92}U^{238}$, a nuclear rotator, the shift is about 1 eV. For $_{82}Pb^{208}$ a vibrator, about 0.1 eV is obtained.
Sapirstein (1989a)	Two-photon exchange graphs for He-like systems in the Furry picture.
Sapirstein (1989b)	Rewiev.
Schmidt et al. (1989)	Muonic 1s - 5g levels in atoms with Z = 30-100. Vacuum polarization to order $\alpha(Z\alpha)^n$, $n\geq3$ included.
Seely (1989)	QED contributions to the 2p-2s transition energies of Li-like ions.
Soff (1989)	Vacuum-polarization charge densities for spherically symmetric external potentials.
Soff and Mohr (1989)	Vacuum-polarization corrections for H-like U.
Yennie (1989a)	Recoil effects in H-like bound-state problems.
Yennie (1989b)	PT approach to many-electron energy levels.
Doncheski et al. (1990, 1991)	Lamb-shift in H-like atoms.
Drake and Swainson (1990)	Bethe logarithms for H, $n\leq20$. Approximations for 2-electron atoms.
Ishikawa (1990a)	Be-like ground states, Z=4...50, with SCF Breit term.
Radozycki (1990a)	The EM virtual cloud in the H atom ground state.
Radozycki (1990b)	The $2p_{3/2}$ state.
Blaive et al. (1991)	Discrete-state contribution to the Lamb shift in hydrogen atoms.
Cheng et al. (1991)	Screened Lamb-shifts for Li-like U, Na-like Pt and Cu-like Au.
Eberlein (1991)	Level shifts for a hydrogen atom between two conducting plates.
Hersbach and Ruijgrok (1991)	Quasipotential theory for H-like atoms.
Indelicato and Mohr (1991ab)	QED effects in atoms. Effect of electron interactions on the self energy.
Mohr (1991)	Lamb shift for the n=4 state of He^+.
Plunien et al. (1991)	Nuclear polarisability and the Lamb shift.
Soff (1991)	Review on few-electron atoms.
Snyderman (1991)	Electron radiative self-energy for large $Z\alpha$. H-like atoms treated, others discussed.
Barut et al. (1992)	Lamb shift in self-field QED.

Table 3.3. (continued).

Blundell (1992)	Screened QED calculations in high-Z ions.
Bukowski et al. (1992)	Bethe logarithm and Lamb shift for H_2^+ and isotopic species.
Drake and Yan (1992)	Rydberg levels of He.
Goldman and Drake (1992)	Rydberg levels of He. Long-range Casimir-Polder effects seen.
Komachiya and Fukuda (1992)	On-shell QED expansions for energy levels of H-, He- and Li-like atoms.
Mohr (1992)	Self-energy correction to n=1, 2 levels of H-like atoms, Z = 5-110.
Mohr and Kim (1992)	Self-energy of excited states (n=3-5) in a strong Coulomb field, Z =10-110.
Zhang and Koniuk (1992)	Muonium FS splitting in a new bound-state formalism.

Table 3.4. Higher-order corrections: interatomic and -molecular interactions.

Reference	Comments
Meath and Hirschfelder (1966b)	Retarded intermolecular forces.
Bernabéu and Tarrach (1976)	Interactions between two particles, at least one of them spinless and at least one neutral.
Golshan and Kobe (1986)	Retarded interactions between nonidentical two-level atoms.
Plunien et al. (1986)	Review on the Casimir effect.
Spruch (1986)	Review on the retarded, Casimir long-range potentials.
Barut and Dowling (1987)	Casimir-Polder forces from "QED without 2nd quantization".
Milonni et al. (1988)	Interpretation of the Casimir force between two parallel, conducting plates.
Mosteepanenko and Trunov (1988)	Review on the Casimir effect.
Babb et al. (1992)	Casimir shifts for He-like low-Z Rydberg ions.
Milonni and Shih (1992)	Source theory of the Casimir force.

4. MULTIELECTRON ATOMS: METHODS

Table 4.1. General methods and basic theory for multielectron atoms. See also Table 1.1. for general references and the Table 3.1 for QED aspects.

Reference	Comments
Fock (1933,1934)	The exchange energy for Dirac electrons.
Johnson Jr. and Breit (1933)	Breit interaction for one electron outside closed shells.
Breit (1937)	Approximately relativistic equations for nuclear particles.
Brown and Schaefer (1956)	Expansion in angular momenta in bound-state PT.
Malli and Saxena (1971)	Spin-spin interactions in f^4 configurations.
Braun and Parera (1985)	Bethe-Salpeter equations for two-electron atoms.
Parera López et al. (1985); Parera and Grave de Peralta (1987)	Relativistic electron-pair theory in multielectron atoms.
Braun and Parera (1986)	TP between degenerate levels in a many-electron atom.
Chang et al. (1986)	Two-component Pauli-like effective Hamiltonians in Dirac theory. Tested on H-like atoms and on rare-gases.
Feldman and Fulton (1986,1987);	Derive the DF, 1/Z and RRPA models from QED.
Fulton (1987,1989)	Fundamentals of basis-set Dirac calculations. Advocates self-consistent treatment of the Breit term.
Grant (1986a)	
Heully et al. (1986a)	Transformed 1- and 2-electron Dirac Hamiltonians for MB calculations.
Heully et al. (1986b)	Unsuitable positive-energy projection operators for 1-electron atoms introduce negative-energy states, don't remove them.
Rosenberg and Spruch (1986)	Extremum principles for relativistic bound-state energies.
Rutkowski (1986c)	New PT approach for two-electron atoms.
Sazdjian (1986)	Relativistic wave equations for systems of two (spin 0 or ½) particles.
Sebastian (1986)	Relativistic invariance of the two-photon transition amplitude of a composite system.
Sucher (1986-1989ab)	Relations of the many-electron Hamiltonian to QED.

Table 4.1. (continued).

Blundell et al. (1987b)	1st, 2nd and 3rd-order MBPT formulae for one-valence-electron atoms.
Brown (1987a)	Review on the relativistic atomic many-body problem.
W.Q. Chen and Cook (1987)	Quasirelativistic treatment of nuclear motion in atoms and molecules.
Crater and van Alstine (1987)	Two-body Dirac equations for particles interacting through world scalar and vector potentials.
Glöckle et al. (1987)	A solvable 2-fermion 1D model with δ-function interaction. Lorentz-boost operator constructed.
Goldman (1987)	Accelerated variational convergence in DF-LCAO calculations by a virial-theorem method.
Grant (1987)	Relativistic electron-electron interaction.
Lindgren (1987a)	Relativistic many-body problems: proceedings of Trieste, 1986.
Lindgren (1987b)	The NR and R coupled-cluster approaches: a review.
Lindgren and Lindroth (1987)	Review on R MBPT.
Lindroth (1987a-c)	The relativistic pair equation.
Lindroth et al. (1987)	A relativistic pair equation, projected onto positive-energy states. Permits higher-order Breit corrections.
L.D. Miller (1987)	FW reduction of the DF equations for atoms.
Olsson and Lindgren (1987)	Analyse the FS and hfs tensor operators needed in relativistic MBPT.
Sapirstein (1987)	QED of many-electron atoms: a Furry picture based on the many-electron field. MBPT. Quotes Mohr (1986), Labzovskii (1970ab).
Broyles (1988)	Relativistic propagator for a multi-electron atom consisting of n spin-½ fermions of different masses.
Coutinho et al. (1988b)	A "two-body Dirac equation" in 1D for various interaction potentials.
Datta and Devaiah (1988)	The minimax technique in relativistic HF calculations.
Drake and Goldman (1988)	Finite-basis-set-methods in atomic physics.
Grant (1988a)	Review on high-Z atoms.
Grant (1988b)	Review on relativistic atomic calculations.
Grant and Quiney (1988)	Foundations of relativistic atomic and molecular theory.
Johnson (1988)	Review on RRPA.
Johnson et al. (1988a)	B-spline Dirac basis.
Liaw et al. (1988b)	QED-based theory for atomic energies and transitions.
Lindgren (1988)	A relativistic MBPT in Feynman gauge.

Table 4.1. (continued).

Lindroth (1988)	Numerical solutions of the relativistic pair equation.
Pilkuhn (1988)	A "Kramers-type" equation for two spin-1/2 particles in an external field derived from the Breit equation.
Quiney (1988); Quiney et al. (1989c, 1990)	Relativistic MBPT.
Shabaev (1988)	Nuclear recoil effects for multicharged ions.
Bagrov and Obukhov (1989)	Separation of DF equation.
Broyles (1989)	Continuum dissolution discussed. Positive-energy states propagate forward in time, $E<-mc^2$ states backwards. Therefore no danger of continuum dissolution.
Chaix and Iracane (1989)	From QED to DF.
Chaix et al. (1989)	Variational stability of the QED vacuum in the mean-field theory.
Chang and Chung (1989)	Electron correlation due to intrashell core excitation and intershell core-valence interaction.
Darewych and Horbatsch (1989,1990)	Two-body Coulomb systems of spin 0, ½.
Datta (1989)	Relativistic equations of motion method.
Hess and Ohno (1989)	The g-Hartree method tested on Be and Ne.
Jankowski and Rutkowski (1989)	Double-PT for relativity and correlation. Applications for He, Be- and Ne-like systems ($Z\le80$) and for Zn^{2+}.
Jansen and Hess (1989b)	Test the 'no pair' Hamiltonian of Hess (1986) on Au.
Johnson et al. (1989b)	Relativistic, QED and PNC effects in atoms: The Santa Barbara 1988 workshop.
Lindgren (1989ab)	Reviews on relativistic MBPT.
Lindgren (1989c,1990)	Derivation of interelectronic potentials from QED. Explains the different results for Feynman and Coulomb gauges in MBPT as missing, irreducible multiphoton exchange, see next entries.
Lindroth and Mårtensson-Pendrill (1989); Lindroth et al. (1989b)	Analysis of the complete Breit interaction, cp. Gorceix and Indelicato (1988). Excitations <u>from</u> $\epsilon<0$ states are allowed, excitations <u>into</u> them not.
Mårtensson-Pendrill (1989ab)	R/NR pair-correlation theory. Numerical solution in a 2D grid. Projection operators needed. Applications discussed.
Mittleman (1989)	Theory for 3-electron atoms from the Bethe-Salpeter equation.
Nicolaides (1989)	"State-specific theory".
Oreg et al. (1989)	Configuration-average energy shift owing to CI. Both ls and lsj coupling.

Table 4.1. (continued).

Polyzou (1989)	Relativistic two-body models.
Roothaan (1989)	Discusses retardation effects and the many-electron Dirac problem.
Rosenberg (1989)	Virtual-pair effects in atomic theory.
Salomonsson and Öster (1989)	Relativistic all-order pair functions from discretized single-particle Dirac functions. N^3 operations needed for N radial lattice points. Convergence to $0.075\alpha^2(1+\frac{1}{2})^{-2}$.
Sampson et al. (1989)	A new DS program for highly charged ions.
Schwarz et al. (1989)	The radial origin of direct and indirect relativistic effects in atoms analysed.
Shadwick et al. (1989)	A method optimizing variationally the relativistic atomic potential. Equivalent to DF.
Tisenko (1989)	Dul'yan-Faustov pseudopotential equation for the He atom FS.
Bijtebier (1990ad)	Three-fermion bound states: separable equations.
Bijtebier (1990b)	Scattering from a relativistic bound state.
Bijtebier (1990c)	Two relativistic fermion bound states in a potential.
Cheung and Li (1990)	A two-body Dirac equation with phenomenological potentials. Reduced to two coupled 2nd-order equations. Klein paradox discussed.
Darewych (1990)	Three-fermion wave function with Coulomb interactions.
Kim (1990)	Relativistic and QED effects in highly-charged ions.
Migdalek (1990)	Book on model potentials for atoms.
Scott et al. (1990)	PT by symbolic computation.
Sneddon and Whittingham (1990)	Radial matrix elements involving continuum Dirac functions.
Styszynski and Karwowski (1990)	Screening constants in MCDF.
Franosch and Soff (1991)	Influence of nuclear shape and of muonic vacuum polarization for strongly bound electrons.
Gaida et al. (1991)	Changing centers-of-mass in the relativistic Lagrangean mechanics of a system of particles.
Grant (1991)	Review on relativistic basis-set calculations.
Hamacher and Hinze (1991)	"Finite-volume variational method" for the Dirac equation.
Johnson (1991, 1992)	Review on R MBPT.
Kagawa et al. (1991)	STO-basis-set CI theory.
Lindgren (1991)	Review on R MBPT.
Liu and Kelly (1991)	MBPT, based on MCDF.
Parpia and Mohanty (1991)	Theory part for the MOTECC 91 atomic program MCGDFB.

Table 4.1. (continued).

Samzow and Hess (1991)	Two-electron terms in the "no-pair" Hamiltonian.
Avgoustoglou et al. (1992)	MBPT formulae through 3rd order for 1-particle-1-hole excited states of closed-shell atoms.
Crater and Van Alstine (1992)	Two-body interactions by classical relativistic field theory.
Grant (1992)	Review on basic theory.
Hersbach (1992)	A momentum-space theory for two spin-½ particles. Arbitrary mass ratio. Reduces to the Breit equation.
Lister and Judd (1992)	Matrix elements of the spin-other-orbit operators in the atomic f shell.
Parpia et al. (1992b)	Nuclear-motion corrections for the GRASP2 package. Ne-Rn.
Scott et al. (1992)	Accurate (18-figure) (1D) FEM solution of the two-body Dirac equation with static Coulomb interactions.
Szász (1992)	Book on HF-(or DF-) level atomic theory.
Uylings (1992)	Second quantization for atoms in various angular-momentum couplings.

Table 4.2. Published programs for atoms. For a more detailed review of pre-1979 programs, see Grant (1979). Further non-relativistic programs with relativistic perturbation are listed there.

Reference	Comments
Rinker (1979)	Energy levels and transition intensities in muonic atoms. Dirac equation solved in finite nucleus potential + Uehling + Källen-Sabry potential.
Fraga et al. (1987)	A Breit-Pauli-level CI program. Uses numerical radial functions from Froese Fischer's program.
Joc and Bojara (1987)	A PC version of Desclaux' code.
Bieron (1988)	PC version of Grant's atomic program.
Band et al. (1989b)	The RAINE package. Includes EM transitions and nuclear processes.
Dyall et al. (1989b); Parpia and Grant (1989)	GRASP: The Oxford MCDF package.
Dzuba (1989)	The Novosibirsk atomic package.
Scott et al. (1989)	Many-particle electrodynamics by symbolic manipulation.
Shadwick et al. (1989)	A program for variational optimization of atomic potentials. Equivalent to DF.
Froese Fischer (1991a)	A MCHF package including Breit-Pauli corrections.
Froese Fischer (1991b)	A BP CI program for the MCHF package.
Froese Fischer and Godefroid (1991)	Transition probability program, LS and LSJ. EL and ML transitions.
Hibbert and Glass (1991)	A program for angular integrals of the Breit-Pauli Hamiltonian.
Mohanty et al. (1991b); Parpia and Mohanty (1991)	The MOTECC 1991 atomic GTO basis programs.
Perger and Karighattan (1991)	A continuum wave function solver for the package of Grant.
Salvat and Mayol (1991)	Accurate numerical solutions of Dirac equations for central fields.
Kagawa (1992)	Program for fractional parentage coefficients of jj-coupled states.

Table 4.3. Numerical, non-statistical four-component methods.

Reference	Comments
Rabin (1982)	Solves the lattice fermion doubling problem.
Stacey (1982)	Elimination of fermion doubling. Used for atoms by Salomonsson and Öster (1989).
Kogut (1983)	"Fermion doubling": spurious states for a discretized Dirac eqation.
Heully (1985)	Thesis on relativistic MB calculations.
Johnson et al. (1988a)	Finite B-spline basis-set for Dirac equation.
Salomonsson and Öster (1989)	Ways to remove spurious states from discretized basis functions.
Durand and Gara (1990)	Matrix methods for the numerical solution of relativistic wave equations.
Jaskólski et al. (1990)	Outward integration procedure for QR radial equations.
J. Wu et al. (1990)	Numerical methods for TD nuclear Dirac equations in 3+1 or 1+1 dimensions.
Perger et al. (1991)	Numerical treatment of continuos functions a frozen DF potential.
Cai et al. (1992)	Extension of GRASP, the Oxford MCDF code, to thousands of configurations.

Table 4.4. Four-component LCAO approaches for many-electron atoms. For one-electron atoms, see Tables 2.3. and 2.4.

Reference	Comments
Goldman and Dalgarno (1986)	A method avoiding spurius roots, variational collapse and continuum dissolution. He, Ne^{8+}; Ne^{6+}; Ne^{4+}. STO-basis.
Grant (1986ab)	Fundamental theory for basis-set work. He-like and Be-like systems, Ne. SCF vs. PT Breit.
Krause (1986)	He-like systems, Z=2-100. Screened hydrogenic basis in a variational treatment.
Y.S. Lee (1986)	An approximation, neglecting small-component Coulomb and exchange contributions. Test on I^{n+}.
Y.S. Lee and Baeck (1986)	Conditions for kinetic balance tested on U^{90+}.
Mark (1986)	GTO basis sets for H-Ne.
Hegarty and Aerts (1987)	Ne.
Matshuoka (1987)	Nuclear attraction integrals for GTOs for a homogeneously charged spherical nucleus.
Matsuoka and Huzinaga (1987)	Well-tempered (geometric series, $\alpha=\alpha_0\beta^{n-1}$, n=1-N) GTO basis calculations for He-Ar, Ag, Xe. Finite nuclei.
Mukoyama and Lin (1987ab)	STO-basis applied for the Dirac equation in a DS potential. One spurious root for kappa > 0.
Mukoyama and Lin (1987c)	Inner-shell ionization in ion-atom collisions.
Quiney (1987); Quiney et al. (1987a)	Finite-basis studies of the Dirac equation.
Quiney et al. (1987b)	SCF treatment of the Breit interaction. He- and Be-like systems.
Baeck and Lee (1988)	Open-shell DF-LCAO. Boron SO splitting.
Datta and Devaiah(1988); Datta and Madhusudana Rao (1988)	He.
Goldman (1988a)	He, Be, C and Ne sequences using a method avoiding spurious roots. Breit terms also included.
Goldman (1988b)	Variational treatment of the Breit terms.
Ishikawa and Sekino(1988)	Be, Ne with various GTO basis sets and finite nuclei.
Goldman (1989a)	Accurate polarizabilities for 1-electron atoms, Z=1...115, by a gauge-invariance method.

Table 4.4. (continued).

Grant (1989)	Assessment of basis-sets for Dirac and QED calculations.
Ishikawa et al. (1989)	Be, Ne^{6+}, Ar^{14+}, Ne. Up to 3rd-order MBPT with a GTO basis.
Y.S. Lee et al. (1989)	STO basis-set recipes for Li-F, Na-Cl, U^{82+}.
Matsuoka (1989)	Comments on "unified" and "separated" basis spinors: equivalent.
Matsuoka and Okada (1989)	Tl-Rn using a well-tempered GTO basis.
Mohanty and Clementi (1989a)	DF-GTO finite-nucleus results for He-Ar, Xe, Rn.
Mohanty and Clementi (1989b)	Reviews on GTO relativistic calcula-lations for atoms.
Okada and Matsuoka (1989)	Well-tempered GTO basis for He-Hg.
Quiney et al. (1989a)	Ar-like systems.
Quiney et al. (1989b)	Review.
Quiney et al. (1989c)	Relativistic MBPT.
Sekino and Ishikawa (1989)	MBPT on Ne.
Baeck and Lee (1990)	Al SO.
Ishikawa (1990a)	Be-like ground states, Z=4, 10, 18 and 50, with SCF Breit term.
Ishikawa (1990b)	He-like ground states, Z=2...50. Ne, Ar. MBPT.
Ishikawa (1990c)	Summarises Ishikawa (1990ab).
Ishikawa and Sekino (1990)	Variational bounds for finite c and finite nuclei. C, Hg^{74+}, Hg.
Ishikawa et al. (1990)	Effects of GTO contraction. Ne, Ar, Ge.
Kumar et al. (1990)	K- and L-shell ionization of Au by H^+, He^{2+} impact.
Matsuoka (1990a)	A finite-nucleus model with constant dV/dr proposed for STO basis. $V(r)=v_0 + v_1 r$, $r< R$. Tests on H, Ag^{46+}, H^{79+}.
Mohanty and Clementi (1990a)	Kinetically balanced geometric (well-tempered) GTO basis sets for Z=2-86. Single-configuration DF results compared to numerical ones.
Okada et al. (1990)	Well-tempered GTO basis for He-Hg. DF + Breit at SCF level.
Quiney (1990ab, 1991)	Reviews on relativistic finite-basis calculations.
Sekino and Bartlett (1990)	Coupled-cluster E for Be-like systems, Z=4...50. No essential contributions beyond MBPT(4).
Grant (1991)	Review on basis-set atomic calculations. Hg with SCF versus PT Breit.
Ishikawa (1991a)	Kr. Up to 3rd-order MBPT with GTO basis.
Ishikawa (1991b)	Review on basis-set DF and MBPT calculations.

Table 4.4. (continued).

Ishikawa (1991c)	Effects of GTO basis-set contraction on relativistic and Breit energies. Be...Sr.
Ishikawa and Binning (1991)	Ar.
Ishikawa et al. (1991)	DFB GTO solutions for He-Xe, Be-Sr.
Kagawa et al. (1991)	STO-basis-set CI theory. Applications on Ne-like ions, Z=18...54.
Mohanty et al. (1991b)	Review on atomic basis-set calculations. Description of the MOTECC 1991 program REATOM.
Mukoyama and Lin (1991)	Virtual orbitals used to describe continuum states in ionization during ion-atom collisions.
Visser et al. (1991)	An open-shell GTO DF program tested on C and Sn atoms.
Wilson (1991)	Review on relativistic MBPT. Gives details on Ar ground-state E.
Datta (1992)	Minimax techniques for two-electron atoms.
Goldman (1992)	Fundamentals of basis-set approach tested on the Dirac-Coulomb case.
Ilyabaev and Kaldor (1992)	E of He-Xe and Be. CC using a no-pair Hamiltonian. GTO basis.
Ishikawa (1992a)	E of Xe. MBPT with well-tempered GTO.
Ishikawa (1992b)	Xe pair correlation energy.
Ishikawa et al. (1992b)	DF-Breit on Yb, Hg, Pb, Rn.
Matsuoka (1992)	I, Au. Contractions tested.
Malli et al. (1992)	He...Xe using a "universal" (18s, 12p, 11d) basis.
Parpia and Mohanty (1992)	Finite-nucleus calculations with a Fermi-distribution for He-like ions, Z=10...160. Hg.
Parpia et al. (1992)	DF and DF-Breit E of $(ns)^2$ and $(np)^6$ systems, He...No; H^-, F^- - At^-.

Table 4.5. Various four-component local-density methods.

Reference	Comments
Liberman (1968)	An exchange potential tested on Ar and Hg.
Euwema and Stukel (1971)	Several density functionals tested on E of He-Xe.
Ramana and Rajagopal (1981c)	Theory of spin polarised inhomogeneous relativistic electron gas.
Datta (1987)	Relativistic extension of the Hohenberg-Kohn theorem.
Dreizler et al. (1987)	Relativistic density functional theory.
Baltin and March (1988)	Relativistic electron gas in a weak local potential.
Y.-F.Guo and Whitehead (1988, 1989ab)	Several LSD potentials implemented at QR (Cowan-Griffin) level.
Glushkov (1989)	LDF model including BP terms.
Sampson et al. (1989)	An accurate DS program for highly-ionized atoms. Big mesh, 2nd-order Dirac equation used. QR option.
Sen et al. (1989b)	A "Z-transition state" concept for energy differences of isoelectronic atoms and molecules. QR. $Z^* = (Z_A + Z_B)/2$.
Dietz and Hess (1990)	Inhomogeneous many-electron systems and the "g-Hartree" method.
Dreizler and Gross (1990)	Book on density functional theory.
Glushkov (1990)	An expression for the electron-electron repulsion potential.
Dietz and Hess (1991)	HF-type equations from QED with non-linear gauge fixing.
Guo and Whitehead (1992)	Ne-Rn treated at NR/QR/R Ξ level.

Table 4.6. Thomas-Fermi calculations

Reference	Comments
Malzacher and Dreizler (1986)	Discuss the non-integrable density of relativistic TF models.
March (1986)	Applies relativistic TF theory on bond lengths of CH_4-PbH_4.
Engel and Dreizler (1987); Engel et al. (1987)	Relativistic density-functional theory including gradient corrections ("TFDW"). Derivation from field theory.
S.H.Hill et al. (1987)	Statistical inclusion of h_d and H_B in RTF theory. Results compared with DF ones for Cu-like systems, Z=29...58 and neutral atoms, Z=20...120.
Engel and Dreizler (1988, 1989)	Solution of the RTFDW model for atoms and positive ions.
Englert (1988)	Semiclassical theory of atoms. Book.
March (1988)	Approximate R electron densities from NR TF theory.
Membrado et al. (1988)	Relativistic TF potential outside a surface.
Leung and Pei (1989)	High-density expansions of the relativistic TF equation.
March (1989)	Relativistic TF theory: a review.
H.Müller et al. (1989)	Extension of the relativistic TFDW model to arbitrary external fields.
Yonei (1989)	Relativistic TF models of compressed atoms: Fe and Au. Significant R/NR differences at high compression, even for light atoms.
Centelles et al. (1990)	Relativistic TF with $\hbar^2$ corrections for nuclei.
Pohlner and Dreizler (1991)	Fourth-order gradient corrections to the relativistic Thomas-Fermi-Weizsäcker model.
Stroucken and Dreizler (1991)	Relativistic TF theory at finite temperatures.
Cinal and Englert (1992)	The Scott (1952) nuclear singularity correction in momentum space. Hg.
Von-Eiff and Weigel (1992)	Nuclear RTF model with quantum corrections.
Von-Eiff et al. (1992)	Relativistic TF model for nucleons in a vector and scalar potential.

Table 4.7. Independent-particle models. See also Table 2.4.

Reference	Comments
Bohr and Coster (1923)	Introduce a screening parameter, δ, for the Sommerfeld relativistic term $E = -[(Z-\delta)/n]^4[n/k-3/4]/2c^2$.
Parker (1986)	Inner-shell energy levels with $V = -Z_{eff}/r$ for every shell, plus corrections.
Zapryagaev et al. (1986)	Relations between the quantum defects, phases and the Green function.
Ivanov and Safronova (1987a)	Screening parameters and relativistic corrections.
Johnson (1987)	Parity violation using several model potentials (Tietz, Green, Norcross).
Lombardi (1987)	Na valence levels. Determines quantum defects and core polarizabilities.
Salvat et al. (1987a)	A 5-parameter MP consisting of three Yukawa potentials for Z=1-92, fit to DS.
Churilov et al. (1988)	n=4, $\bar{n}$=4 transitions of Ni- and Zn-like ions, Z=30 - 50.
Glushkov (1988)	A model for multielectron atoms, based on screened Coulomb functions.
L.N. Ivanov et al. (1988)	Review on multicharged ions.
Johnson et al. (1988a)	B-spline basis used with a Tietz potential for Cs and Tl.
Laughlin and Victor (1988)	Review on model-potential methods.
Malinovskaya (1988)	Cl-, Ar- and K-like ions, Z=19-36.
Rogers et al. (1988)	A parametric potential of Yukawa terms plus long range Coulomb tails, fit to QR calculations. Both neutral and highly ionized atoms considered.
Zapryagaev etal. (1988)	Polarizabilities of Li- and Na-like ions using a quantum defect method.
Bao et al. (1989)	Quantum defects for s, p, d, f and g states of ions with $2 \le Z \le 95$.
Martin (1990); Martin and Karwowski (1991)	Quantum-defect orbitals for valence, Rydberg and continuum atomic states.
Qiu and Li (1990)	Energy levels for a Dirac electron in a spherical TF potential, confined to radius R_W. Fe as example.
Glushkov (1991b)	Ga-like systems, Z = 31-36.
Glushkov (1992ac); Malinovskaya and Glushkov (1992)	K-, Rb- and Cs-like systems.
Glushkov (1992b)	Kr, Xe. Claims to find positive EA.
Loginov (1992)	$5d^9 6p$ levels of Tl IV-Br VI.
Misra et al. (1992)	Parametrized formula for x-ray spin doublets.

Table 4.8. Definitions, reviews and background for effective potential calculations.

Reference	Comments
Marketos (1989)	Discusses theoretical justifications for atomic, Phillips-Kleinmann-type calculations.
Migdalek (1990)	Book on model potential methods for atoms.
Walzer (1990, 1991ab)	Radial and energetic atomic parameters derived from the PP of Bachelet et al. (1982) for classifying binary intermetallic compounds.

Table 4.9. Effective-potential methods. For a comprehensive summary of work on molecules, see Table 7.7. Other tests on atoms are also given in references there and in ch. 5.

Reference	Comments
Taksar et al. (1971)	A simple PP for Na valence transitions.
Durand (1986a)	Core-valence correlation in PP metods: a review.
Grave de Peralta and Boada (1986)	Discuss pseudopotentials for one- valence-electron atoms.
Syromyatnikov and Savicius (1986)	Semiempirical, relativistic method for atomic energy levels.
Durand and Malrieu (1987)	A review on effective-Hamiltonian theory.
Hurley and Christiansen (1987)	PP QMC, applied on Li and K EA.
Mohan and Hibbert (1987)	A 2-electron MP for 6s-6p transitions of Hg.
Silberbach (1987)	Semiempirical PP for Ba, Au, Hg, Tl and Pb fit to three lowest s, d and f states.
Ermler et al. (1988a)	Review.
Fuentealba et al. (1988)	Correlation energy differences between N- and (N-1)-electron systems from a polarization potential.
Gropen (1988)	Review.
Laughlin and Victor (1988)	Review on model potentials for 1- and 2-valence-electron atoms.
Dolg (1989)	Energy-adjusted QR PP for Ln. 4f in core.
Hamann (1989)	Generalized norm-conserving pseudopotentials for arbitrary electron energies. Tested on bulk BaSe. Avoids problems with f-resonances for Ba in a plane-wave expansion and with multiple reference states for various l.
Hibbert (1989)	Reviews model potentials in oscillator -strength calculations.
Bylander and Kleinmann (1990)	4f resonances of Ba using a norm-conserving PP.
Masovic (1990)	A SO PP for band-structure calculations.
Milman et al. (1990)	Relativistic PP for transition-metal atoms. Four-component DFT. M, M^+, M=Sc-Cu, Y-Ag, Lu-Au.
Sellers and Ove (1990)	Anisotropic PP for modelling partially filled d shells.
Mitás et al. (1991)	Nonlocal PP in QMC. Si, Sc and Cu IP and EA.
Fernandez Pacios (1992)	Ar-core PP for Sc-Cu, improved valence excitation energies.

Table 4.9. (continued).

Flad et al. (1992)	Cartesian Gaussian projection operators for quantum Monte Carlo.
Leininger et al. (1992)	K-Cs, Ca-Ba and Fe-Os valence levels as tests of the Barthelat-Durand, Christiansen, Hay-Wadt and Stoll-Preuss PP.
Seijo et al. (1992)	Correct valence orbital nodal structure required for valence E_{corr} of X^-, X, X^+; X = F-I.

Table 4.10. Available relativistic effective potentials. QR = quasirelativistic (one-component), DF = Dirac-Fock, LD = local density, SO: spin-orbit perturbation potential given.

Reference	Comments
Greenside and Schlüter (1983)	$_{21}$Sc-$_{29}$Cu. LSD. SO.
Hurley et al. (1986)	$_{19}$K-$_{36}$Kr. V_{SO} included. Both Ne- and Ar-like cores included for Sc-Kr.
Dolg et al. (1987a)	Sc-Zn with a Ne-like core.
G.Durand et al. (1987)	Hg.
La John et al. (1987)	$_{37}$Rb-$_{54}$Xe, including V_{SO}.
Sakai et al. (1987a)	The nd-transition metals $_{21}$Sc...$_{80}$Hg. Only the valence s and d electrons in valence space.
Sakai et al. (1987b)	Ditto. The semicore p-shell also in valence space.
Schwerdtfeger (1987)	Tl.
Wadt (1987)	Pu.
Igel-Mann et al. (1988a)	Main-group elements in groups 13-17, n=2-5 ($_5$B to $_{53}$I).
Jasien and Stevens (1988)	$_{24}$Cr.
Dolg (1989)	Ln, 4f in core.
Dolg et al. (1989a)	Ln, Ce-Yb. Valence space from $4s^2$ up. SO potential given.
Dolg et al. (1989c)	Ln, La-Lu. Valence space from $5s^2$ up.
Schwerdtfeger et al. (1989b)	R/NR 19-e PP for Au.
Schwerdtfeger et al. (1989c)	R/NR Pb. SO.
Strömberg et al. (1989)	R/NR Cd and Hg.
Andrae et al. (1990)	QR and SO potentials for 4d and 5d metals ($_{39}$Y-$_{48}$Cd,$_{72}$Hf-$_{80}$Hg).
Cristiansen (1990)	Al. SO included. Core polarisation tested.
Kaupp et al. (1990a)	Zn 2-el. PP.
Musolino et al. (1990)	Sb. DFT.
Ross et al. (1990)	$_{55}$Cs-$_{57}$La, $_{72}$Hf-$_{86}$Rn. V_{SO} included.
Sellers (1990a)	Nb.
Sellers and Ove (1990)	Anisotropic PP for modelling partially filled d shells. Co.
Dolg et al. (1991)	$_{58}$Ce (12-el. PP).
Ermler et al. (1991)	$_{87}$Fr-$_{94}$Pu. 6s6p6d5f5g in valence space.
Fernandez Pacios and Botella Olcina (1991a)	Improved Ar-core PP for $_{21}$Sc-$_{29}$Cu.
Kaupp et al. (1991a)	Ca, Sr, Ba. 10-electron PP.
Küchle et al. (1991)	$_{80}$Hg-$_{86}$Rn. Both QR and NR basis sets given. A parametrized core-nucleus repulsion correction.
Wallace et al. (1991)	$_3$Li-$_{18}$Ar basis sets.
Allouche et al. (1992)	Ba. CP and SO potentials.
Barandiarán and Seijo (1992)	$_3$Li-$_{57}$La.
Stevens et al. (1992)	$_{19}$K-$_{57}$La, $_{72}$Hf-$_{86}$Rn.

Table 4.11. One-component and perturbation calculations.

Reference	Comments
Fraga et al. (1986)	Matrix elements of the Breit Hamiltonian in various coupling schemes.
Menendez et al. (1988)	BP terms as tests of the quality of STO atomic wave functions.
Uylings (1988,1989)	Angular formulae for two-electron BP terms.
Vijayakumar et al. (1989)	A LDF method with 1-component, SO-split one-electron functions. IP and EA.

Table 4.12 (1/Z)- and other similar expansions for many-electron atoms. For 1-electron atoms, see table 2.3.

Reference	Comments
Mohr and Safronova (1986)	Perturbation expansions for energies of 2-electron systems.
Safronova (1986a)	1/Z PT for $1s^2 2l2l'$-$1s^2 2l2l''$ TP. BP terms included.
Shestakov (1986)	3rd-order 1/Z PT on $3d^9 4s^2\ ^2D$ FS of Cu-like systems, Z=28-80.
Ivanov and Safronova (1987b)	An αZ expansion for $1s^{n1} 2s^{n2} 2p^{n3}$ systems.
Sakibaev and Safronova (1988)	Comparison of PT in αZ and 1/Z with MCDF for 2 $^3S_1$-^3P_J energies.
Ivanov and Safronova (1989a)	An αZ expansion for $1s^{n1} \ldots 3d^{n6}$ systems. Satellite spectra of 2p-1s and 3p-1s type.
Kim (1989)	Breit correlation for Al-like ions is Z^2 and Z^3. for $3p_{1/2}$ and $3p_{3/2}$, respectively.
Anno (1990)	Regularities of relativistic effects on total energies in 1/Z theory.
Pal'chikov and Safronova (1990)	1/Z analysis of DF theory.
Safronova et al. (1990)	New terms in 1/Z and αZ estimated for Be-like systems (Z = 18, 26, 42) by comparing MCDF and PT.
Shabaev (1990)	PT for relativistic atoms.
Karwowski et al. (1991)	DF + Breit energies expanded in a double series in 1/Z and $(\alpha Z)^2$. Main correlation-relativity crossterm from H_{Br}. He-, Be-, and Ne-like series.
Kagawa and Safronova (1992)	1/Z and αZ expansions for He-like systems. Z = 10-100.

Table 4.13. Relativistic nuclear calculations.

Reference	Comments
L.D.Miller (1974,1975ab)	Local approximations to the exchange potential in nuclear DF theory.
L.D.Miller (1976)	Elastic electron scattering formalism for relativistic nuclear models.
Brockmann (1978)	Nuclear relativistic Hartree-Fock theory: formulation, implementation, results for ^{16}O and ^{40}Ca.
Brockmann and Weise (1981)	SO coupling in nuclei.
Eisenberg (1981)	A relativistic nucleon in scalar and vector potentials.
L.D.Miller (1983ab)	Nucleon-nucleus scattering in a Dirac formalism.
Brown (1986)	Review.
Glöckle et al. (1986)	Relativistic 3-body effects reduce bound-state E by about 3 %, if R and NR 2-body E is held constant.
Gross (1986)	Relativistic equations for nuclear systems.
Kurasawa and Suzuki (1986)	Coulomb sum rule for electron scattering by nuclear matter. RRPA.
Boussy et al. (1987)	Relativistic HF for nuclei. NR limits discussed.
Brown (1987b)	Review.
Brown et al. (1987)	Review.
De Sanctis and Prosperi (1987)	One- and two-photon matrix elements of weakly bound systems.
Glöckle et al. (1987)	A solvable 2-body, 1D model with δ-function interaction. Lorentz-boost operator constructed.
Kurasawa and Suzuki (1989ab)	Coulomb sum rules for electron scattering by nuclear matter: Dirac sea and sigma-omega effects.
Lev (1988)	Relativistic corrections to binding energy for ≥4 nucleons.
Malfliet (1988)	Relativistic nuclear MBPT.
Thies (1988)	Disputes relativistic effects in nuclear physics.
Bijtebier (1989ab,1990a-d)	Few-fermion bound-state and scattering formalism.
De Sanctis (1989)	Relativistic corrections to photon-scattering polarizabilities.
Mc Neil et al. (1989)	Use of B splines (Johnson et al., 1988 a) in Dirac SCF and RPA nuclear models. ^{16}O, ^{40}Ca.
Plunien et al. (1989,1991)	Nuclear polarizability contributions to electronic Lamb shift.
Reinhard (1989)	Relativistic mean-field description of nuclei and nuclear dynamics, a review.
Brockmann and Machleidt (1990)	Relativistic nuclear structure. Includes review.

Table 4.13. (continued.

Centelles et al. (1990)	Relativistic TF with h^2 corrections for nuclei. Tested on a harmonic-oscillator model.
Dawson and Furnstahl (1990)	Relativistic spectral RPA in finite nuclei.
Mishustin (1990)	Instability of the Dirac nucleon sea in relativistic nuclear collissions.
J. Wu et al. (1990)	Numerical methods for time-dependent, 3+1 or 1+1 dimensional nuclear Dirac equations.
De Sanctis and Prosperi (1991)	Two-body interactions for nucleons. Pion exchange.
Keister and Polyzou (1991)	Review on relativistic Hamiltonian dynamics in nuclear and particle physics.
Toyama et al. (1991)	Relation between bound-state radius and scattering length.
Wasson (1991)	Nuclear Dirac-Hartree methodology.
Weigel et al. (1991)	Semiclassical expansions in nuclear DF theory.
Zhang and Onley (1991)	Retardation effects in finite nuclei. Retardation of the pion exchange may change the total binding energy by up to 5%.
Berghammer et al. (1992)	A 1D model for Coulomb dissociation.
Brockmann and Toki (1992)	Relativistic density-dependent Hartree approach for finite nuclei.
Cruz Barrios and Nemes (1992)	Field theoretical Lagrangians for mean-field approximations.
Jin et al. (1992)	Electron Coulomb effects in quasi-relativistic (e, e'p) reactions.
Marcos et al. (1992)	R/NR nuclear HF models.
Nikolaus et al. (1992)	R HF point coupling model for nuclear ground states.
Price et al. (1992)	R RPA for nuclear ground states. Contributions from the negative energy sea at one-loop level included.
van Dijk (1992)	Relativistic effect on the radius of a model deuteron.
Von-Eiff and Weigel (1992)	Relativistic TF model for finite nuclei including quantum corrections.
Von-Eiff et al. (1992)	Relativistic TF model for nucleons in a vector and scalar potential.
Zamick et al. (1992)	Interplay of relativity and core polarization in SO splitting.
X.-Q. Zhu et al. (1992)	One-time relativistic equation for nucleon-nucleon scattering derived from the Bethe-Salpeter equation.

5. MULTIELECTRON ATOMS. RESULTS

'Dielectronic recombination' by an ion means the absorption of one electron, from an electron beam or another atomic or molecular system, and the simultaneous excitation of an existing electron to a higher state. We include papers on it in Table 5.12.

Table 5.1. Tabulations of atomic ground-state properties.

Reference	Comments
Band and Trzhaskovskaya (1986)	Wave function expansions near origin from DS and DF functions.
Chen and Crasemann (1990)	Auger and Coster-Kronig radial matrix elements for Z=6-92.

Table 5.2. Data on atomic energy levels. The ground-state energy is denoted by E. FS = fine structure.

Reference	Comments
Fricke and Greiner (1969)	Ground-state DS configurations searched for Z=160-170. E164 a $7d^{10}(8p_{\frac{1}{2}})^2$ noble gas. Next noble gas is E170, $7d^{10}8p^6 9s^2$.
Safronova et al. (1971)	S XI and S XII, including BP terms.
Safronova and Labzovskii (1971)	BP terms for 1s2s and 1s2p 2-electron systems.
Dehmer (1973)	Atomic SO splitting arises in the K-shell region. Z-dependence analysed.
Beck and Nicolaides (1976)	Alkali atom nd ^2D FS.
Cole and Harmon (1981)	Orbital eigenvalues and E for 4f atoms in various LDF schemes.
Beck and Nicolaides (1984)	Fine structure of Be^- bound states.
Bruch et al. (1984)	B III 1s2pnp ^2P and 1s2pnd ^2D states at BP level.
Dyall and Grant (1984)	Ar K_β x-ray spectrum. MCDF-EAL.
Baluja (1986)	$2s2p^3\ ^5S - 2s^2 2p^2\ ^3P$ TP in the C-sequence, O III - Fe XXI.
Band et al. (1986)	X-ray transitions.
Bauche et al. (1986)	D_1 and D_2 lines of Fr. Isotope shifts for $p_{1/2}$ and $s_{1/2}$ electrons.
Biémont (1986)	Sulfur sequence 3s-3p, 3p-3d and 3p-4s, up to Ti VII.
Biémont and Hansen (1986b)	Forbidden transitions in $3p^4$ and $4p^4$ configurations.
Cogordan and Lunell (1986)	$2p^{-1}3s$, 3p and 3d levels of Ne-like ions, Z=20-54.
Das and Grant (1986)	3s3p FS levels of Mg-like systems.
Fawcett (1986)	Fe XV 3-3 and 3-4 transitions.
Froese Fischer (1986)	$2p^{-1}3s3p$, 3s3d and 3s4s quartet states of Na I and Mg II.
Froese Fischer and Godefroid (1986)	P-like $3p^3\ ^4S$, ^2D, ^2P terms, Z = 16...42.
Frost et al. (1986)	Pb II SO splitting.
Gogava et al. (1986)	Nuclear and QED effects for Li-like ions, Z = 20...92.
Gurchumeliya et al. (1986)	Zn-like series.
Hagelstein (1986a)	Ni-like Gd.
Hagelstein (1986b)	F-like Se.
He et al. (1986)	Positronium anion ground-state energy.
Indelicato (1986)	The 1s binding energies of He-, Ne-, Xe-, and Rn-like ions, Z=100...170.
Indelicato et al. (1986a)	He-like Kr.
Ivanova et al. (1986)	Mg-like systems, Z=25-84.
Johnson and Sapirstein (1986)	Ground-state energies of He-like systems, Z=1-100, using the B-spline MBPT.
Karim and Bhalla (1986)	He-like Cr.

Table 5.2. (continued).

Kotochigova and Tupitsyn (1986a)	MCDF for low-lying levels of Ba.
Maurer and Watson (1986)	Multiple K+L-shell vacancy states, Z=9-18.
Migdalek and Baylis (1986a)	Cd-like series, $5s^2$-5s5p transitions. Limited intravalence CI + polarisability core-valence correlation term.
Migdalek and Baylis (1986b)	$6s^2$-6s6p of Yb.
Mu and Crasemann (1986,1988)	Two-photon transitions in inner shells of Mo, Ag, Xe. DS.
Parera and Lopez Boada (1986)	Valence levels of alkaline earths.
Parker (1986)	Inner-shell energy levels with a $-Z_{eff}/r$ potential for every shell.
Riisager and Rosén (1986)	M x-ray shifts for Dy as function of atomic charge states.
Ross et al. (1986)	SO splitting of B-Tl, F-At. PP.
Rutkowski (1986c)	E of He-like atoms, Z=2-36 using a new PT approach.
Shestakov (1986)	The $3d^9 4s^2\ ^2D$ FS of Cu-like systems, Z=28-80. 1/Z PT.
Syromyatnikov and Savicius (1986)	Semiempirical, relativistic method for atomic energy levels.
Zhao et al. (1986)	X-ray energies of Ac-U. DS.
Babb and Spruch (1987,1988)	Retardation effects on Rydberg states of He.
Beck et al. (1987)	No bound states for Sc^-, several for Cu^-.
Bhattacharya et al. (1987)	M-shell vacancy effects on L-x-ray spectra using screened hydrogenic functions.
Biémont and Hansen (1987a)	2p-3s, 2p-3d, 2s-3p and 3p-3d transition energies for Ne-like ions, K X-Se XXV.
Biémont and Hansen (1987b)	M1 and E2 transitions within $3p^N$ and $4p^N$. Ga I - Xe XXIV.
Carroll et al. (1987)	Sulfur 1s IP.
Chen and Crasemann (1987c)	$1s2s^2 2p$, $1s2s2p^2$ and $1s2p^3$ configurations of Be-like ions, Z=6...26.
Cheng and Wagner (1987)	4s-4p transitions in Cu-like, $4s^2$-4s4p transitions in Zn-like ions, Z=50...92. No significant deviation from experiment.
Combet Farnoux (1987)	A review on giant resonances in heavy elements.
Crasemann et al. (1987)	A bibliography on theoretical atomic-electron binding energies.
Curtis (1987)	Lowest P-term fine structure for the Cu, Zn, Ga and Br series, Z≤92.
Drake (1987)	1s nd levels of He.
Dzuba et al. (1987a)	Au, Tl.

Table 5.2. (continued).

Fawcett (1987a)	C-like series, Z=9-28. Allowed 2-2 and 2-3 transitions.
Fawcett (1987b)	C I, N II, O III.
Francisco and Pueyo (1987)	Fine-structure levels of 3d metals M^{n+}, n=0,2,3,M=Sc-Zn.
Froese Fischer (1987)	2p-3s transitions of O I.
Glushkov et al. (1987)	Al-like systems,Z=23-32, from a model potential.
Gorceix et al. (1987)	Coulomb and magnetic correlation energies for $1s^2$ and 1s2p states of He-like ions, Z=10...60. Magnetic exceeds Coulomb around Z=50 for $1s^2$.
Hagelstein (1987)	Ne-like Se.
Hsuan et al. (1987)	He-, Li- and Be-like Ni.
Indelicato (1987)	Li-like ions. MCDF.
Indelicato et al. (1987ab)	He-like ions at MCDF level. A new treatment of radiative corrections.
Johnson et al. (1987)	Li-Cs valence levels.
Kelleher and Saloman (1987)	Rydberg states with anisotropic core states: Stark effect on $5d_{3/2}8l$ autoionizing states of Ba.
Kim and Greene (1987)	Two-electron excitations in Ca.
Kim and Weiss (1987)	$2s^2$-2s2p energies of Be-like ions.
Kotochigova and Tupizin (1987)	MCDF energy levels of neutral Ba, Ce, Sm, Eu, Yb. R/NR comparisons. The 4f collapse occurs between Ba and Ce.
Kulagin (1987)	Order of occupation of nf and (n+1)p levels in Ln^{q+}, An^{q+}.
Lisini and Decleva (1987a)	ns^{-1} and np^{-1} PES of the alkali atoms.
Lisini and Decleva (1987b)	The 6s to 7s shake-up satellite in Cs $5s^{-1}$ PES.
Lombardi(1987)	Na_2 FS. Model-potential fits.
Migdalek and Baylis (1987a)	$6s^2$-6snp, n=6-20 for Hg.
Migdalek and Baylis (1987b)	$6s^2$-6s6p of Yb-like ions, Z=71-73. Influence of the 5d orbital collapse on 5d6p mixture to 6s6p studied.
Migdalek and Baylis (1987c)	$6s^2$-6s6p of Ba.
Migdalek and Baylis (1987d)	$5s^2$-5s5p of Sr I and Y II.
Migdalek and Bojara (1987, 1988)	Comparison of CI and MC for ns^2 - nsnp transitions of Cd and Hg.
Migdalek and Wyrozumska (1987)	Cs sequence. Collapse of f and d orbitals.
Mohan and Hibbert (1987)	Hg $6s^2$-6s6p using a MP.
Perger and Das (1987b)	Be sequence 2s2p FS.
Radojevic (1987)	Pb $6s^26p^2$ and $6s6p^3$ levels.
Shi et al. (1987)	X-ray energies for Y, Zr, Nb, Mo, Tc, Ru. HFS with relativistic corrections.
Steiger (1987)	Li-like ions, Z=26...94. States up to $5f_{7/2}$.
Veseth (1987a)	Isotope-shifts in C II and N II FS.

Tabel 5.2. (continued).

Wadt (1987)	Pu^{2+} excited states as tests of PP.
Zilitis (1987)	Valence levels of K-like species, Z=19...92. DF.
Band et al. (1988)	Atomic mechanisms for mixed-valence states in Ln, An.
Beck (1988)	Three bound states of Zn^-: $4s4p^2$ ^4P, $4p^3$ ^4S and $3d^9 4s4p^3$ ^6D.
Beck and Cai (1988)	Bound states of Mn^-.
Biémont (1988)	n = 3-4 and 4-4 transitions of Cu-like ions (Z = 38-60).
Bieron and Migdalek (1988)	Ge-like sequence.
Blomberg and Wahlgren (1988)	Pd $4d^{10}$-$4d^9 5s^1$ distance influenced by 4s and 4p orbital relaxation in relativistic, 1st-order PT.
Blundell et al. (1988)	Up to 3rd-order MBPT for 6s, $6p_{1/2}$ and 7s of Cs.
Carlsson (1988)	Cu I. Odd levels. MCHF+BP. Applied on Cu-vapor laser.
Chang and Bryan (1988)	Al II, f-level FS perturbed by 3p3d ^3F.
Chen and Crasemann (1988a)	$1s^{-1}$ states of B-like ions, Z=6...54.
Cowan et al. (1988)	Hg $5d^9 6s^2 6p_j$ levels.
Churilov et al. (1988)	n=4, $\Delta n=0$ transitions of Ni- and Zn-like ions, Z = 30 - 50.
Curtis et al. (1988)	3s, 3p and 3d levels of Ne-like systems, Z=16-18.
Drake (1988a)	Ground states of H^-, He; 1s2s of He.
Drake (1988b)	n=1, 2 states of He-like ions, $Z \leq 100$.
Drake and Makowski (1988)	1s2p states of He.
Froese Fischer (1988)	Li II 2^3S-2^3P, MI ns-np, M=Li, Na, K.
Gogava et al. (1988)	$2s^{-1}$ and $2p^{-1}$ levels of Ne-like systems, Z=18...42.
Gorceix and Indelicato (1988)	Effects of the complete Breit interaction on two-electron ion 1s2s and 1s2p energy levels, Z=6...92.
Iglesias et al. (1988)	Mo III. $4d^4$, $4d^3 5s$, $4d^3 5p$ levels.
L.N. Ivanov et al. (1988)	Review on multicharged ions, especially He-, Li- and Ne-like ones.
Indelicato (1988)	He-like systems, Z=10-92.
Johnson et al. (1988b)	Li-like n=2 levels, Z=3-92.
Johnson et al. (1988c)	Na-like n=3 levels, Z=11-92.
Y.-K. Kim et al. (1988a)	2s3p levels of Be-like ions.
Y.-K. Kim et al. (1988b)	QED contributions to ground-state FS of Al-like ions.
Klobukowski and Fraga (1988)	Accurate ground-state E for He-like systems, Z=2-4.
Liaw et al. (1988a)	Na-Cs and Ne-Xe ground and excited-state energy contributions.
Lindroth (1988)	E for $1s^2$ systems, Z=2...50, from a numerical solution of the relativistic pair equation.
Liu and Li (1988)	Fr exited states. DS.
Malinovskaya (1988)	Cl-, Ar- and K-like ions, Z=19-36. A model-potential fit.

Table 5.2. (continued).

Ohtsuki and Hijikata (1988)	He-like system 2 ^{1,3}P FS, Z≤10.
Rashid et al. (1988)	E of U I to U XCII. DF.
Reed (1988)	The 36 n=3 states of Ne-like Ba.
Reed et al. (1988)	n=2 states of O-like Se.
Rutkowski et al. (1988)	E of closed-shell atoms, He...Xe, using a new PT approach.
Sakibaev and Safronova (1988)	2 $^3S_1$-^3P_J energies. Comparison of MCDF and PT in αZ and 1/Z.
Sapirstein (1988)	n=2 state IP of Li-like ions.
Silver (1988)	Precision spectroscopy on 1- and 2-electron ions of medium and high Z.
Styszynski and Karwowski (1988)	E of $1s^2$ atoms, Z=2-20. MCDF with radiative corrections and relativity-correlation cross terms. Final error estimated < 0.001 a.u.
Sugar et al. (1988)	Al-like ions, Z=29-42.
Theodosiou and Curtis (1988)	IP of Na-like systems, Z=30...92.
Vainshtein and Safronova (1988)	Ne-like ions.
Veseth (1988)	FS of B-F, Al-Cl. MBPT.
Walker et al. (1988)	Excitation energies of Sc...Co as tests of PP.
Zhao and Pan (1988a)	$2p^5$3s, 3p and 3d levels of Ne-like Kr. MCDF.
Zhao and Pan (1988b)	Fe XVII.
Zhao and Wang (1988)	Pb I $6p^2$-6p7s. MCDF.
Zhao et al. (1988)	X-ray energies of Sc-Ni. About 1% accuracy with TS DS.
Anno and Teruya (1989)	Relativistic contributions to E, Z=2-20.
Au (1989ab)	He Rydberg states: QED corrections beyond Breit.
Babb and Spruch (1989)	Retardation effects for a Rydberg electron outside a multielectron core.
Baker et al. (1989)	He energy levels.
Bauche and Bauche-Arnoult (1989)	Mixing of many transition arrays: 30 configurations of Xe^{27+}: 110 000 lines in 80 arrays.
Beck (1989)	The ^{4}P to ^{4}S transition of Mg^-.
Biémont (1989)	n=4-4 transitions in zinc-like ions.
Biémont and Hansen (1989)	3d and $3d^9$ systems.
Biémont et al. (1989)	Zn-like ions.
Blanke et al. (1989)	3d-4f and 3d-5f transitions of doubly-excited three-electron ions, N V...Mg X.
Blundell et al. (1989a)	2s,2p and 3s levels of Li and Be^+. All-order MBPT gives 0.01% accuracy.
Blundell et al. (1989b)	Ground-state energy of He from all-order MBPT. Briand et al. (1989) H-like and He-like Xe. 1s2p $^{1,3}P_1$.
Cai and Beck (1989)	Bound excited states of Cr^-, Fe^-, Co^- and Ni^-.

Table 5.2. (continued).

Chang (1989)	1s4f states of He.
Chung (1989a)	$1s2s2p^2$-$1s2p^3$ ^5S of Be-like systems, Z=4-10.
Crasemann (1989)	A review on relativistic phenomena in atomic physics.
Drake (1989)	Review on He.
Dzuba et al. (1989a)	Tl 6p.
Fraga et al. (1989)	Test the approach of Fraga et al. (1986, 1987) on He ^1S levels and one ^3P level.
Froese Fischer (1989)	M, M^-, M=Mg-Ba. Sc, Sc^+.
Froese Fischer and Chen (1989)	nsnp ^3P states of Rb and Cs. The latter is bound. Thumm and Norcross (1991) later find it unbound.
Glebov et al. (1989)	Ku (E104) ground state a J=2 with mainly $6d7s^27p$. The $6d^27s^2$ lies 0.5eV higher.
Indelicato (1989)	He- and Li-like ions: a review.
Ivanov and Safronova (1989b)	2l4l' states of 2-electron systems.
Jansen and Hess (1989b)	Au ^2D and ^2P levels using the 'no pair' approach.
Johnson (1989)	Li-like atoms. MBPT. The obtained $2p_{1/2}$-2s difference compared to Lamb-shift calculations.
Johnson et al. (1989a)	Negative ions of Pd, Ca-Ba.
Kim (1989)	3p FS of Al-like ions, Z=13-60. The Breit correlation varies as Z^2 and Z^3 for $3p_{1/2}$ and $3p_{3/2}$, respectively.
Lagowski and Vosko (1989)	J-independent relativistic contributions to s-d interconfiguration energies for M and M^+, M=Ca-Cu. LDF.
Migdalek and Stanek (1989)	$4s^2$-4s4p transitions in the Zn sequence.
Mukoyama (1989a)	M-shell x-rays of Dy. DF.
Nicolaides et al.(1989)	Bound exited states of atomic negative ions. New results for A^-, A=Al, Fe, Mn, Cu.
Nilsen (1989)	Dielectronic satellite spectra for Ne-like ions.
Ostrovskii and Sheinerman (1989)	Hg^+ 7p-7s, 6s and nd-ns energies. R/NR HF.
O'Sullivan (1989)	Rb IV - Rb IX. 4p-5s, 4p-4d. DF.
O'Sullivan and Kane (1989)	Ru XII and Rh XIII $4s^24p^3$ and $4s^24p^25s$ levels.
Pacheco and Sañudo (1989)	Observation of "strange matter" in nuclei through change of nuclear radius. 1s-2p levels calculated for Z=80-104.
Polasik (1989ab)	$K\alpha L^n$ lines of Mo, Pd, La, Ho. MCDF + QED.
Polasik (1989c)	$K\alpha L^oM^r$ lines of Mo, Pd, Ho.

Table 5.2. (continued).

Qian et al. (1989)	$2s^2$-2s2p for $Ne^{6+}...U^{88+}$. MCDF.
Raghavachari and Trucks (1989a)	The d^ns^2-$d^{n+1}s$ energies of Sc-Cu.
Rashid et al. (1989)	E of Be-like systems, Z=4-20. MCDF.
Rudzikas and Ryabtsev (1989)	Identification of the spectra of multielectron ions.
Saloman and Kim (1989)	Ground-state terms of S-like ions, Z=16-92. MCDF with Breit and Lamb-shift corrections.
Salomonson and Öster (1989)	E of He using all-order pair functions.
Sampson et al. (1989)	Resonance transitions of Ne- and Ni-like ions, as test of a new DS program.
Sapirstein (1989a)	E of He-like systems.
Schmidt et al. (1989)	Muonic 1s-5g levels in atoms with Z=30-100.
Seely (1989)	2s-2p transitions of Li-like ions, Z=24-54. QED included.
Seely et al. (1989)	Cu-like systems, Z=44-92. All n=4,5 levels. Values corrected semiempirically.
Tisenko (1989)	He atom FS formulated a Dul'yan-Faustov pseudopotential equation.
Tragin et al (1989)	$d^{10}nl$-$d^{10}nl'$ and d^{10}-d^9nf transitions near ionization limit for Cu- and Ni-like ions, respectively. Z=42...83, 35...92, respectively.
Aashamar and Luke (1990)	Cr II sextet-quartet mixing.
Anisimova and Semenov (1990a)	Fine structure of $np^5n's$ configurations.
Anisimova and Semenov (1990b)	Fine structure $nd^9n's$ configurations.
Anisimova and Semenov (1990c)	Fine structure of ps configurations.
Anno (1990)	E from 1/Z theory.
Au and Mesa (1990)	He Rydberg levels: corrections beyond Breit.
Biémont (1990)	Forbidden lines in the K sequence, Cd^{29+}-Ce^{39+}.
Biémont and Quinet (1990)	Ga-like ions, Br V-In XIX.
Blundell et al. (1990a)	n=2 states of Li-like U. Higher-order Breit, mass-polarization and QED included.
Blundell et al. (1990b)	3rd-order MBPT ground-state energies of Cs and Tl.
Bylander and Kleinman (1990)	4f resonances in Ba using a norm-conserving PP.
Chen and Huang (1990)	Anomalous fine-structure splittings in the DF model corrected by proper NR limits. 3p hole states of Ar-Sr as example.
Chung (1990a)	$1s^{-1}$ four-electron resonances of C.
Chung (1990b)	$1s^{-1}2l2l'2l''$ states of O.

Table 5.2. (continued).

Connerade and Karnatak (1990)	Centrifugal-barrier double-well atomic potentials and valence changes. Sm as example.
B.P. Das et al. (1990)	FS intervals for $1s^2 2p$ levels of Li-like ions, Z=5-28.
Drake (1990)	Rydberg states of He.
Greene (1990)	M^-, M = Rb-Fr. See Thumm and Norcross (1991).
Hansen (1990)	Atomic spectra for astrophysics and fusion. Proceedings of Amsterdam 1989.
Indelicato (1990)	K_α-transitions in few-electron systems.
Indelicato and Desclaux (1990)	Li-like ions, Z=15-92.
W.R. Johnson et al. (1990)	Cu like n=4 levels, Z=29-92.
Y.-K. Kim (1990)	Relativistic and QED effects in highly-charged ions.
Koc and Migdalek (1990)	nl-levels of Ag and Au. Inverted FS of Au 5f reproduced.
Lindroth and Salomonsson (1990)	Ar^{16+} 2^3S_1-1^1S_0 energy calculated as 114.072(7) au. R PF.
Marketos and Lambropoulos (1990)	The J=1/2 even spectrum of Sr II.
Mazzoni and Joshi (1990)	As I $3d^9 4p^4$ and $3d^9 4p^3 np$ levels.
Migdalek and Baylis (1990)	Ba $6s^2$-6s5d transitions.
Migdalek and Stanek (1990)	Comparison of 'optimal level' and 'average level' MCDF for ns^2-nsnp transitions of Zn I, Rb VIII, Hg I, and Rn VII. For neutrals, OL clearly better.
Nielsen et al. (1990)	2l4l' states as function of Z.
Ohanessian et al. (1990)	ΔE between $5d^n$, $6s^1 5d^{n-1}$ and $6s^2 5d^{n-2}$ for Ba^+, La^+, Hf^+-Hg^+, as tests of the QR PP.
Parera López (1990)	E of He-like atoms at DF levels using a scaled hydrogenic basis. Includes Breit.
Pararpia and Grant (1990)	Accurate ground-state E for He-like atoms.
Polasik (1990)	$K\alpha L^1 M^r$ x-ray lines of Pd.
Quinet and Biemont (1990)	Resonance transitions in Ni-like ions, Ge V - La XXX.
Safronova et al. (1990)	Be-like ions, Z=18,26,42. Radiation and correlation corrections appended to MCDF. New terms in 1/Z and αZ estimated.
Seely and Wagner (1990)	3s-3p transitions of Na-like ions, Z=39-92. QED included.
Sugar et al. (1990)	Si-like ions, Z=29-42.
Viar and Safronova (1990)	$2s^2 2p^5 3l$ and $2s2p^6 3l$ levels of Ne-like systems, Z=18...92.
Wycech et al. (1990)	FS in the antiprotonic $\bar{p}$-^{174}Yb atom.
X.Y. Xu et al. (1990)	Ru $4d^7 ns$ (n=28-50) and nd (n=24-56) Rydberg levels. DS.

Table 5.2. (continued).

Zhuikov et al. (1990)	E104 energy levels. Ground state s^2pd. MCDF.
Zilitis (1990)	Rydberg s and p levels of K-like ions.
Zou et al. (1990)	XUV spectra of C-like Co.
Aggarwal et al. (1991)	Lowest 20 levels of Ca XV.
Au et al. (1991)	Recoil corrections to Rydberg levels of He.
Band et al. (1991)	4d-nf excitations of Ba^{2+}.
Bauche et al. (1991)	Break-down of jj-coupling. $3d^{10}$-$3d^9nf$ transitions of Ta^{45+}.
Biémont (1991)	Allowed transitions in Al V-Al VII.
Blundell et al. (1991a)	All-order MBPT energies of Cs.
Cai (1991); Cai et al. (1991)	Hg $6p^2$ resonance states.
Cheng et al. (1991)	n=2 states of Li-like U, n=3 of Na-like Pt and n=4 of Cu-like Au. MBPT + QED.
Chou et al. (1991)	Pb-like $^3P_{0-1}$ transitions, Z=82-98.
Chung (1991b)	$1s^{-2}$ four-electron systems: $2s^22p^2$ ^3P>>$2s2p^3$ ^5S, Z=3-10.
Couch and Scott (1991)	Kα x-ray satellites for Z=19-32. R/NR comparisons made.
Davidson et al. (1991)	Ground-state correlation energies for 2- to 10-electron ions, $Z \leq 20$.
Dzuba et al. (1991a)	np-state FS for M^-, M=Ca-Ra. For Ra^-, only $7p_{1/2}$ is bound.
Dzuba et al. (1991b)	M $6p^2$ and np(n+1)s levels, M=Sn, Pb, 114. Bi $6p^3$.
Feldman et al. (1991)	$3d^4$ ^5D levels of Ti-like ions, Z= 42...92.
Glushkov (1991b)	$4p^1$ and $4d^1$ levels of Ga-like systems, Z = 31-36.
Greene and Aymar (1991)	Sr-Ra excited states.
Hsu et al. (1991)	1s2s2p ^4P levels for Li-like ions.
Indelicato and Mohr (1991)	Li-like U $2s$-$2p_{1/2}$.
Ivanova and Gulov (1991)	Ne-like $2p^{-1}3l$, 4l and $2s^{-1}3l$, 4l levels, Z=27-92. Revised results for Na- and F-like ions.
Kagawa et al. (1991)	$2s^{-1}$ and $2p^{-1}$ levels of Ne-like ions, Z=18...54. Finite-basis CI.
Karwowski et al. (1991)	E of He-, Be- and Ne-like series as a double series in 1/Z and $(\alpha Z)^2$.
Y.-K. Kim et al. (1991)	ns-np transitions in Li-, Na- and Cu-like ions, Z≤92.
Litzén and Zeng (1991)	4s-4p transitions in Ga-like ions, Z=44-49.
Liu and Kelly (1991)	E of Be from MCDF-based MBPT.
Mårtensson-Pendrill et al. (1991)	E of Be, now agrees with experiment.
Martin et al. (1991)	Quantum defect orbital study of 3s-3p and 3p-3d in the Na-like sequence, Z=11...99.

Table 5.2. (continued).

McKenzie and Drake (1991)	E of Li -7.478 060 326(10) au, exp. -7.478 060 34(20). also QED corrections for $1s^2 2s$ and $1s^2 2p$ states, Z=7...36.
Migdalek and Baylis (1991a)	Yb $6s^2$, 6s6p and 5d6s levels.
Migdalek and Baylis (1991b)	La II $5d^2$, 5d6s and 5d6p levels.
Perger et al. (1991)	$6p^2$ resonances of Hg.
Quinet and Biemont (1991)	X-ray spectra of Ni-like ions, Ag XX -Pb LV.
Quinet et al. (1991)	Allowed and forbidden transitions in Ne-like ions Ni XIX - U LXXXIII.
Samzow and Hess (1991)	FS of Br as test of the "no-pair" Hamiltonian.
Tayal (1991)	Ga II 4s4p, $4p^2$ and 4s4d level energies. BP PT.
Thumm and Norcross (1991)	The 6s6p $^3P^o_J$ states of Cs^- are unbound, narrow resonances.
Vukajlovic et al. (1991)	E difference of $3d^{n-1}4s^1$ and $3d^{n-2}4s^2$ of Ca-Cu. DFT.
Zilitis (1991)	Rydberg d and f levels of K-like ions. DF.
Ali (1992)	FS of K-like (Z=22-54) and Rb-like (Z = 40-60) ions. MCDF.
Ali and Kim (1992)	K ground-state FS.
Andersson and Roos (1992)	Ni atom d^8s^2,d^9s and d^{10} states. NR correlation at CASPT2 level, $\langle h_m+h_d \rangle$ from Martin and Hay (1981).
K. Ando et al. (1992ab)	$1s^2 2lnl'$ (n=2, 3, 4) levels of Na VIII - S XIII. Comparison of MCDF and 1/Z.
Anno and Teruya (1992)	Comments on E(Z) for He-like systems.
Avgoustoglou et al. (1992)	$2p^{-1}3s$ - $2p^6$ energy of Ne-like Xe.
Babb et al. (1992)	Casimir shifts for He-like Rydberg ions, Z = 2...20.
Band and Trzhaskovskaya (1992)	5d-5f energies of gaseous and metallic U and Th. Single-configuration DF.
Biémont et al. (1992)	Intra $3d^2$, $3d^2$-3d4s and $3d^2$ - 3d4d transitions for Z = 22-47. MCDF, HFR and "SST" compared.
Blundell (1992)	$2s-2p_{1/2}$ of Li-like U, $3s-3p_{3/2}$ of Na-like Pt, $4s-4p_{1/2}$ of Cu-like Bi. Accurate screened QED used.
Bruch et al. (1992)	Ne^{q+} K-shell binding energies, q=4-6. MCDF, 1/Z and a saddle-point method used.
Cai et al. (1992)	Pr^{3+} levels using large-scale MCDF.
Cheng and Huang (1992)	4s-4p transitions of Zn-like ions, Z=31...92. MC RPA.
Chou and Huang (1992a)	6s-6p transitions in Hg-like systems, Z=80-93. MC RPA.
Chou and Huang (1992b)	5s-5p of Cd-like systems, Z = 48-65.

Table 5.2. (continued).

Chung (1992)	E of He- and Li-like systems, Z=11-20.
Connerade and Sarpal (1992)	$3p^{-1}$ levels of Ca I. MCDF.
Deutsch and Kizler (1992)	$1s^{-1}$ levels of Xe.
Dolg et al. (1992a)	Yb^{n+} (n=0-2) energy levels, as test of PP.
Drake and Yan (1992)	Very high precision Rydberg levels for He, $n \leq 10$, $L \leq 7$.
Eissner and Tully (1992)	Two-electron, $2p^2$-2s3p and $2p^2$-2s2p transitions of Be-like systems Z = 6...26.
Goldman and Drake (1992)	He Rydberg levels.
Hessels (1992)	Higher-order relativistic polarisation corrections to He Rydberg states, n = 10.
Ilyabaev and Kaldor (1992)	Ne excited levels. CCSD.
Indelicato and Lindroth (1992)	K_α ($1s^{-1}$ - $2p^{-1}$) energies for Z = 31-100. MCDF+QED. Uncertainty of finite-nucleus corrections is the bottleneck.
Johnson and Sapirstein (1992)	n = 2 triplet states of He-like ions, Z = 10-36.
Koc and Migdalek (1992)	Yb II and Lu III $4f^{14}nl$ levels, especially FS.
Kuplyauskene and Zhukauskas (1992)	Decay of $2p^{-1}3l3l'$ states of Na-like Cl and Ar.
Lao and Christiansen (1992)	Ne ground-state valence E_{corr} from PP Monte Carlo.
Leininger et al. (1992)	Valence levels of K-Cs, Ca-Ba and Fe-Os as test of four different PP.
Lindroth and Hvarfner (1992)	$2\ ^1S$-$2\ ^{1,3}P_1$ transitions of Be-like Fe and Mo. Theoretical uncertainty dominated by radiative effects.
Lindroth et al. (1992)	E of Be. "Experimental NR energy" -14.667 353(2) a.u.
Loginov (1992)	$5d^9 6p$ levels of Tl IV - Bi VI.
Malinovskaya and Glushkov (1992)	K-like ions.
Mäntykenttä et al. (1992)	Satellites in 4d, 5s and 5p PES of Cs and Ba. MCDF.
Marian et al. (1992)	Cu valence levels, especially ^2D FS.
Mei and Davenport (1992)	IP and s-d excitation energies of Ca-Cu. DF and various DFT compared.
Miecznik et al. (1992)	High-spin states of B-like systems: $1s2s2p^3\ ^6S^o$ -to- $1s2s2p^2 3s\ ^6P^e$ and $1s2s2p^2 3d\ ^6P^e$, Z = 4...13.
Misra et al. (1992)	Parametrized formula for x-ray spin doublets.
Mooney et al. (1992)	K, L and M x-ray of Xe.
Parpia and Mohanty (1992)	E for He-like ions, Z = 10...160 and for Hg. A finite nucleus with Fermi-distribution.

Table 5.2. (continued).

Rashid and Ahmad (1992)	Predicted wavelengths for K-like ions with ± 1/3 , ±2/3 additional (quark) charges.
Seijo et al. (1992)	Valence E_{corr} for I^-, I, I^+.
Safronova and Wyart (1992)	$2s^{-1}$ 31 levels of Ne-like ions, Z = 30...92.
Scofield and MacGowan (1992)	4d-4p laser lines of Ni-like Ta-U.
Sugar et al. (1992)	Rb-like $4p^6 4d$ - $4p^5 4d^2$ transitions, Z = 47-60.
Theodosiou et al. (1992)	2s-2p energies of Li-like systems, Z = 3-92.
Vidolova-Angelova (1992)	Rydberg states of Lu.
W.-J. Wang (1992)	N-like ions, Z = 19-22. MCDF.
Z.-W. Wang et al. (1992a)	$1s^2nd$ and $1s^2nf$ levels of Li-like systems, Z = 3-10.
Z.-W. Wang et al. (1992b)	$1s^2ns$ (n=3-5) states of Li-like systems, Z = 3-10.
Wijesundera et al. (1992a)	Hg^- $6s^2 6p_{1/2, 3/2}$ resonances calculated to lie at +0.19, +0.21 eV (exp. +0.63(3)).
Yamamoto et al. (1992)	Energy-levels of Hf.
Yang and Li (1992)	FS of 4f and 3d levels in K-like ions, Z=26, 31, 36.
Y. Zhang et al. (1992a)	Na-like ions, n≤ 15, l ≤ 6, Z = 38-45.
Q.-R. Zhu et al. (1992)	Li-like ions, Z = 20-25.

Table 5.3. Auger and autoionization processes. For dielectronic recombination processes, see Table 5.12.

Reference	Comments
Nicolaides (1983)	Radiative autoionization (RA) and short-wavelength lasers. Li 1s2s2p ^4P undergoes relativistic RA.
Aksela et al. (1986a)	$N_{4,5}OO$ resonance Auger spectra of Xe.
Aksela et al. (1986b)	$M_{4,5}NN$ Auger of Kr.
Aspromallis and Nicolaides 1986)	Relativistic and Coulomb autoionization in $2p^43s\ ^2S$ and $2p^43d\ ^2P$ states of F.
Aspromallis et al. (1986)	Relativistic autoionization of $2s2p^2\ ^4P$ state of Be^-.
Beck and Nicolaides (1986)	Alkaline earth Auger energies.
Cavell and Sodhi (1986)	Auger parameters of phosphorus compounds.
Karim and Bhalla (1986)	2lnl' autoionizing states of He-like Cr.
Kotochigova (1986)	Autoionization treated at MCDF level. 5dnp, n=9, 10 for Ba (J=1).
Levin et al. (1986)	Kr L-MM Auger spectra.
Mannervik and Cederquist (1986)	Relativistic autoionization of Li ^4P_J levels: comments on Nicolaides and Aspromallis (1986).
Nicolaides and Aspromallis (1986)	Relativistic autoionization of Li ^4P_J.
H. Aksela and S. Aksela (1987)	Review on free-atom Auger spectroscopy.
Aksela et al. (1987)	$L_{2,3}MM$ Auger of Ni.
Armen et al. (1987)	Post-collision interactions in inner-shell photoionization. Ar, Xe.
Bruneau (1987)	A relativistic close-coupling approach to the Auger process. Application on the Ne K-LL.
Chen and Crasemann (1987a)	1s2p3d $^4F_{9/2}$ metastable states of Li-like ions, Z=6...42. Breit-induced Auger transitions important.
Chen and Crasemann (1987b)	Auger-decay of $1s^{-1}$ states of Be like ions, Z=6...26.
Chen and Crasemann (1987c)	Auger-decay of $1s^{-1}$ states for B-like systems, Z=6...54.
Radojevic (1987)	Low-energy $6s6p^3$ autoionizing states in Pb.
Tulkki et al. (1987)	Post-collision interactions (Auger and Coster-Kronig) in inner-shell photoionization. Ar, Xe.
Aksela et al. (1988a)	Auger of Ar.
Aksela et al (1988b)	Resonance Auger of Rb.
Chen and Crasemann (1988a)	Auger energies and rates of $1s^{-1}$ states of B-like ions, Z=6...54.
Chen and Crasemann (1988b)	L-shell transitions of Be-like ions, Z= 30-54. Autoionizing resonances.

Table 5.3. (continued).

Cowan et al. (1988)	Hg $5d^9 6s^2 6p_j$ autoionization lifetimes.
Karazija and Rudzikaite (1988)	Mean energy and spin doublets of Auger lines.
Lefebvre-Brion et al. (1988)	SO effects in autoionization of HCl.
Nicolaides and Aspromallis 1988)	Autoionizing states of 2- and 3- electron model atoms with Z=2.05.
Sampson et al. (1988)	Rates of populating excited n=4 levels of a 28-electron, Ni-like ion when a core-excited 29-electron, Cu-like ion decays.
Aksela (1989)	Auger spectra of free atoms: a review.
Aksela et al. (1989a)	Auger of $3d^{-1}np$ states of Kr.
Aksela et al. (1989b)	Auger of $1s^{-1}$ states of Ne.
Campbell and Wang (1989)	L-subshell x-ray emission rates including overlap and exchange effects. Interpolated DF.
Chen (1989a)	Auger and radiative decays of $2p^{-1}$ 3l3l' and $2s^{-1}$3l3l' states of Na-like ions, Z=18...92. Effects of CI and Breit Hamiltonian studied.
Chen (1989b)	L-shell fluorescence yields from Ne- and Na-like low-lying autoionizing states.
Chen (1989c)	Relativistic effects on x-ray and Auger transitions of highly charged ions.
Chen et al. (1989)	The 2p-3p4s Auger spectrum of free Ca atoms: strong correlation effects due to collapse of d AOs.
Chung (1989b); Davis and Chung (1989)	Spin-induced autoionization of Li-like systems. $1s2s2p\ ^4P$, Z=3-18.
Dyubo and Safronova (1989)	Nonradiative decays of Be-like ions.
Froese Fischer and Idrees (1989)	Autoionization of core-exited ^5P states Na^-.
M. Meyer et al. (1989)	Autoionization of the Ca $2p^5 3d$ core resonances.
Safronova (1989)	Autoionization states of atomic systems.
Aksela et al. (1990a)	2p-excited Auger of Cl, applied on HCl.
Aksela et al. (1990b)	Coster-Kronig width of Rb.
Aksela et al. (1990c)	$M_{4,5}NO$ and $M_{4,5}NN$ Auger of Rb.
Aksela et al. (1990d)	Auger decay of Rydberg-excited Ne 1s, Ar 2p, Kr 3d and Xe 4 d.
Chen (1990b)	X-ray and Auger transitions in atoms and ions.
Chen et al. (1990a)	Auger- and Coster-Kronig radial matrix elements for Z=6-92.
Chung (1990c)	Auger-spectra of Be and B.
Ford et al. (1990)	Autoionization and Auger Spectra of Mn and Sm vapours. HFR.
Ivanov et al. (1990)	Autoionization states of 2-electron systems.

Table 5.3. (continued).

Radojevic and Talman (1990)	Beutler-Fano autoionizing resonances in Ne.
Chen (1991b)	Auger rates and fluorescence yields for a double-K-hole state.
Chi et al. (1991)	Autoionizing levels of Be.
Cowan and Wilson (1991a)	Cd $4d^9 5s^2 5p$ autoionization lifetimes.
Crater (1991)	Singlet-positronium decay.
Fritzsche and Zschornak (1991)	L-MM Auger spectra of Kr.
Fritzsche et al. (1991)	MCDF K-LL Auger rates for Ne-like systems, Z=10...92.
Greene and Aymar (1991)	Autoionization of Sr-Ra.
Hayaishi et al. (1991)	Kr 3d and Xe 4d conjugate shake-up satellites in treshold-electron spectra.
Hwang et al. (1991)	Autoionizing levels of Zn. MC-RRPA.
Ivanov and Safronova (1991)	Autoionization 1s2l4l' states.
Vidolova-Angelova and Ivanov (1991)	Autoionizing Rydberg levels of Tm.
Vijayakumar and Gopinathan (1991)	Auger energies from LDF. Ne-Xe, Cr...Cu, Ge.
Whitfield et al. (1991)	Shake-modified resonant autoionization of Mg.
Zimmerer et al. (1991)	Relativistic Auger rates for two-electron ions. Breit effects.
Aksela et al. (1992a)	Resonantly excited $2p^{-1}ns$, md states of Ar, $3d^{-1}np$ of Kr.
Aksela et al. (1992b)	Kr $3d^{-1}$ np, n = 5,6 Auger.
Aksela et al. (1992c)	Decay of $2p^{-1}$ in HCl.
Batkin et al. (1992)	K-shell autoionization in beta decay.
M.H. Chen (1992)	Effect of intermediate coupling on angular distribution of Auger electrons. $M_{4,5}$-NN of Kr, $M_{4,5}$-NN and $N_{4,5}$-OO of Xe. MCDF.
Fritsche et al. (1992)	Zn L_1 level width and Coster-Kronig yields.
Fritzsche and Fricke (1992)	2p Auger spectra of Mg-like ions, Z = 13-36. MCDF.
Hayaishi et al. (1992)	Ar 2p shake-up structures.
Kuplyauskene and Zhukauskas (1992)	Autoionization of Na-like Cl and Ar.
Leväsalmi et al. (1992)	KLL spectrum of Ne.
Lohmann (1992)	Angular distribution and spin polarisation of Auger electrons from Hg.
Szunyogh et al. (1992)	Relativistic theory of Ti or Pd core-core-valence Auger spectra.
Tulkki et al. (1992b)	Ne KLL and Ar LMM Auger rates.
Yamamoto et al. (1992)	X-ray and Auger energies of Hf.

Table 5.4. Ionization potentials and electron affinities. (For a recent summary of the experimental EA for the elements 1-85, see H. Hotop and W.C. Lineberger, J. Phys. Chem. Ref. Data 14 (1985) 731-750).

Reference	Comments
Beck and Nicolaides (1979)	One-electron binding energies: a review of the contributions.
Hess (1986)	Relativistic increase of Ag IP, EA 0.47 and 0.15 eV, respectively. Br ground state. No-pair formalism used at MRD-CI level.
Beck et al (1987)	No bound states found for Sc^-, several for Cu^-.
Dzuba et al. (1987a)	Au, Tl IP.
Hurley and Christiansen (1987)	Li and K EA from PP MC calculations.
Ohno (1987a)	IP of Ca by g-Hartree method.
Ohno (1987b)	IP of He, Be, Ne, Mg, Ar.
Radojevic et al. (1987)	EA of F-I.
Zhao and Pan (1987)	DS IP of K-Cs, Cu, Ag.
Y.-F.Guo and Whitehead (1988)	IP and EA at QR-LSD level for Z= 37...80.
Ohno (1988)	IP of Be-like systems, Z=4-10.
Ortiz (1988)	EA of K-Cs. PP, up to 4th-order MBPT.
Rashid et al. (1988)	IP_n of U.
Dzuba et al. (1989a)	Tl 6p IP.
Froese Fischer (1989)	Sc IP. Ca, Sr, and Ba EA: 0.062 (exp. 0.043(7)), 0.106 and 0.148 eV, respectively. Relativistic EA decrease almost 0.1 eV for Ba.
Froese Fischer and Chen (1989)	Cs 6s6p ^3P predicted to have a positive EA, Rb 5s5p not. See Thumm and Norcross (1991, 1992).
Fuentealba et al. (1989)	EA of Ca, Sr, Ba. Five different density functionals tested.
Y.-F.Guo and Whitehead (1989)	EA of Mg-Ra and An (Ac-Lr) at LSD level.
Y.-F.Guo et al. (1989)	EA of rare gases (He-Rn) and some An (Pu-Es) at LSD level.
Jansen and Hess (1989b)	Au EA and IP using the 'no pair Hamiltonian.
Johnson et al. (1989b)	EA/eV: Pd 0.525, exp. 0.558(8) Ca 0.056 0.043(7) Sr 0.093 Ba 0.192
Raghavachari and Trucks (1989b)	IP of Sc-Cu. Corrections of Martin and Hay (1981) used.
Rubio et al. (1989)	Spherical DS models for the IP_n of Cu_N, Ag_N clusters, N≤25.

Table 5.4. (continued).

K.D.Sen et al. (1989a)	Electronegativities, $\chi = -[\partial E/\partial N]_Z$ (E total energy, N number of electrons) calculated for non-integral (n/3) nuclear charges, Z, near halides. LSD.
Vosko et al. (1989)	EA of Sr, Ba and Ra. from DFT. p^1, d^1 and d^1 configurations predicted, respectively.
Fuentealba et al. (1990)	EA of Ca-Ba. None for Be, Mg.
Illas et al. (1990)	Rh IP and EA.
E.Johnson et al. (1990)	IP_n of Ti-Hf and E104.
Koc and Migdalek (1990)	IP of Ag and Au.
Marian (1990b)	Cu EA increased by 0.02 eV by relativity.
Sundholm and Olsen (1990a)	QR EA of B.
Zhuikov et al. (1990)	E104 (Ku) IP_1-IP_4. MCDF.
Chung (1991a)	IP of Li-like $1s^2 2s$ systems, Z= 3-10.
Cowan and Wilson (1991b)	EA for Be, C-Pb, Mg-Ra and Zn-Hg calculated at HFR level.
Küchle et al. (1991)	EA and IP of Hg-Rn, as tests of new PP. QR/NR.
Mitás et al. (1991)	Si, Sc and Cu IP and EA from QMC.
Nicolaides and Aspromallis (1991)	No EA for He or Xe. Levinson's theorem and MCHF used, respectively.
Nogueira and Guenzburger (1991)	Valence-orbital ionization potentials (VOIP) for 5d elements, Hf-Hg, from DFT. R/NR.
Tupitsyn (1991)	U IP. MCDF.
Vosko et al. (1991a)	Yb EA predicted to be 0.054(27) eV for the $6s^2 6p_{1/2}$ state. The $6p_{3/2}$ is unbound. DFT.
Vosko et al.(1991b)	La and Ac EA from DFT.
Chung (1992)	IP of Li-like systems, Z = 11-20.
Glushkov (1992b)	Obtains positive EA for Kr and Xe in a model potential method.
Ionova et al. (1992)	IP_n of V-105, 103-105. MCDF.
Leininger et al. (1992)	EA and IP of K-Cs, Ca - Ba and Fe -Os using four different PP.
Miecznik et al. (1992)	EA of Be in a $1s2s2p^2$ ^5P state.
Samzow et al. (1992a)	IP and EA fo Ag and Au as tests of a 'no-pair' method.
Vijayakumar and Gopinathan (1992)	IP_1-IP_3 of He - Zn using the relativistic, correlated Ξ method.
Yamamoto et al. (1992)	IP of Hf.

Table 5.5 Supercritical ($Z>137$) systems. For a more thorough compilation, see the treatises of Greiner et al. (1985) and Greiner (1987). Several further calculations on many-electron systems are covered in Ch. 7.

Reference	Comments
Pomeranchuk and Smorodinsky (1945)	Dependence of the critical radius for $Z>137$ on nuclear radius.
Voronkov and Kolesnikov (1960)	ns-states for finite nuclei with $Z>137$.
Klein and Rafelski (1975)	Bose condensation in supercritical external fields: KG particles and $Z \geq 10^3$.
Bosch (1986)	A review on superheavy atoms.
Herath Banda et al. (1986)	δ-electron emission from superheavy collision systems with Z_{tot}=110-171.
Soff et al. (1986,1987)	Exotic phenomena in collisions of very heavy ions.
C.-Y. Wong (1986,1987)	Suggested origin of the positron peaks.
Greiner (1987)	Physics of strong fields. Book.
Rumrich et al. (1987)	Influence of strong **B** (10^{15} G) on positron production in heavy-ion collisions.
B.Müller (1987)	The many facets of the Dirac vacuum.
Bottcher and Strayer (1988)	Pair production in heavy-ion collisions. A one-dimensional TD model solved numerically.
Butler et al (1988)	Positron interactions with element 164: knock-out and positronium formation.
Ionescu et al. (1988)	Nonlinear terms ?
Seba (1988)	Absorption of Dirac eigenvalues by the continuum. 1D and 3D.
B.Müller et al. (1989)	Attempts to explain the GSI positron lines.
Rhoades-Brown et al. (1989)	Pair production associated with electron capture in heavy-ion collisions.
Rumrich et al. (1989)	Continuum states.
Mishustin (1990)	Instability of the Dirac nucleon sea in relativistic nuclear collisions.
Rosenberg (1990)	Dynamics of neutral vacuum decay.
Strayer et al. (1990)	Muon pair production and subsequent μ^- capture into a 1s state. Numerical 3D solution for Au+Au.
Bawin and Cugnon (1991)	"Partially screened solutions" for the Dirac equation in $Z>Z_c$ Coulomb fields.
Graf et al. (1991)	Review on QED in strong Coulomb fields.
Momberger et al. (1991)	Electron-positron pair production in relativistic heavy-ion collisions.
Horbatsch (1992)	Search for narrow resonances due to spin-spin interactions.

Table 5.5. (continued).

Thiel et al. (1992)	Electron-positron pair creation in U^{92+} - U^{91+} collisions.
Wells et al. (1992)	Lepton-pair production in relativistic heavy-ion collisions. Au + Au at 2-100 GeV/nucleon 3 + 1D FEM solution in a Cartesian lattice.

Table 5.6. Electromagnetic transition probabilities.

Reference	Comments
Drake and Dalgarno (1968); Drake et al. (1969)	Two-photon decay of 2 $^3S_1$ He-like systems attributed to singlet-triplet mixing in the intermediate P state.
Taksar et al. (1971)	Na 3p, 3d TP from a simple PP.
Nicolaides (1972)	C I $2s2p^3$ and O I $2p^33s$ $^5S_2$ state lifetimes. SO allowed E1 dominates, $^3P_2$ mixed to upper state.
Lin and Feinberg (1974)	Radiative corrections to M1 of H-like ions.
Barbieri and Sucher (1987)	Radiative corrections to one-photon decay rates.
Borisoglebskii and Yakimovich (1985)	Nuclear E2-induced, $\Delta L=\Delta j=\pm 3$ electronic transitions.
Band et al. (1986)	X-ray transitions.
Biémont (1986)	Sulfur sequence 3s-3p, 3p-3d and 3p-4s TP, up to Ti VII.
Biémont and Hansen (1986a)	M1 and E2 TP in Ge and As sequences.
Biémont and Hansen (1986b)	M1 and E2 TP within $3p^4$ and $4p^4$ configurations.
Braun and Parera (1986)	TP between degenerate levels of the relativistic many-electron atom.
Fawcett (1986)	Fe XV 3-3 and 3-4 transitions.
Froese Fischer (1986)	$2p^{-1}$ 3s3p, 3d and 4s quartet level lifetimes.
Froese Fischer and Godefroid (1986)	E2 and M1 TP for P-like, $3p^3$ systems, Z=16...42.
Fulton and Johnson (1986)	2s-np, 2p-3d TP for Li at DF level using "effective current" approach derived from QED. Length and velocity forms equal.
Karim and Bhalla (1986)	E1 TP for 2l3l' and 2l4l' states of He-like Cr.
Kotochigova and Tupitsyn (1986b)	MCDF calculations of TP of Ba.
Krause (1986)	2-1 TP of He-like systems, Z=2-100, in a screened hydrogenic basis.
Migdalek and Baylis (1986a)	Cd sequence $5s^2$-5s5p TP.
Migdalek and Baylis (1986b)	Yb $6s^2$-6s6p TP.
Mu and Crasemann (1986, 1988)	Two-photon transitions in inner shells of Mo, Ag, Xe, DS.
Safronova (1986a)	$1s^22l2l'$ - $1s^22l3l''$ TP from 1/Z PT.
Sebastian (1986)	Relativistic invariance of the two-photon transition amplitude of a composite system.
Vidolova-Angelova et al. (1986)	Tm $4f^{13}6s$ (n_Rl_R) Rydberg states, n_R=10-50, l_R=0,1.
Biémont and Hansen (1987a)	2p-3s, 2p-3d, 2s-3p and 3p-3d TP for Ne-like ions, K X - Se XXV.

Table 5.6. (continued).

Biémont and Hansen (1987b)	M1 and E2 TP within $3p^N$ and $4p^N$. Ga I - Xe XXIV.
Brage and Litzén (1987)	Ni-like 4s-4p transitions, Zr XIII - Mo XV. MCDF.
Chen and Crasemann (1987a)	1s2p3d $^4F_{9/2}$ metastable states of Li-like ions, Z=6...42. E1, E2 and M2 processes.
Chen and Crasemann (1987b)	Radiative decay of $1s^{-1}$ states in B-like ions, Z=6...54.
Dzuba et al. (1987a)	Au, Tl E1 TP.
Fawcett (1987a)	Allowed 2-2 and 2-3 TP for the C-like ions, Z=9-28.
Fawcett (1987b)	C I, N II, O III.
Fechner et al. (1987)	Eu I $4f^7 6s6p$ TP.
Froese Fischer (1987)	2p-3s TP of O I.
Glushkov et al. (1987)	Al-like systems, Z=23-32. E1 TP. Model potential.
Johnson et al. (1987)	Li-Cs ns-np TP.
Migdalek and Baylis (1987a)	Hg $6s^2$-6snp TP, n=6-20.
Migdalek and Baylis (1987b)	$6s^2$-6s6p TP, Lu II - Ta IV.
Migdalek and Baylis (1987c)	Ba $6s^2$-6s6p TP.
Migdalek and Baylis (1987d)	$5s^2$-5s5p TP, Sr I, Y II.
Migdalek and Bojara (1987, 1988)	ns^2-nsnp TP of Cd and Hg.
Migdalek and Wyrozumska (1987)	Cs sequence. Collapse of f and d orbitals seen in TP.
Mohan and Hibbert (1987)	$6s^2$-6s6p TP of Hg using a MP.
Steiger (1987)	E1, E2, M1 and M2 TP for Li-like ions, up to $5f_{7/2}$ levels. MCDF + Breit + QED. Z=26...94.
Su and Eberly (1987)	"Essential states" in the theory of electron-photon interactions.
Verhey et al. (1987)	M1 $^2P_{1/2}$ - $^2P_{3/2}$ TP for the B-like sequence. 1st-order Breit and QED included.
Ali and Kim (1988)	E2 and M1 TP for the K isoelectronic series. (4s above 3d for Sc^{2+}- Cr^{5+}).
Barut and Salamin (1988)	Relativistic theory of spontaneous emission. Applied on 2p-1s and 2s-1s of hydrogen and muonium.
Biémont (1988)	n = 3-4 and 4-4 TP for Cu-like ions, Z = 38-60.
Bouchiat and Piketty (1988)	M1- and E2-hfs-induced 6s-7s TP for Cs.
Brage (1988)	Thesis on TP calculations.
Carson (1988)	Bound-bound multipole TP for H-like atoms.
Curtis et al. (1988)	3s, 3p and 3d level TP for Ne-like systems, Z=16-18.

Table 5.6. (continued).

Fawcett and Cowan (1988)	Fe I. HFR.
Froese Fischer (1988)	MCHF TP of $2^3S - 2^3P$ Li II. Also the ns-np TP of Li I, Na I, K I.
Luo et al. (1988)	Si II resonance lines.
Mohanty and Sampson (1988)	E1 TP for H-like ions as test of various QR approximations.
Saloman and Kim (1988,	M1 and E2 TP between ground-state terms 1989) of S-like systems, Z=15-92. For heaviest atoms $3p^4$ mixes significantly with p^2d^2 for $^3P_1$, $^1D_2$ and $^1S_0$ states.
Theodosiou and Curtis (1988)	Na-sequence TP, Z=11-54...92. 3p, 3d.
Zhao and Pan (1988a)	$2p^5 3s$, 3p and 3d TP of Ne-like Kr. MCDF.
Zhao and Pan (1988b)	Fe XVII.
Zhao and Wang (1988)	Pd I 6p7s TP. MCDF.
Becker et al. (1989)	M1 and E2 TP for N-like, $2p^3$ levels, Z=9-26.
Biémont (1989)	n=4-4 TP in zinc-like ions.
Biémont and Hansen (1989)	TP in 3d and $3d^9$ configurations.
Biémont et al. (1989)	Zn-like ions.
Blundell et al. (1989a)	All-order E1 TP for 2s, 2p and 3s states of Li and Be^+. Agreement with experiment 0.1 %.
Chen and Crasemann (1989c)	n=2 and 3 E1 TP for O-like ions.
Curtis (1989)	M1 TP for nsnp states in the Be, Mg and Zn series.
Curtis and Theodosius (1989)	4p and 4d TP for Cu-like systems, Z=29-92.
Davis and Chung (1989)	Li-like 1s2s2p ^4P_J TP, Z=3-18.
Drake and van Wijngaarden (1989)	Radiative transitions in one- and two-electron ions.
Dzuba et al. (1989b)	Cs 6p, 7p E1 TP.
Goldman (1989b)	Two-photon TP for 2s H-like systems, Z=1...100.
Gulov and Ivanova (1989ab)	Ne-like ions.
Hibbert (1989)	Reviews model potentials in oscillator-strength calculations. Al II, Zn, Cd-like and Hg ns-np E1 TP.
Ivanova et al. (1989)	Forbidden 2-3 radiative transitions in Ne-like ions.
Jankowski and Polasik (1989)	$K\beta/K\alpha$ x-ray intensity ratios for Ti...Ge. Influence of Breit and QED terms considered.
Migdalek and Stanek (1989)	Zn sequence $4s^2$-4s4p TP.
Mukoyama (1989a)	M-shell x-ray TP of Dy. DF.
Mukoyama and Adachi (1989)	K-shell x-ray TP in superheavy elements.
Ostrovskii and Sheinerman (1989)	Hg^+ E1 TP for 7p-7s, 6s and E2 TP for nd-ns. R/NR HF.

Table 5.6. (continued).

Sampson et al. (1989)	E1 TP for Ne- and Ni-like ions as test of a new DS program.
Sen and Puri (1989a)	E1 TP for ns-np of Be-Ra. LSD. Isoelectronic cations (q=+1 to +3) included.
Sen and Puri (1989b)	E1 TP for ns-np of alkali metal-like systems.
Theodosiou (1989)	TP of Ca^+. ns, np, nd, 4f.
Tong and Li (1989); Tong et al. (1990)	Two-photon decay rates in atomic inner shells. Mo, Ag, Xe. Relativistic decrease analysed.
Tulkki and Åberg (1989)	Photoabsorption of positive heavy ions. Oscillator strength concentrates to nl ->n (l+1) transitions. Origin a phase-shift of π between the wave functions in a Coulomb field.
Vidolova-Angelova et al. (1989)	Tm Rydberg states.
Zilitis (1989)	TP of K-like systems, Z=19...92.
Beck and Cai (1990)	Tl^+ $6s^2$ - 6s6p E1 TP.
Biémont (1990a)	Forbidden lines in the K sequence, Cd^{29+} - Ce^{39+}.
Biémont (1990b)	Review on heavy-element TP.
Biémont et al. (1990)	n=4-4 TP for Ge-like ions Rb VI - Ag XVI.
Biémont and Quinet (1990)	Ga-like systems, Br V - In XIX.
Curtis (1990)	Semiempirical formulations of line strengths using singlet-triplet mixing angles.
B.P. Das and Idrees (1990)	nsnp $^3P_1$ $^1S_0$ TP for B^+ - Tl^+.
B.P. Das et al. (1990)	M1 FS TP between $1s^22p$ levels of Li-like systems, Z=5-28.
Davis and Chung (1990)	Li-like $2p^2np$ $^{4,2}S$ and 1s2pmp $^{4,2}P$ states.
Greene (1990)	nsnp ^3P lifetimes of M^-, M=Rb-Fr.
Guet et al. (1990)	3s-3p TP of Na-like ions. Z = 11-29.
Hansen (1990)	Oscillator strengths for astrophysics and fusion. Proceedings of Amsterdam 1989.
Hartley and Mårtensson-Pendrill (1990)	E1 transitions of Cs and Tl.
Lindroth and Salomonsson (1990)	Ar^{16+} 2 $^3S_1$ - 1 $^1S_0$ transition dominated by M1. Calculated lifetime 203.4(4) ns.
Migdalek and Baylis (1990)	$6s^2$-6s5d E2 TP for Ba.
Migdalek and Stanek (1990)	ns^2-nsnp TP for Zn I, Rb VIII, Hg I and Rn VII. Various approximations compared.
Mohan and LeDourneuf (1990)	E2 and M1 for the J = 3/2 - J = 1/2 FS transition of Si^{5+} $2p^5$.
Nefiodov (1990)	Spectral line shape in non-resonant laser field.
Pal'chikov (1990)	Two-photon transitions.

Table 5.6. (continued).

Sampson et al. (1990)	TP of Na-like ions, Z=22-92.
Tiwary et al. (1990)	Rb I, Sr II, CS I, Ba II ns-np TP at BP level.
M. Wilson (1990)	Radiative decay rates in Hg^+.
Zhu and Pan (1990)	MCDF methodology for E1 TP.
Baik et al. (1991)	E1, E2 and M1 TP for Na-like ions, $56 \le Z \le 92$.
Band et al. (1991)	4d-nf TP of Ba^{2+}.
Biémont (1991)	Allowed transitions in Al V - Al VII.
Bieron et al. (1991)	np^2-np(n+1)s TP for Ge I...Bi II.
Blundell et al. (1991a)	All-order TP for Cs.
Chou et al. (1991)	Pb-like $^3P_{0-1}$ M1 TP. Z=82-98.
Dzuba et al. (1991b)	Bi $6p^3$ M1 TP.
Eissner (1991)	Breit-Pauli calculations of E1 and M1 TP. He-like systems.
Fawcett (1991)	Review on oscillator strengths.
Fawcett and Wilson (1991)	Yb II.
Feldman et al. (1991)	M1 TP between $3d^4\ ^5D$ levels of Ti-like ions, Z=42...92.
Froese Fischer and Godefroid (1991)	MCHF BP program for E1, E2,... and M1, M2... TP.
Glushkov (1991)	E1 TP of Fr-like ions using a MP method.
Guet and Johnson (1991)	ns-np, ns-(n-1)d and np-(n-1)d TP for Ca^+-Ba^+. E1, E2.
Ivanova and Gulov (1991)	E1-E3 and M1-M3 TP for $2s^{-1}nl$ and $2p^{-1}nl$, n=3, 4, levels of Ne-like systems, Z=27-92.
Kagawa et al (1991)	E1 TP for $2s^{-1}$ and $2p^{-1}$ levels of Ne-like systems, Z=18...54. STO-CI.
Karwowski and Martin (1991)	E1 TP for the ns-np transitions in the Li-, Na- and Cu-like sequences using "relativistic quantum defect orbitals".
Martin et al. (1991)	3s-3p and 3p-3d TP for the Na-like series, Z=11...99. QDO.
Migdalek and Baylis (1991a)	Yb $6s^2$-6s6p and -5d6s E1, E2, M1 and M2 TP.
Migdalek and Baylis (1991b)	La II 5d6p lifetime.
Perger et al. (1991)	$6p^2$ resonance width for Hg.
Quinet and Biemont (1991)	X-ray TP of Ni-like ions,Ag XX-Pb LV.
Quinet et al. (1991)	Allowed and forbidden transitions in Ne-like ions, Ni XIX - U LXXXIII.
Reiche and Zschornak (1991)	X-ray emission rates i ncluding electron-exchange corrections. MCDF.
Sampson et al. (1991)	TP of F-like ions, Z= 22-92.
Shabaev (1991b)	QED theory of resonance line shapes in few-electron atoms.
Tayal (1991)	Ga II 4s4p, $4p^2$ and 4s4d-level TP.
Y. Zhang et al. (1991)	Na-like Fe.
Zhu et al. (1991)	n=3 to n=2,3 TP in F-like Se.
Biémont et al. (1992)	Intra-$3d^2$, $3d^2$-3d4s and $3d^2$-3d4d TP (M1, E2) for Z = 22-47. MCDF, HFR and "SST" compared.

Table 5.6. (continued).

Bieron and Migdalek (1992)	As I, Sb I, Bi I $np^2(n+1)s$ TP. MCDF.
Chou and Huang (1992a)	6s-6p E1 TP of Hg-like systems, Z=80-93.
Chou and Huang (1992b)	5s-5p TP of Cd-like systems, Z = 48-65.
Eissner and Tully (1992)	Be-like systems, $Z = 6...26$. $2p^2$-2s2p and two-electron $2p^2$-2s3p TP.
Glushkov (1992ac)	s-p and p-d TP for K-, Rb- and Cs-like systems.
Gorshkov et al. (1992)	QED theory of U^{90+} line profiles.
Grotch et al. (1992)	Relativistic wave-function corrections to E1 TP of bound $b\bar{b}$ systems.
Grozdanov et al. (1992)	Photoexcitation of H-like Rydberg states in microwave fields.
Hibbert et al. (1992)	$2p^5 3l$ lifetimes of S VII-Ar IX.
Kuplyauskene and Zhukauskas (1992)	Decay of $2p^{-1}3l3l'$ states of Na-like Cl and Ar.
Lamoreaux and Fortson (1992)	Electric-field (Stark effect) changes of Hg $6s6p(^3P)$-$6s^2$ TP.
Lavín et al. (1992)	Na-like systems, 3s-np (n=4-6), Z=18...42. Quantum defect theory.
Liaw (1992)	Transition amplitudes in DF theory. B-splines. Li-, B-, Na- and Al-like systems.
Loginov (1992)	$5d^6 6p$ lifetimes of Tl IV - Bi VI.
Martin et al. (1992)	Excited-state p-d and np-6s TP for Cu-like ions, Z = 33...70.
Miecznik et al. (1992)	High-spin ^6P-^6S TP of B-like systems, Z = 5...13.
Pinnington and Baylis (1992)	Hg-like systems, Z = 80, 82, 83.
Safronova and Wyart (1992)	Ne-like systems, Z = 30...92. $2s^{-1}3l$-state TP.
Seke (1992c)	Non-exponential $2p_{1/2}$-1s decay of Dirac hydrogen.
Theodosiou et al. (1992)	2p lifetimes in the Li sequence, Z = 3-92.
Vidolova-Angelova (1992)	Rydberg-state lifetimes of Lu.
Zhang and Sampson (1992)	$\Delta n=0$ transitions of Be-like systems, Z = 8-92.
Y. Zhang et al. (1992a)	Radiative TP for Na-like ions, $n \le 15$, $l = \le 6$, Z = 38-45.
Q.-R. Zhu et al. (1992)	Li-like ions, Z = 20-25.
Zilitis (1992)	Principal series of K-like ions, Z = 22...92.

Table 5.7. Polarizabilities and screening constants. α_L and χ_L stand for electric and magnetic polarizabilities and γ_L and σ_L for electric and magnetic screening constants, respectively.

Reference	Comments
Dzuba et al. (1986b)	Shielding of external electric fields in atoms. Radial dependence for Tl^+.
Lombardi (1987)	Determines core dipole and quadrupole polarizabilities for Na from model-potential fits.
Silberbach (1987)	Polarizabilities of Ba, Au, Hg, Tl and Pb as test of a semiempirical PP.
Gray et al. (1988)	Experimental dipole and quadrupole polarizabilities for Na^+ from Rydberg states.
Zapryagaev et al. (1988)	Polarizabilities of Li- and Na-like ions from the quantum defect method.
Goldman (1989a)	Accurate dipole polarizabilities for H-like systems, Z=1...115.
Pal'chikov and Tkachev (1989)	Electric polarizability and hyperpolarizability of He-like systems.
Zhou and Norcross (1989)	Scalar and tensor polarizabilities α_0 and α_2 for Cs $6p_{3/2}$. Also α_0 for 6s, 7s, $6p_{1/2}$. Model potential.
De Sanctis (1991)	Polarizabilities of exotic atoms. All EL evaluated. E1 relativistic α^2 correction.
Sadlej (1991)	Au α_1. QR/NR.
Sadlej et al. (1991)	Polarizabilities of Ca-Ba. Relativistic decrease analysed.
Kutzner et al. (1992)	Frequency-dependent polarizabilities of H^-, F^- - I^-.

Table 5.8. Electric and magnetic hyperfine properties.
E0: Isotope and Moessbauer isomer shifts, E2: Nuclear electric quadrupole interactions, E4: Nuclear electric hexadecapole interactions, M1: Magnetic dipole and M3: Magnetic octupole hyperfine effects.
For Sternheimer factors see Table 5.7.

Reference	Comments
Schüler and Schmidt (1936)	Angular parts and relativistic corrections for E2 hfs of p electrons.
Desclaux (1971)	M1 and E2 hfs of Sc, Cu, Ga, Br and Li-Fr at DF level.
Huang and Hughes (1979)	M1 hfs of muonic He, $\alpha\mu^- e^-$. Relativistic, recoil and radiative corrections included.
Mårtensson-Pendrill and Salomonsson (1984)	M1 and E2 hfs of Ca^+ 4s, 4p and 3d states.
Band et al. (1985)	M1 hfs in highly excited H-like, Li-like and Na-like atoms, Z ≤ 28.
Bauche et al. (1986)	Cs and Fr isotope shifts.
Baumann and Braun (1986)	Yb 6s6d state E2, M1. Exp.,DF.
Bouazza et al. (1986)	Pb II.
Childs et al. (1986)	$4f^{11}5d6s^2$ of Er I. M1 and E2 hfs. MCDF.
Nielsen and Pfeufer (1986)	E0 hfs of Er I, $4f^{11}5d6s^2$. MCDF.
Swainson and Drake (1986)	M1 and E2 hfs of muonic Li, Be and B.
Vainshtein et al. (1986)	Hfs parameters for Li-like systems.
Blundell et al. (1987a)	Field isotope shifts in atoms. New theoretical formulation. Z=19-95 tabulated.
Druzbick and Williams (1987)	M1 hfs of ^3He 1s2s and $^3He^+$, 1s and 2s.
Dzuba et al. (1987a)	Au, Tl.
Johnson et al. (1987)	Li-Cs valence states. M1.
Martynenko and Faustov (1987)	M1 hfs of muonium.
Nielsen et al. (1987)	M1 and E2 hfs of Er I + II + III MCDF.
Olsson and Lindgren (1987)	Analyse the hfs tensor operators needed in relativistic MBPT. Sandars and Beck (1965) OK.
Wendt et al. (1987)	E0, E2 and M1 hfs of Ra I and Ra II.
M. Wilson (1987)	QR calculations of charge densities at nucleus for Re as function of configuration.
Yelkhovsky (1987)	Comments on the nuclear M1-induced hydrogen-atom 1s (F=3/2) E2. See also Pilkuhn (1987).
Zimmermann (1987)	Effect of s-electron normalization on M1 hfs.
Amus'ya et al. (1988bc)	M1 hfs in the He^{++}-μ-e atoms.
Bouazza and Bauche (1988)	Pt II, Au II, Bi II M1 data interpreted at DF level.
Childs and Nielsen (1988)	La I $(5d+6s)^3$ M1, E2 hfs. MCDF.

Table 5.8. (continued).

Grexa et al. (1988)	Tl np state A, n=6-8.
Kutzelnigg (1988)	Origin and meaning of the "Fermi contact" (s-AO M1 hfs) interaction.
Millack (1988)	M1 hfs for a single valence electron, Li-Cs, Cu-Au, B-Tl. The g-Hartree method used.
Trusov and Eglais (1988)	E0 of Fe.
Young et al. (1988)	Sc II hfs. MCDF.
Aaen Andersen et al (1989)	M1 and E2 hfs of $4f^{12}5d$ levels of Er II. MCDF with core exitations and CI.
Blundell et al. (1989a)	M1 and E2 for Li and Be^+ from all-order MBPT.
Dzuba et al. (1989b)	Cs ns, np (n=6, 7) M1 hfs.
Kuten' and Rapoport (1989)	An atomic octupole (M3) moment is induced in the F=3/2 $1s_{1/2}$ and $2p_{1/2}$ states of H-like atoms by the M1 hfs.
Mansour et al. (1989)	Sc II. E2 well predicted by MCDF, M1 hfs less well.
Millack (1989)	M1 hfs of Li and B using "2nd-order g-Hartree" approach.
Panigrahy et al. (1989)	M1 hfs in Be^+-Ne^{7+}, Bi^{80+}. Radiative corrections included.
M.-K.Chen (1990)	M1 hfs of muonic He.
Hartley and Mårtensson-Pendrill (1990)	Hfs of Cs and Tl.
Milstein and Yelkhovsky (1990)	Vacuum-polarization "dressing" of nuclear Q.
Neugart et al. (1990)	M1 hfs of Ra isotopes.
Sen et al. (1990)	Eu $4f^7 5d$ and $4f^7 6p$ field isotope shifts, E0.
Shabaev et al. (1990)	Nuclear-size effects, based on nuclear $\langle r^2 \rangle$ and $\langle r^4 \rangle$ estimated for multiply charged ions, $Z \leq 100$. Precision $\leq 0.2\%$.
Sundholm and Olsen (1990b)	QR corrections to E2 hfs of S^- $(^2P_{3/2})$.
Beck and Datta (1991)	M1 hfs of the $5s^1$ Ag I series, Ag I-Sb V.
Blundell et al. (1991a)	All-order MBPT M1 hfs for Cs.
Elkhovskii and Mil'stein 1991)	Vacuum polarization, induced by nuclear M1 and E2 multipole moments.
Hartley and Mårtensson-Pendrill (1991)	Volume and mass isotope shifts of Cs and Tl.
Panigrahy et al. (1991a)	M1 hfs of $^{213}Ra^+$.
Panigrahy et al. (1991b)	M1 hfs of M^+, M=Zn-Hg.
Andriessen et al. (1992)	Ra II 7s and 7p M1 and E2 hfs. MCDF versus MBPT.
Beck (1992)	Y II $4d^2$ M1 hfs. MCDF + CI using H-like orbitals.
M.-K. Chen (1992)	M_1 hfs of $(1s)_e$ $(2s)_\mu$ muonic helium. NR Hylleraas + 1st-order R and QED corrections.
Indelicato et al. (1992)	Decay of He-like Gd $2\ ^3P_0$ to $1\ ^1S_0$ due to M1 hyperfine mixing with $2\ ^3P_1$.

Table 5.8. (continued)

Mårtensson-Pendrill and Ynnerman (1992)	E0, M1 and E2 hfs of Ba^{+} (6s, 6p).
Mårtensson-Pendrill et al. (1992)	Isotope shifts of Ca.
Pindzola et al. (1992)	Dielectronic recombination between ground-state hyperfine levels of Bi^{82+}.
Shabaeva and Shabaev (1992)	M1 hfs of Li-like Fe^{23+}.
Villemoes et al. (1992)	Sc II 3d4s, $3d^2$ and 3d4p-level M1 and E2 hfs. MCDF.

Table 5.9. Average radii $<r^n>$ and magnetic g-factors.

Reference	Comments
Karaziya (1981)	Excited-electron orbit collapse and atomic spectra. The d shells collapse from outer to inner well around Z=20, 38, 56, 88 and 120, f shells around 57, 90 and 122, g at 124. Effective potentials sketched.
Amus'ya (1987)	Quadrupole moments of atoms with J=½.
Veseth (1987b)	g_J for Al-Cl.
S. B. Zhang et al. (1987)	"Nodal radii" (radius to last node) calculated for s, p and d orbitals of Li-La, Hf-Rn. Relativistic effects up to 10% for s, show no "gold maximum".
Cortona (1989)	$<r^n>$ for Gd^{3+} in a spin-polarized LD model.
Raspini (1989)	Relativistic Zeeman effect in positronium, n=2.
Sai Sunder and Sen (1989)	Ionic radii (r_m) of Be^--Ba^- and some metastable atoms. QR LDF.
Sen and Politzer (1989)	Radii r_m for 3d, 4d and 5d metal anions. LSD.
E. Johnson et al. (1990)	Hf and E104 r_m.
Zhuikov et al. (1990)	Atomic and ionic radii of E104.
Cortona (1991)	Spin-polarized Ce^{3+} $<r^n>$.
Demkov and Rebane (1991)	Propose a direct atomic or molecular experiment for g-2.
Dzuba et al. (1991b)	Pb $6p^2$, Bi $6p^3$ level g-factors.
Fawcett and Wilson (1991)	Yb II g-factors.
E. Johnson and Fricke (1991)	r_m for M^{n+}, M = Ti-E104, n=1-4. Used for estimating thermodynamic properties.
Kohout et al. (1991)	Search for shell structure in plots of density Laplacian or local kinetic-energy density. Be...Pb.
Aktas et al. (1992)	SO-corrections to g_J of Li.
Sundholm and Olsen (1992a)	Atomic quadrupole moment of Ca 3d4s ^1D. QR.

Table 5.10. Compton profiles, momentum distributions and spin densities. For basic theory of the Compton effect, see also Table 5.14.

Reference	Comments
Dutta et al. (1986)	Compton profiles for Kr-Rn. Screened hydrogenic functions used.
Frost et al. (1986)	Pb 6s and 6p momentum distribution. MCDF.
Glass (1987a)	Spin magnetic form factor for V. Atomic correlation and SO included.
Glass (1987b)	Ditto for Co and Ni.
Cortona (1989)	Spin densities for Gd^{3+} in a spin-polarized LD model.
Cortona (1991)	Spin-polarized Ce^{3+} spin densities.
H.-S. Li et al. (1991)	Intra-atomic exchange integrals between 4f, 5d and 6s in Ln^{3+}. Magnetic properties of rare-earth intermetallics.

Table 5.11. X-ray scattering factors.

Reference	Comments
Kane et al. (1986)	Review on elastic scattering of gamma- and x-rays by atoms.
M. S. Wang (1986)	Relativistic dispersion relation to order $\alpha^2 Z^2$ for x-ray anomalous scattering factors.
Smith (1987)	Anomalous x-ray scattering. Rayleigh scattering reduced for bound electrons from the free-electron value by relativistic effects (mainly h_m). Higher multipoles and retardation must be included for heavier elements.
Saloman et al. (1988)	Total x-ray absorption cross-sections for E = 0.1-100 keV and Z=1-92. DS.
Ehlotzky (1989)	Scattering of x-rays by relativistic electrons in a strong laser field.
Fox et al. (1989)	X-ray scattering factors for atoms, Z=2-98.
Kahane et al. (1989)	Gamma-ray Rayleigh scattering from U, E = 0.4-3 MeV.
Sasaki (1989)	Tables of anomalous scattering factors.
Wang and Chia (1989)	X-ray atomic anomalous scattering factors for Zr, Nb, Mo.
Kissel and Pratt (1990)	Z-dependent, energy-independent corrections to anomalous scattering factors. Higher multipoles important. Z=1-100.
Namikawa et al. (1990)	X-ray magnetic scattering from atomic inner-shell electrons.
Kefi et al. (1992)	X-ray anomalous scattering factors of Pd...Xe. R correction to the dispersion term.

Table 5.12 Electron and positron scattering.

Reference	Comments
Rosen (1963)	Relativistic Coulomb scattering.
Rotenberg (1966)	Small relativistic effects in a model potential study of slow-electron scattering by Na and Cs.
Spruch (1966)	Discusses R/NR comparisons for slow-electron scattering.
Meister and Weiss (1968)	Scattering of E<200 eV electrons from Ar, Xe, Hg.
Scofield (1978)	K- and L-shell ionization by relativistic electrons, Z=18-92, E=50 keV -1 GeV. DS.
Kemper et al. (1985)	Elastic scattering of slow electrons from Ar-Xe. Two-channel approach.
Bartschat and Burke (1986)	Relativistic effects on low-energy electron scattering from Hg.
Chen (1986a)	Dielectronic recombination of He-like systems, Z=6...42.
Chen (1986b)	Dielectronic recombination of Ne-like systems, Z=18...54.
Chen (1986c)	Dielectronic recombination of F- and Ne-like Se.
Eichler (1986)	Relativistic "eikonal" theory of electron capture.
Huberland and Fritsche (1986,1987)	Low-energy elastic electron scattering from Hg-Bi using a Kohn-Sham approach. Spin-polarization emphasized.
Hasenburg et al. (1986)	Relativistic effects in elastic electron scattering from heavy atoms.
Jaskolski and Karwowski (1986)	QR approach to low-energy electron scattering from closed-shell atoms.
L.H. Kim et al. (1986)	Ratio of positron to electron bremsstrahlung energy loss.
McEachran and Stauffer (1986)	Spin polarization of electrons, elastically scattered from Xe.
Rosenberg (1986)	Relativistic scattering in a slowly time-dependent electric field.
Eichler (1987)	Relativistic charge exchange with Coulomb boundary conditions.
Jaskólski et al. (1987)	Low-energy electron scattering from Xe.
Kim and Pratt (1987)	Electron bremsstrahlung from Coulomb and DS potentials, Z=13...79, E= 1 eV-0.5 MeV.
Kobus and Jaskólski (1987)	Low-energy positron scattering from Hg as test for the QR approach.
Mc Carroll (1987)	Review on relativistic effects in electron capture.
McEachran and Stauffer (1987a)	Elastic and inelastic electron scattering from Xe.

Table 5.12. (continued).

McEachran and Stauffer (1987b)	Elastic electron scattering from Hg.
Norrington and Grant (1987)	Low-energy electron scattering by Fe XXIII and Fe VII using the Dirac R-matrix method.
Salvat et al. (1987b)	Elastic electron scattering of NR electrons from a MP fit to DS.
Sienkiewicz and Baylis (1987)	Elastic electron scattering from Ar.
Sud et al. (1987)	L-shell ionization by relativistic electrons.
Zygelman and Dalgarno (1987)	Electron-impact excitation of H-like n=2 FS levels.
Butler et al. (1988)	Positron interactions with Th, (164). Knock-out and positronium formation discussed.
Chen (1988a)	Dielectronic recombination for F-like ions, Z=26...63.
Chen (1988b)	Dielectronic recombination of H-like ions, Z=8...54. 1s+e -> nln'l' -> 1s n"l" + h .
Chen (1988c)	Effect of Koster-Kronig transition on dielectric recombination of He-like ions.
Chen and Crasemann (1988c)	Dielectronic recombination for $1s^2 2l2l'$ configurations of Be-like ions, Z=30...54.
Ehlotzky (1988)	Scattering of (spinless) electrons from a potential in a strong laser field. Sum rule.
Pindzola and Buie (1988); Pindzola et al. (1988)	Electron-impact ionization of h eavy atomic ions.
Reed (1988)	Electron-collision strengths for the 36 n=3 states of Ne-like Ba.
Reed et al. (1988)	Autoionizing resonances in electron-impact excitation of O-like Se. MCDF.
Ross and Fink (1988)	Spin-polarized atomic scattering factors for electrons. DF. E = 40 keV and Au as examples.
Stauffer (1988)	Relativistic effects in electron-atom scattering: a review.
Chen and Crasemann (1989a)	Contribution of autoionizing resonances to the n=2-2 electron- collisional excitation rates for Be-like ions.
Chen and Reed (1989)	Electron-impact excitation of n=2 and 3 transitions in Ne like ions, Z=15...54.
Fox et al. (1989)	Electron scattering factors for atoms, Z=2-98.
Jaskolski (1989)	QR (1-component but SO split) approach to many-channel electron scattering from atoms. Ca^+, Ba^+.

Table 5.12. (continued).

Pindzola and Buie (1989)	Electron-impact ionization of U^{16+} and U^{82+}. A HF/HFR/DF comparison.
Pindzola et al. (1989)	Electron-impact ionization of highly-charged ions in lowest-order QED.
Qian et al. (1989)	Electron-impact $2s^2$-2s2p excitations of Be-like systems, Ne^{6+}...U^{88+}.
Rhoades-Brown et al. (1989)	Electron capture associated with pair production in heavy-ion collisions.
Ritchie (1989)	Relativistic theory for electron-ion scattering. One continuum electron + Dirac-Coulomb.
Talwar and Wright (1989)	The radiation tail in electron scattering from finite nuclei.
Zhang and Sampson (1989)	Collision strengths for excitation to the n=3 and 4 levels of Ne-like ions, Z = 22-92.
Zhang et al. (1989)	Fully relativistic and QR distorted-wave methods for collision strengths of highly charged ions. He-, Li, Ne, Na and Ni-like ions.
Bartschat (1990)	Spin-orbit and interference asymmetries in elastic electron scattering from heavy one-electron ions.
Chen (1990a)	Resonant excitation in Ca^{q+} collisions with H_2 and He targets.
Chen and Reed (1990)	Effect of autoionizing resonances on n=3-3 electron-impact excitation rates in Ne like Se.
Chen et al. (1990b)	Contribution of resonant excitation double autoionization to electron-impact ionization ot Fe^{15+}.
Czyzewski et al. (1990)	Electron elastic scattering cross sections, $20eV \leq \epsilon \leq 20keV$, $Z \leq 94$.
Dasgupta and Whitney (1990)	Dielectronic recombination data in F-like ions, Z=18-34. HFR.
Pindzola and Badnell (1990)	Dielectronic capture processes in $U^{89,90+}$, colliding with H_2.
Reed et al. (1990a)	Excitation-autoionization processes in electron-impact ionization of Au^{68+}.
Reed et al. (1990b)	Xe^{43+}. Relativistic distorted-wave calculation of the cross section for n=2 -> 3,4.
Rumrich et al. (1990)	Basis-set truncation violates gauge invariance in the coupled-channel representation.
Sienkiewicz (1990)	Low-energy electron scattering from Hg.
Yuan and Zhang (1990)	Low-energy (E = 0.01-5 eV) electron scattering with Ca-Ra. QR.
Zhang and Sampson (1990)	Ionization cross sections for highly charged ions. H, He, Li, Ne and Na-like cases.

Table 5.12. (continued).

Zhang et al. (1990a)	Collision strengths and oscillator strengths for Ni-like ions with Z=60-92.
Zhang et al. (1990b)	Li-like ions, Z=8-92.
Zhang et al. (1990c)	Cu-like ions, Z=60-92.
Badnell and Pindzola (1991)	Dielectronic recombination of Bi^{80+} and Bi^{82+}.
Bartschat (1991)	Spin-dependent effects in elastic and inelastic low-energy electron scattering from Na.
Chen (1991a)	U^{q+} collisions with light targets.
Chen (1991c)	Dielectronic recombination of Li-like systems, Z=6...54.
Chen and Crasemann (1991)	Resonance enhancement of electron-impact excitation rates for Ar-like ions.
Decker and Eichler (1991a)	Exact 2nd-order Born calculations for relativistic electron capture.
Decker and Eichler (1991b)	Electron capture in U^{92+} + U^{91+} collisions.
Dzuba et al. (1991c)	Scattering of very slow electrons by Ba and Ra. Resonance enhancement.
Hussein et al. (1991)	Symmetrical representation of incoming and outgoing wave functions.
Kato and Safronova (1991)	$2s^2$-2s2p-$2p^2$ electron excitation cross sections of Be-like systems, Z=8...42.
Kessler (1991)	Review on electron-polarization phenomena in electron-atom collisions.
P. Li and Q.-Q. Gou (1991)	Elastic scattering of slow electrons by Hg. R/NR.
Mian (1991)	Low-energy electron scattering from Hg.
Nahar (1991)	Cross sections and spin polarizations for $e^{\pm}$ scattering from Cd.
Nahar and Wadehra (1991)	Ditto for Ar.
Pindzola et al. (1991)	Electron scattering theory for highly charged ions: ionization, excitation and dielectronic recombination.
Reed et al. (1991)	Electron excitation-autoionization in Na-like ions, Z=26-92.
Rosenberg (1991)	Relativistic multiphoton bremsstrahlung.
Salvat (1991)	Elastic scattering of fast electrons and positrons from atoms. V(r) sum of Yukawa terms. Al, Ag, Au.
Sienkiewicz and Baylis (1991a)	Elastic electron scattering from Xe.
Sienkiewicz and Baylis (1991b)	Elastic positron scattering on Hg. Negative-energy DF treatment.
Sienkiewicz and Baylis (1991c)	Kr.
Szmytkowski (1991)	Relativistic polarized-orbital theory for elastic electron and positron scattering from closed-shell atoms.

Table 5.12. (continued).

Thumm and Norcross (1991)	The 6 $^3P^o_J$ states of Cs^- are unbound, narrow resonances. Model potential.
Wijesundera et al. (1991a)	Electron scattering by Ne-like Se using the Dirac R-matrix method.
Wijesundera et al. (1991b)	O-like Kr.
Yuan and Zhang (1991a)	Low-energy electron scattering from Xe. QR. E= 0.01-20 eV.
Yuan and Zhang (1991b)	Mg, Zn and Cd.
Zhang et al. (1991)	Electron collision strengths and oscillator strengths for Ni-like ions, Z=60-92.
Zuo et al. (1991)	Electron-impact excitation of Xe.
Badnell and Pindzola (1992)	Dielectronic and radiative-recombination cross sections for U^{90+} near KLL resonances.
Bartschat et al. (1992)	Polarization of 0-2.7 eV light, emitted by electron impact on Cs. Dirac R-matrix framework; BP not sufficient.
Chen and Reed (1992)	Electron-impact ionization of highly-charged Li-like ions, Z = 18...54.
Fritsche et al. (1992)	Consistent theory of electron scattering by atoms.
Hsu et al. (1992)	Positron-impact ionization of H.
Jakubassa-Amundsen (1992)	Relativistic (e, 2e)collisions. K-shell ionization of Cu-Au, E=300, 500 keV.
Kao et al. (1992)	Electron-impact ionization of H-like atoms, Z=1...47. R/NR.
Karasiov et al. (1992)	Electron recombination with H-like U. Resonance effects with doubly excited, $2s^2$ or $(2p^*)^2$ states.
Kheifets and Amusia (1992)	Xe 5s (e, 2e) process.
McEachran and Stauffer (1992)	Spin polarization of electrons elastically scattered from Cd and Zn. DF.
Mikhailov and Porsev (1992)	Double ionization of the atomic K shell in positron annihilation with a K electron.
Moiseiwitsch (1992)	α-expansions for electron capture.
Moores and Pindzola (1992)	Electron impact ionization of highly charged hydrogenic ions. Fully relativistic disorted-wave approximation, including Mφller interaction.
Pindzola (1992)	Positron-impact ionization of highly-charged ions in lowest-order QED.
Pindzola et al. (1992)	Dielectronic recombination between ground-state hyperfine levels of Bi^{82+}.
Reed and Chen (1992)	Electron-impact ionization of Li-like ions, Z = 6-9.
Safronova et al. (1992a)	Electron-impact $2s^2$-2s2p-$2p^2$ transitions of Be-like O, Si, Fe and Mo.
Spies et al. (1992)	Dielectronic and radiative recombination of Li-like Au.

Table 5.12. (continued).

Srivastava et al. (1992a)	Electron impact 5s-5p excitation of Cd.
Srivastava et al. (1992b)	Ditto for Hg.
Srivastava et al. (1992c)	Electron-impact excitation of Ca and Ba D states.
Srivastava et al. (1992d)	Electron-impact excitation of P states of Ca-Ba.
Thumm and Norcross (1992)	Electron scattering from Cs. Very narrow ^3P_J resonances found.
Weigold (1992)	A review on (e, 2e) electron momentum spectroscopy of atoms, molecules and solids. Relativistic effects seen for Xe and Pb valence shells.
Wijesundera et al.(1992a)	$6s^26p$ and $6s6p^2$ resonances in electron scattering by neutral Hg.
Wijesundera et al. (1992b)	Electron scattering from Pb. The $6s^26p^3$ states claimed to be resonances, despite of known Pb EA.
H.-Y. Yang et al. (1992)	Dielectronic recombination of F-like and Ne-like Ge.
Yuan and Zhang (1992a)	Spin polarization of electrons elastically scattered from Ba.
Yuan and Zhang (1992b)	Ditto, Hg.
Zuo et al. (1992)	Electron-excitation of Ar-Xe.

Table 5.13. Particle-atom collisions.
For processes involving hydrogen-like atoms, see also Table 2.3.

Reference	Comments
Au (1986)	Two-photon exchange in electron-atom scattering.
Becker et al. (1986)	U^{92+}-U^{91+} collisions.
Chen and Crasemann (1986)	N-shell ionization of Bi and U by protons.
Hagelstein (1986a)	Electron inelastic scattering from Ni-like Gd.
Hagelstein (1986b)	Electron inelastic scattering from F-like Se.
Leung et al. (1986)	Relativistic α^2 corrections to the Bethe sum rule.
Moiseiwitsch (1986,1987)	Relativistic eikonal approximation for electron capture.
Mukoyama (1986)	Excitation cross sections of H-like ions by heavy charged particles.
Hagelstein (1987)	Dielectronic capture into excited levels of Ne-like Se.
Hagelstein and Jung (1987)	Electron-collision cross sections for Ne-like ions.
Hino and Watanabe (1987,1989)	Radiative electron capture from target atom (Be) to relativistic, naked nucleus (Xe^{54+}, U^{92+}).
Kim and Desclaux (1987)	Relativistic effects in electron-atom collisions. Li-like ions.
Mukoyama and Lin (1987c)	Inner-shell ionization in ion-atom collisions using an analytic atomic basis.
Pacher et al. (1987)	Retardation effects in radiative electron capture. Xe^{54+}+e, H^+ + H(1s).
Pindzola (1987)	Giant resonances in elecron-impact double ionization of Fr^+.
Zygelman and Dalgarno (1987)	Proton-impact excitation of H-like n=2 FS levels.
Bar-Shalom et al. (1988)	Electron collision excitations of heavy ions in a "distorted-wave approximation". He-like Al XII, Ni-like Gd XXXVII.
Kim and Desclaux (1988)	Energy-dependence of electron-ion collision cross sections. Fe^{22+} $2s^2$-2s2p excitations.
Mehler (1988); Mehler et al. (1989)	Inner-shell ionization in ion-atom collisions.
Moiseiwitsch (1988)	Second-order OBK approximation for electron capture.
Momberger et al. (1988)	Angular distributions of electrons emitted from 1s and 2s states in heavy-ion collisions. Z+Au, Z+U.

Table 5.13. (continued).

Mukoyama and Sarkadi (1988)	Dirac-Coulomb continuum function at zero kinetic energy in ion-atom collisions.
Rustgi et al. (1988b)	Ionization of K-shell electrons by highly relativistic protons.
Struensee and Cohen (1988)	n=2 FS transitions in H+p collisions.
Bell (1989)	Double and triple differential cross sections for K-shell ionization by relativistic electrons.
Chen and Crasemann (1989b)	K-, L- and M-shell cross-sections for ionization by protons, Z=22...92. DS.
El-Ashry et al. (1989)	K-shell ionization.
Jakubassa-Amundsen (1989)	K-shell ionization by fast electrons. Ag and Ta targets.
Leung (1989)	Bethe stopping theory for heavy target atoms (Al...Au).
Mukoyama (1989b)	K-shell ionization by protons: screened Dirac-Coulomb and DS wave functions compared.
Grant (1990)	Electron-atom collisions: a review.
Jakubassa-Amundsen (1990)	Relativistic 2nd-order Born theory for electron capture.
Kumar et al. (1990)	K- and L-shell ionization cross-sections of Au by proton or alpha-particle impact. DF-LCAO.
Mayol and Salvat (1990)	Cross-sections for K-shell ionization by electron impact.
Mohan and LeDourneuf (1990)	Electron-impact collision strength for the FS excitation of Si^{5+} $2p^5$ ^2P.
Moiseiwitsch (1990)	Electron capture at relativistic energies.
Moores and Pindzola (1990a,1991)	Electron-impact ionization of highly-charged hydrogenic ions.
Moores and Pindzola (1990b)	Electron-impact ionization of highly-charged ions. MCDF + Breit. Se^{24+}, Xe^{44+}, $U^{85,86+}$.
Mukoyama et al. (1990)	K-shell ionization by mesons.
Aggarwal et al.(1991)	Electron impact excitation of Ca XV. Lowest 20 levels.
Decker (1991)	Electron-positron pair production in relativistic heavy-ion collisions (p-U, U-U).
Eby (1991)	Production of low-energy electron-positron pairs by relativistic heavy ions.
Kürpick et al. (1991ab)	S^{16+} collisions on Ar.
Legrand et al. (1991)	L-shell ionization in p+Sm collisions, E=1 MeV. Dirac-Coulomb. Semiclassical.
Lindhard (1991)	Radiation from ultrarelativistic electrons, moving parallel to a crystal axis. Discusses the Dirac equation background.

Table 5.13. (continued).

Miller and Manson (1991)	Secondary electron energy-spectra in electron-impact ionization. He, E ≤ 5 MeV.
Moores (1991)	Electron-impact ionization cross-sections of highly-charged ions.
Mukoyama et al. (1991)	Relativistic effect on the ionization probability in the geometrical model.
Mukoyama and Lin (1991)	Inner-shell ionization cross sections in ion-atom collisions using discretized states with E>0 for the continuum.
Smit (1991, 1992)	Semiclassical approach to K-shell ionisation using H-like wave functions.
Sud and Moattar (1991)	K-shell ionization by relativistic protons.
Yudin (1991)	Unitary semiclassical theory of Coulomb excitation of atoms. Dirac equation for target electrons.
Amusia (1992)	A new "polarizational" Bremsstrahlung, due to dipole polarization of a colliding atom.
Glass et al. (1992)	The 'distorted wave model' of charge exchange must have correct spinor structure. Otherwise spurious spin effects result at NR limit.
Hofstetter et al. (1992)	Ionization of hydrogen by a relativistic heavy projectile.
Khriplovich and Pospelov (1992)	High-frequency asymptotics of optical activity and energy difference of chiral molecules.
Moiseiwitsch (1992b)	Electron capture from H-like ions by ultrarelativistic charged particle.

Table 5.14. Photon scattering and photoionization.

Reference	Comments
Waller (1929, 1930ab, 1934)	Photon scattering by bound and free Dirac electrons.
Seiler (1936)	X-ray dispersion and absorption in Dirac theory.
Pratt et al. (1964a)	K-shell photoionization cross sections, E=200 keV-2MeV. Dirac-Coulomb functions.
Pratt et al. (1964b)	Spin-polarized photoelectrons.
Mikhailov (1969)	K-shell photoionization near threshold.
Nicolaides and Beck (1978)	MB theory of photoabsorption in atoms and molecules.
Bartschat et al. (1986)	Photoionization of Ba. 5d states of Ba^+ more populated than 6s or 6p.
Deshmukh and Manson (1986)	RRPA 4f, 5s and 5p photoionization parameters of Yb.
Deshmukh et al. (1986a)	Cooper minima in Rn. RRPA.
Deshmukh et al. (1986b)	Multiple minima in the 7s photoionization of Ra.
Deshmukh et al. (1986c)	Review on Cooper minima.
Fink and Johnson (1986)	Low-energy photoionization of Li-Cs. RRPA.
Schmitt and Arenhoevel (1986)	Integrated photoabsorption cross section for a Dirac particle.
Silfvast et al. (1986)	Multielectron photoionization from $3p^6$ Ar to $3p^4 4p$ Ar^+.
Svensson et al. (1986)	Yb 4f and 6s photoionization ß parameter. RRPA.
Doolen and Liberman (1987)	Photoionization and photoabsorption, H-Lr, h < 1keV.
Goldberg and Pratt (1987b)	Photoionization of the 6d state of Cs. A deep minimum found near threshold. DS.
W.T. Hill et al. (1987)	Photoabsorption of Ba^{2+}, $5p^6$ to $5p^5nl$.
Kim and Greene (1987)	Photoionization of Ca.
Manakov et al. (1987)	Photon scattering by a bound electron using the Dirac-Coulomb Green function.
Radojevic et al. (1987)	Photodetachment cross sections of X^-, X=F-I.
Sacchetti (1987)	Relativistic effects on x-ray scattering. Solid-state applications.
Svensson et al. (1987)	Cross sections for Ar in the $3s3p^6np$ ^1P region.
Tulkki et al. (1987)	Post-collision interactions in inner-shell photoionization. Ar, Xe.
Yin et al. (1987)	High-energy relativistic photoionization cross sections from ns, $np_{1/2}$ and $np_{3/2}$ shells. A power-series, screened potential used.

Table 5.14. (continued).

Yin and Pratt (1987a)	Survey of relativistic Cooper minima.
Yin and Pratt (1987b)	Ionic and point-Coulomb relativistic Cooper minima.
Bagrov et al. (1988)	Polarization mechanism of photoelectrons.
Deshmukh et al. (1988)	Photoionization of Al^+. RRPA.
Guo and Åberg (1988)	QED approach to multiphoton ionization in the high-intensity limit.
Holm (1988)	Compton scattering from bound, relativistic atoms.
L. Liu et al. (1988)	Photoionization of excited Fe atoms.
Shanthi et al. (1988a)	Near-threshold photoionization of Kr 3p.
Shanthi et al. (1988b)	Photoionization cross section, angular distribution and spin polarization of Kr 3p and 3d.
Zilitis (1988)	Photoionization cross sections for K-like ions, Z=19-74. DF.
Aymar and Lecomte (1989)	Photoionization from Rydberg states of Sr.
Bechler and Pratt (1989)	Higher retardation and multipole corrections to the dipole angular distribution of 1s photoelectrons.
Dong et al. (1989)	Photoionization, Ne, Ar, Fe, U.
Nasreen et al. (1989)	Photoionization of inner shells in the Mg^{q+} and Ar^{q+} sequences. RRPA.
Radojevic et al. (1989)	4d photoionization of Ba.
Scofield (1989,1990)	Angular distribution of photoelectrons from polarized x-rays.
Shanthi and Deshmukh (1989)	Xenon 4p photoionization near the 4d Cooper minimum. RRPA.
Tulkki (1989)	Multiple excitation at Xe 5s photoionization threshold. Multichannel MCDF.
Veltman (1989)	Radiative corrections to polarized Compton scattering.
Wague (1989)	n=2 resonant photoionization of He-like ions.
Aymar (1990)	Odd-parity spectrum of Ba.
Band et al. (1990b)	K-shell photoionization cross sections.
Cuéllar et al. (1990)	SO effects on Cs ns (n=8-12) photoelectron angular distribution.
Fink and Johnson (1990)	Two-photon ionization of closed-shell atoms. TD DF. He, Ne-Xe.
Florescu et al. (1990)	Low-frequency Rayleigh scattering including spin flip for H-like atoms.
Glotov et al. (1990)	5p and 4f photoabsorption of Yb using LDF methods.
Greene (1990)	Photoabsorption of M^-, M=Rb-Fr.
Greene and Theodosiou (1990)	Photoionization of Ba 6s6p $^1P_1$.
Kelly (1990)	Review on photoionization.

Table 5.14. (continued).

Kutzner et al. (1990)	4d photoionization of Ba.
Mikhailov (1990)	K-shell photoionization near threshold: analytic solution for a screened nuclear field.
Reiss (1990)	Relativistic strong-field photoionization (photon field stronger than atomic one).
Rozsnyai (1990)	Inner-shell photoionization of Ba^{n+}, n=0-2, Xe, Kr.
Schäfers et al. (1990)	Hg 5d and 6s photoionization. RRPA.
Shanthi and Deshmukh (1990)	Photoionization of Pd.
Tulkki et al. (1990)	Photoelectron recapture in post-collision interactions. Ar.
Zhou et al. (1990)	Anomalous photon scattering factors for Ne^{q+}, q= 0...7. Photoionization cross sections. DS.
Chi and Huang (1991)	Photoionization of Be.
Connerade (1991)	Photoionization: a review.
Greene and Aymar (1991)	Photoabsorption of Sr-Ra.
Grochmalicki et al. (1991)	Relativistic effects may inhibit stabilization of atoms against ionization in intense laser fields.
Klimenko and Klimenko (1991)	Elastic scattering amplitude for an electron in a plane wave EM field.
Liu and Li (1991)	High-energy photoionization of atomic ions.
Rosenberg and Zhou (1991)	Soft-photon approximation for bound-state Compton scattering.
Suric et al. (1991)	Compton scattering of photons by 1s shell electrons. Al, Ho, Pb.
Zapryagaev and Nefedov (1991)	Photoionization of H-like ions. E1, M1, E2 approximations. Z = 10-150.
Altun and Radojevic (1992)	3d photoionization of Kr and Br^-. R RPA.
Bergstrom et al. (1992)	Whole atom Compton scattering of unpolarized photons.
Choe et al. (1992)	Photoionization cross sections for the $6s^2$ and 6s6p levels of Yb using DFT.
Connerade (1992)	Review on "giant resonances".
Deshmukh et al. (1992)	Outer-shell photoionization of Rn and Ra. RRPA.
Johnson and Cheng (1992)	4d photoionization of Xe.
Kheifets and Amusia (1992)	5s photoionization of Xe.
Kim et al. (1992)	ns photoionization of U, E = 1 eV-100 keV. DS field.
Krainov and Roshupkin (1992)	Spin effects on angular distribution of electron emission by strong EM fields.
Radojevic and Kelly (1992)	Photodetachment of I^-.

Table 5.14. (continued).

Reiss (1992)	Interaction of atomic electrons with strong EM fields (up to 10^6eV interactions with 1 eV photons).
Scherdin et al. (1992)	Delbrück scattering (photon scattering by nuclear Coulomb potential) for large Z.
Tulkki et al. (1992)	Kr 4p, 4s and 3d photoionisation, E <260eV.
C.-M. Wu et al. (1992)	Sr photoelectron angular distribution. Double-excitation effects. MC-RRPA.
Zhou and Pratt (1992)	Spurious structures in photoionisation calculations in the independent-particle approximation.
Zhou et al. (1992)	Anomalous scattering factors for Rayleigh scattering from Ar.

Table 5.15. Atom-atom collisions and interatomic potentials.
See also Table 7.10.

Reference	Comments
Knipp (1938)	Angular parts for the R^{-5} atomic quadrupole-quadrupole forces for light-heavy (large SO) and heavy-heavy combinations.
Merzbacher (1985)	Fundamental interactions in atomic collisions.
Anholt (1986)	Relativistic atomic collisions.
Anholt and Gould (1986)	Relativistic heavy-ion-atom collisions.
Anholt and Meyerhof (1986)	Relativistic heavy-ion collisions with solid targets.
Anholt et al. (1986)	Radiative processes in relativistic heavy-ion collisions with atoms.
Amus'ya et al. (1987a)	Polarizational Bremsstrahlung.
Amus'ya et al. (1987b, 1988a)	Atom-atom Bremsstrahlung.
Anholt and Becker (1987)	Ultrarelativistic heavy-ion collisions. Electron capture and 1s-ionization.
Anholt et al. (1987)	Uranium ion charge-state studies.
Deco and Rivarola (1987abc)	Charge transfer at relativistic ion-atom collisions.
Denisenko and Mrówczynski (1987)	Interaction of various relativistic elementary (H-like) atoms with matter (C-Pb).
Hino and Watanabe (1987, 1989)	Radiative electron capture for relativistic bare-nucleus atom collision, Xe^{54+} or U^{92+} on Be.
Mrówczynski (1987)	Interaction of relativistic elementary atoms with matter.
Lin (1988)	Heavy-atom collisions: a review.
Toshima and Eichler (1988ab)	Xe^{54}-Ag, -Au and U^{92+}-U^{91+} collisions: electronic excitations and charge transfer.
Aquilanti and Vecchiocattivi (1989)	Fine-structure effects on collision integrals of O and F atoms with He and Xe.
Deco and Grün (1989)	Ionization of a bound relativistic electron by impact of a bare, relativistic ion.
Deco and Rivarola (1989a)	Pair production with electron capture in relativistic heavy-ion collisions.
Deco and Rivarola (1989b)	Two-center effects in relativistic radiative electron capture.
Hutton et al. (1989)	Double and single electron capture in slow collisions of $Ar^{8+, 9+}$ with He.
Toshima and Eichler (1989)	Inclusion of "pseudostates" to describe the continuum.
Amus'ya (1990)	Atomic Bremsstrahlung.

Table 5.15. (continued)

Amus'ya and Solov'ev (1990)	Frequency-angle distribution of radiation in relativistic atomic collisions.
Decker (1990)	Electron capture in C^{6+}-Au and Ar^{18+}-Ag collisions. Negative-energy intermediate states included.
Nakatsuji and Ehara (1990)	CsXe. Collision-induced 6s-5d transitions in a quasistatic approximation.
Toshima and Eichler (1990a)	Distorted-wave approximations for relativistic atomic collisions.
Toshima and Eichler (1990b)	Necessity of Coulomb boundary conditions in coupled-channel calculations. $U^{92}+U^{91}$.
Decker and Eichler (1991a)	Exact 2nd-order Born calculations for relativistic electron capture.
Decker and Eichler (1991b)	Electron capture in U^{92+}-U^{91+} collisions.
Yudin (1991b)	Stopping power of fast multicharged ions.
Bransden and McDowell (1992)	Charge exchange in ion-atom collisions. Book.
Kürpick et al. (1992)	K-vacancy transfer in collisions of S^{15+} on Ar.
Legrand et al. (1992)	L-shell ionization in ion-atom collisions. He+Pt, Ne+Pt, Ne+Yb, S+Au.

Table 5.16. Nuclear processes involving electronic wave functions.

Reference	Comments
Green and Rose (1951)	Nuclear-structure effects in internal conversion.
Rose and Holmes (1951)	Effect of finite nuclear size in beta-decay.
Band and Trzhaskovskaya (1986)	Wave-function expansions near origin from DS and DF functions.
Band et al. (1987)	Beta-decay probabilities.
Z.-H.Chen et al. (1987)	Bound-state versus continuum beta decay of ^{187}Re.
Batkin et al. (1988)	Inner-shell ionization in beta decay.
Band et al. (1989a)	The ^{193m}Ir decay.
Trusov and Eglais (1989)	Contact charge density of Fe.
Band (1990a)	Internal conversion coefficients.
Kolomietz and Kondratjev (1990)	Nuclear gamma-transitions via the accompanying atomic excitations: nuclear M4 transitions for ^{93}Nb and ^{193}Ir. Both elastic and inelastic electronic contributions.
Band and Trzhaskovskaya (1991ab)	Internal-conversion coefficients for $_{41}$Nb ... $_{97}$Bk. DF. Various E_L, M_L.
Jakubassa-Amundsen (1991)	Transfer of electron shells during nuclear reactions.
Schober et al. (1991)	Internal-conversion coefficients for valence-electrons of Tc and U.
Batkin et al. (1992)	K-shell autoionization in beta decay.
Harston and Pyper (1992)	Exchange effects in β decays. ^{106}Ru, ^{241}Pu. ^{14}C, ^{35}S.
Jin et al. (1992)	Electron Coulomb effects in quasielastic (e, e'p) reactions.
Németh (1992)	Comparison of DF EL, ML (L=3,4) internal conversion coefficients with experiment.

Table 5.17. Parity-violation effects in atoms and molecules. Only the time-even, parity-odd or 'TE-PO' parity violation has so far been experimentally unequivocally confirmed. Upper limits exist for the TO-PO case.

Reference	Comments
Salpeter (1958)	H-like atoms with electron dipole moment.
Gajzágó and Marx (1974)	SO required for an energy difference between mirror molecules.
Hegstrom et al. (1980)	PNC energy difference between mirror-image molecules is of order $E^{(2)} = \Sigma' \langle 0 \| V_{PNC} \| n \rangle \langle n \| h_{SO} \| 0 \rangle / (E_0 - E_n)$.
Barra et al. (1986,1988)	PNC effects on NMR chemical shifts in enantiomeric molecules estimated.
Dzuba et al. (1986a)	P and T nonconservation in rare-earth atoms.
Johnson et al. (1986)	PNC effects for monovalent atoms using DF or model potentials.
Botham et al.(1987)	PNC "equations of motion". Application on Pb.
B.P.Das (1987)	PNC TP for 6s-7s of Cs and $6p^*$-$7p^*$ of Tl.
Dzuba et al. (1987b)	PNC amplitudes of Tl.
Dzuba et al. (1987c)	Cs, Tl.
Johnson (1987)	P- and CP-violation in Rb, Cs, Au, Tl.
Kozlov (1987,1988); Kozlov et al. (1987)	PNC effects in PbF. X $^2\pi_{1/2}$.
Mårtensson-Pendrill (1987)	Cs, Tl.
Mårtensson-Pendrill and Öster (1987)	Atomic electric dipole moments. Cs, Hg, Xe ground states.
Perger and Das (1987a)	Cs, Tl.
Sandars (1987)	Review.
Bouchiat and Piketty (1988)	Cs 6s-7s PNC calibration from M1- and E2-hfs-induced TP.
Dzuba et al. (1988)	PNC in Pb and Bi.
Johnson et al. (1988d)	PNC 6s-7s transition in Cs.
Khriplovich (1988)	Book on PNC in atoms.
Kraftmakher (1988a)	PNC in Tl.
Kraftmakher (1988b)	PNC in Cs.
Parpia et al. (1988)	PNC in Cs and Tl.
Wiesenfeld (1988)	Effect of Z on PNC energy difference between enantiomers.
Blundell (1989)	MBPT calculation of PNC in Cs.
B.P.Das (1989)	Review on PNC effects in atoms.
Dzuba et al. (1989c)	Cs 6s-7s PNC E1 amplitude.
Khriplovich (1989)	T-violation in atoms.
Lindroth et al. (1989a)	Atomic (E1) dipole moments, due to an electron dipole moment. Leading terms relativistic, of order α^2. Cs, Tl.
Lindroth et al. (1989b)	Effects of H_{Br} on the Tl E1 moment.
Schäfer et al. (1989)	Proposed PNC two-photon laser experiment on U^{90+}.

Table 5.17. (continued).

Blundell et al. (1990c)	The 6s-7s PNC transition of Cs.
Boston and Sandars (1990)	The P-violating, T-invariant L=1 "anapole moment" of hydrogenic atoms.
Dzyuba et al. (1990)	PNC of Cs.
Hartley et al. (1990)	PNC and electric dipole moments of Cs and Tl.
Onishchuk (1990)	Chirality of valence electrons at nuclei in asymmetrical molecules. No effect on nuclear electron capture.
Blundell et al. (1991b)	The 6s-7s PNC transition of Cs.
Chou and Huang (1991)	MCDF theory including PNC.
Hartley and Sandars (1991)	Review on PNC effects in atoms.
Khriplovich (1991)	Book on PNC.
Flambaum (1992)	Long-range PNC interaction.
Mårtensson-Pendrill and Lindroth (1991)	P- and T-violating scalar-pseudoscalar electron-nucleon interaction. Cs, Tl.
Dmitriev et al. (1992)	P- and PT-odd terms for HgF and PbF.
Karasiev et al. (1992)	PNC effects in radiation spectra of He-like ^{235}U. MCDF.
Liu and Kelly (1992)	E1 moment of Tl.
Mårtensson-Pendrill (1992ab)	Reviews on atomic effects of nuclear PNC effects.
S. Sen and Gautam (1992)	NR aspects of SU2*U(1) PNC hamiltonians in atoms.

Table 6.1. Group theory and symmetry aspects.

Reference	Comments
Elliott (1954)	Spin-orbit coupling in band theory. Character tables.
Altmann (1986)	Rotations, quaternions and double groups.
Eschrig and Richter (1986)	Kramers' theorem. Cp. Rösch (1983a), Dongarra et al. (1984ab).
J. Meyer (1988)	Molecular basis functions for C_n, C_{nh}, S_n, C_{nv}, D_n, D_{nd}, D_{nh}, T, T_h, T_d, O and O_h. Projection operators used.
J. Meyer et al. (1989)	A program for constructing relativistic, symmetry-adapted molecular basis functions for 36 double groups.
Pitzer and Winter (1988)	A real Hamiltonian can be obtained in a $\|lm_lm_s\rangle$ basis for C_{2v}, D_2, D_{2h} or higher symmetry. Applications on SO PP.
Smirnov and Evarestov (1988)	Symmetry of localised orbitals in molecules with SO interaction. Example: T_d Bi_4.

7. MOLECULAR CALCULATIONS

Table 7.1. One-electron systems. For overcritical fields, see also Table 5.5.

Reference	Comments
Klaus (1980b)	General properties of Dirac operators for several Coulomb singularities.
Becker et al. (1986)	U^{92+}-U^{91+} collisions at 1 GeV/amu. TD finite-difference solution of the Dirac equation.
Wietschorke (1986); Wietschorke et al. (1987)	Two-center continuum.
Wolniewicz and Poll (1986)	Higher ro-vibrational states of HD^+ and H_2^+
Ishikawa et al. (1987, 1988)	H_2^+ using momentum-space GTOs.
Mark and Becker (1987)	H_2^+ energy correction at R=2 a.u. using a GTO basis.
Rutkowski and Rutkowska (1987)	H_2^+ at R=2 a.u. using a new PT approach. Terms of order c^n, n=2, 4, 6 obtained.
Sundholm et al. (1987); Sundholm (1988)	Fully numerical Dirac results for H_2^+. V(R), Q_2, Q_4.
Bakalov (1988)	Mesic molecules.
Ixaru and Pantea (1988)	The two-centre Dirac equation and positron creation. A double-square-well model in 1D.
Toshima and Eichler (1988ab)	U^{92+} - U^{91+} collision system. Excitation and capture.
Tóth et al. (1988)	Two-centre Dirac equation with separable potentials.
Dietz and Roemer (1989)	Avoided level crossings and pair production in strong Coulomb fields.
Moss and Sadler (1989)	High J, low v levels of the HD^+ electronic ground state.
Petelenz and Smith (1989)	Vacuum-polarization in muonic molecules, ppμ, ddμ etc.
Rumrich et al. (1989)	Continuum states of the two-center Dirac equation. Heavy collision systems.
Aissing et al. (1990); Aissing and Monkhorst (1990)	Binding energies of muonic molecules.
Balint-Kurti et al. (1990)	Rovibrational levels of HD^+, v = 0-21, J=0,1.
Howells and Kennedy (1990)	Ground and first excited states of H_2^+ and its isotopic species.
Moss (1990ab)	H_2^+, D_2^+ rovibrational states.

Table 7.1. (continued).

Rutkowski and Schwarz (1990)	H_2^+ properties using a new PT.
Strayer et al. (1990)	Muon-pair production and negative-muon capture into the 1σ state in a superheavy collision system. Numerical, B-spline solution in 3D. Au+Au.
Toshima and Eichler (1990b)	Necessity of Coulomb boundary conditions in coupled-channel calculations. $U^{92}+U^{91}$.
Decker and Eichler (1991b)	Electron capture in U^{92+} + U^{91+} collisions.
L. Yang et al. (1991)	Accurate FEM solution of the Dirac equation for H_2^+ (R=2 a.u.).
Bukowski et al. (1992)	Bethe logarithm and Lamb shift for H_2^+ and isotopic species.
Franke and Kutzelnigg (1992)	Very accurate, "direct perturbation theory" E_0, E_2, E_4, E_6 for H_2^+, HeH^{2+}.
LaJohn and Talman (1992)	Minimax variational solution of the Dirac equation for H_2^+, Th_2^{179+}.
Rutkowski et al. (1992)	H_2^+ using double PT. Effects on R, k, D_e.
Thiel et al. (1992)	Finite-difference treatment of U^{92+} + U^{91+} collisions. Creation of electron-positron pairs.

Table 7.2. LCAO-DF calculations on molecules. For LCAO calculations on one- and many-electron atoms, see the Table 2.3. or 2.4. and 4.4., respectively.

Reference	Comments
Aerts (1986)	DF-LCAO methodology.
Y.S. Lee (1986)	An approximation neglecting small-component Coulomb and exchange contributions. Tested on HI.
Malli and Pyper (1986)	AuH. Both DF and CI using numerical, atomic DF basis, completed with STOs. See comments by Balasubramanian et al. (1991a) and reply by Malli et al. (1991).
Hegarty and Aerts (1987)	Kinetically balanced F_2 and GeH_4.
Lee and McLean (1987)	AgH and AuH. SCF-level STO MO energies. Effects on E_{corr} mentioned.
Baeck and Lee (1988)	Open-shell DF-LCAO. CH SO splitting.
Datta and Devaiah (1988)	H_2 using a minimax technique.
Laaksonen et al. (1988)	HCl by GTO MCDF. Energies and dipole moment given.
Ramos et al. (1988)	A four-component numerical DF basis, completed with STO:s. Bonding and dipole moments in HgH^+, TlH, PbH^+, BiH, TlI, PbTe. R/NR. (See Balasubramanian et al. 1991a, Malli et al. 1991).
Roszak and Chojnacki (1988)	Molecular calculations using a modified H-like basis.
Wilson (1988)	Review.
Baeck and Lee (1990)	Diatomic open-shell hydrides of B-F, Si, S.
Wilson (1990)	Universal basis sets and Cholesky decomposition of the two-electron integral matrix.
Mohanty and Clementi (1990b, 1991c); Mohanty et al. (1991a)	Working formulae for the authors' basis-set DF method for molecules.
Dyall et al. (1991a)	Polyatomic Gaussian-basis DF calculations reviewed. H_2O.
Dyall et al. (1991b)	CH_4 - PbH_4 at DF level. Results compared with PP and PT.
Glushkov (1991c)	Formalism for TD MCDF theory for molecules.
Visscher et al. (1991ab)	Kinetic balance and basis-set contraction tested on SiH_4, SnH_4 and atoms.
Wilson (1991)	Review on relativistic MBPT.
Baeck et al. (1992)	CO, CO^+.
Bauschlicher et al. (1992c)	KO. $^2\Sigma$ - $^2\pi$ separation at DF level.
Dyall (1992a)	XH_2, XH_4, X = Si-Pb. DF-LCAO compared with PT and PP.
Dyall (1992b)	Omission of (SS|SS) Coulomb integrals for the small component S studied on Hg^{2+}, Hg and PbO.
Fujimura and Matsuoka (1992)	Breit interaction integrals over Laguerre Gaussians.

Table 7.2. (continued).

Goldman (1992)	Spurious-root elimination tested on Dirac-Coulomb.
Ishikawa et al. (1992a)	AuH^+ ground state. First open-shell application of the Malli-Pyper method.
Malli (1992)	ThPt. Predicted SCF R_e 274 pm. R/NR. Unbound by 1.38 eV.
Matsuoka (1992b)	HX, HX^+, X=I, At.
Mohanty (1992)	Formalism for molecular Dirac-Fock-Breit. 79 references.
Visser (1992)	Thesis on group 14 MH_4 and EuO_6^{9-}.
Visser et al. (1992a)	MH_4, M = C-Pb. Effect of H_{Br} studied.
Visser et al. (1992b)	EuO_6^{9-} crystal-field levels.
Yang et al. (1992)	Numerical 2D-DS model solved for H_2, Li_2, BH with 10-figure accuracy by FEM. R/NR at $R_{exp.}$.

Table 7.3. Molecular all-electron calculations using "no-pair" or other transformed Dirac Hamiltonians.

Reference	Comments
Hess et al. (1986)	Variational use of the Pauli Hamiltonian leads to problems.
Hess (1986)	Derives the "no-pair" method. Results on Br, Ag.
Hess and Chandra (1987)	AgH X $^1\Sigma$ and A $^1\Sigma$.
Marian et al. (1988)	ThO. Bonding analysed.
Sundholm (1988)	Fully numerical solution of the DS equations (α=0.70) for H_2, LiH, Li_2, BH and CH^+. R/NR for V(R), Q_1, Q_2.
Carnell et al. (1989)	CsH. Lowest 17 states at MRD CI level using Hess' "no-pair" operator. X $^1\Sigma$ R_e contraction 3.0 and 1.4 pm at SCF and CI level, respectively. For A $^1\Sigma$ at CI level, 19.6 pm.
Jansen and Hess (1989b)	Tests on Au atom IP, EA and excitation energies.
Jansen and Hess (1989c)	AuH lowest states.
Marian et al. (1989); Marian (1990a)	NiH and NiD. Vibronic spectra using the "no-pair" Hamiltonian or PT. Includes SO.
Kellö and Sadlej (1990a)	A multiple-perturbation theory for correlation and the Cowan-Griffin QR Hamiltonian. HX, X=Cl-I. AgH.
Kellö and Sadlej (1990b)	EFG in HX, X=Cl-I. Nuclear Q for X obtained.
Knappe (1990)	GTO basis QR LDF approach.
Knappe and Rösch (1990)	The "no-pair" Hamiltonian implemented with GTOs and LDF. Tests on Ce and Pb atoms.
Marian (1990b)	The Cu atom using the "no-pair" Hamiltonian.
Kellö and Sadlej (1991)	MH, M=Cu-Au. Molecular Q.
Marian (1991ab)	CuH, CuD and other transition-metal diatomics. "No-pair" Hamiltonian.
Sadlej (1991)	Dipole moment and polarisability and Q_2 of AuH.
Sadlej and Urban (1991)	Dipole moment of AgH.
Sadlej et al. (1991)	Polarisabilities of Ca-Ba.
Samzow and Hess (1991)	FS of Br as test of two-electron terms in the "no-pair" Hamiltonian.
Bueckert et al. (1992)	2nd-order Dirac eq. used in variational Monte Carlo. E_{rel} of Li and LiH obtained from fit to various c, down to 137./10.
Hippe and Peyerimhoff (1992)	CuO ground and excited states using 'no pair'.
Häberlen and Rösch (1992)	AuH, AuCl, Au_2 using a GTO-LDF approach with a transformed QR Hamiltonian.
Häberlen et al. (1992)	M@C_{28}, M@$C_{28}H_4$ endohedral M(IV) compounds, M=Ti, Zr, Ce.

Table 7.3. (continued).

Ilyabaev and Kaldor (1992a)	He-Xe and Be. CC using a no-pair Hamiltonian.
Ilyabaev and Kaldor (1992b)	No-pair CCSD for excitation energies of the Ne atom.
Kellö and Sadlej (1992a)	XCN, X = F-I. QR CI.
Kellö and Sadlej (1992b)	PbH_4, BiH_3, PoH_2, AtH. Dipole moments and polarisabilities.
Marian et al. (1992)	Cu valence levels using the 'no pair' Hamiltonian. R/NR.
Rösch and Häberlen (1992)	Reply to Samzow et al. (1992b).
Sadlej (1992a)	SbH_3, H_2Te, HI. SCF-level R/NR.
Sadlej (1992b)	XX', X = F-I.
Samzow et al. (1992a)	Two-electron terms of the no-pair Hamiltonian. Ag, Au.
Samzow et al. (1992b)	Comments on Knappe and Rösch (1990). For reply, see Rösch and Häberlen (1992).
Wilkin and Salahub (1992)	SCF treatment of h_m+h_d and h_{SO} found possible using suitable (NR!) innermost contraction coefficients. Cu, Ag, Au.

Table 7.4. Molecules treated by the DS-DVM method.

Reference	Comments
Fricke and Sepp (1981)	Kr^{2+}-Kr^{2+} correlation diagram.
Ellis and Holland (1986)	Review on cluster compound calculations.
Ruscic et al. (1986)	M_2O, M=Al-Tl. PES interpreted. TlCl.
Sepp et al. (1986)	Improve numerical accuracy to obtain V(R) of diatomics. N_2, Ne_2 tests.
Thies et al. (1986)	Ne^+-Ne, Ne^+-O collision systems.
Hartung et al. (1987)	Ar-Ar^+ potential curve for laboratory energies of 15-400 keV and MO correlation diagram. The new code of Sepp et al. (1986) is used.
Hohl et al. (1987)	$M(BH_4)_4$, M=Zr, U. Bonding and PES. Results compared with QR-MS ones of Hohl and Rösch (1986).
Ryzhkov and Gubanov (1987)	EuX_6^{10-}, X=O-Se. LnS_6^{10-}, Ln=Nd, Sm, Eu, Gd. $Sm_2S_{10}^{16-}$. Bonding.
Sepp (1987); Sepp and Fricke (1987)	Positive-energy projection operators for basis-set DS calculations.
J. Guo et al. (1989)	Use DS-DVM potentials for MS calculations of the M-edge x-ray spectra of UO_2(s) cluster model and molecular UCl_4.
Rösch et al. (1989)	XAu_5^+, XAu_6^{2+}, X=nothing, C. $C(Au\ SH_2)_6^{2+}$. R/NR. Bonding analysed.
Thies et al. (1989ab)	F^{8+}-Ne and F^{6+}-Ne collision system.
DeKock et al. (1990)	CeO_2, $YbCl_2$. Both found bent, mainly due to 5d.
Goodman and Soderholm (1990)	$MBa_2Cu_3O_7$ cluster models, M=Y, Pr, Nd, Ho, Cm.
Kimura et al. (1990)	$[I(H_2O)_6]^-$ in a polarized dielectric medium. Charge Transfer To Solvent (CTTS) transitions calculated. I^- SO splitting 1.02 eV (exp. 0.88).
Ryzhkov and Gubanov (1990)	Uranyl complexes. Energy-level diagrams given.
Cheng and Ellis (1991)	Pt_n, n=2-4.
Görling et al. (1991)	$\{[(H_3P)Au]_6\ X\ \}^{m+}$, X=B, C, N. Also Au_6^{2+}, $[(H_3P)Au]_6^{2+}$. Bonding analyzed. 5d and 6s equally important. Both Au-Au, Au-X and Au-PH_3 interactions must be included.
C.-X. Guo et al. (1991)	Cluster models for Sm^{2+} in BaF_2.
Kürpick et al. (1991ab)	S^{16+} scattering from Ar.
Pierloot et al. (1991)	AnX_4, An=Th, U; X=F, Cl. PES interpreted.
Bastug et al. (1992)	Al_n (n=2-8).
Bowmaker et al. (1992)	$(RN)U(NR_2)_3$, R = H, SiH_3. Bonding and PES.

Table 7.4. (continued).

Kürpick et al. (1992)	K-vacancy transfer in S^{15+} collisions on Ar.
Nakamatsu et al. (1992)	CF_4, SF_6, UF_4, UF_6.
Onoe et al. (1992ab)	UV excited states of UF_6. XPS.
Pershina et al. (1992a)	MCl_5, M = V, Nb, Ta, E105 and Pa. Bonding analysed. E105 atomic ground state $6d^3 7s^2$, as for Nb.
Pershina et al. (1992b)	MBr_5, M = Nb, Ta, Pa, E105.
Pershina et al. (1992c)	$MOCl_5$, $MOBr_5$, M = V-E105, Pa.
Ryzhkov et al. (1992)	MCl_4, M = Ti-Hf, 104; Pb.

Table 7.5. Molecules treated by the DS-MS-Xα method.

Reference	Comments
Arratia-Perez and Malli (1986a)	Au_6^{2+}. S-d hybridization discussed.
Arratia-Perez and Malli (1986b)	Ag_3^{2+}, Au_3^{q+}, Au_4^{q+} (q=1, 2). S-d hybridization, g-tensors and hfs discussed.
Arratia-Perez and Malli (1986c)	Ag_5^{q+} (q=0, 2-4).
Case (1986)	SO effects in metalloporphyrins. M=Cu, Fe.
Mukoyama and Adachi (1986)	CO, UO. R/NR comparisons for orbital energies.
Zuloaga and Arratia-Pérez (1986)	Cis-Pt$(NH_3)_2Cl_2$. R/NR. Donation to 6p found more important than to 5d.
Arratia-Perez et al. (1987)	$Cu(CO)_3$ bonding, optical, magnetic properties.
Arratia-Perez and Malli (1987)	Ag_4^{q+} (q=1, 3) Bonding, optical and magnetic properties.
Arratia-Perez and Marynick (1987)	$Ag(CO)_3$. Bonding, optical and magnetic properties.
Ramos et al. (1987)	Au_{13} (I_h). R/NR. IP, optical spectra. SO splitting of the 5d band 2.2 eV. Relativity increases s-d hybridisation.
Arratia-Perez and Marynick (1988)	$PaCl_6^{2-}$. Energy levels, g, hfs.
Arratia-Perez et al. (1989)	Au_{13}(O_h and I_h). Bonding, g, hfs.
J. Guo et al. (1989)	M-x-ray spectra for UCl_4 and a cluster model of UO_2 using a DS-DVM potential.
W. Mueller (1989)	$Y_2Ba_2Cu_{12}O_{18}$ cluster model for solid $YBa_2Cu_3O_7$. Cu 3d state SO splitting about 0.3 eV. Bonding both ionic and covalent.
W. Müller (1989)	Cluster models for BaO/W(100): $(BaO)_nW_m$, n=1, 2, m=0...14. Also W-Os mixed clusters. Result: "a Ba salt of the O/W substrate".
Glebov et al. (1990)	U_6 octahedral and distorted clusters as models for the fcc and bcc structures. U-U bonding 40 % 6d, 30 % 6p, 20 % 5f, 3 % 7s.
Zhuikov et al. (1990)	$(104)Cl_4$, hybridization $7s^{0.34}$ $6p^{5.81}$ $7p^{0.21}6d^{1.34}5f^{13.98}$. Results compared with MCl_4, M=Ti-Hf.
Arratia-Perez (1991)	$Os_2Cl_8^{2-}$.
Kalinkin et al. (1991)	Cluster models for S-doped Bi_2Sr_2Ca Cu_2O_8. Sr 4p SO effects on Bi $6p_{1/2}$ levels seen.
Chermette (1992)	Review on MS Xα.

Table 7.6. Molecules treated by the quasirelativistic DS-MS Xα or "RXα" approach. See Table 2.2. of Pyykkö (1986a) for the QR 1-electron equation.

Reference	Comments
Agaskar et al. (1986)	$Os_2Cl_8^{2-}$. Evidence for a $\sigma^2\pi^4\delta^2\delta^{*2}d^5$-$d^5$ triple bond.
Andersen et al. (1986)	$Ln(cp)_2$, Ln = Sm, Eu, Yb. PES interpreted.
Hofmann and Rösch (1986); Hofmann et al. (1986)	$[(\eta^5\text{-}C_5H_5)_3Re_3(\mu^2\text{-}O)_6]^{2+}$. Expected $(1e')^2\ {}^3A_2$ ground state confirmed. Also the $Re_3O_6^{5+}$ fragment.
Hohl and Rösch (1986)	$M(BH_4)_4$, M=Zr, Hf, Th, U. Bonding and PES.
Katrib et al. (1986)	$Re_2X_8^{n-}$, X=Cl, Br, n=2,4,6. Re(4f) binding energies and multiple bonding.
Kovba and Topol' (1986ab)	Low-lying states of AlO-InO.
Loder (1986); Loder et al. (1988)	TlX, X=F-I, NO_3. Tl_2O, Tl_2S. PES interpreted: 6s character in two MO:s, more in the lower.
Mattar and Ozin (1986)	Ag_3^{q+}, q=0-2. D_{3h}. Eigenvalues, spin densities, IP.
Moine et al. (1986)	$(MCl_6)^{5-}$, M=Cu, Ag, as models for impurities in LiCl. SO in PT. Geometries and excitation energies.
Axe and Marynick (1987)	$HgTe_2^{2-}$, $PbTe_2$. Relativistic reduction of 5s-$5d_{z^2}$ σ_g repulsion found.
Bice (1987)	Lead (II) halide PES.
Bursten and Clark (1987)	$M_2(O_2CH)_4$, M=W. A review on multiple M-M bonds.
Bursten and Novo-Gradac (1987)	$(cp)_2M$-$Ru(cp)(CO)_2$, M=Zr, Th. Th metal character $s^{0.097}p^{0.036}d^{0.718}f^{0.155}$.
Bursten and Strittmatter (1987)	$(cp)_3UL$, L=CO and OC. U-C preferred to U-O.
El-Issa et al. (1987)	_Cis_ and _trans_ $PtCl_2(NH_3)_2$.
Goursot et al. (1987a)	$PtCl_5^{2-}$. Photoreduction of $PtCl_6^{2-}$ discussed.
Goursot et al. (1987b)	$[Ta_2Cl_6(NH_3)_4]^{q-}$, q=0,1. Electronic transitions.
Green et al. (1987)	$Ln(C_5H_5)_2$, Ln=Eu, Yb. Bonding and PES.
Kondratenko et al. (1987)	EXAFS and XANES spectra of Pt in $PtCl_n^{2-}$, n=4, 6.
Makhyoun (1987)	$NpO_2Cl_4^{2-}$, $NpO_2(NO_3)_3^-$. The $5f^1$ levels studied, including SO. Effective ligand-field parameters obtained.
Makhyoun et al. (1987)	$Th(\eta^5\text{-}C_5H_5)_2\ (\mu\text{-}PH_2)_2\ Ni(CO)_2$. Evidence for a Th-Ni dσ-dσ bond.
Scioly et al. (1987)	$Ta_2Cl_4(PH_3)_4H_n$, (n=2, 4), $Ta_2Cl_4(PH_3)_4$. Free and hindered rotation of the hydride ions for n=4 and 2, respectively.
Topol' et al. (1987)	RuO_4, OsO_4 PES.

Table 7.6. (continued).

Akrivos and Aldridge (1988)	$PtCl_2(OH)_2(NH_3)_2$.
Axe and Marynick (1988)	20-electron Zintl anions $Sn_2Bi_2^{2-}$, $Pb_2Sb_2^{2-}$ ("tetrahedral"), $Tl_2Te_2^{2-}$ (boat).
Berksoy and Whitehead (1988a)	Al_2X_6, X=F-I. NQCC interpreted.
Berksoy and Whitehead (1988b)	$SnCl_2Ph_2$ monomer and dimer NQCC, IS.
Cayton (1988)	Organometallic "piano-stool" systems $(cp)ML_n$, n=3, 4. M=Mo, W, Re,..; $U_2(OH)_{10}^{2+}$, $(cp)_2U_2O_4$.
Dai and Ma (1988a)	$MoCl_2^{2+}$, $MoCl_4$.
El-Issa et al. (1988)	IrX_6^{n+}, n=1-4, x=Cl-I. IrX_n, n=1-4.
Makhyoun (1988a)	$[W(WS_4)_2]^{2-}$. Confirms W(II) and W(VI) oxidation states for middle and end atoms, respectively. W-W bond order 0.5.
Makhyoun (1988b)	$5g^1$ complexes EO_2^{2+} and EF_6 of E= element 125. NpO_2^{+2}.
Makhyoun and El-Ezaby (1988)	UOX_5^{2-}, X =F-Br. Axial compression of the <u>trans</u> U-X bond suggested. Ligand-field parameters for the $5f^1$ levels derived.
Makhyoun and Hasanein (1988)	$UO_2Cl_4^{2-}$. Dependence on U-O studied.
Bowmaker et al. (1989)	X_2^n, n=-1, 0, +1, X=Cl-I. NQCC interpreted.
Bursten et al. (1989a)	$(cp)_3An$, An=Th-Pu.
Bursten et al. (1989b)	$(cp)_3AnL$, L=H, CO, NO, OH, Cl, cp.
Bursten et al. (1989c)	$(cp)_3AnL$ versus Cl_3AnL, An=Th-Cf.
Cotton and Feng (1989)	$Ru_2(HNNNH)_4$, $Rh_2(HNCHNH)_4$. The M-M δ^* and π^* levels discussed.
Daul (1989)	Inclusion of multiplet states and SO splitting in MS Xα calculations.
El-Issa et al. (1989)	Bonding in $MoS_nO_{4-n}^{2-}$.
Goursot et al. (1989)	Pt(III) complexes containing aquo and chloro ligands: $PtCl_kO_l(OH)_m(H_2O)_n$, CN=4-6. Bonding and CT transitions.
King et al. (1989)	$[Au(H_2PCH_2PH_2)]_2^{2+}$. Bonding discussed: Origin of the Au(I)...Au(I) attraction not understood.
Makhyoun and Sawan (1989)	$Cu_3AsSe_6^{6-}$ as model for Cu_3AsSe_4.
R.M. Miller et al. (1989)	CrO_4^{2-}, CrO_3X^-, CrO_2X_2, X= F, Cl. Excited states studied.
Morin et al. (1989)	Inclusion of SO effects in MS-LSD methods by a j-dependent QR method. Results for Pt_n, n=2, 4, 5, compared with full Dirac calculations.
Pershina et al. (1989)	$HfCl_4$, $(E104)Cl_4$. Bonding analysed.
D.-S. Yang et al. (1989)	<u>Cis</u>-$PtMe_2(C_2H_4)_2$. PES.

Table 7.6. (continued).

Novo-Gradac (1990)	Bimetallic organoactinide compounds, $Cp_nMM'Cp(CO)_2$, n=1, 2, M=Zr, Ru, Th, U. Also various M_2L_6, L=CH_3, Cp. 5f contributions increase along Th, U, Pu. $U_2(OMe)_{10}$. Lack of U-U bonds discussed.
Rogemond et al. (1990)	BiO_6^{9-}, GeO_4^{4-} and GeO_6^{8-} cluster models for $Bi_4Ge_3O_{12}$.
Tahri et al. (1990)	AnX_8^{4-}, An=Th, Pa, X=Cl, Br. The f-f and f-d optical transitions interpreted, including SO.
D.-S. Yang et al. (1990a)	*Cis*-$PtMe_2(EH_3)_2$ (E=N-As), $PtMe_2(C_2H_4)_2$, $PtMe_2(CNH)_2$. NR comparison for *cis*-$PtMe_2(PH_3)_2$. Only small effects on orbital energies.
D.-S. Yang et al. (1990b)	$PtMe_3$(HC(O)CH COH). PES. Extensive 5d delocalization to the ß-diketonato and methyl groups.
D.-S. Yang et al. (1990c)	Cp Pt Me_3.
D.-S. Yang et al. (1990d)	$Pt(CF_3)_2$, $Pt(CF_3)_2L_2$ (L=C_2H_4, PH_3). Significant C 2s character in the Pt-CF_3 bonds. Little interaction between d_π and C-F σ^*.
Berksoy-Bölük (1991)	$PbCl_2Ph_2$ NQCC, IS.
Cayton et al. (1991)	Dinuclear U(V) systems. U_2H_{10}, $U_2(OH)_{10}$ models for known $U_2(OR)_{10}$. Also $(cp)_2U_2O_4$. U-U bonding searched.
Chermette et al. (1991)	$[Cr(NH_3)_6]^{3+}$. SO-induced intersystem crossings studied. LSD potential chosen. Weighted average of monoatomic SO used.
Cotton and Feng (1991)	$[W_3OCl_6(O_2CH)_3]^-$ energy evels and bonding.
El-Issa et al. (1991)	ML_5, M=Sb, Bi; L=-H,-C≡CH models for $BiPh_5$. MCl_6^{2-}, M=Sn, Pb: LUMO stabilization important for colour. MS_4^{2-}, M=Mo, W: LUMO destabilization not important.
El-Issa and Zeedan (1991)	$[M'(MS_4)_2]^{2-}$, M'=Ni-Pt, M=Mo, W. Optical transitions and bonding.
Kryuchkov et al. (1991)	M_2^{n+}, M=Mo, Tc, Ru, Re, n=0-6, as models for multiply-bonded dimers.
Ren (1991)	Diruthenium and diosmium organometallics
Schneider (1991)	Symmetric and asymmetric, multiple M-M bonds, M=Mo, Ta-Re. AnX_3, An=Th-Pu, X=H, CH_3, NH_2, OH. Also UX_4.
Strittmatter (1991); Strittmatter and Bursten (1991)	$An(cp)_3$ compounds An = Th-Cf. Also $Ln(cp)_3$, Ln=Ce-Dy. PES for $Zr(cp)_3Cl$.
Subramanian et al. (1991)	MCl_4, M=Ti-Hf. PES, EA.
Chermette (1992)	Review on MS Xα.
Jolly and Marynick (1992)	Sb_7^{3-}. 5p bonding. NR/QR.

Table 7.6. (continued).

Gallego-Planas and Whitehead (1992)	Pentastanna [1.1.1.] propellane, bicyclo [1.1.1] pentastannane. QR/NR study of bonding.
Nour et al. (1992)	PaX_8^{4-}. QR + SO. Optical spectrum of $ThX_4:Pa^{4+}$ (X = Cl, Br) assigned.
Schwerdtfeger et al. (1992b)	NQCC of AuCl, $AuCl_2^-$, AuX_4^-, X=F-I. R/NR.

Table 7.7. Molecules treated by pseudopotential methods.

Reference	Comments
Cohen and Basch (1983)	Ethene and ethyne complexes of Ag.
Low and Goddard (1984b)	Pd insertion to H-H, H-C and C-C bonds. Existence of H-Pd-H a relativistic effect.
Flad et al. (1985ab)	H adsorption on cluster models of Cu and Ag.
Flad et al. (1985c)	Comparisons between the group 1 and 11 clusters of Li-K and Cu, Ag.
Sakaki et al. (1985, 1987)	Nucleophilic attack by water on a Hg(II)-ethylene complex.
Balasubramanian (1986a)	BiH^+. Eight low-lying states. Relativistic avoided crossings found. Bi^+ SO lowers D_e.
Balasubramanian (1986b)	BiH. 10 low-lying states.
Balasubramanian (1986c)	BiF.
Balasubramanian (1986d)	PbS.
Balasubramanian (1986e)	Sn_3.
Balasubramanian (1986f)	Ga_2.
Balasubramanian (1986g)	SnH_2.
Balasubramanian et al. (1986)	Br^+-Kr collisions.
Basch et al. (1986a)	A cis-$Pt(NH_3)_2(CH_2OH)(OCH_3)$ model for ascorbic acid complexes. Pt-C bond energy higher than Pt-O one. Strong trans influence for the NH_3.
Basch et al. (1986b)	$Pt(NH_3)_3^{2+}$ complexes with guanine. adenine, cytosine and thymine.
Bauschlicher et al. (1986)	CsOH, BaOH.
Baykara et al. (1986)	Hydrogen atoms on and in PdH_n cluster models.
Bernier et al. (1986)	HgH, lowest states including SO.
Bertoncello et al. (1986)	$SnL_nL'_{4-n}$, L = CH_3, Cl, NCS. PES interpreted.
Boland (1986)	AgH, AuH at NR and QR HF level.
Caballol et al. (1986)	XH_4^+, X = Si-Sn. Geometries.
Cabot et al. (1986)	MO_2, M = Ga-Tl, superoxide.
Christiansen (1986)	A core-valence correlation potential tested for K.
Di Bella et al. (1986)	$Pt(acac)_2$ PES (acac=acetylacetonate).
Fernandez et al. (1986a)	SnH_4 VUV spectrum.
Fernandez et al. (1986b)	$SnCl_4$ and $Sn(CH_3)_4$ VUV spectrum and PES.
Gavezzotti et al. (1986)	Pt_2, PtH and PtCO.
Hay et al. (1986)	Cl_2 Th-$(\mu$-$PH_2)_2Pt(PH_3)$: a compound with a Th-Pt bond. Mainly 5d-to-6d donation in a σ HOMO.
Hliwa et al. (1986)	Low-lying states of CuH.
Igel-Mann et al. (1986)	Alkali dimers, XY, X,Y=Li-Cs. Ground-state properties.

Table 7.7. (continued).

Illas et al. (1986)	Geometrical isomers of AgO_2: C_s 0.4 kcal/mol below C_{2v}.
Jarque et al. (1986)	PdH_2. Side-on preferred to head-on.
Koga and Morokuma (1986)	$M(CH_3)(H)(CO)(PH_3)$, M=Pd, Pt. The carbonyl insertion proceeds via methyl-group migration. TS established. Pd less endothermic, has lower $E^{\neq}$, due to weaker M-C bonds.
Low and Goddard (1986a)	MR_1R_2, M=Pd, Pt; R=H, CH_3. Models for reductive coupling on a metal surface.
Low and Goddard (1986b)	$M(CH_3)_2(PH_3)_2$, M=Pd, Pt. $Pt(CH_3)_2(Cl)_2(PH_3)_2$. $Pt(Cl)_2(PH_3)_2$. $Pt(H)(CH_3)(PH_3)_2$. Models for reductive elimination from bis(phosphine) complexes. Explains why Pd(II) favours reductive elimination but Pt(0) favours oxidative addition.
McMichael Rohlfing et al. (1986)	M, MH, M=Pd, Pt. X $^2\Sigma$ and $^2\Delta$, respectively.
Nicolas and Barthelat (1986)	$Cu-C_2H_4$. Bonding and optical spectrum.
Novaro et al. (1986)	$M-H_2$ systems, M=Li, Cu, Pd, Ag and Pt. C_{2v} only.
Pettersson and Langhoff (1986)	GeH.
Poulain et al. (1986)	$Pt-H_2$. The excited $^1A_1$ state is binding.
Rappé (1986)	TeF_6 EA. Ground-state properties of TeF_6 and TeF_6^-.
Ricart et al. (1986)	SnX_2, X=Cl-I. Ground-state properties and IP.
Ross et al. (1986)	SO splitting of group 13 and 17 atoms, B-Tl, F-At.
Ruiz et al. (1986)	CuH+H->Cu + H_2 proceeds without barrier; CuH_2 is unstable, in agreement with matrix data.
Schwerdtfeger (1986)	Thesis on relativistic pseudopotential methods. Dimers and hydrides of Br, I, Au-Pb. Ions included.
Schwerdtfeger et al. (1986)	X, HX, X_2, X=Br, I.
Selmani et al. (1986)	Ag_n + O, O_2; n=2, 4.
Silberbach et al. (1986)	M_2^+, M=Rb and Cs.
Walch et al. (1986)	Dimers and trimers of Cu, Ag, Au.
Wisner et al. (1986)	$M(CH_3)_2(PH_3)_2$, $M(PH_3)_2$, M=Pd, Pt. CH_3 bonds strongly to Pd, while PH_3 prefers Pt. See Low and Goddard (1986ab).
Zangrande et al. (1986)	<u>Trans</u>-$Pt(PH_3)_2X_2$, X=Cl-I. PES.
Alvarez et al. (1987)	Br_4, Br_4^{2-}, Br_4K_2.
Andzelm et al. (1987a)	As_2, As_4 geometry and PES.
Andzelm et al. (1987b)	MM', MM'=Si-Sn. Geometry, ground electronic state.
Balasubramanian (1987a)	ScH_2. $X^2\Sigma_g^+$.
Balasubramanian (1987b)	GaAs, $GaAs^+$. $X^3\Sigma^-$ and $X^4\Sigma^-$, respectively, predicted.
Balasubramanian (1987c)	Ge_2. 14 bound low-lying states found.

Table 7.7. (continued).

Balasubramanian (1987d)	SnS.
Balasubramanian (1987e)	Se_2.
Balasubramanian (1987f)	PtH_2. X^1A_1 (H-Pt-H 85°). Pt, PtH^+ tests.
Balasubramanian (1987g)	$GaAs_2$. X^2B.
Balasubramanian (1987h)	Pt_2.
Balasubramanian et al. (1987e)	PdH.
Balasubramanian et al. (1987a)	TeH.
Balasubramanian and M.Z. Liao (1987)	Au_3
Balasubramanian et al. (1987c)	SeH, SeH^+.
Balasubramanian et al. (1987b)	$KrHe^+$.
Balasubramanian and Pitzer (1987)	Review.
Balasubramanian and Ravimohan (1987)	Te_2.
Balasubramanian et al. (1987d)	SbH.
Bernier and Millié (1987)	HgH_2 potential-energy surfaces.
Caballol et al. (1987)	$Cu(CO_2)$ geometry.
Chapman et al. (1987a)	HI. SO decreases the dipole moment by 3.6 % at R_e.
Chapman et al. (1987b)	HBr. SO decreases D_e by 0.16 eV, contracts R_e by 0.5 pm.
Czuchaj et al. (1987)	MHe, MNe, M = Mg, Cd, Hg.
Dolg et al. (1987b)	MO, M = Sc-Zn. FeH.
G. Durand et al. (1987)	Mg_2^{2+}, Hg_2^{2+}. SO added for Hg.
Fernandez et al. (1987)	SnH_4^+ geometry, SO splitting.
Flad et al. (1987)	Cluster models for silver halide surfaces.
Francisco and Pueyo (1987)	Fine-structure of M^{n+}, n = 0, 2, 3; M = Sc-Zn using Huzinaga-type NR PP.
Fuentealba and Reyes (1987)	MH^+, M = Be-Ba.
Fuentealba et al. (1987)	MH, M = Li-Cs, Be-Ba. Ground-state properties.
Hay (1987)	Classical $M(H)_2$ and non-classical M-(η^2-H_2) complexes of five-coordinate W(0), d^6. L=PH_3, CO.
Hurley and Christiansen (1987)	Li and K EA from PP MC.
Igel-Mann et al. (1987)	CaO. Ab initio and semiempirical pseudopotentials compared.
Jarque et al. (1987)	$Pd-C_2H_4$ as model for heterogeneous catalysis.
Kozlov et al. (1987)	PNC effects in PbF. A QR PP calculation completed with monoatomic PNC matrix elements.

Table 7.7. (continued).

Langhoff et al . (1987)	MH, M=Y-Ag. Three atomic asymptotes important in bonding: $4d^n5s^2$, $4d^{n+1}5s^1$ and $4d^{n+2}5s^0$.
Lespes et al. (1987)	CH_3SnH_3. VUV and PES.
R.L. Martin (1987)	AgH, Ag_2, AgO. Ground-state properties.
McKee (1987)	O, O_2, C_2H_4 interactions with Ag_2 and Ag_2^+: models for surface-adsorbate interactions.
Merchán et al. (1987)	CuCO, $CuCO^+$.
Nakatsuji et al. (1987)	$Pd-H_2$, Pd_2-H_2 models for hydrogen chemiabsorption on Pd. The latter shows catalytic activity. Bonding to 4d primary, to 5s secondary. Excited states unimportant.
Neisler (1987); Neisler and Pitzer (1987)	Hg_2^{2+}. Stabilised by relativity by 1.0 eV wrt $2Hg^+$. Cavity models for hydration.
Nizam et al. (1987)	PbX_2, X=F-I.
Radzio and Andzelm (1987)	Non-additive interactions in Xe_3.
Ross et al. (1987)	RbH. Lowest states.
Russo et al. (1987)	PdX, X=C-Sn; RhC. Ground-state properties. LSD.
Sakai and Miyoshi (1987)	MF_6^{n-}, M=Cr-W, n=0-2. EA deduced.
Sánchez Marcos et al. (1987)	$Mo(CO_2)_2(PH_3)_4$, $Mo(CO_2)_2(CNH)$ $(PH_3)_3$. Perpendicular carbon dioxides preferred.
Savin et al. (1987)	IP of a Cs_4O cluster. Decrease by oxygen explained as 2s repulsion of a_1 HOMO.
Schilling et al. (1987a)	SrH^+-CdH^+. Bonding analysed.
Schilling et al. (1987b)	MCH_3^+, M = Sc...Zn, Y...Cd. $D(M^+-CH_3)>D(M^+-H)$ by about 6 kcal/mol on the average. Bonding similar.
Schwerdtfeger (1987)	TlH^+, TlH, TlH_3. Ground-state properties. R/NR.
Schwerdtfeger et al. (1987)	HX^n, X=Br, I; n= -1, 0, +1.
Smith and Goddard (1987)	$Re_2Cl_8^{2-}$. Re-Re quadruple bond energy 85(5) kcal/mol (exp. 97(12)). Rotational barrier 3.0 intrinsic strength of bond 6(3) kcal/mol.
Wadt (1987)	5f-5f excitations of PuF_6.
Walch (1987)	Excited states of Ag_3. Ag_3^+.
Winkler et al. (1987)	Mössbauer isomer shifts in SnX_4, X=F-I, H, Me. Atomic DF integrals grafted to QR PP.
Anslyn et al. (1988)	$Cl_3MoC_3H_3$. Metallatetrahedrane preferred to metallacyclobutadiene.
Balasubramanian (1988a)	Br_2, Br_2^+.
Balasubramanian (1988b)	PbCl.
Balasubramanian (1988c)	MH_2, M = Ge-Pb.
Balasubramanian (1988d)	Pd_2.
Balasubramanian (1988e)	SnCl.

Table 7.7. (continued).

Balasubramanian (1988f)	MH_2, M = Co. Results compared with M = Sc, Y, Pd, Pt.
Balasubramanian and Feng (1988)	Ga_3, Ga_3^+.
Balasubramanian et al. (1988a)	$Ar-Cl^+$ collision system.
Balasubramanian et al. (1988b)	Pd + H_2 potential-energy surface, PdH^+. The Pd-H bonds are weaker, more ionic than Pt-H ones.
Balasubramanian and Li (1988a)	GeH. Dipole moment 0.098 D (exp. "1.24D").
Balasubramanian and Li (1988b)	In_2. Calculated D_e 0.87 eV, larger than that of Tl_2.
Balasubramanian and D.-W. Liao (1988a)	RhH_2 potential-energy surface.
Balasubramanian and D.-W. Liao (1988b)	RhH.
Balasubramanian and M.-Z. Liao (1988a)	AuH_2.
Balasubramanian and M.-Z. Liao (1988b)	SeH_2, SeH_2^+.
Balasubramanian and M.-Z. Liao (1988c)	Ag_3, Au_3.
Balasubramanian and Ravimohan (1988)	YH_2.
Basch et al. (1988)	TeO_2. A lower, open C_{2v} and higher (2.04 eV) ring geometry found.
Cauletti et al. (1988)	Sn-S and Sn-Se bonding in $(Me_3Sn)_2X$, $(Me_2SnX)_3$, X = S, Se.
Chapman et al. (1988a)	HX^-, X = Br, I. The HI EA predicted to be > 0.
Chapman et al. (1988b)	MH, M = Ge-Pb. Spectroscopic constants and dipole moments. Significant SO differences between $X^2\pi_{1/2}$ and $^2\pi_{3/2}$.
Daniel et al. (1988)	Olefin hydrogenation by the Wilkinson catalyst $RhCl(PR_3)_3$. Comparisons with RhH, Pd and Pt systems.
de Loth et al. (1988)	The Cu(II) - V(II) exchange interaction evaluated in a model complex for CuVO $(fsa)_2$ en · CH_3OH.
Di Bella et al. (1988)	Glyoximate complex of Pt. PES.
Ermler et al. (1988a)	Review.
Ermler et al. (1988b)	Anisotropic semiconductor clusters: PbI_2, GaAs, GaAs:S.
Fernandez Pacios et al. (1988)	Core-polarization corrections to PP. Na_2, K_2, NaK.
Fernandez Pacios and Gomez Calzada (1988)	Core-valence partition discussed for 3d metals.
Francisco and Pueyo (1988)	SO splittings for 3d ions in crystals.
Gavezzotti et al. (1988)	CO + Pt_n, n = 1, 3, 4, 7.
Gropen (1988)	Review.

Table 7.7. (continued).

Igel-Mann et al. (1988b)	XH, XO, X=K, Ca,Ga-Br. Ground-state properties.
Illas et al. (1988)	H chemisorption on M_5 models for M(100), M=Cu, Ag.
Jasien and Stevens (1988)	CrO, CrO^+.
Jeung and Koutecky (1988)	ScX, X=H, C, N, O.
Koga et al. (1988)	$Rh(H)(C_2H_4)(CO)_2(PH_3)$. Berry pseudorotation and olefin insertion around d^8 Rh(I).
Krauss et al. (1988ab)	Binding of Pt di-and triammines to oligonucleotides. Also $[Pt(NH_3)_4^{2+}]$ models bound to H_2O or $H_2PO_4^-$.
Langhoff and Bauschlicher (1988a)	MO, MS^+, M=Y, Zr, Nb.
Langhoff et al. (1988)	YX, X=F-Br.
Lee et al. (1988)	Mo_2, Mo_2^+. LDF. Ionization shortens and strengthens the bond.
Z.-X. Ma and S.-S. Dai (1988a)	Mo_2. Ground state.
Z.-X. Ma and S.-S. Dai (1988b)	KX^-, X=Cl-I.
McAdon and Goddard (1988ab)	M_N ring models for 1D metals, M=Cu-Au, Li, Na. N=2-12. Also M_2^+, M_3^+.
McKee and Worley (1988)	$Rh(CO)_n^{q+}$, n=1, 2; q=0,1.
P.K. Miller et al. (1988)	$OTeF_5^{q-}$, q = 0,1. $NaOTeF_5$.
Miralles-Sabater et al. (1988)	$M^+ + C_2H_2$, M=Cu, Ag. A model for metal-olefin complexes.
Miyoshi and Sakai (1988a)	$ZrCl_4$, AgH, AuH, ZrO, Mo_2.
Miyoshi and Sakai (1988b)	AuF_6^{q-}, q = 0-3.
Miyoshi et al. (1988)	MF_6^{q-}, q=0,1; M=Cr-W. The EA of WF_6 is 1.52 eV smaller than that of MoF_6.
Nakamura and Morokuma (1988)	$Pt(PH_3)_2(CH_2=C=O)$ and $Pt(PH_3)_2(CH_2)(CO)$. TS from ketene to carbene carbonyl searched. Thermolysis of the ketene complex discussed.
Nakatsuji et al. (1988)	H_2-Pt_n, n=1-3, as models for hydrogen chemisorption on Pt. Pt and Pt_3 react but Pt_2 has a high barrier. SO small for Pt-H_2 X $^1A_1$ but important at large separation, mainly due to Pt atom ^3D_J effects. Pd$\neq$ Pt.
Ortiz (1988)	EA of K-Cs.
Pitzer and Winter (1988)	Inclusion of SO PP for molecules. A real Hamiltonian obtained for C_{2v} D_2, D_{2h}, or higher symmetry in a $\|lm_lm_s\rangle$ basis.
Planelles et al. (1988)	$CuCH_2$. X $^2B_1$ (d^{10}, $\sigma^2\pi$) ionic state with a strong Cu-C bond.
Pouchan et al. (1988)	CH_3SnH_3 vibrational spectrum. MP2.
Rasch and Tobisch (1988)	Pt, PtH, PtH_2.

Table 7.7. (continued).

Saalfrank et al. (1988)	Te_∞ chains of trigonal tellurium. Cohesive properties, including inter-chain interactions.
Sakai and Miyoshi (1988)	EX_3, E=As, Sb; X=F-I. Geometries and dipole moments.
Selmani and Salahub (1988a)	XH_3, XH_4, X=C-Sn. LSD.
Selman and Salahub (1988b)	XH_2, X=Si-Sn. Singlet-triplet splitting at LSD level.
Tadjeddine et al. (1988a)	CH_3I. IP including SO.
Tadjeddine et al. (1988b)	CH_3I^+. Molecular states related to photofragmentation.
Tanpipat (1988)	MX, M=Sc, Lu, Lr; X=H, O, F, Cl. Lr resembles Lu (cp. Pyykkö (1979c)).
Walker et al. (1988)	Excited states of Sc...Co for various core sizes.
Andreocci et al. (1989, 1990)	Alkyl acetylides of tin and lead. PES.
Bagus et al. (1989)	Halogens (F-Br) bound to Ag_n surface models (n = 4, 7, 10). Essentially ionic bond. "Image charge" important.
Balasubramanian (1989a)	Pd_3. $X\ ^1A_2$.
Balasubramanian (1989b)	MH_2^+, M = As-Bi. AsH_2.
Balasubramanian (1989c)	Review on relativity and chemical bonding.
Balasubramanian (1989d)	GaH_n, GaH_n^+, n = 1-3.
Balasubramanian (1989e)	Review on heavy p-block diatomic hydrides, halides and chalconides.
Balasubramanian (1989f)	Review on coinage-metal clusters.
Balasubramanian and Feng (1989)	M_n, M_n^+ (M = Ag, Au; n = 1-4). IP and AE.
Balasubramanian et al. (1989)	Au_4. X^1A_1 rhombus.
Balasubramanian and Li (1989)	Sb_2. $X^1\Sigma_g^+$ has a D_e of 2.17 eV, SO increased by 0.1 eV.
Balasubramanian and D.-W. Liao (1989)	Rh_2.
Balasubramanian and M.-Z. Liao (1989)	AgH_2.
Balasubramanian and Nannegari (1989)	AsH.
Balasubramanian and Ravimohan (1989)	Nb + H_2 potential-energy surface.
Balasubramanian and Wang (1989a)	ZrH.
Balasubramanian and Wang (1989b)	YH. $X^1\Sigma$ gross populations $6s^{1.55}$ $6d^{0.45}5d^{0.69}$.
Balasubramanian and Wang (1989c)	Tc + H_2 and Ru + H_2 reactions.
Basch (1989)	X_3, X_3^-, X = Se, Te.
Bauschlicher (1989)	EA of Au_3.
Bauschlicher et al.	$LaFe^{2+}$, $LaRu^{2+}$, YRu^{2+}. Multiple d-d'

Table 7.7. (continued).

Langhoff (1989)	bonds of order 2-3.
Bauschlicher et al. (1989a)	M_2, M_3, MM',MM'M",M = Cu-Au: structures and EA.
Bauschlicher et al. (1989b)	MCH_3^{n+}, M = Y-Ag.
Bowmaker et al. (1989)	X_2^n, n = -1, 0, +1, X = Br, I. Ground-state properties, EA, IP.
Brusich (1989)	Organometallic Th chemistry. L = H, CH_3, Cl.
Burton et al. (1989)	$Pd(acac)_2$, acac = acetylacetonate, O=CH=CHCH=O. Geometry, bonding and PES.
Castillo et al. (1989)	Cu + CH_4 -> $HCuCH_3$ -> CuH + CH_3. CI.
Cave et al. (1989)	$X_2Y_2^{2-}$ Zintl ions. $Pb_2Sb_2^{2-}$ and Sb_2Bi_2 are pseudo-T_d, $Tl_2Te_2^{2-2}$ butterfly-like C_{2v}. Also isoelectronic neutrals.
Chang and Pitzer (1989)	$U(COT)_2$. Green color due to 5f -> 6d excitations. Charge $U^{+0.98}$, $6d^{1.98}$ $5f^{2.50}$.
Chapman (1989)	HX, HX^-, X = Br, I. MH, M = Ge-Pb. Electric dipole and transition moments.
Chapman et al. (1989)	$Ar-Cl^+$ collision system.
Christiansen (1989)	PP MC reviewed.
Crépin and Millié (1989)	$Hg-N_2$ potential curves. A $Hg(^3P)$ exciplex found.
Czuchaj et al. (1989)	CaHe, CaNe.
Daoudi et al. (1989)	MCO, M = Ru, Os.
K.K. Das and Balasubramanian (1989a)	YH_2^+, ZrH_2^+.
K.K. Das and Balasubramanian (1989b)	NbH_2^+, MoH_2^+.
Dolg and Stoll (1989)	LnH, LnO. LnF, Ln = La-Lu. The 4f AO in core. For comments on ground-states, see McDonald et al., JCP 93 (1990)7676.
Feng and Balasubramanian (1989)	In_3.
Fernandez Sanz and Márquez (1989)	H_3Sn-XH_3, X = C-Sn. Structure and vibrational spectrum.
Fernandez Sanz et al. (1989)	SnH_4 vibrational spectrum.
Fuentealba et al. (1989)	EA for Ca, Sr, Ba.
Garcia-Cuesta et al. (1989)	Pd/C_2H_2. A weakly-bonded π complex.
Igel-Mann et al. (1989)	X_2, XH, XO, XF, XS, X=In, Sn, Sb. Ground-state properties.
Illas et al. (1989)	H and O chemisorption on a Cu_5 cluster model for a Cu(100) surface.
Kim and Balasubramanian (1989)	GaH.
Kober and Hay (1989)	Transition-metal dihydrogen chemistry. $WL_5(H_2)$, L=CO, PR_3.$Cr(CO)_4(H_2)_2$.
Koga and Morokuma (1989)	Olefin hydrogenation by the Wilkinson catalyst. $H_2RhCl(PR_3)_3$. Also fragments: RhCl, H_2RhCl, etc.

Table 7.7. (continued).

Krogh-Jespersen et al. (1989)	$(NH_3)_5Ru(III)L$, $L=NH_3$, $S(CH_3)_2$.
Kudo and Nagase (1989)	PbH_4^+ Jahn-Teller distortions. Ready decomposition to $PbH_2^+ + H_2$.
Langhoff et al. (1989)	MO, MS, M=Mo, Tc.
Lee et al. (1989)	Pd_2. LDF.
Li and Balasubramanian (1989)	I_2, I_2^+, 30 and 13 electronic states, respectively, as functions of R. V_{SO} included.
Z.-X. Ma and S.-S. Dai (1989)	HgX_2, X=Cl-I.
Marino et al.(1989a)	Be_n surface models (n=19-45) for H and Cs adsorption.
Marino et al. (1989b)	Bi_6I_{18} model for BiI_3. Optical absorption spectra and geometry discussed.
Maseras et al. (1989)	$[Rh(PH_3)_4H_2]^+$: classical or nonclassical?
Miralles-Sabater et al. (1989)	Cu + H_2O^{q+}, q=0,1.
Moullet et al. (1989)	Cs_2, Cs_2^+ using DFT. R/NR comparisons. 5d important.
Nagase (1989)	X_nH_n, X=Ge-Pb, n=4, 6, 8 (tetrahedrane, prismane, cubane).
Nizam et al. (1989); Nizam and Bouteiller (1990)	Cluster models for Pb(II) dihalides. X=F-I. Geometries of PbXX' optimized.
Novaro (1989)	Review on activation of small molecules by transition-metal atoms.
Pavolini et al. (1989)	RbCs. 30 electronic states.
Planelles et al. (1989)	$CuCH_2^+$ X $^1A_1$.
Ramírez-Solís and Daudey (1989)	CuF low-lying levels.
Ramondo et al. (1989)	MXX', M=Sn, Pb; X=F-I. Structure and vibrational spectra.
Ross (1989)	Effects of SO on atoms, molecules and clusters.
Schwerdtfeger (1989)	AuX_2^-, AuX_4^-, X=F-I. Larger relative stability of Au(III) halide complexes, compared to Au(I) ones, attributed to relativity.
Schwerdtfeger et al. (1989a)	TlR_n^{m+}, n=1-4, $R=CH_3$, H, m=0, 1. TlR_2Br. TlR', $R'=C_2H_3$, C_2H. TlBr, Tl_2. Explain why organometallic chemistry uses Tl(III), inorganic Tl(I). $TlCH_3$ predicted to be stable.
Schwerdtfeger et al. (1989b)	AuL, L=H, F-I, Li, Na, S. Ions of Au_2, AuH. Largest D_e decrease (-86 kJ/mol) for AuF, largest D_e increase (+174 kJ/mol for AuLi. All contract.
Schwerdtfeger et al. (1989c)	PbH^+, PbH_n, n=1, 2, 4. Ground-state properties.

Table 7.7. (continued).

Sodupe et al. (1989a)	$Cl_4W=CH_2$. Conformational preferences.
Sodupe et al. (1989b)	Ground and low-lying states of FeH^+. X $^5\pi$.
Sohn et al. (1989)	Nb_2. X $^3\Sigma_g^-$. $^3\Delta_g$ slightly higher.
Spiegelmann et al. (1989)	Rb_2.
Strömberg et al. (1989)	MH^+, MCl_2, M=Cd, Hg. $Hg(OH)_2$, HgO (bad D_e). Consequences on global "mercury cycle" quoted.
Wang and Balasubramanian (1989)	TcH.
Balasubramanian (1990a)	GaAs, $GaAs^+$, $GaAs^-$.
Balasubramanian (1990b)	Review on heavy p-block dimers and trimers.
Balasubramanian (1990c)	InSb, $InSb^+$, $InSb^-$.
Balasubramanian (1990d)	$(GaAs)_2$. X^1A_g rhombus.
Balasubramanian (1990e)	InH, InH^+.
Balasubramanian (1990f)	MH, M = Y-Cd.
Balasubramanian and Dai (1990)	The Ir + H_2 and Ir^+ + H_2 potential energy surfaces. The IrH_2^+ X^3A_2 bond angle increased by 9° by SO.
Balasubramanian and Das (1990)	Reply to comments by Bauschlicher et al. (1990a) on coinage-metal tetramers.
Balasubramanian and Feng (1990a)	Pt_2 + H and Pt + H potential-energy surfaces.
Balasubramanian and Feng (1990b)	Cu_4, Ag_4.Ag-Ag 287pm, compared to Au-Au = 269 pm in Au_4.
Balasubramanian and Li (1990)	MoH, MoH^+.
Balasubramanian and Ravimohan (1990)	Zr_2.
Balasubramanian and Wang (1990)	RuH.
Ballone and Galli (1990)	LD-pseudopotential results on M_2, M_2^+, M=Cu-Au. D_e overestimated by a factor of 3, R_e underestimated by 10.
Barandiaran et al. (1990)	YO, TcO, AgO, AgH. Rel. contractions given.
Barnes et al. (1990)	Mono- and dicarbonyl ions of 3d and 4d metals. MCO^+, $M(CO)_2^+$, M = Sc-Cu, Y-Ag.
Barone et al. (1990)	X_4H_9 and X_5H_{12} cluster models for diamond-like crystals, X = C-Ge.
Bauschlicher and Langhoff (1990)	Transition-metal cations: a review on 3d- and 4d-metal Ar, $(H_2O)_n$, CO, H, CH_3,... complexes.
Bauschlicher et al. (1990a)	Comparisons with Balasubramanian and Feng (1990b) on M_4, M = Cu-Au.
Bauschlicher et al. (1990b)	M_4, M_5, M = Cu-Au.
Bauschlicher et al. (1990c)	AgO, AgS.

Table 7.7. (continued)

Bowmaker and Schwerdtfeger (1990)	ZnX, X = F-I. Ground-state properties and UV spectra. Monotonous k_2 decrease suggested for MX, M = Zn-Hg.
Bravo et al. (1990)	AgO_2 potential-energy surfaces. O_2^--production in olefin oxidation explained.
Bruno et al. (1990)	PtL_2, L =$-CH_2-PH_2-BH_2-PH_2CH_2-$. Bonding and PES.
Christiansen (1990)	Al^{n+}, n=0-2. Core polarization potentials. PP MC.
Dai and Balasubramanian (1990)	EH_3^{n+}, E = As-Bi, n = 0, 1. Also EH, EH_2.
K.K. Das and Balasubramanian (1990a)	Rh_3. X^2A_3 (D_{3h}).
K.K. Das and Balasubramanian (1990b)	Au_2 low-lying states. X 0_g^+ is 96% $^1\Sigma_g^+$. R_e 251.8 pm, ω_e 193 cm^{-1}, D_e 2.05 eV (exp. 247, 191, 2.30).
K.K. Das and Balasubramanian (1990c)	$Tc^+ + H_2$, $Ru^+ + H_2$.
K.K. Das and Balasubramanian (1990d)	LaH. Low-lying states.
K.K. Das and Balasubramanian (1990e)	GeH_n^{q+}, n = 1-3, q = 0,1.
K.K. Das and Balasubramanian (1990f)	YNH, NYH. Linear $X^2\Sigma^+$ YNH lowest.
K.K. Das and Balasubramanian (1990g)	NbH. X $^5\Delta$.
K.K. Das and Balasubramanian (1990h)	Rh_3.
Dolg et al. (1990a)	GdO. $X^9\Sigma^-$. Bond-length expansion 1.6 pm.
Dolg et al. (1990b)	EuO. $X^8\Sigma^-$.
Feng et al. (1990)	$Xe-I^+$ collision system. Transfer of translational and electronic energy calculated.
Fernandez Pacios (1990)	Modified Ar-core PP for 3d metals. Improved $s^2d^n-s^1d^{n+1}$ HF-level energies.
Flad et al. (1990)	Adsorption of Ag_n^{q+} on AgBr surfaces.
Haberlandt (1990)	NiH.
Igel-Mann et al. (1990)	$(XY_m)_n$, X=In, Sn, Sb; Y=H, O, F, S; n≤6. $In(C_5H_5)$. Ground-state properties.
Illas et al. (1990)	Rh, RhH, Rh_2.
Jarret-Sprague et al. (1990)	HXO_3, X=As, Sb. Structure and frequencies.
Karna et al. (1990)	CHX_3, X=F-I. Static polarizability α, first and 2nd hyperpolarizabilities β and γ. Geometries, dipole moments.
Katsuki (1990)	Ag, AgH using a spectral representation PP.
Kaupp et al. (1990a)	$M(CH_3)_n$, M=Mg, Zn, n=1, 2. Ground-state properties. Chemical properties of Mg and Zn compared.

Table 7.7. (continued).

Knight et al. (1990)	PdH_2^+. X $^2A_1$.
Koga and Morokuma (1990)	C-H activation in CH_4 by $RhCl(PH_3)_2$.
Krauss and Stevens (1990)	Rb_2, Cs_2 and their anions.
Kyle et al. (1990)	$Cu_4I_4(NH_3)_4$ as model for luminescence in $Cu_4I_4(C_5H_5N)_4$.
LaJohn (1990)	M_2, M_2^+, M=Li-Cs. Includes core-valence correlation potentials.
Langhoff and Bauschlicher (1990)	NbN. X $^3\Delta$. Results compared with isoelectronic ZrO, Nb is $5s^1$ and has more stable 4d orbitals.
Li and Balasubramanian (1990)	Mo + H_2. Insertion (C_{2v}) mode easier than collinear ($C_{\infty v}$) one.
Lin and Ortiz (1990)	CrH^-, MnH^-, FeH^-. IP.
Z.-X. Ma and S.-S. Dai (1990a)	BaCl. X $^2\Sigma$ and A states.
Z.-X. Ma and S.-S. Dai (1990b)	$TiCl_2$.
Z.-X. Ma et al. (1990)	BX_3, X=F-I. PES reassigned, including SO.
Marino et al. (1990)	Cluster models for PbI_2.
Merchan et al. (1990)	Cu_2, CuH.
Musolino et al. (1990)	Sb_2, Sb_4 using several DFT PP. PES interpreted.
Nagase and Kudo (1990)	Bicyclo [2.2.0] hexaplumbane. M_6H_{10}, M=Ge-Pb. Geometries (C_2) and frequencies.
Nagase et al. (1990)	HM=MH, M=As-Bi. Double bonds found, even for Sb and Bi. Geometries (C_{2h}) and frequencies.
Nakatsuji and Ehara (1990)	CsXe. Collision-induced 6s-5d absorption spectra from a quasistatic model.
Nakatsuji and Nakai (1990)	O_2-Ag_n, n=2, 4, as models for oxygen chemiabsorption on silver.
Nakatsuji and Saito (1990)	Excited states of $[MoO_{4-n}S_n]^{2-}$ and $MoSe_4^{2-}$.
Nakatsuji et al. (1990)	^{95}Mo NMR chemical shifts in $[MoX_{4-n}Y_n]^{2-}$, X, Y = O, S, Se. E(d-d*) important. No SO.
Nicolas and Spiegelmann (1990)	M-C_2H_4, M = Cu-Au. Bonding and excited states.
Ohanessian et al. (1990)	MH^+, M = Ba, La, Hf-Hg. Bonding analysed, compared with 3d and 4d analogs.
Ohanessian and Goddard (1990)	Review on bonding in MH^+, M = 3d, 4d or 5d metal.
Pacchioni and Bagus (1990)	CO/Pd_n, n = 2, 8 as models for chemisorption. R PP of Hay and Wadt (1985ab) compared with NR PP of Huzinaga.

Table 7.7. (continued).

Paniagua and Illas (1990)	H/Rh_n, n = 3, 4, as models for chemisorption.
Partridge et al. (1990)	$(MM')^q$, $(MM'M'')^q$, q = -1, 0, +1, M = Cu-Au IP and EA.
Pepper and Bursten (1990)	U_2. X $^{13}\Delta_g$ at R_e = 300 pm, D_e = 160 kJ/mol.
Plass et al. (1990)	Cu_2X_2, Cu_4X_4, X = Si, Sn.
Poulain et al. (1990)	M + H_2, M = Ni, Ru, Rh, Pt.
Ramírez-Solís and Daudey (1990)	CuCl.
Ramírez-Solís et al. (1990)	CuI. SO included.
Rochefort et al. (1990)	Chemisorption and diffusion of H in and on cluster models of Pd, Rh and bimetallic PdSn, RhSn and RhZn catalysts. LSD potential.
Rosi and Bauschlicher (1990)	Bonding of La^+ and La^{2+} to C_2H_2, C_2H_4 and C_3H_6.
Rosi et al. (1990)	M^{n+}-C_2H_6 and $M(CH_3)_2^{n+}$, M = Y - Ag, n = 0,1.
Sawamura and Ermler (1990)	$(PbI_2)_7$ cluster models. Quantum confinement effects in layered clusters.
Schwerdtfeger (1990)	$M(CH_3)_2^{n+}$, M = Au - Pb, n = -1 - +2. R/NR. Force constant maximum at $HgMe_2$. Decrease at $TlMe_2^+$ relativistic, further decrease at $PbMe_2^{2+}$ both R and NR.
Schwerdtfeger et al. (1990a)	TlMe, $TlMe_2^+$, $TlMe_3$, $TlMe_4^+$, $TlMe_2Br$. Frequencies reported here.
Schwerdtfeger et al. (1990c)	Au(I) complexes. AuL, AuL_2^-, L = H, F-I, CN, SCN, CH_3 PH_3^+. Two-coordination, 5d participation and Cl NQCC of $AuCl_2^-$ increased by relativity.
Sellers (1990a)	Nb_n, n = 1-4, Nb-N_2. Cations included.
Sellers (1990b, 1991)	Pd_n + H, O, N, S; n = 10...22. Chemisorption models.
Sellers and Ove (1990)	Anisotropic PP for modelling partially filled d shells. CoH, Co_2.
Sodupe et al. (1990)	Ground and low-lying states of FeH. X $^4\Delta$.
Spiegelmann et al. (1990)	The ArKr* exciplex. SO included.
Strömberg et al. (1990)	$M(H_2O)_6^{2+}$, M = Cd, Hg. $Hg(H_2S)_6^{2+}$. Quasioctahedral, T_h energy minima.
Sumathi and Balasubramanian (1990)	H_2Te, H_2Po and their cations. TeH and PoH. SO important for Po. Dipole moment changes sign from H_2Te to H_2Po.
Tobisch and Rasch (1990)	Pt, PtH, PtH_2.
Trinquier (1990)	XH_n (n = 1, 2, 4), X_2H_4, X = Sn, Pb. <u>Trans</u>-bridged forms, rather than $H_2X = XH_2$ predicted to be stable.
Trinquier and Barthelat (1990)	XF_2, X_2F_4, X = Sn, Pb. Dito.

Table 7.7. (continued).

J.-Z. Wang et al. (1990)	Zr + H_2.
L.-S. Wang et al. (1990a)	E_2, E_2^+, E = As-Bi.
L.-S. Wang et al. (1990b)	AB, AB^+, A = Sn, Pb; B = Se, Te.
Winkler and Vetter (1990ab)	Mössbauer IS and NQCC in $SnX_n(CH_3)_{4-n}$, X = F-I.
Andrae et al. (1991)	Ag_2, Au_2, RuH, OsH as tests of the potentials of Andrae et al. (1990). Also M_2^+, M_2^-, M = Ag, Au. No SO. (Cp. Benavides-Garcia and Balasubramanian (1991)
Andreocci et al. (1991)	Cyclic Me_2 Si(N-Me)$_2$ M, M = Sn, Pb. (Me_2Sn) (NMe)$_2$ ($SnMe_2$). PES.
Audouard and Spiegelmann (1991)	Kr_2^* and Kr_2^+ potential energy curves at MR CI level including SO.
Balasubramanian (1991a)	Pt_2H_2 potential-energy surface.
Balasubramanian et al. (1991a)	Compare the small-basis DF-LCAO calculations of Ramos et al. (1988) with large-basis correlated PP ones. For reply, see Malli et al. (1991).
Balasubramanian and Das (1991a)	HfH. Strong SO mixing found. $X^2\Delta_{3/2}$. The $^4\Phi$ states have more d, less s and go up, compared to ZrH.
Balasubramanian and Das (1991b)	M_4^+, M = Cu-Au. X $^2B_{1u}$, rhombus.
Balasubramanian and Das (1991c)	Au_3 excited states.
Balasubramanian and Feng (1991a)	Ga_4, In_4.
Balasubramanian and D.-W. Liao (1991a)	Au_6 isomers. C_{5v} is 0.73 eV below D_{6h}.
Balasubramanian and D.-W. Liao (1991b)	Bi_2, Bi_2^-. Bi_2 $X^1\Sigma$ R_e 276 pm (exp. 265.96). Important SO increase of the EA.
Balasubramanian and Ma (1991)	WH_2, WH_2^+ potential surfaces. WH_2 more stable than MoH_2 by 0.5 eV, compared to M(^{7}S) + H_2.
Balasubramanian et al. (1991b)	E_3, E_3^+, E = P-Bi. The different photo fragmentation patterns of Sb_n and Bi_n explained: IP(Bi) < IP(Bi_2) but IP(Sb) > IP(Sb_2).
Balasubramanian and Tao (1991)	EH_n, EH_n^+, E = In, Tl; n = 1-3.
Balasubramanian et al. (1991c)	GaAs + X, InSb + X, X = Cl, Br. Models for halogen-etching.
Barthelat et al. (1991)	NbH_3 and TaH_3 non-classical (dihydrogen) and classical, respectively.
Basch (1991)	M_2^{2+}, M = Pt, Au, Hg. $R_e(Hg_2^{2+})$ 285 pm!
Bauschlicher and Langhoff (1991)	M^+, M^{2+} complexes (M = Y, La) with C_2H_2, C_2H_4, C_3H_6.
Bauschlicher et al. (1991a)	Bonding of Nb^{3+} to CH_2, C_2H_2 and C_2H_4.

Table 7.7. (continued).

Bauschlicher et al. (1991b)	Zr_2. New spectral assignments.
Benavides-Garcia and Balasubramanian (1991)	OsH. $X(^4\Pi + {}^4\Delta)$.
Bock et al. (1991)	SiI_2 PES.
Breidung et al. (1991)	AsH_nF_{3-n}, AsH_nF_{5-n} vibrational spectra.
Castillo et al. (1991)	The $HCuCH_3$ reaction intermediate.
Cauletti et al. (1991)	$SnCl_2$. PES.
Chang (1991)	$C_{60}M$, M = Cs, Ba, La, Eu, U. $(COT)_2An$, An = U, Np, Pu, including SO. Magnetic moment and optical spectra calculated.
Chang et al. (1991)	$C_{60}M$, $C_{60}M^+$.
Cheng and Balasubramanian (1991)	TaH.
Christiansen (1991)	PP MC on Sc and Y.
Czuchaj et al. (1991)	NaHg potential curves, TP.
Dai and Balasubramanian (1991a)	$Os^+ + H_2$.
Dai and Balasubramanian (1991b)	$Re + H_2$, $Re^+ + H_2$.
Dai and Balasubramanian (1991c)	IrH. Results compared with RhH.
Dai et al. (1991a)	$Ta + H_2$, $Ta^+ + H_2$.
Dai et al. (1991b)	Pd_3^+. X^4B_2, nearly D_{3h}.
K.K. Das and Balasubramanian (1991a)	Ag_n, Au_n EA calculated, n = 3,4.
K.K. Das and Balasubramanian (1991b)	LaH_2. X^2A_1.
K.K. Das and Balasubramanian (1991c)	LaH^+, LaH_2^+. $X^2\Delta$, X^1A_1, respectively.
K.K. Das and Balasubramanian (1991d)	RhH^+, $X^2\Delta$.
K.K. Das and Balasubramanian (1991e)	HfH_2. X^1A_1. Results compared with ZrH_2.
K.K. Das and Balasubramanian (1991f)	$InSb_2$, $SbIn_2$, $GaAs_2$, $AsGa_2$ and their cations.
K.K. Das and Balasubramanian (1991g)	Ru_2. $X^7\Delta_u$.
K.K. Das and Balasubramanian (1991h)	RhH_2^+.
Dolg et al. (1991a)	$Ce(C_8H_8)_2$. X $4f^1e_{2u}^3$ singlet. First triplet 0.5 eV higher.
Dolg et al. (1991b)	HgH_n(n = 1,2), TlH_n(n = 1,3), PbH_n(n = 1,2,4), BiH_n (n = 1,3,5), PoH_n(n = 1,2), AtH. RnF_n(n = 2,4,6,8). NR/QR/SO comparisons reported. Relative chemical stability for various n analysed.
Dolg et al. (1991c)	CeO. 4f in core or in valence space.
Dolg et al. (1991d)	LnX_2, LnX_3, Ln = La... Lu, X = F-I. Atomization energies compared to exp.
Dolg et al. (1991e)	$LnHe^{3+}$, Ln = La...Lu. The La, Gd, Lu ones stable, Eu and Yb ones metastable.
Faegri and Bauschlicher (1991)	ScNi, YNi, ScPd, YPd. $X^2\Sigma$.

Table 7.7. (continued).

Fernandez Pacios and Botella Olcina (1991ab)	Improved Ar-core PP for Sc-Cu.
Flurer and Busch (1991)	M_2X^+ ions, M = Cu, Ag, X = H, CH_3, O, CO. Mass spectroscopic data interpreted.
Focher et al. (1991)	The eigenvalue for the last valence electron compared to experimental IP_n, Z = 1-94. LDF-based QRPP of Bachelet et al, (1982) used.
Gayen et al. (1991)	Cluster models for GaAs surface passivation by sulfur.
Hay (1991a)	$Pt(PH_3)_2$ + H_2 revisited. If 5s and 5p shells are in core, too large E_{corr} obtained, due to missing 6s, 6p nodes.
Huzinaga (1991)	Review.
Illas et al. (1991)	Cu_5O as model for O chemisorption on Cu(100). 3d correlation decreases the interaction.
Kaupp et al. (1991a)	MH, MH^+, MH_2, M=Ca-Ba. CaH_2 linear, SrH_2 and BaH_2 bent. Both core polarization and direct d-AO contributions to the bending.
Kaupp et al. (1991b)	MX_2, M=Be-Ba, X=F-I. All Be and Mg halides, $CaCl_2$-CaI_2 and SrI_2 linear. CaF_2, $SrCl_2$, $SrBr_2$ and BaI_2 are borderline. SrF_2 and BaF_2-$BaBr_2$ bent. Bending force constant (> or < 0) correlates with d character.
Klobukowski (1991)	XH, X=N-Sb. $^3\Sigma^-$ and $^1\Sigma^-$ states. Compared to the nodeless PP, the model potential with nodal wave functions is useful for SO.
Koga and Morokuma (1991a)	Review on catalytic elementary reactions and catalytic cycles of transition-metal complexes.
Koga and Morokuma (1991b)	CO insertion into the Pd-R bond of $HPd(R)(CO)(PH_3)$, R = H, CH_3.
Jarid et al. (1991)	CrF_4, CrF_5, RuF_4 and RuF_5. High-spin T_d, D_{3h}, D_{4h} and C_{4v} geometries predicted, respectively, at CI level.
Lee and Lee (1991)	MP2 with V_{SO} for HX, XX', X=Br, I. PbH_4.
Ma and Balasubramanian (1991)	WH. SO effects separated. Results compared with MoH.
Mains and White (1991)	Rh_n, $Rh_x(CO)_y$, RhH, RhH_3, RhC, RhCH, RhO, RhO_2. PP and AE results compared.
Marino (1991)	Summary of the cluster-model work on PbI_2, BiI_3 and Cs-activated Be-surfaces.
Marino and Ermler (1991)	$Be_{45}X_2$, X=Cs, OCs, HCs surface models. Cesiation lowers surface work function by 2.2 eV.
Martinez-Magadan et al. (1991)	ZnH_2^{n+}, n=0-2. Reaction surfaces.

Table 7.7. (continued).

Morokuma and Borden (1991)	$(H_3P)_2Pt(C_2H_4)$: a study of alkene pyramidalization.
Nakatsuji and Saito (1991)	RuO_4, OsO_4 excited states. Ru-O bond weaker than Os-O one. E(LUMO) -4.1 and -2.9 eV, respectively, explaining why RuO_4 is a stronger oxidation agent. First allowed excitation energies 3.22 and 3.90 eV, respectively.
Nguyen et al. (1991)	M_2E_3 and $M_2E_3H_2$ propellanes, M = C - Sn, E = O, S.
Novaro and Jarque (1991)	$Pd-H_2$. Results compared with matrix spectroscopy. R/NR.
U. Meier et al. (1991)	GaAs, $GaAs_2$, Ga_2As_2, As_2 and As_4 at MRD-CI level. Ions included.
Ortiz and Ballone (1991a)	PP for various non-local density functionals applied on M, M^+ and M_2, M = Be-Ba, Zn-Hg. IP_n and atomic excitation energies stable. D_e of M_2 vary strongly.
Ortiz and Ballone (1991b)	M_2^{2+}, M = Ag, Cd, Au, Hg. DFT PP. Gradient corrections remove binding for Au. M = Cd and Hg predicted to be metastable.
Ortiz and Ballone (1991c)	M_2, M = Be-Ba, Cd, Hg. LDA PP.
Pak et al. (1991)	LiBe, with full SO CI.
Pitzer and Winter (1991)	SO integrals.
Pyykkö and Zhao (1991a)	Bonding trends in XAu_n^{m+}, X = B-N, Al-S; n = 4 - 6. CAg_6^{2+}. At QR HF level, radial bonding dominant.
Pyykkö and Zhao (1991b)	$(Cl\ M\ PH_3)_2$, M = Ag, Au. The weak, "aurophilic" Au...Au attraction ascribed to correlation effects, not hybridization, by a HF/MP2/MP3 comparison. BSSE not considered. [See Li and Pyykkö (1992)].
Pyykkö and Zhao (1991c)	The large range of uranyl bond lengths interpreted as a frozen soft mode of a UO_6^{6-} octahedron. The D_{2h} structure of a $(OAnO)_2^{n+}$ dimer reproduced.
Sakaki and Ieki (1991)	Oxidative addition of SiH_4 to $Pt(PH_3)_2$. MP4.
Savin et al. (1991)	Visualisation of the electron density in Sn(II) compounds: SnO, Sn_2O_2, Sn_2H_4, Cu_4Sn_4. The $5s^2$ lone pair seen.
Schwerdtfeger (1991a)	$(TlH)_2$ model for $(Tlcp)_2$. A weak Tl(I)...Tl(I) attraction found at correlated level, none at HF level. TlH studied at seven different levels, up to QCI.
Schwerdtfeger (1991b)	Au and Au_2. Au EA increased 1.07 eV by relativity. Total EA 2.133 (exp. 2.309(5)) eV, IP_1 8.930 (9.226) eV. Au-Au 253.7 (247.2)pm.

Table 7.7. (continued).

Schwerdtfeger and Dolg (1991)	AuLa, AuLu. D_e tripled by relativity. In AuLa, Au $5d^{9.91}$, not much d transfer. Cp. Watson et al. (1987) for solid AuLu.
Schwerdtfeger et al. (1991)	PdH, PdF.
Seijo et al. (1991)	MX_2, M = Mg-Ba, X = F-I. Geometries and frequencies obtained at HF level.
Smith and Carter (1991)	Interactions of NO and CO with Pd and Pt atoms. MCO linear, X $^1\Sigma$. MNO bent. Pd and Pt prefer CO and NO, respectively.
Sodupe and Bauschlicher (1991)	$M^+ + C_2H_2$, M = Y-Ag. Y^+-Mo^+ insert into the π bond, Tc^+-Ag^+ have electrostatic bonding.
Sodupe et al. (1991b)	$Cl_4W{=}CH_2 + C_2H_4$. Formation of the metallacyclobutane exothermic.
Sodupe et al. (1991c)	Collinear reaction of Fe^+ with H_2.
Song and Hall (1991)	$WOCl_2(NH_3)_3$, $MoOCl_2(PH_3)_3$: no evidence for bond-stretch isomers.
Trinquier (1991)	X_2H_4, X = Sn, Pb. A singly-bridged local minimum.
Wang and Ermler (1991)	$Cs_{11}O_3^{n+}$ cluster model for cesium suboxide, n = -1, 0, +1.
Wexler et al. (1991)	$(C_6H_6)Ta(OH)_2$. Structure and bonding.
Albright and Tang (1992)	$Ta(CH_3)_5$, TaH_5. C_{4v} preferred to D_{3h}.
Allouche et al. (1992)	BaH. SO included.
Anchell et al. (1992)	PdH, Pd_nH_2 (n = 2,6).
Andrae et al. (1992)	OsH. Results compared with Benavides-Garcia and Balasubramanian (1991).
Andreocci et al. (1992)	$Sn(C{\equiv}CMe)_4$. PES.
Armstrong et al. (1992)	$LiMPh_3$, M=Sn, Pb. Sn-Li>Pb-Li.
Balasubramanian et al. (1992)	Hg_2. 12-VE PP. Many excited states given.
Balasubramanian and Feng (1992a)	M_2 + H, M = Ni, Pd. C_{2v} ground state. SO for Pd. Cp. B and F (1990a) on Pt_2H.
Barandiarán and Seijo (1992)	E_2, E = N-Sb. ScO, ScS. MnO, CuO.
Bauschlicher et al. (1992b)	MCH_2^+, M=Y-Ag.
Benavides-Garcia and Balasubramanian (1992)	GeX_2, GeHX, X=Cl-I. X $^1A_1$. Singlet-triplet separation.
Bender et al. (1992)	$Os(CO)_4(C_2H_4)$, an osmacyclopropane. GVB. $Os(CO)_4$, $H_2Os(CO)_4$.
Blomberg et al. (1992b)	$CH_4 + Pd_3$.
Boatz et al. (1992)	CdAr potential-energy surfaces.
Boehm and Banerjee (1992)	ReO_2, ReO_4^{n-}, n = 0,1.
Boldyrev and Simons (1992)	TeF_8^{2-} electron detachment energy 5 eV.
Boldyrev et al. (1992)	ICCH, ICCI, $IF_n(CCH)_{3-n}$, $I_4(CC)_4^{4+}$, $I_4(CC)_4F_4$. HF. New iodonium compounds predicted.

Table 7.7. (continued).

Breidung and Thiel (1992)	MI_3, MI_5, M = Sb, Bi. Ground-state properties.
Casarin et al. (1992)	$M(TAA)^{p+}$, M=Ni, Pd; TAA = tetraaza-annulene. PES.
Colmenares-Landin et al. (1992)	Ru + H_2.
Cundari (1992a)	TM imido complexes, L_nM=NR. M=Ti-Hf, ..., Rh, Ir.
Cundari (1992b)	$(NH_2)_2Zr$=NH, **1**. **1** · H_2.
Cundari (1992c)	H_2M=NH + CH_4 reaction, M = Ti-Hf. X_2Zr =NH, X = $-NH_2$, -Cl.
Cundari and Gordon (1992a)	Olefin metathesis catalysts: $M(OH)_2(NH)(=CHR)$, M = Mo, W, Re; R = H, CH_3. $MCl_2(XH)(CH_2)$, $M(OH)_2(XH)(CH_2)$.
Cundari and Gordon (1992b)	Transition metal alkylidenes: H_2M=CHX, M = Ti-Hf. H_3M = CHX, M = Nb, Ta.
Czuchaj et al. (1992a)	CdHg potential-energy curves and transition moments.
Czuchaj et al. (1992b)	KHg potential-energy curves and transition moments.
Dai and Balasubramanian (1992a)	M_3 + H, M = Pd, Pt. X $^2A_2$, C_{3v}. SO
Dai and Balasubramanian (1992b)	Zr_4. X D_{2h} rhombus, ^1A_g or $^3B_{1u}$, 5s-to-4d and 5p promotion observed.
Dai and Balasubramanian (1992c)	Rh_3. X S=6 C_{2v}. Low barrier to D_{3h}.
Dai and Balasubramanian (1992d)	M_4, M = Ge-Pb. X D_{2h} rhombus ^1A_g. SO for Pb_4, increases R by 0.5 pm, lowers E by 0.2 eV, changes atomization energy from 148 to 56 kcal/mol. Pb $6s^{1.94}$, nearly "inert pair".
Dai and Balasubramanian (1992e)	Os + H_2. X $^3B_1$, C_{2v}. (cp. linear RuH_2 quintet states). SO effects on bond angle up to 10°. Os ^5D ground state doesn't insert into H_2, the ^3F does.
Dai and Balasubramanian (1992f)	M_4^+, M=Ge-Pb. SO.
Dai et al. (1992a)	Ge_3: X $^1A_1$ or $^3B_2$.
Dai et al. (1992b)	IrH^+.
Dixon et al. (1992)	CH_2X-CH_2X, X = Br, I.
Dmitriev et al. (1992)	P- and PT- odd terms for HgF and PbF.
Dolg et al. (1992a)	Yb and YbO. An $f^{13}\sigma^1$ 0^- ground state still found, instead of f^{14} $^1\Sigma^+$.
Dolg et al. (1992b)	YbH, YbF. SO included.
Dolg et al. (1992c)	Ln_2, Ln=La-Lu. The 4f AO in core, 5s5p5d6s AOs in valence space.
Eckert et al. (1992)	H_2 rotation barriers in $M(CO)_3(PR_3)_2H_2$, M = Mo, W.
Fanizzi et al. (1992)	Pt(II) coordination in $Pt(N-N)(L)X_2$ (N-N a bidentate N-donor ligand, L and X monodentate): 5-coordinate tbp for L = strong π-acceptor. Otherwise CN = 4.
Fernandez Pacios (1992)	New PP with [Ar] core for Sc-Cu: improved $s^2d^n-s^1d^{n+1}-d^{n+2}$ energies.

Table 7.7. (continued).

Flad et al. (1992)	Cartesian Gaussian projection operators for quantum Monte Carlo. Li-Cs, Be-Ba, MM' (M = Li-K).
Foucrault et al. (1992)	Rb_2, Cs_2 at 1-VE level, including core-valence correlation potentials.
Gayen and Ermler (1992)	Cluster models for K adsorption on Ag.
Goursot et al. (1992)	M_2, M_4, M_4CO, M = Rh, Pd. Chemisorption analysed.
Gropen et al. (1992)	PtH. Two lowest states. R/NR. Results compared with PT ones.
Hagelberg et al. (1992)	Sb_4.
Hay (1992)	Oxidative addition and reductive elimination of H_2 and alkanes.
Haynes et al. (1992)	MH_7L_2, M = Tc, Re; L = PH_3. Classical (M-H) hydrides preferred by Re, non-classical (M-$H_2(\eta^2)$) ones by Tc. Cp. Pyykkö (1988a), p. 585.
Hickman et al. (1992)	XeHe and XeAr excited states and cations. SO. The ab initio CI curves improved by fits to empirical parameters.
Igel-Mann et al. (1992)	X_2, XH, XY, XY_m, X = In, Sn, Sb; Y = F, O, S. $(XY_m)_n$. Effect of the X $5s^2$ inert pair discussed. Electron-density pictures.
Jonas et al. (1992)	TiH_6^{2-} D_{3h} < Oh by 3 kcal/mol. VH_6^- C_{3v} < D_{3h} << O_h. MPn unreliable. QCISD(T) or CCSD(T) preferred.
Kalinkin et al. (1992)	Cluster model for Bi_{12} Ge O_{20}.
Kaupp and Schleyer (1992)	MX_2, M = Ca-Ba; X = Li, BeH, BH_2, CH_3, NH_2, OH, F. Most Ca compounds linear and Ba compounds bent at SCF level. QR SCF.
Kaupp and v. R. Schleyer (1992b)	Shapes of group 1-3 metal compounds $M^{n+}L_m$, M = K-Cs, Ca-Ba, Sc-La; L= NH_3, H_2O, HF. d AOs important.
Kaupp and v. R. Schleyer (1992c)	$(CH_3)_n$ PbF_{4-n}, $(CH_3)_nPbF_{2-n}$. Why are organic Pb(IV) and inorganic Pb(II) more stable?
Kaupp et al. (1992a)	MCp_2^+, M = Sc, La. Bending attributed to $d\sigma$. Also MH_2^+, MF_2^+ models.
Kaupp et al. (1992b)	MCp_2, M = Ca-Ba, Sm, Eu, Yb. M = Ba linear in HF, bent in MP2. All systems floppy.
Kaupp et al. (1992c)	$CaX_2(L)$, X=H, F; L=Ne-Xe, CO, N_2.
Kawamura-Kuribayashi et al. (1992)	$(SiH_2\ Cp_2)$ ZrR^+: Silylene-bridged zirconocene catalyst for olefin polymerization. R = CH_3, C_2H_4, C_3H_7.
Kim and Balasubramanian (1992a)	GaBr. 15 states.
Kim and Balasubramanian (1992b)	GaCl
Kippels et al. (1992)	Harmonic force fields for $M(CH_3)_2$, M = Zn-Hg. HF and MP2.

Table 7.7. (continued).

Klobukowski (1992a)	$M^{2+}(H_2O)_n$, M = Ca, Sr; n = 1-6. SCF + MP2. Experimental gas-phase hydratisation energies reproduced.
Klobukowski (1992b)	HX, X_2, X - Cl, Br. ECP methods (without inner nodes) and model potentials (with inner nodes) compared.
Krogh-Jespersen et al. (1992)	$(NH_3)_5Ru(II)S(CH_3)_2$. HF, MP2, MP3. Bonding and electronic spectrum.
Kudo and Nagase (1992)	$X_4H_6^{m+}$, X = Ge-Pb, m = 0,1.
Laasonen et al. (1992)	$La@C_{82}$.
Lao and Christiansen (1992)	Ne ground state, quantum Monte Carlo. Valence E_{corr} 0.313(3)a.u.
Lee and Lee (1992)	Polyatomic program including double-group symmetry. HX, X_2 (X, X=Br, I), CH_3I. PES.
Li and Pyykkö (1992)	Au_2^{2+}, $(ClAuPH_3)_2$, $SeAu_2$, $Se(AuPH_3)_2$. The "aurophilic" Au(I)...Au(I) attraction still attributed to correlation, strengthened by relativity.
Liao and Balasubramanian (1992a)	Group III-V tetramers and their cations: Al_2P_2, Ga_2As_2, In_2Sb_2 and Tl_2Bi_2. Also AlP - TlBi. SO. TlBi D_0^0 only 0.53 eV.
Liao and Balasubramanian (1992b)	M_6, M_6^+, M = Cu-Au. All M_6 ground states capped pentagonal, C_{5v}.
Lin and Hall (1992a)	Classical and nonclassical hydrides, IrH_5L_2(L = Be, PH_3, PMe_3). Correlation important, MP4 close to MRCI, CISD worse.
Lin and Hall (1992b)	Dito, $ML_{7-n}H_n$, $ML_{8-n}H_n$ (n = 2-7, M = Mo, W, Tc, Re, Ru, Os, Rh, Ir; L = PH_3, CO.
Lin and Hall (1992c)	$M(CO)_6$ + CO, M = Ta, W. SCF.
Lin and Hall (1992d)	$M(CO)_6$ + G, G = CO, PH_3; M = Cr-W, V-Ta. Substitution reactions.
Lin and Hall (1992e)	$ML_{8-n}H_n$, n = 4-7, L=PH_3. Stability of non-classical isomers.
Lin and Hall (1992f)	M_8C_{12}, M = Y.
Lundell and Kunttu (1992)	EHE^+, E = Kr, Xe. $D_{\infty h}$. IR frequencies given.
Marino et al. (1992)	LiBe valence and Rydberg states. SO.
Márquez and Fernández Sanz (1992a)	$(CO)_5Mo{=}MH_2$, M = Ge, Sn. CASSCF.
Márquez and Fernández Sanz (1992)	$MoSnH_2$.
Maseras et al. (1992)	$Rh(PH_3)_3Cl$.
Maseras and Morokuma (1992)	NiH_2, $[Os(PH_3)_4H(H_2)]^+$. Effect of the formally empty metal np AOs on population analysis. "Natural population analysis" compared to Mulliken one.
Matsunaga et al. (1992)	Heteroatom 6π benzene analogues $X_3Y_3H_6$, X, Y = B-Ga, C-Ge, N-As, O-S, Zn.

Table 7.7. (continued).

Miralles et al. (1992)	$[Cu_2Cl_6]^{2-}$ singlet-triplet gap.
Mösges et al. (1992)	$M(NH_2)_2(HF)_4$, M = Ca-Ba. H $Ba(C_4H_4N)$. $M(C_4H_4N)$, M = Na-Cs.
Müller and Reinhold (1992)	$R_3P...HX$, R=H, CH_3; X = Cl-I. HF, MP2. No trace of ion-pair complexes $R_3PH^+...X^-$.
Nakamura et al. (1992)	MeCu, Me_2Cu^- addition to acetylenes.
Nakatsuji et al. (1992a)	Dative metal-metal bonds found in $(OC)_5Os-M(CO)_5$, M = Cr, W. Results compared with $(OC)_5M-M(CO)_5$, M = Mn, Re.
Nakatsuji and Nakai (1992)	End-on O_2-Ag "dipped adcluster" chemisorption model.
Ortiz et al. (1992)	$An(CH_3)_3$, An = U-Pu. Pyramidal, C_{3v} structures found, even for the free molecule. 6d important.
M. Otto et al. (1992)	XN_3, X = F-I.
P. Otto et al. (1992)	$(MI)_n$ cluster models (M = In, Tl) for the orthorhombic crystal. $(CaSi)_n$ and $(RbI)_2$ also considered.
Pacchioni and Bagus (1992)	$Pd-PX_3$, $L-Pd-PX_3$, X = H, -Me, -OMe. $(CO)_3Pd-PX_3$. σ donation, π^* back donation analysed.
Pápai et al. (1992)	$M(CO)_n^{m+}$, n = 1, 2; m = 0,1. DFT.
Perry et al. (1992)	MR_1R_2, M=Co-Ir, R=H, CH_3.
Probst (1992)	Hydration of M^{2+}, M = Zn-Hg, Mg. $[M(H_2O)_n)]^{2+}$, n = 1-6.
Ramirez-Solis et al. (1992)	CuCl. Diagonal and transition dipole moments.
Reinhold et al. (1992)	$L-HgCH_3$, L = C≡CH, $CH=CH_2$, CH_2-CH_3, CH_3. Weak C-C bond or longer chain (C_2 versus C_1) weaken the <u>trans</u> C-Hg bond. Correlations with $^1J(HgC)$ and group electronegativity.
Reynolds (1992)	$C_2H_4Br^+$, $C_4H_8Br^+$ isomers.
Riehl et al. (1992)	$Ir(III)L_2ZZ'X$, L = PH_3, Z = H, CH_3; X = H, Cl, NH_2, OR. Stability and bonding analysed.
Saethre and Gropen (1992)	X_4^{n+}, n = 0-2, X = S-Te. d AOs required for agreement with experimental (n=2) geometries. SCF.
Sakai et al. (1992)	NaI ground and first excited state.
Sargent and Hall (1992)	Oxidative addition of H_2 to <u>trans</u>-$IrX(CO)(PR_3)_2$, X=H^-, Cl^-, Ph^-; R = H, Me.
Sargent et al. (1992)	Oxidative addition of H_2 to IrCl(CO)(dppe), dppe = 1, 2 - bis(diphenylphosphino) ethane.
Schleyer et al. (1992)	$H_3X-X'H_3$, X = C-Pb. No systematic relativistic effect on the rotation barrier.
Schmidt et al. (1992)	R-C≡E, E = N-Sb, R = H, F, Me, Ph.
Schneider and Thiel (1992)	MH_3X, M = C-Sn, X = F-I. Anharmonic force fields. HF.

Table 7.7. (continued).

Schröder et al. (1992)	FeF^+ X $^5\Sigma$.
Schwerdtfeger (1992a)	Comments on MH_n^{m+}, M = In, Tl; n = 1-3; m = 0,1.
Schwerdtfeger and Boyd (1992)	H_3PAu-$AuPH_3$.
Schwerdtfeger et al. (1992a)	XH_3 (X = N-Bi) inversion barriers and other ground-state properties. HF, MP2, R/NR and QCI for Bi.
Schwerdtfeger et al. (1992b)	AuL_3, AuL_4^-, Au_2L_6, L = H, F-I. AuF_6^-. Relative stability of AuL_4^- over AuL_2^- increased by relativity, as suggested by Pyykkö (1978).
Schwerdtfeger et al. (1992c)	MX, MX_3 (M=B-Tl; X = H, F-I). AuH, HgH_2. PbX_2, PbX_4; BiX, BiX_3, BiX_5 (X=H, F). Energetics of the valency change from Pb(IV) to Pb(II) etc. R/NR. Relativity energetically favours the lower valency, however the $6s^2$ pair is not more "inert" than ns^2, n<6.
Sellers and Sudhakar (1992)	Hg clusters and their interaction with H_2O. Water-Hg, Hg-Hg potentials.
Seijo et al. (1992)	Valence E_{corr} for I^-, I and I^+. CISD. Correct nodal structure required.
Seitz-Beywl et al. (1992)	Li^+ and I^- interactions with Pt_n surface models, n = 5, 9.
Sodupe et al. (1992)	$[Sr(CO_2)_n]^+$, n = 1, 2.
Takeda and Shiraishi (1992)	$(SnH_2)_n$ polystannane.
Tornieporth-Oetting et al. (1992)	I_2^{m+}, m = -1, 0, +1. $I_2N_3^+$. A I-I-N_3 chain structure lowest, a C_{2v} one next. MP2.
Trinquier (1992)	$X_2H_5^+$, X = Ge-Pb. Doubly bridged, $H_2XH_2XH^+$ isomer lowest for Sn and Pb.
Vitale et al. (1992)	$Cu_4I_4L_4$, L = NH_3, C_5H_5N. I-to-pyridine and I-to-Cu transitions studied.
Vijayakumar and Balasubramanian (1992)	Tl_n^{m+}, n=2-5, m=0,1. Including SO, Tl_4 is D_{4h}, X $^1A_1$. Tl_3 is near D_{3h}, X $^4A_2$.
Wallace (1992)	SO CI potential curves for CuF, CuCl and AuF.
Wang and Carter (1992)	Pt_3. Ten states within 6.6 kcal/mol: one linear, two D_{3h}, seven distorted D_{3h}.
Wehman-Ooyevaar et al. (1992)	MCl(H)(Me)$_2$(NH_3)...NH_3 and MCl(Me)$_2$(NH_3)...NH_4^+, M = Pd, Pt.
Wiest et al. (1992)	L-ReO_3, L= cp, CH_3, CF_3, F, Cl.
H.-X. Zhang and Balasubramanian (1992a)	M^++H_2, M = Ni-Pt. For Pt, SO stabilizes the small-angle minimum.
H.-X. Zhang and Balasubramanian (1992b)	E_4, E = P-Bi. SO effects stabilize Bi_4 by 0.82 eV.

Table 7.8. Molecules treated by the Perturbative Hartree-Fock-Slater (P-HFS method).

Reference	Comments
Ziegler (1985c)	Proton affinity of d^8 metal pentacarbonyls (Cr^{2-}, Mn^--Re^-, Fe-Os, Co^+) and d^{10} tetracarbonyls (Fe^{2-}, Co^--Ir^-, Ni-Pt, Cu^+).
Dyke et al. (1986a)	As_2, As_4 PES.
Dyke et al. (1986b)	Sb_2, Sb_4 PES.
Louwen et al. (1986)	Trans-$MCl_2(EH_3)_2$ PES. M=Pd, E=N, P; M=Pt, E=N.
Boerrigter (1987)	Thesis. A summary on SO effects in X_2^+, X=F-Br, and bonding in $An(COT)_2$ and AnX_4.
Boerrigter et al. (1987)	SO effects in excited states of X_2^+, X=F-Br. A localised hole found for the $^2\pi_u$ state of Br_2^+.
Dyke et al. (1987); Fehér (1988)	MO, MO^+, M=Sr, Ba. PES.
Ziegler (1987)	Review on periodic trends. Dimer D_e is $Mo_2>W_2$ and $Mo_2<W_2$ for NR and R, respectively. Becke's non-local correction used.
Ziegler et al. (1987a)	Multiple M-M bonds in M_2, $M_2(OH)_6$ (M=Cr-W), $M_2Cl_4(PH_3)_4$ (M=V-Mn, Nb-Tc, Ta-Re). Relative bond strengths discussed.
Ziegler et al. (1987b)	Relative bond strengths of M-R bonds, R=H, CH_3. $RRe(CO)_5$, $RIr(CO)_4$ R/NR.
Ziegler et al. (1987c)	$M(CO)_6$ (M=Cr-W), $M(CO)_5$ (M=Fe-Os), $M(CO)_4$ and $M(PF_3)_4$ (M=Ni-Pt). M-CO bond strengths 4d<5d<3d and 5d<4d<3d for R and NR, respectively.
Allen et al. (1988); Fehér (1988)	UO, UO_2. PES interpreted. Bond-length **expansion** by 7.5pm found for UO_2.
Boerrigter et al. (1988a)	Bonding in $An(COT)_2$, An = Th-Pu, COT = C_8H_8. Relative importance of $6d\pi$, $5f\pi$ and $5f\delta$ bonding discussed. Rel. contractions of An-C reported. PES interpreted.
Boerrigter et al. (1988b)	AnX_4, An=Th, U; X=F, Cl. T_d geometry found. Rel. contractions reported. PES interpreted.
Ziegler et al. (1988a)	Relative strengths of M-R bonds, R=H, CH_3, in cp_2 MR, M=Y, La, Tc, Re. For early metals H≈Me, for late metals H>Me.
Ziegler et al. (1988b)	M-L bond strengths (L=OH, OCH_3, SH, NH_2, PH_2, CH_3, SiH_3, CN and H) in Cl_3 ML (M=Ti-Hf). $LCo(CO)_4$.
Ball et al. (1989)	$W(SH)_6^{2-}$, $Cu(PH_3)_2^{2+}$ and their complex. Metal-metal interaction 100kJ/mol of total 1233.

Table 7.8. (continued).

Grabandt et al. (1989)	InF, InF^+. PES and ground-state properties.
Tschinke (1989)	W_2. Cl_3MR, R=H, CH_3; M=Th, U; Ti-Hf. $RM(CO)_5$, M=Mn-Re, $RM(CO)_4$, M=Co-Ir.
Ziegler et al. (1989a)	$[Pt(CN)_4]^{2-}$, $Tl_2Pt(CN)_4$. Bonding and optical spectrum. NR/QR/QR+SO. Tl behaves as a 6s donor and $6p_z$ acceptor.
Ziegler et al. (1989bc)	Reviews.
Ziegler et al. (1989d)	Cp ML (M=Rh, Ir; L=CO, PH_3), $M(CO)_4$ (M=Ru, Os and their insertion into CH_4. 5d bonds stronger due to relativity.
Lelj and Rosa (1990)	Bonding in MIn, M=Cu, Ag, Au. Au-In $<$ Ag-In.
Folga et al. (1991)	L_2 Lu-H + H_2 exchange reaction. Experimental L = cp or cp^* modelled by L=Cl. Activation energy, enthalpy and entropy obtained.
Vanquickenborne et al. (1991)	Reassignment of AnX_4 PES, An=Th,U; X=F,Cl. (Modification of Pierloot et al. (1991) for the solids).
van Wezenbeek et al. (1991)	Relativistic bond lengthening of UO_2^{2+} and UO_2, because depopulation of 6p upon bond shortening leads to smaller $\langle h_m \rangle$.
Folga and Ziegler (1992)	X_2Lu-H + D_2 -> X_2LuD + HD, X_2LuCH_3 + H_2 -> X_2LuH + CH_4. $(Cp^*)_2$ Lu modelled by Cl_2Lu. Four-centre transition state.
Kirchner et al. (1992)	High-energy O_2-Pt(111) scattering. PtO. R/NR, significant effects found.
van Daelen et al. (1992)	Adsorption of S on Ir_4. Influence of Mg^{2+} studied.
van Wezenbeek (1992)	Thesis on uranium compounds.

Table 7.9. First-order pertubation theory on molecules.

Reference	Comments
Meath (1966)	BP contributions to interatomic forces for degenerate states. An α^2R^{-3} dispersion term found.
Meath and Hirschfelder (1966a)	BP contributions to intermolecular forces.
Chang (1967)	Moderately long-range intermolecular forces. SO effects included.
Walker and Richards (1970)	SO of diatomics at HF level. UHF calculations of core-polarisation effects in atoms.
Barron and Buckingham (1973)	SO coupling allows, besides E1, also M1 and E2 transitions of molecules. New singlet-triplet intensity term found.
Beck et al. (1974)	Fine structure of the a $^3\Sigma_u^+$ state of He_2.
Bendazzoli and Palmieri (1974)	GTO SO calculations for polyatomics. CO_2^+, N_3, H_2CO. SOO (spin-other-orbit) important.
Abegg (1975)	GTO SO for diatomics.
Dunning et al. (1978)	GaKr, $GaKr^+$ potential curves and transition moments, including an SO term.
Michels et al. (1978)	Noble-gas dimer ions. SO effects near dissociation limit.
Gribov et al. (1984)	PT for relativistic effects in molecules.
Almlof and Faegri (1986)	ZR_4, Z=C-Pb, R= H, $-CH_3$. Bond-length contraction studied.
Bauschlicher and Langhoff (1986)	ZnO, ZnS. D_e increased by 0.02 eV.
Chong et al. (1986)	CuH, ZnH spectroscopic constants.
Langhoff and Bauschlicher (1986)	CuO and CuS. X $^2\pi$ and $A^2\Sigma^+$.
Langhoff et al. (1986)	Cu_2, Cu_3 ground-state properties.
Pettersson and Langhoff (1986)	GeH.
Thakkar (1986)	$\langle h_m \rangle$, $\langle h_d \rangle$ for diatomics, H_2...RbF.
Werner and Buckingham (1986)	GeH. Effects on R_e and dipole moment.
Yarkony (1986)	Line strengths of spin forbidden transitions at BP level. Example NF b $^1\Sigma$-$X^3\Sigma$.
Alvarez Collado (1987)	RbF HF potential curve.
Grein and Peyerimhoff (1987)	Ne_2 low excited states with semiempirical SO.
Grein et al. (1987)	Ne_2 with ab initio SO.
Matsushita et al. (1987)	SeO. Radiative lifetimes from $a^1\Delta$ and b $^1\Sigma$ to $X^3\Sigma$.
Pettersson et al. (1987)	MH^+, M=Sc-Cu, Y-Ag.
Raseev et al. (1987)	Spin polarisation of molecular photoelectrons from HBr and HI.

Table 7.9. (continued).

Alvarez-Collado and Buenker (1988)	SeS transition probabilities. MRD-CI.
Banichevich et al. (1988)	Potential curves and predissociation of HBr^{2+} and DBr^{2+}.
Bauschlicher et al. (1988)	Cu_n^{m-}, n=1-3, m=0,1. (NR reference).
Blomberg et al. (1988)	PdCO, Pd_2CO. All-electron CPF level. Relativistic contributions to CO binding 11 and 14 kcal/mol, respectively, originate from smaller $4d^{10}$-$4d^95s$ splitting. Pd and PdH tests.
Blomberg and Wahlgren (1988)	Pd atom. 4s and 4p orbital relaxation shown to be important in 1st-order PT estimates of relativistic effects.
Buenker and Chandra (1988)	CI calculations including Breit-Pauli terms. NBr, SeS.
King and Furlani (1988)	Computation of 1- and 2-electron SO integrals for GTO:s.
Langhoff et al. (1988)	ScX, X=F-Br.
Lefebvre-Brion et al. (1988)	SO effects in HCl autoionization.
Senekowitsch et al. (1988)	H_2Se vibrational transition probabilities.
Vrbik et al. (1988)	Quantum Monte Carlo including $h_m + h_d$. Ground-state energy of LiH.
Blomberg et al. (1989)	$LPdH_2$, $L=C_2H_4$, CO, N_2, PH_3, H_2O, Cl^-. Ligand effect on Pd-H bonding investigated. Only relativistic H-Pd-H exists: Low and Goddard (1984b), Balasubramanian et al. (1988b).
Carlacci (1989)	SO effects in organic molecules.
Frey and Davidson (1989)	ScCO, Sc atomic states.
Marian et al. (1989)	NiH.
Pan and Zhao (1989)	The usual mass-velocity term, h_m, is OK in 1st-order PT, contrary to claims of Farazdel and Smith (1986).
Thümmel et al. (1989)	MgO. Spin-forbidden transitions and BO breakdown.
Yarkony (1989)	Quenching of He 2 ^3S in helium gas.
Balasubramanian (1990g)	Stability of He_n^+, n=3,4, including differential relativistic effects.
Daskalopoulou et al. (1990)	Xe_2^+, Xe_3^+ MRCI.
Gemein and Peyerimhoff (1990)	$HeAr^+$ potential curves for X $^2\Sigma$ and A $^2\pi$.
Graff and Wagner (1990)	Long-range forces for $O(^3P) + OH(^2\pi)$.
Kölmel and Ahlrichs (1990)	$(CuX)_n$, X=H, F, Cl, n=1,2. Cu_2, Cu_2^+. $[Cu(HNNNH)]_2$, $[Cu(HN_5H)]_3$: models for "supershort" (245, 235pm) Cu-Cu distances.
Matsuoka (1990b)	BP integrals for spherical Gauss-Laguerre functions.
Schwerdtfeger et al. (1990b)	CuF ^{63}Cu NQCC.
Staemmler (1990)	$HeAr^+$. SO.

Table 7.9 (continued).

Strömberg and Wahlgren (1990)	Au_2, Hg_2^{2+}. Ground-state properties.
Yarkony (1990)	The decomposition reaction $HN_3(S=0) \rightarrow N_2(S=0) + NH(S=0,1)$. Large CI with BP terms. Landau-Zener treatment at the crossing.
Banichevich et al. (1991)	Potential curves and predissociation of HBr^{2+} and DBr^{2+}.
Blomberg et al. (1991a)	M +H-CH_3, M+CH_3-CH_3, M=Fe, Co, Ni, Rh, Pd. E_{tot} at minima and at TS lowered by 2-4 and 6-10 kcal/mol for 3d and 4d metals, respectively.
Blomberg et al. (1991b)	$LRhH_2$ (L=Cl^-, PH_3, CO), (OC)MH_2 (M=Rh, Ru, Ru^+), M-CH_4 (M=Rh, Rh^+): Importance of the ligand in activating a C-H bond.
Bär and Ahlrichs (1991)	AlCu. R_e contraction 0.08a.u., D_e increase 0.15eV.
Ciullo et al. (1991)	NiN_2, $Ni(PH_3)_2N_2$: end-on and side-on preferred, respectively.
Faegri and Bauschlicher (1991)	YNi, ScPd, YPd: results compared with PP.
Ishii et al. (1991)	Free-radical g-values by FPT.
Parlant et al. (1991)	Spin-forbidden decay of HS^{2+}.
Roth et al. (1991)	$MgFe^+$ and its reactions with hydrocarbons. $X\,^6\Sigma^-_+$.
Sodupe et al. (1991a)	$Sr(H_2O)_n^+$. Model for gas-phase clusters.
Åkeby et al. (1992)	Sc_2 X $^5\Sigma$.
Andersson and Roos (1992)	Ni atom valence states. CASPT2 + $\langle h_m + h_d \rangle$ from Martin and Hay (1981).
Banichevich et al. (1992)	Predissociation of HBr^+ and DBr^+. BP MRD-CI.
Bauschlicher et al. (1992a)	Ni_2, Ni_2^+. D_0 calculated at MCPF level.
Bauschlicher et al. (1992b)	MCH_2^+, M = Sc-Cu.
Blomberg and Siegbahn (1992)	CrN, CrO, FeN. Ground-state properties.
Blomberg et al. (1992a)	M + CH_4, M = Y-Pd. Lowest repulsion toward methane in the d^{n+2} state, strongest bonds to the $d^{n+1}s$ state. For Rh both are low. Only Pd has a bound η^2 precursor.
Blomberg et al. (1992b)	Pd_2 + CH_4, Pd_2 + H_2.
Blomberg et al. (1992c)	MC_2H_4, M = Y-Pd.
Gropen et al. (1992)	PtH. Results compared with PP ones.
Hippe and Peyerimhoff (1992)	CuO ground and excited state SO splittings. Occupation of Cu 3d states crucial.
Koseki et al. (1992)	The effective charges in $h_{SO}=(\alpha^2/2)Z_{eff}r^{-3}\mathbf{l}\cdot\mathbf{s}$ determined for Li...Cl, used for diatomics.

Table 7.9. (continued).

Partridge et al. (1992)	Metal ions bound to He, Ne, Ar. SO-corrected NR. MAr^+ D_0 for M = V, Cr, Co and Ni about 20 % below experiment. Also ML_2^+.
Reuter and Peyerimhoff (1992)	Vibrational structure of X $^2A_1$ and A $^2B_1$ electronic states of CH_2^+.
Siegbahn and Blomberg (1992)	C-C bond activation in ethane, cyclopropane and cyclobutane by Y-Pd atoms.
Vahtras et al. (1992a)	SO in O_2, CH_2.
Vahtras et al. (1992b)	2nd-order SO effects in Cr_2.

Table 7.10. Density functional calculations. PP calculations are treated in Table 7.7. GTO calculations using transformed Dirac Hamiltonians are listed in Table 7.3.

Reference	Comments
March (1986)	Discusses bond lengths of CH_4-PbH_4 in relativistic TF theory.
Dietz et al. (1992)	A mean-field model for Be and LiH, derived from QED.

Table 7.11. Semiempirical methods.

Reference	Comments
Minaev and Terpugova (1969)	Quote inclusion of SO at CNDO or EHT level for CT complexes.
Hinkley et al. (1972)	SO splittings from MINDO. Diatomics.
Lohr (1972)	Spin-forbidden electronic excitations in transition-metal complexes.
Wittel (1972)	Different π_g and π_u SO splittings in I_2^+ attributed to σ_g repulsion from below. See also Jungen (1972).
Masmanidis et al. (1975)	SO coupling in organic molecules using CNDO/S.
Mingos (1976)	EHT on Au_n^{q+}, n = 6...13, and $[Au_6(PPh_3)_6]^{2+}$. Predicts the stability of the unknown $[CAu_6(PPh_3)_6]^{2+}$, made in 1988 by Schmidbaur et al.
Minaev (1980)	INDO-level TP for O_2 M1 transitions a $^1\Delta$ - X $^3\Sigma$ and b $^1\Sigma$ - X $^3\Sigma$. External heavy-atom effects discussed.
Minaev and Muldakhmetov (1984)	INDO/S + SO CI interpretation of forbidden transitions in CBr, HOO and O_2.
Dewar et al. (1985a)	MNDO parameters for Hg.
Dewar et al. (1985b)	MNDO parameters for Pb.
Dewar and Jie (1985)	AM1 parameters for Hg.
Krogh-Jespersen (1985)	REX calculations on cis-platinum hydrolysis. MCD intensities calculated via PT.
Barra et al. (1986, 1988)	PNC effects on NMR cemical shifts. REX.
Boudreaux et al. (1986)	$Pt_2(P_2O_5H_2)_4^{4-}$ bonding and spectra.
Gleghorn (1986)	Reaction of C_6H_6 with MCl^+ or M^{2+}, M=Cd, Hg. REX.
Jostes et al. (1986)	$MO_{4-n}S_n^{2-}$, M=Mo, W. Spectra interpreted. IEHT.
Larsson and Pyykkö (1986)	"ITEREX": a charge-iterative version of the Relativistic Extended Hückel program REX. XeF_n, n=2, 4, 6. UF_6. $UO_2X_4^{2-}$, X=F, Cl.
Lohr and Jia (1986)	REX on LnX_3. DZ fits given for Ln=La, Nd, Gd, Lu and X, X=F-I.
Lyudchik et al. (1986)	Bond angles discussed. Inclusion of SO at EHT level using a $h_{SO}=c\mathbf{l}\cdot\mathbf{s}$. $PtCl_6^{2-}$.
Z.-X. Ma and S.-S. Dai (1986)	HX, X_2, X = F-I. REX energies.
Pyykkö (1986b, 1988b)	ITEREX: a charge-iterative REX. Diskette included in (1988b).
Rösch and Pyykkö (1986)	Nuclear spin-spin coupling tensors in XeF_n, n = 2, 4, 6, MF_6, M = S-Te, U. ITEREX.
Trunschke and Miessner (1986)	CO + Pt_n, n = 1, 3, 4. REX.
van der Waals (1986)	SO coupling in lower excited states of d^0 MO_4^{n-} anions.

Table 7.11. (continued).

Wilson et al. (1986)	NMR spin-spin coupling in Zintl anions. $Sn_2Bi_2^{2-}$, Sn_8Pb^{4-} etc. REX.
Boca (1987)	A QR CNDO/1. Geometries and IP for 50 compounds, from H_2 to $PtCl_4^{2-}$.
Culberson et al.(1987)	An INDO model for LnX_n, n=2-4, and $Ce(NO_3)_6^{2-}$. Geometries.
Dixon and Tuckett (1987)	SO splitting in $(t_2)^5\ {}^2T_2$ states of MX_4^+ ions M=C-Ge; X=F-I. MNDO.
Edlund et al. (1987)	REX interpretation of ^{207}Pb NMR chemical shifts in a PbH_3^- model.
Firth and Grice (1987)	SO interaction for the $H+X_2$ surface, X=F, Cl. DIM (diatomics in molecules).
Kertesz and Guloy (1987)	EHT band-structure study on chains, close-packed layers and honeycomb networks of Hg^q, as function of the charge q. Break-up into Hg_4^{2+}.
Krogh-Jespersen and Altonen (1987)	Hydrolysis of cis-platinum and acetamide. REX.
Lohr (1987)	A crystal version, "REXBAND", of the relativistic extended Hückel method REX. Te and Po.
Pyykkö et al. (1987)	Interpretation of the relativistic contribution to the NMR chemical shift $\sigma(^1HX)$, X = F-I, successful. CH_3X, HC≡CX only qualitatively understood. REX.
Schall et al. (1987)	LaF low-lying states. A ligand-field model.
Tatsumi and Nakamura (1987a)	Unusual Th-C-C angles in $(cp)_2Th(CH_2CH_3)_2$. "Agostic" Th...H reproduced. QR EHT.
Tatsumi and Nakamura (1987bc)	$(cp)_2AnR_2$, An = Th, U. Also -metallacyclopentadiene and -cyclobutadiene complexes.
Boca (1988a)	The QR-CNDO/1 applied on $[Co(NH_3)_6]^q$, $[MCl_4]^q$ (M=Ni-Pt) and their solvation.
Boca (1988b)	A QR-INDO/1 method, including the monocentric repulsion terms. MCl_4^{2-}, M=Ni, Pd, Pt, E110. Geometries for hydrides and halides.
Faiez and Ruiz Lopez (1988)	SO coupling constants from INDO-level finite-perturbation theory. ClF_4.
Gavezzotti et al. (1988)	Compares REX and EHT results with PP ones for Pt_n, n=1-7.
Jokisaari et al. (1988)	REX calculations of σ and J tensors in CX, XCX', X=O-Te.
Jostes (1988)	Iterative EHT energy parameters for Hf-Hg.
Marian et al. (1988)	ThO. REX compared with ab initio.
Pyykkö (1988b)	Review on semiempirical relativistic methods. ITEREX diskette included.
S. Wang et al. (1988)	A qualitative MO dimer model for organometallic $(AuTl)_\infty$ chains.

Table 7.11. (continued).

Azuma et al. (1989)	NbN X $^3\Delta$ -$^1\Delta$ cross terms from SO and hfs.
Boca (1989)	CNDO applied on XY diatomics (X,Y group 1, 11 elements).
Böhm et al. (1989)	Cluster models for magnetic anisotropy of amorphous iron, Fe_n, n=12-14, treated at ZDO level, including SO.
Borkovskii and Lyudchik (1989)	SO effects on uranyl complexes: $UO_2F_5^{3-}$, $UO_2X_4^{2-}$, X=Cl, Br.
Janiak and Hoffmann (1989a)	Tl(I)-Tl(I) interaction in an H-Tl... Tl-H model. EHT.
Janiak and Hoffmann (1989b)	1-dimensional, mixed-valent, halogen-bridged gold chains. EHT.
Kotzian et al. (1989)	A SO-INDO/S-CI method, tested on diatomic hydrides and oxides of C, N, O; HF^+-HI^+, TiO.
Minaev et al. (1989)	MINDO/3 - CI + SO. EH^q, q = -1, 0, +1, E=B...S. Intensity-borrowing in SO-allowed electronic lines discussed.
Pellow and Vala (1989)	SO splitting of alkali and noble metal atoms in rare-gas matrices.
Pyykkö et al. (1989)	Chemically realistic ITEREX and REX parameters for An = Th-Np. Applications on $An(C_8H_8)_2$, $An(cp)_4$, ThO_2 versus UO_2^{2+}.
Anderson et al. (1990, 1991)	QR INDO/1 model for 4d transition metals. Geometries for hydrides, oxides, halides etc.
Boca (1990a)	CNDO/1, including SO. Applications on AH, AH_2, AH_3, MH, InX and HgI_2.
Boca (1990b)	Jahn-Teller stability or instability of octahedral copper and gold complexes.
Boca and Breza (1990)	Models for Y-Ba-Cu-O superconductors.
Borkovskii et al. (1990)	$[Ce(H_2O)_n]^{3+}$, n=8, 9. Iterative MWH + SO for the f-d bands.
Breza and Boca (1990a)	Cluster models for $YBa_2Cu_3O_y$ (y=6,7). QR INDO.
Breza and Boca (1990b)	QR-INDO on the Tl-Ba-Ca-Cu-O system cluster models $[Tl_4Ba_{12}Ca_{6n-6}Cu_{2n}O_{16+7n}]^{2n-8}$, n=1-3.
Dai and Ma (1990)	REX on HX, XX', InX, X=F-I.
Gleghorn and Hammond (1990)	Erratum to earlier REX work on M_2H_6, M=Ge-Pb.
Janiak and Hoffmann (1990)	M(I)-M(I) interactions, M=In, Tl, in molecules and solids.
Mingos and Kanters (1990)	EHT on bonding in mono- and dicarbido gold clusters, $[X(AuPH_3)_m]$, X = B-N, C_2.
Mukherjee et al. (1990)	Au_2^{2+} at tight-binding (EHT) level: X $^1\Sigma_g$ ground state repulsive, Σ_u excited state bound. Longer Au_m^{2+} chains, n = 3-6, also considered.
Nizam and Bouteiller (1990)	MNDO on cluster models of PbF_2.

Table 7.11. (continued).

Novak (1990)	REX/EHT on five mixed, chiral halomethanes CXYZU, X- = H, F-I.
Pedash et al. (1990)	Inclusion of SO in the π-electron approximation. Singlet-triplet splittings for aromatics.
Pineda de Castro and Müller (1990)	REX on M_6X_{12} and FeM_6X_{12} cluster models, M = Zr, Th; X = Br, I. Bonding analysed.
Stewart (1990)	Review on MOPAC. Includes MNDO up to Pb, PM3 up to I.
Tatsumi et al. (1990)	The U-S bonds in $[U(SCH_2CH_2S)_4]^{2-}$ involve U 7s, 7p and 6d but little 5f and weak U-S π interactions. QR EHT.
Assefa et al. (1991); Assefa (1992)	Models for $Tl[Au(CN)_2]$, $Tl_2Pt(CN)_4$. Bonding and luminescence.
Aucar and Contreras (1991)	MNDO-RPA calculations of nuclear spin-spin coupling constants. Atomic DF $\langle r^{-3}\rangle$ and $\langle\delta(r)\rangle$ used. XH_4, X = C-Pb. Also XH_3^-, XMe_4, H_3XXH_3,...
Jové et al. (1991); Pyykkö and Jové (1991)	Mössbauer NQCC in neptunyl compounds interpreted by REX.
Kotzian and Rösch (1991)	INDO/CI studies of low-lying levels of PrO and TmO.
Kotzian et al. (1991)	INDO/S-CI on $[Ce(H_2O)_6]^{3+}$.
S. Lee (1991)	Structures of heavy main-group elements from a EHT-level band structure 2nd moment model.
Lohr (1991)	REXBAND on DOS of UB_n, UC, UBC, UPt_3.
Nogueira and Guenzburger (1991)	Valence-orbital IP expansions as functions of atomic charge for Hf-Hg from atomic DFT. R/NR.
Pastor et al. (1991)	Metastability of Pb_3^{2+} understood using a parametrized Hamiltonian.
Boca and Pelikán (1992)	Review on quantum chemistry of coordination compounds.
Boca et al. (1992a)	QR INDO/1 cluster models for $(GaAs)_n$, n ≤ 80. Comparisons with EHT band structures and ESCA.
Boca et al. (1992b)	Ge_x, Ge_xGa_y and Ge_xAs_y cluster models.
Breza (1992)	Cluster models for H_2 in Pd. QR CNDO.
Estiu and Zerner (1992)	$(CO)Rh_n$, n = 1-15. INDO. Transition from bridge to on top CO with larger n, related to charge shift.
Jahns et al. (1992)	$[(Cp^*_2M)_2 C_6H_4]$, M = Sc, Lu. Agostic (M...H) interactions found for M = Lu. 5d important. INDO.
Janiak and Hoffmann (1992)	In(I)...In(I) and Tl(I)...Tl(I) attractions in molecules and solids. Assigned to hybridisation. EHT.
King (1992)	Discusses 6d-5f hybridisation of An=Th-Pu in actinyls and actinocenes.
Kotzian et al. (1992)	LnO, Ln=La...Lu. Energies, oscillator strengths, SO splittings. INDO. 5d population constant, 1.1 electrons.

Table 7.11. (continued).

Kotzian and Rösch (1992)	$[Ce(H_2O)_n]^{3+}$, n = 8,9. The 4f-5d spectrum interpreted.
Nogueira et al. (1992)	Bonding and g-tensor for $[Ir(II)(CN)_5]^{3-}$, $[Ir(II)(CN)_{6-x}Cl_x]^{4-}$, x = 1, 2. IEHT. VOIP for Ir given.
Pastor and Reindl (1992)	Ground states of Pb_n^{2+} clusters using a parametrized many-electron Hamiltonian.
Roszak and Lipinski (1992)	INDO including SO. Triplet lifetime of formaldehyde, pyridine etc. SO splitting of X_2, X = Cl-I.
Schwerdtfeger et al. (1992a)	XH_3 inversion barriers (E_a), X = N-Bi. REX/EHT, MP2. 6s radial contraction at C_{3v} increases E_a.
Solà et al. (1992)	ZnO, HgO. Semiempirical VB.
Thiel and Voityuk (1992)	MNDO for d orbitals. Cl-I.
G. S. Wu et al. (1992)	QR CNDO applied on main-group diatomics, LnX_3, Ln = La-Lu, X = F-I.

Table 7.12. Relativistic crystal field theory.

Reference	Comments
Gerasyuk et al. (1986)	O_h and C_{4v} sites for Ce^{3+} in CaF_2.
Pastusiak and Kuriata (1986)	Relativistic CFT for $4f^7$ ions. Interpretation of tetragonal 2nd-order spin-lattice coefficients.
Andriessen and Postma (1987)	Crystal-field model for Tl^0 and Pb^+ $6p^1$ centers.
Makhyoun (1987); Makhyoun and El-Ezaby (1988)	Derives effective ligand-field parameters for $5f^1$ systems from SO-perturbed QR MS $X\alpha$ calculations.
Schall et al. (1987)	Diatomic LaF. A ligand-field model for the low-lying states.
Tuszynski and Dixon (1987)	Gd^{3+} in C_{3h} symmetry. A study of crystal fields, SO and residual interactions.
Siu (1988ab)	S ground-state splitting of Mn^{2+}.
Siu (1988c)	Covalency effects in relativistic crystal fields.
Vokhmin and Ionova (1988)	Electrostatic and SO parameters for f electrons.
Ammerlaan and van Oosten (1989)	Model for Pt(-I), $5d^96s6p$, in Si.
Bartram et al. (1989)	Pb(0) in SrF_2.
Chatterjee and Buckmaster (1989)	New electric-field effective operators.
Tuszynski and Dixon (1989a)	Relativistic effects in 4f and 3d complexes
Tuszynski and Dixon (1989b)	Zero-field splittings in 4f and 3d complexes.
Yu (1989)	Cubic zero-field splitting of a d^5 ^6S-state ion.
Yu (1990)	Rank-four spin-Hamiltonian parameters of a $3d^5$ ion.
Blanchard and Girault (1991)	Cation vacancies in II-VI compounds.
Boulanger et al. (1991)	TP of Mn^{2+} d^5 $^4T_1$-$^6A_1$ transitions in II-VI compounds. Ligand SO important.
Chatterjee and Buckmaster (1991)	Relativistic effective Hamiltonians for S-state ions. EPR of Gd^{3+}.
Goodman et al. (1991)	Crystal-field levels for Ln = Pr, Nd, Ho in Ln Ba_2 Cu_3 O_7.
Wang and Yu (1991)	Zero-field splitting of a ^6S-ion in tetragonal symmetry.
Ning et al. (1991)	Cubic zero-field splitting of CaF_2 : Mn^{2+}.
Balcar and Lovesey (1992)	Magnetic inelastic neutron scattering intensities for Pr^{3+} and Tm^{3+} (f^2 and f^{12}, respectively).

Table 7.12. (continued).

Gajek and Mulak (1992)	Ab initio fundamentals of $5f^n$ ion angular overlap model (AOM) for halide ligands. Both σ, π and δ contributions necessary.
Spain and Morse (1992)	Ligand-field theory for $d^n\sigma^m$ states of MM', M = Ni, Cu, Pt, H.
Yu and Rudowicz (1992)	Zero-field splitting of ^6S-state Mn^{2+} and Fe^{3+} in cubic and tetragonal fields.

Table 7.13. Relativistic theories of molecular properties. For parity non-conservation effects, see table 5.17.

Reference	Comments
Hameka and Goodman (1965); Lohr (1966)	Raise the question of correct operators for spin forbidden molecular transitions. At Pauli level, $\langle \mathbf{r} \rangle$ is OK, $\langle \mathbf{p} \rangle$ must be replaced by $\langle \boldsymbol{\pi} \rangle$.
Englman (1966)	Operators for spin-forbidden transitions. The Dirac level.
Chiu (1968)	Dirac- and Pauli-level operators for radiative transitions of molecules. A thorough analysis.
Ginsburg and Goodman (1968)	Two-electron SO terms in S-T TP. Formaldehyde at CNDO level.
McGlynn et al. (1969)	SO effects in triplet-state spectroscopy. Book.
Tterlikkis et al. (1969)	Knight shifts and nuclear spin-spin coupling tensors in a Dirac OPW basis.
Walker and Richards (1970)	SO of diatomics at HF level.
Barron and Buckingham (1973)	SO-allowed M1 and E2 transitions of molecules.
Bendazzoli and Palmieri (1974)	SO of CO_2^+, NNN, H_2CO.
Abegg (1975)	SO of diatomics.
Turro and Devaquet (1975)	S-T transitions in pericyclic organic reactions.
Masmanidis et al. (1975)	SO coupling in organic molecules. CNDO/S.
Teo and Lee (1979)	Relativistic effects in the EXAFS of W.
Lefebvre-Brion et al. (1985)	SO autoionization effects in HI.
Barra et al. (1986,1988)	PNC effects on NMR chemical shifts.
Dodelson (1986)	Ortho-para H_2 transition rate 10^{-12} a^{-1} in vacuum.
Rösch and Pyykkö (1986)	Time-reversal symmetry aspects in the relativistic theory of nuclear spin-spin coupling.
W.Q.Chen and Cook (1987)	Quasirelativistic treatment of nuclear motion in atoms and molecules.
Dagdigian and Campbell (1987)	Spin-orbit effects in gas-phase chemical reactions.
Pyper (1988)	Relativistic analogues of the NR hfs operators.
Z. C. Zhang and Pyper (1988)	Relativistic analogues of NR hfs operators. Assumes that nuclear magnetization resides on nuclear surface. Individual contributions diverge, their sum not.

Table 7.13. (continued).

Rutkowski and Schwarz (1990)	PT appproach to molecular structures, force constants or polarisabilities.
Rehr et al. (1991)	Theoretical XAFS standards for K-shell Cu, Pt, $GeCl_4$.
Avery and Antonsen (1992)	Magnetic interactions and 4-currents.
Heifets (1992)	3rd-order PT formulae for g-tensors.

8. SOLID-STATE THEORY

The chemical bonds in solids are not unrelated to those in molecules. Therefore we give below a summary of relativistic solid-state calculations on electronic and geometrical structures in Table 8.1. Table 8.2. gives an idea of relativistic theories of various solid-state phenomena.

Table 8.1. Band-structure calculations.

Reference	Comments
Loucks (1966)	R/NR Fermi surfaces for W. Differences from Mo mainly relativistic.
Koelling and Freeman (1976)	Review on An, An intermetallics.
Overhof et al. (1976)	Band-structure study of the optical gap in expanded fluid mercury. Closes at 5.5 and 9.3 g/cm^3 for sc and fcc, respectively. R KKR.
Elyashar and Koelling (1977)	Nb.
Overhof et al. (1978)	RbAu, CsAu. R KKR. "Significant charge transfer to gold found". PES interpreted.
Weinberger (1982b)	Au, Pt and $d^n s^1$ impurities in them. SO-split DOS given for the impurity.
Koelling et al. (1983)	CeO_2, PrO_2. Metal d and f character found in the O2p band.
Dakshinamoorthy and Iyakutti (1984)	Ac Fermi surface, superconductivity and resistivity at high p. RAPW.
Bachelet and Christensen (1985)	GaAs, Ge. Direct gap reduced to half by relativity. Atomic effect depressed <u>or</u> amplified in solids.
Bhattacharyya et al. (1985)	$MoSi_2$, WSi_2. E_{coh} and ν_{opt} larger for W.
Ebert et al. (1985a)	Ag_xPt_{1-x} alloys. SO splitting of the d shells discussed.
Rabe and Joannopoulos (1985)	SnTe, PbTe. Rhombohedral (distorted NaCl) and NaCl structures, respectively, reproduced. R/QR/NR comparisons. NR SnTe a semimetal.
Springelkamp et al. (1985)	ß-NaSn.Contains tin tetrahedra, isoelectronic to P_4: a $(Na^+)_4(Sn_4^{4-})$ semiconductor.
Wei and Krakauer (1985)	Pressure-induced metallization of BaSe and BaTe. QR/NR.
Brooks et al. (1986)	Review on Ac-Am, ThN-AmN.

Table 8.1. (continued).

Christensen et al. (1986)	Pb. NR structure diamond, R : observed fcc. LMTO. SO effects on bcc/hcp/fcc also considered, about ± 1 mRy. The fcc- hcp transition raised by 100 kbar by SO.
Hybertsen and Louie (1986)	Inclusion of V_{SO} in PP work. Atomic splittings are enhanced for highest occupied p states of Ge, Ar, Kr, Xe.
Koelling (1986)	Why does density-functional theory work for excitation energies? A: The orbitals are localized. Ce, U systems.
Kubo and Yamashita (1986)	Pb under high pressure.
Moriarty (1986)	Phonon spectrum of bcc Ba. Partial filling of 5d band essential.
Pyper (1986); Wood and Pyper (1986)	A relativistic ab initio method for ionic solids. Free-ion DF basis. MgO, NaCl, LiF, NaF, AgF, PbF_2. Dispersion forces appended. Good agreement with experiment for D_e, R_e, bonding.
Schadler et al. (1986)	Ce, Pu. Fully relativistic calculation including spin polarization. Zeeman splitting in B_{ext} included.
Watson et al. (1986)	5d-5d metal alloys, isoelectronic with Re: WOs, TaIr, HfPt. E_{coh} increases with Z_2-Z_1. Right structure obtained.
Weinert et al. (1986)	4f core hole energies in Au.
S.C. Wu et al. (1986)	Ag 4d and Au 5d SO splittings only weakly affected in Ag-Au alloys.
Boring et al. (1987)	UMn_2, UFe_2. Magnetic behaviour.
Brooks (1987)	PuE, E = S, Se, Te predicted to be "relativistic [intrinsic] semiconductors" with a filled $5f_{5/2}$ and empty $5f_{7/2}$ band.
Cortona and Sommers (1987)	Spin-polarized calculations on rare earths.
Fritsche et al. (1987)	Ferromagnetic Fe, Ni, Pd.
Goalwin and Kunz (1987)	MI excitonic spectra, M=Li-Cs.
Harrison and Straub (1987)	Discuss bonding mechanisms in d- and f-shell solids. AnAs, An = Th-Cm.
Hashizume (1987)	Solid Ar.
Jaffe and Singh (1987)	Inclusion of h_{SO} in tight-binding calculations.
Johansson et al. (1987, 1990)	An metals and intermetallic compounds.
Kelly and Brooks (1987)	Cohesive properties of AnO_2, An=Th, U, Pu. LMTO.
Kertesz and Guloy (1987)	Stability of Hg^q chains, close-packed layers and honeycomb networks as function of the charge q.
Kubo (1987)	Sr, Yb under normal and high pressure.
Lohr (1987)	A crystal version of REX. Te and Po. Cubic structure of the latter attributed to SO.

Table 8.1. (continued).

Nemoshkalenko et al. (1987)	Pseudopotentials for transition metals.
Norman (1987)	DFT of f electron systems.
Ohlsén and Calais (1987)	Effect of SO found dominant on g-tensor anisotropy in Rh, Pd, Ir and Pt.
Ostanin et al. (1987)	Mo.
Pyper et al. (1987)	Relativistic effects on interionic potentials in PbF_2. The Pb^{2+}-Pb^{2+} interaction is reduced by a factor of 3. The electron-gas method gives too large relativistic corrections by 2-3.
Qi et al. (1987)	Experimental evidence for no 5d holes in Au, $5d_{5/2}$ holes in Pt. For violet $AuAl_2$ and $PtAl_2$, comparable $d_{3/2}$ and $d_{5/2}$ and dominant $d_{3/2}$ holes, respectively.
Schadler (1987)	Fully relativistic magnetic moments. Gamma-Ce, fcc-Pu.
Terakura et al. (1987)	Alloy phase stability of Cu-Ag, Cu-Au and Ag-Au.
Tréglia and Ducastelle (1987)	PtNi. SO induced ordering of the alloy.
Watson et al. (1987)	Au-5d metal alloys.
Weinberger and Gonis (1987)	Actinides.
Zhu et al. (1987)	E_{coh} of Mo, W, $MoSi_2$ and WSi_2. W>Mo but relativistic effects were not separated.
Albers (1988)	Linear Pt-Br chains. LMTO.
Boring et al. (1988)	Np.
Cardona et al. (1988)	Zinc-blende-type semiconductors. SO splitting effects classified. LMTO.
Davenport et al. (1988)	Cu_3Au. Non-overlapping 3d and 5d bands. SO effects discussed.
Ebert (1988); Ebert et al. (1988a)	Spin-polarized relativistic LMTO calculations.
Eppenga et al. (1988)	Semiconductor SO splittings from Augmented Spherical Wave calculations.
Ginatempo and Staunton (1988)	Disordered alloys containing heavy elements. Cu_3Au.
Gonze et al. (1988)	Bi, including SO coupling.
Herbst and Wilkins (1988)	2p and 3d excitation energies in Ac-Am.
Jeong et al. (1988)	Relativistic (SO) effects on the surface electronic structure of Mo(011).
Liu et al. (1988)	WC. Cohesive properties calculated: both W-C bonds and a metallic component.
Mazin and Savrasov (1988)	Xe. Pressure-dependence of the gap.
McAdon and Goddard (1988ab)	1D Cu-Au, Li, Na. Charge or spin density waves, Peiers distortions.

Table 8.1. (continued).

Moriarty (1988)	High-pressure phases of Hg, to 1TPa = 10 Mbar. NR Hg is hcp, like Zn or Cd, R Hg has the observed ß-Hg structure. E_{coh} 1.21, 0.27 and 0.67 eV for NR, R, exp., respectively.
Nizam et al. (1988)	Periodic-HF on ß-PbF_2. QR PP. DOS compared with PES.
Pospelov and Novikov (1988)	SO in As and Sb.
Saalfrank et al. (1988)	Te_∞ chains of trigonal tellurium. Cohesive properties, including interchain interactions.
Weinberger et al. (1988)	Ordered and disordered Cu_3Au. Itinerant Au 5d states in ordered phase.
Yasui (1988)	Orbital and SO magnetic susceptibilities in bcc V, Nb and W.
Anisimov and Korotin (1989)	FePd, FePt, CoPt, MnAl.
Ebert (1989)	Core states in spin-polarized LMTO calculations. Fe, Co, Ni.
Eriksson et al. (1989a)	$AnIr_2$, An = Th-Am. SO-induced volume-increase for Pu is substantial, due to filling of narrow $5f_{5/2}$ band.
Eriksson et al. (1989b)	$AnRh_3$, An = Ac-Cm.
Fernando et al. (1989)	Full-potential linear augmented STO method. Bcc W, fcc Pt.
Florio et al. (1989)	Random Pd-Pt alloys.
Godwal and Jeanloz (1989)	Equation of state of gold up to 200 GPa and 5000 K. LMTO.
Hafner (1989)	Review on solid and liquid sp-bonded systems, including binary alloys and Zintl compounds.
Hamann (1989)	BaSe, as test for generalized norm-conserving pseudopotentials,
Jeong et al. (1989)	Symmetry effects in hydrogen chemisorption electronic structure.
Kim et al. (1989a)	Optical properties of Mo.
Kim et al. (1989b)	Optical properties of Nb.
Krewer and Feder (1989)	Scattering of Dirac electrons from non- spherical potentials. Application on LEED of W(001).
Krutzen and Springelkamp (1989)	Spin-polarized fully relativistic calculations. Omits the orbital contribution. Several LDF used. Ni and Gd.
Liberman and Zangwill (1989)	4p-4f resonance transitions in Ce.
B. Lindgren (1989)	Reviews calculations of magnetic hfs in Fe, its surfaces and compounds. Several LDF used. Sresses the importance of the correct hyperfine integral.

Table 8.1. (continued).

Z.W. Lu et al. (1989a)	La, fcc and bcc. The 5p must be included in valence space. Several LDF tested.
Z.W. Lu et al. (1989b)	HgSe, HgTe. Semimetals, in contrast to the semiconducting Zn, Cd chalcogenides. Cohesive properties reproduced.
Masovic (1989, 1990)	The phase V of CuI. Includes SO.
Nemoshkalenko et al. (1989)	Lu. Energy spectrum and optical properties of single crystals.
Oja et al. (1989)	Indirect nuclear spin-spin coupling in Cu. QR.
Ostanin and Shirokovskii (1989)	Optical absorption of Mo.
Ostanin and Shirokovskii (1989b)	Optical absorption in Nb and Ta.
Richter and Eschrig (1989)	Spin-polarized relativistic LCAO formalism. Ni.
Saalfrank et al. (1989)	Pb (100), (110) and (111) surfaces at HF level. No SO.
Smelyanskii et al. (1989)	U band structure.
Solov'ev and Likhtenshtein (1989)	Linearized approaches to spin-polarised calculations for An.
Solov'ev et al. (1989)	Spin-polarized relativistic calculations for Fe-Ni, Pu, Am.
Sticht (1989)	Band structures for heavy-fermion systems.
Strange et al. (1989)	Relativistic spin-polarized multiple-scattering theory. Ferromagnetic Fe.
Svane (1989)	LMTO and Green-function calculations of ^{119}Sn Mössbauer isomer shifts. Minimum for Sn s character for a tin impurity in the middle of d-transition-metal series. Maximum for sp metals.
Takeuchi et al. (1989)	Equilibrium properties of Ag and Au. R/NR.
Takizawa et al. (1989); Takizawa (1990)	Phase stability of the nine AB alloys, A=Ni-Pt, B = Cu-Au. Relativistic 5d-to-6s population shifts important.
Tamura and Feder (1989)	A layer-KKR method for crystal surfaces. Pt(111).
Tegze and Hafner (1989)	Metallic and semiconducting alkali metal-lead compounds. The composition range $Na_{15}Pb_4$, NaPb, $NaPb_3$ considered. The orbitals of the Pb_4^{4-} Zintl anion are strongly delocalized. Low ionicity.
Turner and Harmon (1989)	HgI_2. Covalent, rather than ionic bonding, Hg charge < +0.1. Directional Hg6s-I5p_z bonds give a small elastic constant in the ab plane.
X.W Wang et al. (1989)	PdH_n, n=1-4.
Watson et al. (1989)	Alloys of Ti-Hf with Tc-Pd or Re-Pt.

Table 8.1. (continued).

Wei and Zunger (1989)	Negative SO bowing in semiconductor alloys.
S.C. Wu et al. (1989)	Relativistically allowed photoemission from Cu {001} interpreted.
J.H. Xu et al. (1989)	The rutile dioxides MO_2, M = Nb, Ru, Ir.
Arola et al. (1990a)	Disordered $Au_{70}Pd_{30}$.
Arola et al. (1990b)	Disordered $Cu_{75}Au_{25}$.
Blaha et al. (1990)	A full-potential, linearized augmented plane wave program.
Elsässer et al. (1990)	Ground-state properties of the 4d and 5d metals Rh-Ag, Ir-Au. NR/QR a_o, B_o and E_{coh} given.
Eriksson et al. (1990)	Cohesive properties of Ln and Ln-like An elements.
Fernando et al. (1990)	Cohesive properties of 5d metals, Hf-Au. Full versus muffin-tin potentials. Relative energies of fcc, bcc and hcp.
Gonze et al. (1990)	As, Sb, Bi.
Hafner and Jank (1990)	Liquid Al-Tl. Relativistic 6s stabilization leads to a pseudogap between 6s and 6p.
Jank and Hafner (1990a)	Liquid Si-Pb. R and NR interatomic potentials compared. Photoemission spectra compared with DOS.
Jank and Hafner (1990b)	Liquid Ca-Ba, Zn-Hg. Incipient d-band effects seen in alkaline earths. Hg pair correlation function clearly requires relativity.
Jank and Hafner (1990c)	Review on polyvalent liquid metals.
K.J. Kim et al. (1990)	$AuGa_2$, $PtGa_2$. Optical properties discussed.
Krutzen and Inglesfield (1990)	Calaverite, $AuTe_2$: both the periodic average structure and the incommensurate one. Te p - Au d hybridisation.
Leiberich et al. (1990)	In NQCC.
MacLaren et al. (1990)	Fe, using several LDF. Some give correct bcc structure. LMTO.
Nemoshkalenko et al. (1990)	Th.
Ostanin and Shirokovskii (1990)	Simplified method for ferromagnets.
Pyper (1990)	Ab initio calculation of interionic potentials in solids. DF ionic functions.
Sakisaka et al. (1990)	SO effects in bulk Fe. Comparison with PES. Experimental zeta parameter 0.110 eV band-structure value 0.045.
Schmid et al. (1990)	Si, Ge, GeSi. Effect of inversion asymmetry on SO splitting.
Sklyadneva and Chulkov (1990)	Hf. Surface and bulk electronic structures.
Smelyansky et al. (1990)	Os Fermi surface.

Table 8.1. (continued).

Söderlind et al. (1990)	Relativistic effects on the thermal expansion of the actinides, Th-Pu. A 5f SO-induced increase. 5f contribution to cohesion separated.
Solov'ev et al. (1990)	α-δ transition of Pu.
Tamura and Feder (1990)	Pd(110) surface.
Walzer (1990,1991ab)	Binary intermetallic compounds classified using radial and energy quantities derived from PP.
Wang and Davenport (1990)	The s-d promotion energy and spin-polarization energy of transition metals.
Wei and Zunger (1990)	Atomic and diatomic hydrogen in fcc Pd.
Weinberger et al. (1990)	Fe in Au.
Yamagami and Hasegawa (1990)	α-U.
Zhong et al. (1990)	Two-step (QR+SO) inclusion of relativistic effects in orthogonalized LCAO calculations with a GTO basis. Ni, Nb and Ce band structure.
Ahuja et al. (1991)	Pt Fermi surface. R/NR.
Aprá et al. (1991)	AgCl. QR-PP-HF binding energy, a_0, bulk modulus and electron density.
Chulkov et al. (1991)	Metal surfaces.
Cortona (1991)	Spin-polarized LD calculations.
Costa Cabral et al. (1991)	Monte Carlo simulation of CsAu melt
Deiseroth et al. (1991)	NaHg. Contains chains of square-planar 6p-bound Hg_4 units. Nearly neutral atoms.
Eriksson et al. (1991)	UNiSn. Magnetic properties.
G.-Y. Guo et al. (1991)	Fe, fcc ground state still predicted. Experimental : bcc.
Gyorffy et al. (1991)	Review on relativistic density functional theory for solids. Mentions magneto-crystalline anisotropy, magneto-optic (Faraday and Kerr) effects, Fano effects, corrections to London moment in superconductivity and new magnetic phase transitions.
Hjelm and Calais (1991, 1992)	Zeeman splitting for Pd and Pt in external **B**.
Hörmandinger et al. (1991)	Spin-orbit effects in $Bi_2Sr_2CaCu_2O_8$.
Kokko (1991)	LMTO calculations on monovalent metals and their alloys. K, Ag-Au, Cu-Au.
Kübler and Harmon (1991)	"Electronic structure in the 1990s": Proceedings of the A. J. Freeman seminar. 37 contributions.
S. Lee (1991)	Structures of heavy main-group elements from an EHT, 2nd-moment argument.
Y. Li et al. (1991)	Noble gases Ne-Xe, and NaCl. QR.

Table 8.1. (continued).

Liu et al. (1991)	High-pressure phases of Pb: fcc -> hcp around 14 GPa, hcp -> bcc around 109 GPa reproduced. Method of Hybertsen and Louie (1986). SO lowers the bcc and raises the hcp wrt fcc, however less than in Christensen et al. (1986).
Lohr (1991)	Uranium borides, carbides UPt_3. REXBAND.
Z.W. Lu et al. (1991)	Random $Ag_{0.5}Pd_{0.5}$ and $Ag_{0.5}Au_{0.5}$ alloys. In the latter, s charge transfer $Ag^{+0.025}Au^{-0.033}$ but d charge transfer $Ag^{-0.015}Au^{+0.012}$ (signs corresponding to ionic charge).
Mansfield and Needs (1991)	Surface energy and stress of Pb (111) and (110) surfaces.
Méndez and Dominguez-Adame (1991)	A numerical method for 1D relativistic band structures. Mathieu potential $V(x) = V_0 \cos 2x$.
Min and Jang (1991)	Effect of SO on the electronic structure of magnetic materials. Fe-Ni, Ce-Gd.
Nikolaev and Ionova (1991)	Th-Am, fcc and bcc.
Ostanin and Shirokovskii (1991)	Simplified method for ferrimagnets and antiferromagnets. Dirac-KKR. Theory only.
Schober et al. (1991)	Inter-conversion coefficients for valence electrons of hcp-Tc and bcc-U.
Solovyev et al. (1991)	Spin-polarized LMTO. Volume-dependent electronic sructure and magnetic moment of Pu.
Takizawa et al. (1991)	CuPd, CuPt alloys. The bcc, respectively $L1_1$ structures reproduced.
Tománek et al. (1991)	H on Pd (001), Pd (110) and bulk Pd. PP used for both Pd and H!
Wang and Chou (1991)	Structural properties of hcp Y.
Watson et al. (1991ab)	s-d promotion and E_{coh} in d transition-metals.
Wills et al. (1991)	High-pressure phases of Ce. Both 4f and 5d contributions found important.
S.C. Wu et al. (1991)	Relativistic effects on the Cu(111) surface electronic structure.
Yamagami and Hasegawa (1991)	Itinerant 4f electrons in CeNi.
Yasui et al. (1991)	Orbital and SO magnetic susceptibilities in fcc Rh, Ir, Ni-Pt and Cu-Au.
Agrawal and Agrawal (1992)	AlAs and CdTe. Elastic properties and PES.
Alouani et al. (1992)	$Pt_2Br_6(NH_3)_6$. Dimerisation of the halogen-bridged linear chains into Pt(II) and Pt(IV).
Blaha and Schwarz (1992)	Solid Cl_2, Br_2 and I_2. Geometry optimised. NQCC.

Table 8.1. (continued).

Chang and James (1992)	HgI_2.
Cortona (1992)	Cohesive properties of the alkali halides, MX. R/NR for RbBr, CsX, MI. Of the CsI dissociation energy (to atoms), 7 % is "relativistic".
Ebert and Akai (1992)	Spin-polarized relativistic treatment of dilute and concentrated, disordered $Fe_x\,Co_{1-x}$ and $Co_x\,Pt_{1-x}$.
Eriksson and Wills (1992)	Am. High-pressure 5f delocalization.
Eriksson et al. (1992a)	Th. High-pressure fcc-bct transition.
Eriksson et al. (1992b)	Ce crystal structures under pressure. The α' phase stabilised by 4f, the bct also by 5d electrons.
Ettema et al. (1992)	SnS. α and ß phases.
Felser-Wenz et al. (1992)	Ba, BaO, BaS and BaF_2. Ba 5d population larger than 6s or 6p one.
García et al. (1992)	Gradient-corrected functionals give no consistent total-energy improvement over LDF for Ge, GaAs, Nb and Pd. R/NR for Ge, Nb, Pd.
Gordienko et al. (1992)	Silver azide, AgN_3.
G.Y. Guo et al. (1991b)	Magnetocrystalline anisotropy of a Fe monolayer on Cu, Ag or Au.
Hasegawa and Yamagami (1992)	PuE, E = S, Se, Te.
Katsnel'son et al. (1992)	α-δ transition of Pu as a Mott transition in the 5f system.
Körling and Hägglund (1992)	Cohesive and electronic properties of 3d, 4d and 5d metals. LSD and gradient-corrected potentials compared.
Markowsky and Podgorny (1992)	Optical properties of MTe, M = Zn-Hg.
Mattheiss (1992)	MSi_2, M = Cr-W. Structures.
Methfessel et al. (1992)	Surface energies of 4d metals, Y-Ag. NR/QR.
Neumann et al. (1992)	KMO_3, M = Nb, Ta.
Orita et al. (1992)	I_2 under pressure. Mechanism of insulator-to-metal transition.
Orlando et al. (1992)	Tetragonal and cubic ZrO_2 using a PP version of the crystal HF program.
Rao et al. (1992)	Th : a 5f metal at ultrahigh pressures. The 6p band must be in valence space.
Schick et al. (1992)	Fully relativistic spin-polarized technique for isolated impurities in crystals. Applications on U, Np, Pu in Pd.
Sigalas et al. (1992)	Band structure and total energy of fcc and bcc 3d, 4d and 5d metals. Relativistic effects worsen agreement with experiment (due to cancellation of errors).

Table 8.1. (continued).

Takeda and Shiraishi (1992)	$(SnH_2)_n$ polystannane.
Tamura (1992)	Dirac-Green functions for general, magnetically polarized cell potentials.
Tegze and Hafner (1992)	A_3B, A = Li-Cs, B = Sb, Bi. Small ionic gap. The ionic ASb compounds contain Sb spirals. The ABi, ABi_2 are metallic.
Trubitsin et al. (1992)	Relativistic, spin-polarized treatment of antiferromagnetic Cr.
Walzer (1992ab)	Pseudopotential approach for binary intermetallic compounds. Structures, melting points.
X.-D. Wang et al. (1992)	Relativistic MS formalism for space-filling potentials.
Wills and Eriksson (1992)	Th, Pa, U. Strong 5f effects on the structure. QR approximation increases the volume by about 5%. This SO effect was attributed to 6p.
Wills et al. (1992)	Elastic constants of $_{55}Cs$ - $_{80}Hg$.

Table 8.2. Relativistic theories of solid- and liquid-state phenomena.

Reference	Comments
Mott (1949)	Interprets earlier x-ray spectra of Pt as evidence of 5d-band SO in the metal. The holes near top are $5d_{5/2}$, can receive electrons from $2p_{3/2}$ ($\Delta j=+1$) but not from $2p_{1/2}$. Hence L_3 seen, L_2 not. See van der Laan and Thöle (1988).
Elliott (1954)	Spin-orbit coupling in band theory.
Jortner (1962)	Halide-ion SO splitting in KI or CsI and in solution.
Ruvalds and Falicov (1968)	SO hybridisation in ferromagnetic Ni.
Tterlikkis et al. (1969)	Knight shifts and indirect nuclear spin-spin coupling tensors in Rb and Cs. Dirac OPW.
Shiba (1976)	SO effects, due to the surface, on the Knight shift and susceptibility of normal and superconducting small particles. Without SO, no shift for even number of electrons.
Sone (1977)	SO effects on spin susceptibility of small particles.
Feder et al. (1983)	Relativistic multiple scattering of electrons by ferromagnets.
Ebert et al. (1985b)	Relativistic theory of nuclear spin-lattice relaxation in metals. Ag_xPt_{1-x}.
Anholt and Meyerhof (1986)	Relativistic heavy-ion collisions with solid targets.
Szpunar (1986)	Method for obtaining orbital moments in complex crystal structures. SO. Fe, Co, Ni.
van der Laan et al. (1986)	Experimental identification of SO-split states in a solid. Ce materials.
Blügel et al. (1987)	3d and 4d impurities in Ni. Quasirelativistic analogs to the M1 hfs terms derived.
Gaylord and Kevan (1987)	SO-induced surface resonance on W(011).
Kaveh and Mott (1987)	SO effects in Si:Sb induce a low-temperature metallic phase, separated from the high-temperature phase.
Lee (1987)	Relativistic corrections to density of states and conductivity of s-dimensional semiconductors. Electron-gas model.

Table 8.2. (continued).

Lefebvre et al. (1987)	Scheme for possible stereochemical rules of Sb $5s^2$ lone pairs: i) Low-lying inactive, ii) Hybridizes with chalcogen ligand p-AOs, active. iii) High-lying inactive, partially depleted.
Misra and Callaway (1987)	Relativistic theory of neutron form factors. The Dirac expression reduced to spin, orbital and spin-orbit terms.
Pankratov (1987)	Dirac theory for semiconductors with heterogeneous structures.
Pilkuhn and Soerensen (1987)	Dirac equation for electron channeling.
Sacchetti (1987)	Relativistic and solid-state effects on x-ray scattering.
Thole and van der Laan (1987,1988)	Systematics between valence band SO and x-ray absorption branching ratio.
Abramovich and Nemtsov (1988)	Multiple-wave scattering and transition radiation of relativistic electrons in crystals.
Band et al. (1988)	Atomic mechanisms for mixed-valence states in Ln, An.
Bychkov (1988)	Energy spectra for 2D systems with SO interaction in an inclined magnetic field.
K.C. Das et al. (1988)	Spin and SO contributions to magnetic susceptibility of divalent hcp metals.
Ebert and Gyorffy (1988)	Scattering solutions of the Dirac equation for nonspherical targets.
Ebert et al. (1988b)	Circularly polarized x-ray absorption by ferromagnetic Fe.
Ebert et al. (1988c)	Relativistic effects on magnetic moments and hyperfine fields of Fe, Co and Ni.
Jansen (1988ab, 1989)	Magnetic anisotropy in transition metals using density-functional theory. Both 1- electron SO and h_{Br}-derived terms considered.
Rashba and Sherman (1988)	SO splitting in 2D quantum wells, due to k^3 terms in the bulk dispersion law. Estimate for GaAs.
Shilkova and Shirokovskii (1988a)	Dipole transition matrix elements in crystals. Outer parts of Wigner-Seitz cell reduced to muffin-tin sphere surface by Green's theorem for the Dirac operator.
Shilkova and Shirokovskii (1988b)	Electron velocity in crystals. Vanadium as example.

Table 8.2. (continued).

van der Laan and Thole (1988); van der Laan (1989)	Core-valence x-ray absorption branching ratio related to valence-state SO. A "local probe" for metal compounds and alloys. Basic idea: absence of a $2p_{1/2}$ edge indicates absence of $5d_{3/2}$ holes in the valence band, transitions to $5d_{5/2}$ being forbidden.
Ando (1989)	Spin-orbit effects on electron localization in two-dimensional square lattices.
Diener and Grafenstein (1989)	Relativistic Kohn-Sham formalism and the microscopic stress tensor.
Ebert et al. (1989)	Relativistic description of hyperfine interaction in magnetic systems.
Feng (1989)	Conductance fluctuations in disordered metals with SO scattering and Zeeman splitting.
Ginatempo et al. (1989)	Relativistic theory of photoemission from crystalline metals and alloys. Cu, Ag, Au, ordered and disordered Cu_3Au.
Reizer (1989ab)	Effects of relativistic electron-electron interaction on superconductivity, specific heat etc.
Strange et al. (1989b)	A first-principles theory of magnetocrystalline anisotropy in metals. Fcc Ni.
Tripathi et al. (1989, (1990)	Effect of electron-phonon interaction on the Knight shift. Spin, orbital and SO-contributions.
Caldas et al. (1990)	Anion antisite defects As_{Ga} and Sb_{Ga} in GaAs.
Connerade and Karnatak (1990)	Centrifugal-barrier double-well atomic potentials and valence changes in solids.
Ferrari and Russo (1990)	Zitterbewegung-like discussion of two-band systems.
Krewer et al. (1990)	Spin-polarized electron diffraction from Pb(110).
Matsumoto et al. (1990, 1991)	Formalism for paramagnetic spin susceptibility of metals. SO-induced anisotropy. Applications on Sc, Co, Y.
Mishra et al. (1990)	The Knight shift in simple metals (Li-Cs, Be, Mg, Zn, Al, Ga) separated into spin, orbital and SO contributions, using the theory of Tripathi et al. (1982).
Bates et al. (1991)	Vibronic reduction factors for second-order spin-orbit coupling.
Brouder and Hikam (1991)	Multiple-scattering theory of magnetic x-ray circular dichroism. SO and spin-polarization included. Application on Fe.

Table 8.2. (continued).

Coffey et al. (1991)	The Dzyaloshinskii-Moriya interaction in cuprates (SO-induced anisotropic exchange interaction between electron spins).
K.C. Das et al. (1991)	Total magnetic susceptibility in metals.
Ebert et al. (1991b)	Theory of magnetic x-ray dischroism.
Ebert et al. (1991a)	Magnetic moments and hyperfine fields of 5d-impurity atoms in ferromagnetic Fe.
Gay and Richter (1991)	SO-induced orbital angular momentum in monolayers.
Ginatempo and Gyorffy (1991)	Spin-polarized photoemission from non-magnetic metals.
Hota and Tripathi (1991a)	Theory of spin, orbital and spin-orbit Knight shifts in PbTe.
Hota and Tripathi (1991b)	g-factors of $Pb_{1-x}Sn_xTe$.
Jablonski (1991)	Elastic electron backscattering from gold. A NR TF potential compared to a R DS one.
Jagannathan (1991)	Multiple SO scattering and anisotropic spin interactions in metals.
Jiang et al. (1991)	The superconducting $\langle I^2 \rangle$ parameter for hcp transition metals Sc...Os.
Kulakowski and Lacroix (1991)	SO contribution to the magnetic susceptibility in degenerate narrow bands.
Lander (1991)	Spin and orbital magnetic moments in An compounds. $V_{CF} \approx V_{SO}$. (CF=crystal field).
H.-S.Li et al. (1991)	Intra-atomic exchange integrals between 4f, 5d and 6s in Ln^{3+}. Magnetic Ln-Ln and Ln-M(3d) couplings.
Lindhard (1991)	Radiation from ultrarelativistic electrons moving parallel to a crystal axis.
Lipnitskii et al. (1991)	The Pb(001) surface.
Macedo and Chalker (1991)	SO effects in disordered conductors. Quaternions used.
Movaghar and Cochrane (1991)	Connection between Hall coefficient and magnetic susceptibility. SO.
Mustre de Leon et al. (1991)	Theory of XAFS (x-ray absorption fine structure). Applications on Cu, Ge, Pt, Br_2, $GeCl_4$.
Rozing et al. (1991)	Two-photon momentum distribution for positron annihilation. W.
Shachal and Manassen (1991, 1992)	SO-induced scanning tunneling microscopy observation of individual paramagnetic spins.
Sinkovic et al. (1991)	Diffraction of spin-polarized photoelectrons in magnetic materials.
Staunton et al (1991)	Theory of magnetocrystalline anisotropy.

Table 8.2. (continued).

Stern and Entin-Wohlman (1991)	Interplay of SO and Zeeman leads to damped conductivity oscillations as function of B_{ext}.
Strange et al. (1991)	Dichroic x-ray fluorescence. Fully relativistic multiple-scattering theory for Fe.
Ando and Tamura (1992)	Conductance fluctuations in quantum wires with SO scattering.
Blügel (1992)	Ferromagnetism of 5d-metal monolayers, e.g. Ir on Ag or Au, or Os on Ag.
Bonesteel et al. (1992)	Role of SO coupling in anisotropic superexchange (Dzyaloshinskii-Moriya interaction). La_2CuO_4.
Bouchaud and Sornette (1992)	SO effects on magnetoconductance.
Chao and Chuang (1992)	SO effects on strained semiconductor quantum wells.
Chatterjee et al. (1992)	Tight binding model within a fibre bundle formalism.
Ebert (1992)	Review on (SO-induced) relativistic effects in magnetic solids.
Ebert et al. (1992)	Spin-polarized relativistic treatment of disordered alloys. Fe_xCo_{1-x}, Co_xPt_{1-x}.
Edelstein (1992)	SO contribution to the Hall effect.
Hafner and Jank (1992)	Liquid As, Sb and Bi. Relativistic effects lead to s-p splitting in the DOS. Structure factors S(q) calculated.
Hernandez and Sanquer (1992)	SO decreases magnetoconductance.
Hörmandinger and Weinberger (1992)	Spin polarized relativistic approach to magnetocrystalline anisotropy energy of Fe.
Jagannathan (1992)	Nature of middle-range spin-spin interactions in metallic alloys. Anisotropic SO-induced terms included.
Kurdak et al. (1992)	SO effects on quantum interference effects in quasi-1D wires and rings.
Oreg and Entin-Wohlman (1992)	SO scattering in low-D disordered mesoscopic systems.
Papanikolaou et al. (1992)	SO scattering cross sections for 4sp and 5sp impurities in a Mg host. For Cu and Ag impurities, d contributions found.
Romanov (1992)	SO splitting in cylindrical quantum wires.
Sahnoune et al. (1992)	SO scattering effects on magnetoresistance. Scales between Au and Ag in Ca-Al as Z^6, instead of theoretical Z^8.
Shen et al. (1992a)	Ground-state wavefunction for a gas of spin-($\frac{1}{2}$) anyons in a **B**.
Shen et al. (1992b)	Second virial coefficient for a 2D spin-($\frac{1}{2}$) anyon gas.

Table 8.2.(continued).

Smith et al. (1992)	MCD of L_2 and L_3 edges of Fe and Ni.
Strange et al. (1992)	Relativistic theory of x-ray fluorescence from magnets.
Tan and Thouless (1992)	Total energy for fermions an a 2D lattice in a **B**.
Xia and Chang (1992)	Magnetic coupling in Gd_mY_n superlattices.
Zeyada (1992)	Effect of SO coupling on diamagnetic susceptibility of thin Cu films.

9. RELATIVISTIC EFFECTS AND HEAVY-ELEMENT CHEMISTRY

Perhaps the most dramatic impact of relativity on chemical thought is the insight that the chemical differences between row 5 (Z=41-54) and row 6 (Z=73-86) contain large, if not dominant, relativistic contributions. The further development of this story is outlined in Table 9.1.

As to particular properties, references on bond lengths and other ground-state properties are summarized in Table 9.2. The magnetic-resonance parameters are discussed in Table 9.3. and the molecular electric dipole and higher multipole moments, charge distributions and polarizabilities in Table 9.4. Data on molecular energy levels and energy-transfer are summarized in Table 9.5. and those on molecular ionization potentials and electron affinities in Table 9.6.

In principle, only the papers comparing the corresponding relativistic and non-relativistic calculations are included in this chapter. For the other relativistic molecular calculations, see Ch. 7.

Table 9.1. "Relativity and the periodic system": Periodic trends, reviews and pedagogical papers.

Reference	Comments
Bohr and Coster (1923)	X-ray spectra and the Periodic System. A screened Sommerfeld relativistic term introduced.
Sugiura and Urey (1926)	Comment on FS. While the lanthanides were known on row 6, the actinides were not known on row 7 in 1926. This was explained by lack of occupied 5f states, first predicted to occur at Z=95.
Wu and Goudsmit (1933)	Contrary to Sugiura and Urey (1926), the (NR) 5f one-electron states in a TF potential become occupied already at Z=92, 93, suggesting a new f-element series.
Goeppert Mayer (1941)	(NR) solution in a TF potential predicts occupied 5f states around Z=91-93. Explains similarity of $_{93}$Np to $_{92}$U.
Fricke and Greiner (1969)	Atomic ground states for Z=160-170. E164 a $(8p_{1/2})^2$ noble gas, E170 a $8p^6 9s^2$ noble gas.
Dehmer (1973)	Atomic SO splitting arises in the K-shell region. Z-dependence analysed.
Keller Jr. (1976)	Predicted properties for superheavy elements.
Mann (1976)	Review on superheavy elements E103-E172.
Desclaux and Pyykkö (1980b)	Popular review on relativistic effects in chemistry.
Ellis (1985); Ellis and Holland (1986)	Reviews on molecular cluster compound calculations.
Fazzio et al. (1985)	Discuss the trends for Cu, Ag and Au impurities in Si.
Rabe and Joannopoulos (1985)	SnTe is rhombohedral (distorted NaCl), PbTe is 'NaCl'.
Romanovskaya (1986)	History of the QM interpretation of periodicity. Relativistic effects discussed on pp. 80-85, 95-107.
Watson et al. (1986)	ZrPd bonds less strongly than HfPt.
Balasubramanian and Pitzer 1987)	Review on relativistic quantum chemistry.
Kotochigova and Tupizin (1987)	MCDF R/NR studies of the 4f-shell collapse between Ba and Ce.
Kutzelnigg (1987)	Reviews methodology for relativistic molecular work.
Lohr (1987)	Te and Po. Cubic structure of the latter attributed to large SO. "REXBAND".
Norman (1987)	DFT of f electron systems.
Weinberger and Gonis (1987)	Review on actinides.

Table 9.1. (continued).

Wulfsberg (1987)	Inorganic chemistry textbook. Ch.11.7. "Relativistic effects on orbitals".
S. B. Zhang et al. (1987)	"Nodal radii" for s, p and d orbitals of Li-La, Hf-Rn. Relativistic contraction up to 10.4 and 5.9% for s and p, respectively. No "gold maximum".
Zhu et al. (1987)	E_{coh} for Mo and W 7.8 and 9.7 eV (exp. 6.82 and 8.90, respectively). No NR reference calculation.
Ziegler (1987)	Review on periodic trends.
Dai and Ma (1988b)	Reviews on relativistic quantum chemistry.
Ermler et al. (1988a)	Review on relativistic effects in atoms and molecules. Especially SO effects and PP methods.
Langhoff and Bauschlicher (1988b)	Review on transition-metal hydrides, halides, oxides, dimers and trimers.
Malli (1988)	Review on relativistic quantum chemistry.
Marian et al. (1988)	Bonding in diatomic ThO analysed.
Miyoshi et al. (1988)	The EA of WF_6 1.52 eV smaller than that of MoF_6.
Moriarty (1988)	NR Hg is hcp like Zn and Cd, R Hg is ß-Hg. See Table 8.1.
Pyykkö (1988a)	Relativistic effects in structural chemistry.
Zel'dovich et al. (1988)	Magnetic or nuclear-spin effects in chemistry.
Balasubramanian (1989c)	Review on relativity and chemical bonding.
Balasubramanian (1989e)	Review on heavy p-block diatomic hydrides, halides and chalconides.
Balasubramanian (1989f)	Coinage metal clusters.
Burns and Bursten (1989)	Review on f-element organometallics.
Dolg et al. (1989b)	Ln compounds.
Glebov et al. (1989)	Element 104 ground state $6d7s^27p$. The $6d^27s^2$ lies 0.5 eV higher.
Kratz et al. (1989)	Chemical properties of E105. The anionic halide complexes rather resemble those of Nb and Pa than Ta.
Z.W.Lu et al. (1989b)	HgSe, HgTe are semimetals but the Cd chalcogenides semiconductors.
Mackay and Mackay (1989)	Introduction to inorganic chemistry. Ch. 16.10.: Relativistic effects.
Malli (1989)	Review on 5d and 5f elements.
Novaro(1989)	Activation of small molecules by transition-metal atoms: a review.
Schwarz et al. (1989)	Analyse the radial distribution of direct and indirect relativistic effects. Direct ones come from the K-shell region. Indirect ones arise from all shells and influence all shells. The "gold maximum" of 6s contraction understood.

Table 9.1. (continued.)

Schwerdtfeger (1989)	AuX_2^- and AuX_4^- complexes compared. Larger stability of Au(III) attributed to relativity.
Sen et al. (1989b)	A "Z-transition state" concept for LDF energy differences of isoelectronic atoms and molecules.
Takeuchi et al. (1989)	Equilibrium properties of silver and gold. R/NR.
Takizawa et al. (1989)	Phase stability of AB, A=Ni-Pt, B=Cu-Au. The absence of Pt-based alloys, like AgPt, related to more s, less d (larger d hole) and hence larger Pt-Pt cohesion. NR AgPt would exist. Au is a counterexample with large E_{coh} and many alloys.
Ziegler et al. (1989bc)	Reviews on periodic trends, especially among nd and 5f metal compounds.
Balasubramanian (1990b)	Review on heavy p-block dimers and trimers.
Baerends et al. (1990)	Suggest the use of different radii, r and r_{FW}, for discussing relativistic orbital contraction.
Elsässer et al. (1990)	Ground-state properties of 4d and 5d metals.
Eriksson et al. (1990)	Cohesive properties of Ln and Ln-like An. Lr found to be the first 6d metal.
Hoffman (1990)	Chemistry of the transactinides.
E.Johnson et al. (1990)	IP_n and radii of Ti-Hf and E104. MCDF.
Massey (1990)	Inorganic chemistry text. Relativistic effects quoted on pp. 58, 61, 158, 180, 206.
Milman et al (1990)	Trends amond d transition metal atoms discussed using 4-component PP DFT.
Ohanessian and Goddard (1990)	Review on bonding in MH^+, M=3d, 4d or 5d metal.
Schwarz (1990)	Relativistic effects in chemistry.
Zhuikov et al. (1990)	E104 atomic ground state s^2pd.
Bonacic-Koutecky et al. (1991)	A review on small group 1, 2 and 11 clusters.
Bursten and Strittmatter (1991)	A review on bonding in An-cp complexes.
Ellis et al. (1991)	A review on TM clusters and their interaction with H and O.
Jolly (1991)	Inorganic chemistry text. Relativistic effects on pp. 22-23.
Koga and Morokuma (1991)	Homogeneous catalysis by transition-metal complexes.
S.Lee (1991)	Structures of heavy main-group elements. EHT-level band-structure 2nd-moment model.
Liu et al. (1991)	Role of SO in the fcc->hcp->bcc high-pressure transition. Smaller than in Christensen et al. (1986).

Table 9.1. (continued).

Norrby (1991)	A pedagogical review on relativity and chemistry.
Pepper and Bursten (1991)	Review on theoretical actinide chemistry: hydrides, halides, oxides and organometallics.
Pyykkö (1991a)	Relativistic effects on periodic trends.
Sabirov and Bagatur'yants (1991)	Review bonding of f-element molecules.
Schwarz et al. (1991)	A nonsingular PT analysis of relativistic changes of molecular structure.
Smith and Carter (1991)	Atomic Pd likes CO but Pt likes NO.
Takizawa et al. (1991)	The different structures of CuPd and CuPt reproduced.
Balakrishnan et al. (1992)	D_0 of H_2, H_2^+. Latest experiment, supports theory.
Bauschlicher and Langhoff (1992)	Review on 3d and 4d TM mono- and dihydrides.
Hafner and Jank (1992)	Structure factors and DOS of liquid Sb and Bi.
Hay (1992)	Oxidative addition and reductive elimination of hydrogen and alkanes.
Ionova and Gerasimova (1992)	Review on actinyl chemistry.
Ionova et al. (1992)	Redox potentials of M^{n+}(aq) ions, M = V-105, 103-105, estimated from MCDF.
Kaupp and v.R. Schleyer (1992c)	Why are organic Pb(IV) and inorganic Pb(II) the more stable oxidation states?
Koga and Morokuma (1992)	Review on hydrogen transfer from transition metal hydrides.
Mamaev et al. (1992)	Reviews calculations on CH_4 activation by group 10 metals. Claims tunneling dynamics relevant.
Neumann et al. (1992)	KMO_3, M=Nb, Ta. The latter has a larger gap and larger E_{coh}.
Ryzhkov et al. (1992)	Is E104 a member of group 4?
Schmidbaur et al. (1992)	Review on the "aurophilic", Au(I) ... Au(I) attraction.
Schädel (1992)	Chemical properties of elements 103-105.
Tegze and Hafner (1992)	ASb (A=Na-Cs) are ionic, contain Sb spiral chains. ABi (A=Li, Na) are metallic. Difference attributed to p AO size.
Tyson et al. (1992)	Experimental evidence for Au 5d loss, 6s gain and Ag 4d gain, 5s loss in Au-Ag alloys.

Table 9.2. Bond lengths. Many of these papers also give data on k_2 and D_e. C is the relativistic bond-length contraction, $C = R_{NR} - R_R$.

Reference	Comments
Huzinaga et al. (1984)	I_2, XeF_2 (see volume 1).
Chong et al. (1986)	CuH, ZnH.
March (1986)	Bond lengths of CH_4 to PbH_4 in R TF theory.
Petterson et al. (1987)	C for MH^+, M=Sc-Cu.
Sakai and Miyoshi (1987)	WF_6^{n-} C/pm 2.7, 2.5 and 1.6 for n=0, 1 and 2, respectively. QR PP HF.
Schwarz (1987)	Analysis of relativistic effects on bond lengths.
Schwerdtfeger et al. (1987)	C for HX^{n+}, X=Br, I; n=-1, 0, +1.
Allen et al. (1988)	UO_2 bond length **expansion**, C=-7.5 pm.
Langhoff et al. (1988)	ScF, C=0.1 pm ScCl, 0.16 pm ScBr, 0.74 pm (1st. order PT).
Pyykkö (1988a)	Table IV gives existing values of C until early 1988.
Jansen and Hess (1989c)	AuH X $^1\Sigma$ C=20 pm.
Marian et al. (1989)	NiH X $^2\Delta$; C=1.5 pm.
Moullet et al. (1989)	Cs_2, Cs_2^+. C=16 and 21 pm, respectively.
Schwerdtfeger et al. (1989a)	C for Tl-H and some Tl-C bonds.
Schwerdtfeger et al. (1989b)	AuL, L=H, F-I, Li, Na, S. Ions of Au_2, AuH. Largest C (109 pm) for AuH^{+2}, smallest (16 pm) for AuF.
Schwerdtfeger et al. (1989c)	PbH^+, PbH_2, PbH_4. C=4.7, 3.7, and 7.0 pm.
Strömberg et al. (1989)	CdH^+ HF C=5 pm, HgH^+ 30, 31 and 31 pm for HF, CAS and CCI, respectively. MCl_2 HF C 4 and 14 pm for M=Cd, Hg.
Tschinke (1989)	W_2 C=4 pm.
Dolg et al. (1990a)	GdO. Rel. expansion, C=-1.6 pm.
Dolg et al. (1990b)	EuO. Small expansion.
Katsuki (1990)	AgH. C=4.2 pm.
Kellö and Sadlej (1990a)	HX, X=Br, I. C=0.26 and 0.90 pm, respectively. Also frequencies.
Lelj and Rosa (1990)	CuIn, AgIn, AuIn C=2, 11 and 32 pm, respectively. P-HFS.
Schwerdtfeger (1990)	$AuMe_2^-$ - $PbMe_2^{2+}$. C=14.8, 14.4, 10.5 and 1.6 pm, respectively. Effects on k_2 and D_e also discussed.
Schwerdtfeger et al. (1990c)	AuL, AuL_2^-, L=H, F-I, CH_3, CN, SCN. $Au(PH_3)_2^+$. C vary from 20 to 30 pm.
Strömberg and Wahlgren (1990)	At PT level, Au_2 C is 22 and 20 pm for HF and CPF, respectively. For Hg_2^{2+}, 30 pm for both.

Table 9.2. (continued).

Dolg et al. (1991d)	MH_n, M=Hg-At. RnF_n.
Dolg et al. (1991a)	CeO. QR/NR expansion 4 pm.
Dyall et al. (1991b)	MH_4, M=C-Pb. DF-LCAO.
Faegri and Bauschlicher (1991)	ScNi, YPd. CASSCF C= 1 and 9 pm, respectively. SCF 5 and 8 pm.
Lee and Lee (1991)	SO effects on R_e for HX, XX' and PbH_4. C_{SO}=-0.3 pm for HI, -1.8 pm for I_2.
Schwerdtfeger (1991a)	TlH C using seven levels of correlation. HF 2.9, QCI 2.2 pm.
Schwerdtfeger (1991b)	Au_2 C=25 pm.
Schwerdtfeger and Dolg (1991)	AuLa, AuLu. C=41 and 36 pm respectively.
van Wezenbeek et al. (1991)	UO_2, UO_2^{2+} -7.4 and -2.6 pm, respectively. Bond **expansion** due to $<h_m>$ of the (diagonal and off-diagonal) 6p hole.
Åkeby et al. (1992)	Sc_2 $X^5\Sigma_u$. Sign(C) depends on approximation.
Barandiarán and Seijo (1992)	HF C for As_2, Sb_2, MnO and CuO 0.3, 3.2, 0.2 and 1.9 pm, respectively.
Dai and Balasubramanian (1992e)	OsH_2. SO effects on the bond angle up to 10°.
Dyall (1992a)	XH_2 C = 0.004, 0.30, 1.10 and 4.44 pm for X = Si, Ge, Sn and Pb, XH_4 C = 0.7, 2.1, 7.3 pm for X = Ge-Pb, respectively. Bond angle effects on XH_2 and vibrational frequencies also studied.
Gropen et al. (1992)	PtH two lowest states. PT C for $^2\Sigma$ and $^2\Delta$ levels.
Häberlen and Rösch (1992)	AuH, AuCl, Au_2.
Ishikawa et al. (1992a)	AuH^+. SCF-level C 101 pm!
Kirchner et al. (1992)	PtO C for various states.
Lee and Lee (1992)	SO effect on ground-state properties of HX, XX' (X=Br, I): C_{SO} < 0.
Malli (1992)	ThPt C 89 pm.
Matsuoka (1992b)	HX, HX^+, H = I, At.
Rutkowski et al. (1992)	H_2^+ using double-PT. Effects on R, k, D_e.
Schwerdtfeger et al. (1992b)	R/NR R, k, D_e for AuL, AuL_2^-, AuL_3, AuL_4^-, Au_2L_6 (L = H, F-I). AuF_6^-.
Schwerdtfeger et al. (1992c)	TlX, TlX_3, X = H, F-I. AuH, HgH_2. PbX_2, PbX_4, BiX, BiX_3, BiX_5, X=H, F.
Visser et al. (1992a)	MH_4, M = C-Pb. Effect of H_{Br} studied.

Table 9.3. Magnetic resonance parameters.

Reference	Comments
Tterlikkis et al. (1969)	Knight shifts and spin-spin coupling tensors in Rb and Cs metals.
Shiba (1976)	SO effects on small-particle Knight shifts.
Ebert et al. (1985b)	Relativistic theory of nuclear spin-lattice relaxation in metals.
Barra et al. (1986,1988)	PNC effects on NMR chemical shifts estimated by REX.
Larsson and Pyykkö (1986)	Assign for the first time the An NQCC in actinyl ions, $OAnO^{n+}$, to the hole in the 6pσ semicore orbital. REX.
Rösch and Pyykkö (1986)	Time-reversal aspects of nuclear spin-spin coupling. Applications on heavy-element fluorides.
Wilson et al. (1986)	Spin-spin coupling in Zintl anions using REX.
Edlund et al. (1987)	A heavy-atom shift of the heavy atom ("HAHA") contribution identified for ^{207}Pb NMR chemical shift in a PbH_3^- model using REX.
Ohlsén and Calais (1987)	Effect of SO dominant in g-tensor anisotropy of platinum-group metals.
Pyykkö et al. (1987)	The NMR 'heavy-atom chemical shift' in HX ascribed to SO-induced spin-density in π and Zeeman effects between σ and σ^*. The perpendicular shift, $\sigma_\perp$ dominant. REX.
Winkler et al. (1987)	Mössbauer isomer shifts of ^{119}Sn in halides, hydride and $SnMe_4$.
Berksoy and Whitehead (1988b)	$SnCl_2Ph_2$. Sn and Cl NQCC.
Blügel et al. (1988)	3d and 4d impurities in nickel. Derive the quasirelativistic analogues to the M1 hyperfine contact, orbital and dipolar terms.
Jokisaari et al. (1988)	NMR parameters for CX and XCX', X=O-Te by REX.
Kutzelnigg (1988)	Origin and meaning of the Fermi-contact interaction.
Pyper (1988)	Relativistic analogues of the NR hfs operators.
Angstl (1989)	Effects of mass-velocity corrections on molecular g-tensors.
Bowmaker et al. (1989)	X_2^n NQCC, n=-1,0,+1, X=Cl-I.
Oja et al. (1989)	Indirect nuclear spin-spin coupling calculated for metallic copper. Seen as magnetic ordering at 58 nK. QR.
Svane (1989)	^{119}Sn Mössbauer isomer shifts for tin impurities in d- and sp-metals and tin compounds.

Table 9.3. (continued).

Kellö and Sadlej (1990b)	q_X in HX, X=Cl-I. Nuclear Q obtained.
Mishra et al. (1990)	Knight shift in simple metals.
Nakatsuji et al. (1990)	^{95}Mo NMR chemical shifts.
Schwerdtfeger et al. (1990c)	$AuCl_2^-$ ^{35}Cl NQCC increased by 85%.
Winkler and Vetter (1990ab)	Mössbauer IS and NQCC in $SnX_n(CH_3)_{4-n}$, X = F-I.
Aucar and Contreras (1991)	MNDO-RPA calculations of nuclear spin-spin coupling constants for group 14 (C-Pb) hydrides, halides and methyl compounds.
Hota and Tripathi (1991a)	Theory of spin, orbital and spin-orbit Knight shifts of ^{207}Pb in PbTe.
Hota and Tripathi (1991b)	g-factors of $Pb_{1-x}Sn_xTe$.
Ishii et al. (1991)	Free-radical g-values by FPT.
Jové et al. (1991), Pyykkö and Jové (1991)	Mössbauer NQCC in neptunyl compounds interpreted by REX. The 6pσ semicore hole important.
Schwerdtfeger et al. (1991b)	CuF q_{Cu} decreased 19%.
Blaha and Schwarz (1992)	NQCC of solid Cl_2, Br_2 and I_2. Intermolecular effects on ^{n}.
Heifets (1992)	3rd-order PT formulae for g-tensors.
Schwerdtfeger et al. (1992b)	MS Xα NQCC for AuCl, $AuCl_2^-$, AuX_4^-, X = F-I.

Table 9.4. Electric dipole moments, Q_1, higher multipole moments, Q_L, and molecular charge distributions.

Reference	Comments
Werner and Buckingham (1986)	GeH dipole moment.
Chapman et al. (1987a)	SO decreases the HI dipole moment at R_e by 3.6%.
Lee and McLean (1987)	AgH, AuH dipole moments decreased by relativity.
Schwerdtfeger (1987)	TlH Q_1, Q_2.
Chapman et al. (1988b)	Dipole moments of MH, M=Ge-Pb. μ(PbH)/D 0.937 and 0.637 for X $^2\pi_{1/2}$ and $^2\pi_{3/2}$, respectively.
Laaksonen et al. (1988)	HCl dipole moment increased by 0.015 D. Cp. Kellö and Sadlej (1990a).
Sundholm (1988)	Fully numerical DS R/NR Q_1, Q_2 for H_2^+, H_2, LiH, Li_2, BH, CH^+.
Marian et al. (1989)	NiH dipole moment decreased -0.3D.
Schwerdtfeger et al. (1989b)	Q_1 of AuL, L = H, Li, Na, F-I, O, S. R/NR.
Schwerdtfeger et al. (1989c)	Q_1 of PbH, PbH_2. R/NR.
Baeck and Lee (1990)	Relativistic effects on Q_1 of BH^+-FH^+ by DF-LCAO.
Kellö and Sadlej (1990a)	Dipole moments of HX, X=Cl-I and of AgH, decreased by relativity. Effects on dipole polarizability also studied.
Sumathi and Balasubramanian (1990)	Signs of H_2Te and H_2Po dipole moments opposite.
Dolg (1991d)	Q_1 of MH, M=Hg-At. NR/QR.
Kellö and Sadlej (1991)	Molecular Q_2 of MH, M=Cu-Au.
Sadlej (1991)	AuH. Relativity decreases Q_1 from 1.21 to 0.62 a.u. Polarizability and Q_2 also studied.
Sadlej and Urban (1991)	AgH. Dipole moment.
Schwerdtfeger and Dolg (1991)	AuLa. Opposite Q_1! Relativistic changes about +1.04 and +0.79 D, respectively.
Dyall (1992a)	XH_2 dipole moments, X = Si-Pb. $(\delta Q_1/D)/Z^2 = 0.00010$.
Häberlen and Rösch (1992)	AuH, AuCl Q_1 decreased by relativity.
Kellö and Sadlej (1992a)	XCN Q_1, Q_2 and α, X = F-I. QR CI. R/NR.
Kellö and Sadlej (1992b)	PbH_4, BiH_3, PoH_2, AtH Q_1 and α_1.
Sadlej (1992a)	Q_1 and α_1 of SbH_3, H_2Te, HI. SCF-level R/NR + NR-MBPT.
Sadlej (1992b)	Q_1 and α_{zz} of XX', X = F-I.

Table 9.5. Molecular energy levels and energy transfer. Chemical reactions.

Reference	Comments
Lohr (1972)	Spin-forbidden electronic excitations in transition metal complexes.
Salem and Rowland (1972)	Intersystem crossing in diradical reactions discussed.
Shaik and Epiotis (1980)	Spin inversion in triplet Diels-Alder reactions discussed.
Z. Wu et al. (1985)	Spin-rotation coupling in MX (M=K-Cs, X=Ar-Xe) is mainly due to noble-gas SO.
Fisher and Michl (1987)	External and internal heavy-atom effects on spin-forbidden proton tunneling in 1,3-perinaphtadiyl.
Zel'dovich et al. (1988)	Magnetic or nuclear-spin effects in chemistry.
Yarkony (1990)	The decomposition reaction HN_3 (S=0) -> N_2 (S=0) + NH (S=0,1). Landau-Zener treatment at the singlet-triplet crossing.

Table 9.6. Molecular ionization potentials and electron affinities.

Reference	Comments
Sakai and Miyoshi (1987)	First and second EA of MF_6, M = Cr-W. Relativity decreases both for WF_6 by 1.1 and 1.2 eV, respectively; the electrons go mainly into a 5d AO on W.
Rubio et al. (1989)	Spherical DS models for the IP_n of Cu_N and Ag_N clusters, N≤25.
Boldyrev and Simons (1992)	TeF_8^{2-} electron detachment energy 5 eV.

Corrections to Volume I

P. 13, Table 2.2., entry for Dirac (1928a), reads $R = B^{-1/2}P$, should read $R = r^{-1}B^{1/2}P$.

P. 24, Table 2.4.: Lapidus (1983) treats a "hydrogen atom" with a 1D δ-function potential.

P. 25, Table 2.4.: Spector and Lee (1985) treat a 1D KG hydrogenic atom

P. 43, Table 4.5.: Band et al. (1984) optimised the Slater α for DS.

P. 75, Table 5.4.: Add Dzuba et al. (1983a) and (1983b) for the IP of Cs and Fr, respectively.

P. 117, Table 7.3.: Add Pyykkö and Desclaux (1977b) for $(114)H_4$.

P. 126, Table 7.6.: Bursten et al. (1985): $An(C_5H_5)_4$.

P. 134, Table 7.7.: Andzelm et al. (1984): reads 'diminish' should read 'increase'.

P. 141, Table 7.8.: The reference Louwen et al. (1984b) is the present Louwen et al. (1986).

Reference 828 (Dirac 1979) has been published as abridged version in Europhys. News 8(1) (1977) 1-4.

Reference 1167 (Goldman and Drake 1984): p. **L**197.

Reference 1278: The authors are: Gurchumeliya, A.D., Ivanova, E.P., Tsirekidze, M.A. and Tsirekidze, T.A. (1985b).

Reference 1591 (Kahn 1984): Volume 25.

Reference 1910 (Lyudchik and Borkovskii 1984): The journal is Dokl. Akad. Nauk **Belor.** SSR.

Reference 2038 (Mendoza and Zeippen 1982): $3p^2$.

Reference 2073 (Migdalek and Baylis 1984a): Page **L**459.

Reference 2593 (Scharf et al. 1985): Page **447**.

Some comments on notations and terminology

In the Bibliography, Smith (1975) and Smith (1975a) are synonymous. The references are in alphabetic order according to:
1) The first author (including initials), 2) The second author, 3) The third author, etc., 4) For each of these, the year, 5) For the same author(s) and year, the letters a, b, c, ...

We have adopted the IUPAC notations 1 to 18 for the groups in the Periodic System (1 for H and Li, 4 for Ti, 12 for Zn, 14 for C, 18 for Ne). The elements Li - F are called "2nd-row" ones, Na - Ar "3rd-row" ones etc.

No apologies are offered for using for ions both the notation Cu^+ and Cu II in a work on the borderline of chemistry and physics.

List of acronyms and symbols.

A	Magnetic vector potential.
$\underline{A}$	Hyperfine coupling tensor.
$\boldsymbol{\alpha}$	Dirac matrix.
α	Fine structure constant (1/137.035 989 5(61), 1986).
AE	All-electron.
AL	Average level.
α_L	Electric polarizability of multipolarity L.
An	An actinide (Ac - Lr).
AO	Atomic orbital.
APW	Augmented plane-wave method.
B	Magnetic field.
β	Dirac matrix.
BP	Breit-Pauli.
c	Speed of light.
CASSCF	Complete active space SCF.
CC	Coupled cluster.
CCSD	Coupled cluster method with single and double substitutions.
CEPA	Coupled electron-pair approximation.
CFT	Crystal field theory.
CG	Cowan-Griffin (approximation).
CI	Configuration interaction.
CN	Coordination number.
CNDO	Complete neglect of differential overlap.
CP	Core polarization.
Cp	Cyclopentadienyl, C_5H_5.
Cp^*	$C_5(CH_3)_5$.
CT	Charge transfer.
D_e	Dissociation energy.
DF	Dirac-Fock. (<u>Not</u> "density-functional").
DFT	Density functional theory.
DOS	Density of states.
DS	Dirac-Slater.
DVM	Discrete Variational Method.
DZ	Double-zeta (radial function).
E	Electric field.
EA	Electron affinity.
EAL	Extended average level.
ECP	Effective core potential.
EHT	Extended Hückel theory.
EM	Electromagnetic.
ESR	Electron spin resonance.
EXAFS	Extended x-ray absorption fine structure.
FAKE	Fast, Accurate Kinetic Energy method.
FEM	Finite-element method.
FFT	Fast Fourier transform.
FPT	Finite perturbation theory.
FS	Fine structure.
FW	Foldy-Wouthuysen.
g	The electron g-factor.
g	The electron g-tensor (in ESR).
γ_L	Electric screening constant of multipolarity L.
GTO	Gaussian type orbital.
GVB	Generalized valence bond.

h_{Br}	Breit Hamiltonian.
h_D	Dirac Hamiltonian (see eq.(2.4)).
h_d	Darwin term at the NR limit.
h_m	The mass-velocity term.
h_p	Pauli Hamiltonian.
HF	Hartree-Fock.
HFR	Hartree-Fock with relativistic correction terms.
Hfs	hyperfine structure.
HFS	Hartree-Fock-Slater.
HOMO	Highest occupied MO.
IEHT	Iterative EHT
INDO	Intermediate neglect of differential overlap.
IP	Ionization potential.
IS	Isomer shift.
χ_L	Magnetic susceptibility of multipolarity L.
KG	Klein-Gordon.
KKR	Korringa-Kohn-Rostocker band-structure method.
L	Multipolarity; orbital angular momentum.
Λ	Electronic orbital angular momentum around the molecular axis for linear molecules.
LCAO	Linear combination of atomic orbitals.
LDA	Local density approximation.
LDF	Local density functional.
LEED	Low-energy electron diffraction.
Ln	A lanthanide (La - Lu).
LMTO	Linear Muffin-Tin Orbital method.
LSD	Local spin density.
LUMO	Lowest unoccupied MO.
MBPT	Many-body perturbation theory.
MCD	Magnetic circular dichroism.
MCDF	Multiconfiguration Dirac-Fock.
MCHF	Multiconfiguration Hartree-Fock.
Me	Methyl (CH_3-).
MINDO	Modified INDO.
MNDO	Modified NDO.
MO	Molecular orbital.
MP(n)	Møller-Plesset PT (of order n).
MRD-CI	Multireference CI.
MRCI	Multireference CI.
MS	Multiple Scattering.
MWH	Mulliken-Wolfsberg-Helmholtz (= EHT).
N	Number of electrons.
NMR	Nuclear magnetic resonance.
NQCC	Nuclear quadrupole coupling constant.
NR	Non-relativistic.
OCE	One-centre expansion.
ODC	Optimised Double-Configuration method.
ω_e	Vibrational frequency.
P	Parity inversion.
PES	Photoelectron spectrum, -spectroscopy.
PF	Pair function.
PK	Phillips-Kleinman.
PM3	"Parametric model 3"
PNC	Parity non-conservation.
PP	Pseudopotential.
PT	Perturbation theory.
QCI	Quadratic CI.
QED	Quantum electrodynamics.
QM	Quantum mechanics.
QMC	Quantum Monte-Carlo.

QR	Quasirelativistic ("one-component").
ρ	Electron density.
RASSCF	Restricted active space SCF.
R_e	Bond length.
REX	Relativistic Extended Hückel.
RPA	Random phase approximation.
S	Overlap matrix.
SCF	Self-consistent field.
σ_L	Magnetic screening constant of multipolarity L.
SO	Spin-orbit.
STO	Slater-type orbital.
SZ	Single-zeta (basis).
T	Time inversion.
TD	Time-dependent.
T_e	Electronic excitation energy.
TF	Thomas-Fermi.
TFDW	Thomas-Fermi-Dirac-Weizsäcker.
TM	Transition metal.
TP	Transition probability.
TS	Transition state.
V	A potential.
VB	Valence bond.
n-VE	n-Valence-electron (/atom) PP.
WKB	Wentzel-Kramers-Brillouin ("semiclassical") approximation.
X	Electronic ground state of a molecule.
$X\alpha$	Slater's $X\alpha$ (= HFS) method.
Z	Nuclear charge.

BIBLIOGRAPHY

3120 Aaen Andersen, L.U., Childs, W.J., Nielsen, U. and Poulsen, O. (1989). Phys. Rev. A, 40, 6281-6285.
Hyperfine structures in ^{167}Er II : Measurements of high $4f^{12}5d$ levels and ab initio multiconfiguration Dirac-Fock calculations

3121 Aashamar, K. and Luke, T.M. (1990). J. Phys. B, 23, L733-L738.
An efficient method for computing relativistic mixing in complex spectra: application to sextet-quartet mixing in Cr II

3122 Abe, S. and Fujita, T. (1987). Nucl. Phys., A475, 657-662.
A simple analytic solution of the Dirac equation with a scalar linear potential

3123*Abegg, P.W. (1975). Mol. Phys., 30, 579-596.
Ab initio calculation of spin-orbit coupling constants for gaussian lobe and gaussian-type wave functions

3124 Abounadi, J. and Darkhosh, T. (1987). J. Math. Phys., 28, 1505-1507.
Dirac wave equation in the presence of an external field

3125 Abramovich, B.S. and Nemtsov, B.E. (1988). Izv. Vyssh. Uchebn. Zaved., Radiofiz., 31, 1477-1481.
Effect of multiple wave scattering on transition radiation of relativistic electrons in crystal (in Russian)

3126 Adkins, G.S. (1989).
Application of the bound state formalism to positronium In "Relativistic, Quantum Electrodynamic and Weak Interaction Effects in Atoms", ed. W. Johnson, P. Mohr and J. Sucher, AIP Conf. Proc., Vol. 189, pp. 65-92.

3127 Adkins, G.S., Bui, M.H.T. and Zhu, D. (1988). Phys. Rev. A, 37, 4071-4078.
New calculation of the three-photon-annihilation contribution to the positronium hyperfine interval

3128 Adler, R.J. and Martin, R.A. (1992). Am. J. Phys., 60, 837-839.
The electron g factor and factorization of the Pauli equation

3129 Aerts, P.J.C. (1986).
Towards Relativistic Quantum Chemistry, Thesis, Groningen.

3130 Aerts, P.J.C. and Nieuwpoort, W.C. (1986). Chem. Phys. Lett., 125, 83-90.
On the optimization of Gaussian basis sets for Hartree-Fock-Dirac calculations

3131 Agaskar, P.A., Cotton, F.A., Dunbar, K.R., Falvello, L.R., Tetrick, S.M. and Walton, R.A. (1986). J. Am. Chem. Soc.,

108, 4850-4855.
The multiply bonded octahalodiosmate(III) anions.
2. Structure and bonding

3132 Aggarwal, K.M., Berrington, K.A., Eissner, W.B. and Norrington, P.H. (1991). J. Phys. B: At. Mol. Opt. Phys., 24, 2201-2211.
Collision strengths for electron impact excitation with Ca XV: a relativistic approach

3133 Agrawal, B.K. and Agrawal, S. (1992). Phys. Rev. B, 45, 8321-8327.
Ab initio calculation of the electronic, structural, and dynamical properties of AlAs and CdTe

3134 Ahuja, R., Solanki, A.K. and Auluck, S. (1991). Phys. Stat. Sol. B, 168, 509-518.
Fermi surface properties of platinum

3135 Aissing, G. (1991). Phys. Rev. A, 44, R2765-R2768.
Monte Carlo variance minimization for the Dirac equation

3136 Aissing, G., Bakalov, D.D. and Monkhorst, H.J. (1990). Phys. Rev. A, 42, 116-126.
Relativistic and quantum eletrodynamics corrections to the binding energies of the (J=1,v=1) states of ddμ and dtμ

3137 Aissing, G. and Monkhorst, H.J. (1990). Phys. Rev. A, 42, 3789-3794.
Relativistic corrections to binding energies of muonic molecules

3138*Aitchison, I.J.R. (1972).
Relativistic Quantum Mechanics
Macmillan, London, 260p.

3139 Åkeby, H., Pettersson, L.G.M. and Siegbahn, P.E.M. (1992). J. Chem. Phys., 97, 1850-1857.
Core correlation and the binding energy of Sc_2

3140 Akrivos, P.D. and Aldridge, L.P. (1988). Inorg. Chim. Acta, 145, 63-69.
A comparative MO study of three dichlorodiamminedihydroxo Pt(IV) isomers

3141 Aksela, H. (1989).
Auger electron spectroscopy of free atoms: theoretical analysis
In "X-ray Spectroscopy in Atomic and Solid State Physics" ed. J.G. Ferreira and M.T. Ramos, NATO ASI, 187, Plenum, New York.

3142 Aksela, H. and Aksela, S. (1987). J. Physique, 48, C9-565-577.
Recent developments in Auger spectroscopy of free atoms

3143 Aksela, H., Aksela, S., Ala-Korpela, M., Sairanen, O-P., Hotokka, M., Bancroft, G.M., Tan, K.H. and Tulkki, J. (1990a). Phys. Rev. A, 41, 6000-6005.
Decay channels of core-excited HCl

3144 Aksela, H., Aksela, S., Bancroft, G.M., Tan, K.H. and Pulkkinen, H. (1986a). Phys. Rev. A, 33, 3867-3875.
$N_{4,5}OO$ resonance Auger spectra of Xe studied with selective excitation by synchrotron radiation

3145 Aksela, H., Aksela, S., Hotokka, M., Yagishita, A. and Shigemasa, E. (1992c). J. Phys. B: At. Mol. Opt. Phys., 25, 3357-3370.
Decay channels of core-excited and ionized HCl studied by time-of-flight mass spectroscopy

3146 Aksela, H., Aksela, S. and Lakanen, R. (1990b). Phys. Rev. A, 42, 1791-1794.
Coster-Kronig decay width of Rb

3147 Aksela, H., Aksela, S., Lakanen, R., Tulkki, J. and Åberg, T. (1990c). Phys. Rev. A, 42, 5193-5200.
Initial- and final-state correlation in the valence-shell Auger spectrum of Rb

3148 Aksela, H., Aksela, S., Mäntykenttä, A., Tulkki, J., Shigemasa, E., Yagishita, A. and Furusawa, Y. (1992a). Phys. Scr., T41, 113-118.
Decay channels of resonantly excited $2p^{-1}ns$, md states of Ar and $3d^{-1}np$ states of Kr

3149 Aksela, H., Aksela, S., Pekkala, T. and Wallenius, M. (1987). Phys. Rev. A, 35, 1522-1527.
Electronic structure of free nickel atoms studied by the $L_{2,3}MM$ Auger decay

3150 Aksela, H., Aksela, S., Pulkkinen, H., Bancroft, G.M. and Tan, K.H. (1986b). Phys. Rev. A, 33, 3876-3884.
Resonance Auger decay after selective excitation of Kr 3d states by synchrotron radiation

3151 Aksela, H., Aksela, S., Pulkkinen, H., Bancroft, G.M. and Tan, K.H. (1988a). Phys. Rev. A, 37, 1798-1801.
Anomalously strong shake-up processes in Auger decay of the resonantly excited $2p^5 3s^2 3p^6 nl$ states of Ar

3152 Aksela, H., Aksela, S., Pulkkinen, H., Kivimäki, A. and Sairanen, O.-P. (1990d). Phys. Scr., 41, 425-428.
Shake processes in Auger decay of resonantly excited states of rare gases

3153 Aksela, H., Aksela, S., Pulkkinen, H. and Yagishita, A. (1989a). Phys. Rev. A, 40, 6275-6280.
Shake processes in the Auger decay of resonantly excited $3d^9 4s^2 4p^6 np$ states of Kr

3154 Aksela, H., Aksela, S., Tulkki, J., Åberg, T., Bancroft, G.M. and Tan, K.H. (1989b).
Phys. Rev. A, 39, 3401-3405.
Auger emission from the resonantly excited $1s^1 2s^2 2p^6 3p$ state of Ne

3155 Aksela, H., Bancroft, G.M., and Olsson, B. (1992b).
Phys. Rev. A, 46, 1345-1348.

Multiplet-dependent shake-up structures in high-resolution Auger spectra at $3d^{-1}4s^{2}4p^{6}np$, n=5,6 resonances of Kr

3156 Aksela, H., Lakanen, R., Aksela, S., Sairanen, O.-P., Yagishita, A., Meyer, M., Prescher, Th., von Raven, E., Richter, M. and Sonntag, B. (1988b). Phys. Rev. A, 38, 3395-3399.
Resonance Auger spectra of free Rb atoms

3157 Aktas, H., Aktekin, N. and Ozmen, A. (1992). Doga Turkish J. Phys., 16, 425-427.
Computation of spin-orbit correction to g_J-factor of lithium atom

3158 Alan, A.T. and Barut, A.O. (1992). Europhys. Lett., 17, 119-122.
Generalization of the Volkov solution for relativistic particles with anomalous magnetic moments

3159 Albers, R.C. (1989). Synthetic Metals, 29, 169-174.
LMTO electronic band-structure calculations for linear Pt-Br chains

3160*Albeverio, S. (1972). Ann. Phys.(N.Y.), 71, 167-276.
On bound states in the continuum of N-body systems and the virial theorem

3161 Albright, T.A. and Tang, H. (1992). Angew. Chem., 104, 1532-1534.
Die Struktur von Pentamethyltantal

3162 Ali, M.A. (1992). J. Mol. Struct.(Theochem), 261, 175-185.
Fine structure splittings of potassium-like and rubidium-like ions by the multiconfiguration Dirac-Fock method

3163 Ali, M.A. and Kim, Y.-K. (1988). Phys. Rev. A, 38, 3992-3997.
Electric quadrupole and magnetic dipole transition probabilities in the potassium isoelectronic sequence

3164 Ali, M.A. and Kim, Y.-K. (1992). J. Opt. Soc. Am. B, 9, 185-190.
Improved calculation of fine-structure splittings in the ground state of potassiumlike ions

3165 Al-Jaber, S.M., Zhu, X.-S. and Henneberger, W.C. (1991). Eur. J. Phys., 12, 268-270.
Interaction of a moving magnetic dipole with a static electric field

3166 Allen, G.C., Baerends, E.J., Vernooijs, P. Dyke, J.M., Ellis, A.M., Fehér, M. and Morris, A. (1988). J. Chem. Phys., 89, 5363-5372.
High temperature photoelectron spectroscopy: A study of U, UO, and UO_2

3167 Allouche, A.R., Nicolas, G., Barthelat, J.C. and Spiegelmann, F. (1992). J. Chem. Phys., 96, 7646-7655.
Theoretical study of the electronic structure of the BaH molecule

3168 Almlof, J. and Faegri Jr., K. (1986). Theor. Chim. Acta, 69, 437-446.
Relativistic effects on molecular structure: methyl derivatives of the fourth main-group elements

3169 Alouani, M., Albers, R.C., Wills, J.M. and Springborg, M. (1992). Phys. Rev. Lett., 69, 3104-3107.
Ab initio calculation of the dimerization in the halogen-bridged transition-metal linear-chain compound $Pt_2Br_6(NH_3)_4$

3170 Altmann, S.L. (1986).
Rotations, Quaternions, and Double Groups
Clarendon Press, Oxford.

3171 Altun, Z. and Radojevic, V. (1992). J. Phys. B: At. Mol. Opt. Phys., 25, 3757-3764.
Relaxation effects on the 3d photoionization of atomic krypton and the bromine negative ion

3172 Alvarez, S., Mota, F. and Novoa, J. (1987). J. Am. Chem. Soc., 109, 6586-6591.
Structure and stability of Br_4 and Br_4^{2-} and their interaction with cations and transition metals

3173 Alvarez Collado, J.R. (1987). An. Quim., 83(5), 5-8.
Principales efectos relativistas para sistemas moleculares de interes quimico. Aplicacion al estudio de la molecula de FRb

3174 Alvarez Collado, J.R. and Buenker, R.J. (1988). Mol. Phys., 64, 725-738.
Theoretical study of the transition probabilities amond the three lowest electronic states of the SeS molecule

3175 Aly, H.H. and Barut, A.O. (1990). Phys. Lett. A, 145, 299-303.
Methods of solution of the Schrödinger equations for a class of anharmonic and singular potentials

3176 Ambjørn, J., Durhuus, B. and Jonsson, T. (1990). Nucl. Phys., B330, 509-522.
A random walk representation of the Dirac propagator

3177 Ammerlaan, C.A.J. and van Oosten, A.B. (1989). Phys. Scr., T25, 342-347.
Electronic structure of platinum in silicon

3178 Amus'ya, M.Ya. (1987). Comm. At. Mol. Phys., 19, 261-276.
The quadrupole moment of atoms with total spin j=1/2

3179 Amus'ya, M.Ya. (1990). Comm. At. Mol. Phys., 24, 251-268.
Polarizational radiation ('atomic' bremsstrahlung)

3180 Amus'ya, M.Ya. (1992). Nucl. Instr. Meth. Phys. Res., A314, 225-234.
Polarizational radiation or "atomic" bremsstrahlung

3181 Amus'ya, M.Ya., Korol', A.V. and Solov'ev, A.V. (1987a).

Polarizational bremsstrahlung in the collisions of relativistic atoms and relativistic particles with atoms In "Polarizational Bremsstrahlung of Atoms and Particles", Moscow, "Nauka", pp. 226-258.

3182 Amus'ya, M.Ya., Kuchiev, M.Yu. and Solov'ev, A.V. (1987b). Zh. Tekh. Fiz., 57, 820-822.
Bremsstrahlung in the atom-atom collisions at relativistic velocities (in Russian)

3183 Amus'ya, M.Ya., Kuchiev, M.Yu. and Solov'ev, A.V. (1988a). Zh. Eksp. Teor. Fiz., 94, 74-86;
Sov. Phys. JETP, 67, 41-48.
Bremsstrahlung in collisions of atomic particles at relativistic velocities

3184 Amus'ya, M.Ya., Kuchiev, M.Yu. and Yakhontov, V.L. (1988b). Yad. Fiz. 47, 877-878; Sov. J. Nucl. Phys., 47, 556.
Relativistic correction to hyperfine splitting in $(^{3,4}He^{++}-\mu^{-}e^{-})^{0}$ atoms

3185 Amus'ya, M.Ya., Kuchiev, M.Yu. and Yakhontov, V.L. (1988c). Muon Catal. Fusion, 3, 351-356.
High accuracy calculations of the hyperfine splitting in the ground state of the $(^{4}He^{++}\mu^{-}e^{-})^{0}$ and $(^{3}He\mu^{-}e^{-})^{0}$ atoms

3186 Amus'ya, M.Ya. and Solov'ev, A.V. (1990). Zh. Eksp. Teor. Fiz., 97, 745-756; Sov. Phys. JETP, 70, 416-422.
Frequency-angle distribution of radiation in collisions between relativistic atomic particles

3187 Anchell, J., Gutowski, M., Nichols, J. and Simons, J. (1992). Int. J. Quantum Chem., 41, 793-810.
Self-consistent-field potential-energy surfaces for hydrogen atom pairs within small palladium clusters

3188 Anders, I.A. and Kotlyarov, V.P. (1991). Teor. Mat. Fiz., 88, 72-84; Theor. Math. Phys., 88, 725-734.
Characterization of the scattering data of the Schrodinger and Dirac operators

3189 Andersen, R.A., Boncella, J.M., Burns, C.J., Green, J.C., Hohl, D. and Rösch, N. (1986). J. Chem. Soc., Chem. Commun., 405-407.
Electronic structure of bis(cyclopentadienyl) lanthanide compounds: photoelectron spectra and molecular orbital calculations

3190 Anderson, A. (1991). Phys. Rev. A, 43, 4602-4610.
Intertwining of exactly solvable Dirac equations with one-dimensional potentials

3191 Anderson, W.P., Cundari, T.R., Drago, R.S. and Zerner, M.C. (1990). Inorg. Chem., 29, 1-3.
Utility of the semiempirical INDO/1 method for the calculation of the geometries of second-row transition-metal species

3192 Anderson, W.P., Cundari, T.R. and Zerner, M.C. (1991). Int. J. Quantum Chem., 39, 31-45.

An intermediate neglect of differential overlap model for second-row transition metal species

3193 Andersson, K. and Roos, B.O. (1992). Chem. Phys. Lett., 191, 507-514.
Excitation energies in the nickel atom studied with the complete active space SCF method and second-order perturbation theory

3194 Ando, K., Safronova, U.I. and Tolstikhina, I.Y. (1992a). Phys. Scr., 46, 107-114.
Energy levels $1s^2 2lnl'$ (n=2,3,4) of Na VIII-S XIII ions. Comparison of two calculation methods: MCDF and MZ

3195 Ando, K., Safronova, U.I. and Tolstikhina, I.Yu. (1992b). Opt. Spectrosc., 72, 285-289; Opt. Spektrosk., 72, 528-534.
Energies of the $1s^2 2lnl'$ (n=2,3,4) ionic levels. Comparison of two methods of calculation: MCDF and MZ

3196 Ando, T. (1989). Phys. Rev. B, 40, 5325-5339.
Numerical study of symmetry effects on localization in two dimensions

3197 Ando, T. and Tamura, H. (1992). Phys. Rev. B, 46, 2332-2338.
Conductance fluctuations in quantum wires with spin-orbit and boundary-roughness scatter

3198 Andrae, D., Dolg, M., Stoll, H. and Preuss, H. (1992). J. Mol. Spectrosc., 155, 430-432.
Comment on "spectroscopic constants and potential energy curves of OsH" by M. Benavides-Garcia and K. Balasubramanian

3199 Andrae, D., Häussermann, U., Dolg, M., Stoll, H. and Preuss, H. (1990). Theor. Chim. Acta, 77, 123-141.
Energy-adjusted ab initio pseudopotentials for the second and third row transition elements

3200 Andrae, D., Häussermann, U., Dolg, M., Stoll, H. and Preuss, H. (1991). Theor. Chim. Acta, 78, 247-266.
Energy-adjusted ab initio pseudopotentials for the second and third row transition elements: molecular test for M_2 (M=Ag,Au) and MH (M=Ru,Os)

3201 Andreocci, M.V., Bossa, M., Cauletti, C., Stranges, S., Horchler, K. and Wrackmeyer, B. (1992). J. Mol. Struct. (Theochem), 254, 171-176.
Electronic structure of tetrahedral tin acetylides by pseudopotential ab initio calculation and gas-phase UV photoelectron spectroscopy

3202 Andreocci, M.V., Bossa, M., Cauletti, C., Stranges, S., Wrackmeyer, B. and Horchler, K. (1989). Inorg. Chim. Acta, 162, 83-87.
UV photoelectron spectra and pseudopotential ab initio calculations of $(CH_3)_3PbC\,CCH_3$

3203 Andreocci, M.V., Bossa, M., Cauletti, C., Stranges, S., Wrackmeyer, B. and Horchler, K. (1990). Phys. Scr., 41,

800-804.
Electronic structure of alkyl-lead(IV) acetylides by gas-phase UV photoelectron spectroscopy and pseudopotential calculations

3204 Andreocci, M.V., Cauletti, C., Stranges, S., Wrackmeyer, B. and Stader, C. (1991). Z. Naturforsch., 46b, 39-46.
UV photoelectron spectra and pseudopotential "ab initio" calculations of some 4-membered cyclic amides of group XIV elements

3205 Andriessen, J. and Postma, H. (1987). Hyperfine Interactions, 35, 655-657.
Theoretical study of Tl^0 and Pb^+ centers in alkali halide and alkaline earth fluoride type crystals

3206 Andriessen, J., Postma, H., van den Brink, A.M. and Das, T.P. (1992). Phys. Rev. A, 45, 1389-1398.
Many-body multiconfigurational Dirac-Fock calculation of hyperfine interaction in 7s, $7p_{1/2}$, and $7p_{3/2}$ states of Ra II

3207 Andzelm, J., Russo, N. and Salahub, D.R. (1987a). Chem. Phys. Lett., 142, 169-174.
Model potential for As and LCGTO MP LSD calculation of geometry, electronic structure and photoelectron spectra for As_2 and As_4

3208 Andzelm, J., Russo, N. and Salahub, D.R. (1987b). J. Chem. Phys., 87, 6562-6572.
Ground and excited states of group IVA diatomics from local-spin-density calculations:Model potentials for Si, Ge, and Sn

3209 Anghel, N. (1989). Diss. Abstr. Int., 50, 1445-B; Comm. Math. Phys., 128, 77-97(1990).
L^2-index theorems for perturbed Dirac operators

3210 Anghel, N. (1990b). J. Math. Phys., 31, 2091-2093.
The two-dimensional magnetic field problem revisited

3211 Angstl, R. (1989). Chem. Phys. 132, 435-442.
Contribution of the relativistic mass correction to the g tensor of molecules

3212 Anholt, R. (1986).
Relativistic atomic collisions
In "Electronic and Atomic collisions", ed. D.C. Lorents, W. Meyerhof and J.R. Peterson, North-Holland, Amsterdam, pp. 245-255.

3213 Anholt, R. and Becker, U. (1987). Phys. Rev. A, 36, 4628-4636.
Atomic collisions with relativistic heavy ions. IX. Ultrarelativistic collisions

3214 Anholt, R. and Gould, H. (1986). Adv. At. Mol. Phys., 22, 315-386.
Relativistic heavy-ion--atom collisions

3215 Anholt, R. and Meyerhof, W.E. (1986). Phys. Rev. A, 33, 1556-1568.
Atomic collisions with relativistic heavy ions. V. The states of ions in matter

3216 Anholt, R., Meyerhof, W.E., Xu, X.Y., Gould, H., Feinberg, B., McDonald, R.J., Wegner, H.E. and Thieberger, P. (1987). Phys. Rev. A, 36, 1586-1600.
Atomic collisions with relativistic heavy ions. VIII. Charge-state studies of relativistic uranium ions

3217 Anholt, R., Stoller, Ch., Molitoris, J.D., Spooner, D.W., Morenzoni, E., Andriamonje, S.A., Meyerhof, W.E., Bowman, H., Xu, J.-S., Xu, Z.-Z., Rasmussen, J.O. and Hoffmann, D.H.H. (1986). Phys. Rev. A, 33, 2270-2280.
Atomic collisions with relativistic heavy ions. VI. Radiative processes

3218 Anisimov, V.I. and Korotin, M.A. (1989). Fiz. Met. Metalloved., 68, 474-480.
Electron structure and magnetic properties of FePd, FePt, CoPt, and MnAl alloys

3219 Anisimova, G.P. and Semenov, R.I. (1990a). Opt. Spektrosk., 68, 1232-1236; Opt. Spectrosc., 68, 721-724.
Determination of the fine-structure parameters of $np^5n's$ configurations from experimental data

3220 Anisimova, G.P. and Semenov, R.I. (1990b). Opt. Spektrosk., 69, 5-7; Opt. Spectrosc., 69, 2-4.
Calculation of the fine-structure parameters of $nd^9n's$ configurations

3221 Anisimova, G.P. and Semenov, R.I. (1990c). Opt. Spektrosk., 69, 734-737; Opt. Spectrosc., 69, 437-439.
Semiempirical calculation of the fine-structure parameters of ps configurations

3222 Anno, T. (1990). Mem. Fac. Sci., Kyushu Univ., C17(2), 177-188.
Regularities among relativistic effects upon total energies of atoms and ions with reference to the Z-expansion theory

3223 Anno, T. and Teruya, H. (1989). J. Chem. Phys., 91, 4738-4744.
Relativistic effect on total energies for determinatiom of correlation energies of atoms from their experimental total energies

3224 Anno, T. and Teruya, H. (1992). J. Chem. Phys., 97, 2174.
Erratum: Relativistic effect on total energies for determination of correlation energies of atoms from their experimental total energies [J. Chem. Phys. 91, 4738 (1989)]

3225 Anslyn, E.V., Brusich, M.J. and Goddard III, W.A. (1988). Organometallics, 7, 98-105.
Metallacyclobutadiene versus metallatetrahedrane structures for $Cl_3MoC_3H_3$ complexes

3226 Aprà, E., Stefanovich, E., Dovesi, R. and Roetti, C. (1991). Chem. Phys. Lett., 186, 329-335.
An ab initio Hartree-Fock study of silver chloride

3227 Aquilanti, V. and Vecchiocattivi, F. (1989). Chem. Phys. Lett., 156, 109-114.
Transport properties of open-shell systems: fine structure effects on collision integrals for oxygen and fluorine atoms with rare gases

3228 Arbuzov, B.A., Boos, E.E., Savrin, V.I. and Shichanin, S.A. (1989). Pis'ma Zh. Eksp. Teor. Fiz., 50, 236-238; JETP Lett., 50, 262-264.
Relativistic Coulomb problem and bound states in continuum

3229 Arbuzov, B.A., Boos, E.E., Savrin, V.I. and Shichanin, S.A. (1990a). Teor. Mat. Fiz., 83, 175-185.
Relativistic Coulomb quasipotential and new narrow resonances in systems of charged particles

3230 Arbuzov, B.A., Shichanin, S.A., Boos, E.E. and Savrin, V.I. (1990b). Mod. Phys. Lett. A, 5, 1441-1449.
On a possible correspondence of the new narrow resonances in charged particle systems to the relativistic Coulomb levels in continuum

3231 Arbuzov, B.A., Shichanin, S.A. and Savrin, V.I. (1992). Phys. Lett. B, 275, 144-148.
On a mechanism of GSI resonances production

3232 Arik, M. and Mungan, M. (1992). Phys. Lett. B, 282, 101-104.
q-oscillators and relativistic position operators

3233 Armen, G.B., Tulkki, J., Åberg, T. and Crasemann, B. (1987). Phys. Rev. A, 36, 5606-5614.
Quantum theory of post-collision interaction in inner-shell photoionization: Final-state interaction between two continuum electrons

3234*Armstrong, B.H. (1963). Phys. Rev., 130, 2506-2519.
Anomalous solutions to the Dirac and Schrödinger equations for the Coulomb potential

3235 Armstrong, D.R., Davidson, M.G., Moncrieff, D., Stalke, D. and Wright, D.S. (1992). J. Chem. Soc., Chem. Commun., 1413-1415.
The observation of a Pb-Li bond; synthesis, structure and model molecular orbital(MO) calculations on the monomeric Ph_3Pb-Li·(pmdeta) complex [pmdeta=$(Me_2NCH_2CH_2)_2NMe$]

3236 Arodz, H. and Ruijgrok, Th.W. (1988). Acta Phys. Pol. B, 19, 99-140.
On a classical limit of the Dirac equation with an external electromagnetic field. Part 1. Proper-time evolution, Lorentz covariant expectation values and classical equations of motion

3237 Arola, E., Barnes, C.J., Rao, R.S. and Bansil, A. (1990a).

Phys. Rev. B, 42, 8820-8826.
Electronic structure of $Au_{70}Pd_{30}$ disordered alloy

3238 Arola, E., Rao, R.S., Salokatve, A. and Bansil, A. (1990b). Phys. Rev. B, 41, 7361-7369.
Electronic structure of $Cu_{75}Au_{25}$ disordered alloy

3239 Arratia-Perez, R. (1991). J. Phys. Chem., 95, 7239-7244.
Relativistic electronic structure of staggered and eclipsed conformations of octachlorodiosmate(III)

3240 Arratia-Perez, R., Axe, F.U. and Marynick, D.S. (1987). J. Phys. Chem., 91, 5177-5183.
Geometry, bonding, and optical and magnetic properties of $Cu(CO)_3$. A theoretical study

3241 Arratia-Perez, R. and Malli, G.L. (1986a). Chem. Phys. Lett., 125, 143-148.
Bonding in the octahedral Au_6^{2+} cluster

3242 Arratia-Perez, R. and Malli, G.L. (1986b). J. Chem. Phys., 84, 5891-5897.
Dirac scattered-wave calculations for Ag_3^{2+}, Au_3^{q+}, and Au_4^{q+} (q=1,2) clusters

3243 Arratia-Perez, R. and Malli, G.L. (1986c). J. Chem. Phys., 85, 6610-6622.
Dirac scattered-wave study of trigonal bipyramidal silver clusters Ag_5^{q+} (q=0,2-4)

3244 Arratia-Perez, R. and Malli, G.L. (1987). J. Magn. Res., 73, 134-142.
Bonding, optical, and magnetic properties of paramagnetic Ag_4^{1+} and Ag_4^{3+} clusters

3245 Arratia-Perez, R. and Marynick, D.S. (1987). J. Chem. Phys., 87, 4644-4650.
Molecular hyperfine interactions in $Ag(CO)_3$

3246 Arratia-Perez, R. and Marynick, D.S. (1988). Phys. Rev. B., 37, 4893-4899.
Paramagnetic resonance hyperfine structure of hexachloroprotactinate(IV)

3247 Arratia-Perez, R., Ramos, A.F. and Malli, G.L. (1989). Phys. Rev. B, 39, 3005-3009.
Calculated electronic structure of Au_{13} clusters.

3248 Arredondo Ruiz, J.H. (1992). J. Phys. A, 25, 4705-4722.
The Klein-Gordon equation with indefinite form

3249 Arshansky, R. and Horwitz, L.P. (1988). Phys. Lett. A, 131, 222-226.
Covariant phase shift analysis for relativistic potential scattering

3250 Arshansky, R. and Horwitz, L.P. (1989a). J. Math. Phys., 30, 66-80.
The quantum relativistic two-body bound state.
I. The spectrum

3251 Arshansky, R. and Horwitz, L.P. (1989b). J. Math. Phys., 30, 213-218.
Relativistic potential scattering and phase shift analysis

3252 Arshansky, R. and Horwitz, L.P. (1989c). J. Math. Phys., 30, 380-392.
The quantum relativistic two-body bound state. II. The induced representation of SL(2,C)

3253 Arshansky, R., Horwitz, L.P. and Saad, D. (1989). Nucl. Phys. B, Proceedings Supplements, 6, 275-277.
Quantum relativistic two-body bound states with invariant direct action potentials and the electromagnetic interaction

3254 Aspromallis, G. and Nicolaides, C.A. (1986). J. Phys. B: At. Mol. Phys., 19, L13-L18.
Relativistic and Coulomb autoionisation in low-lying excited stated of fluorine

3255 Aspromallis, G., Nicolaides, C.A. and Beck, D.R. (1986). J. Phys. B: At. Mol. Phys., 19, 1713-1717.
Relativistic autoionisation of bound states of negative ions: Be^-

3256 Assefa, Z. (1992). Diss. Abstr. Int., 53, 2303-B-2304-B.
Photoluminescence studies of lanthanide ion complexes of gold and silver dicyanides: A new low dimensional solid state class for non-radiative excited state energy transfer

3257 Assefa, Z., DeStefano, F., Garepapaghi, M.A., LaCasce Jr., J.H., Ouellete, S., Corson, M.R., Nagle, J.K. and Patterson, H.H. (1991). Inorg. Chem., 30, 2868-2876.
Photoluminescence and electronic structure of $Tl[Au(CN)_2]$: Evidence for relativistic effects in thallium-gold and gold-gold interactions

3258 Atag, S. (1989). J. Math. Phys., 30, 696-699.
Large-N iterative solution of the Dirac equation

3259*Atiyah, M. and Singer, I. (1984). Proc. Nat. Acad. Sci. USA, 81, 2597-2590.
Dirac operator coupled to vector potentials

3260 Au, C.K. (1986). Phys. Rev. A, 34, 3568-3579.
Coulomb-gauge electrodynamics analysis of two-photon exchange in eletron-atom scattering

3261 Au, C.K. (1988). Phys. Rev. A, 38, 5571-5577.
Unified analysis of multipole and finite-mass corrections in long-range Coulombic interactions

3262 Au, C.K. (1989a). Phys. Rev. A, 39, 2789-2795.
Effects of transverse photon exchange in helium Rydberg states: Corrections beyond the Coulomb-Breit interaction

3263 Au, C.K. (1989b).
Time-ordered diagrammatic approch to the two-photon exchange potential: an examination of the Casimir

effects in long range EM interactions
In "Relativistic, Quantum Electrodynamic and Weak Interaction Effects in Atoms", ed. W. Johnson, P. Mohr and J. Sucher, AIP Conf. Proc., Vol. 189, 165-195.

3264 Au, C.K., Feinberg, G. and Sucher, J. (1991). Phys. Rev. A, 43, 561-564.
Recoil corrections to Rydberg energy levels in helium

3265 Au, C.K. and Mesa, M.A. (1990). Phys. Rev. A, 41, 2848-2850.
Addendum to "Effects of transverse photon exchange in helium Rydberg states: Corrections beyond the Coulomb-Breit interactions"

3266 Aucar, G.A. and Contreras, R.H. (1991). J. Magn. Res., 93, 413-418.
New method to study spin-spin coupling constants involving heavy nuclei at the MNDO-RPA level with localized orbitals

3267 Audouard, E. and Spiegelmann, F. (1991). J. Chem. Phys., 94, 6102-6123.
Theoretical study of the excited states of the Kr_2^* excimer: Potential curves including ab initio spin-orbit coupling

3268 Avery, J. and Antonsen, F. (1992). J. Mol. Struct. (Theochem), 261, 69-78.
Magnetic interactions and 4-currents

3269 Avgoustoglou, E., Johnson, W.R., Plante, D.R., Sapirstein, J., Sheinerman, S. and Blundell, S.A. (1992). Phys. Rev. A, 46, 5478-5488.
Many-body perturbation-theory formulas for energy levels of excited states of closed-shell atoms

3270 Axe, U. and Marynick, D.S. (1987). Inorg. Chem., 26, 1658-1664.
Electronic structure of the ditelluromercurate(II) Zintl anion. A scattered-wave-Xα study

3271 Axe, U. and Marynick, D.S. (1988). Inorg. Chem., 27, 1426-1431.
Relationship between bonding and structure in tetranuclear heteroatomic Zintl anions containing 20 valence electrons

3272 Aymar, M. (1990). J. Phys. B: At. Mol. Opt. Phys., 23, 2697-2716.
Eigenchannel R-matrix calculation of the J=1 odd-parity spectum of barium

3273 Aymar, M. and Lecomte, J.M. (1989). J. Phys. B: At. Mol. Opt. Phys., 22, 223-245.
R-matrix calculation of photoionisation from Rydberg states in strontium

3274 Ayres, V.M. (1986). Diss. Abstr. Int. B, 46, 3890.
Relativistic bound states of fermions and antifermions

3275 Azuma, Y., Barry, J.A., Lyne, M.P.J., Merer, A.J., Schröder, J.O. and Féménias, J.-L. (1989). J. Chem. Phys.,

91, 1-12.
Spin-orbit distortion of the hyperfine structure in heavier molecules: Breakdown of the case (a_β) formalism in the B $^3\Phi$ - X $^3\Delta$ system of gaseous NbN

3276 Babb, J.F., Habs, D., Spruch, L. and Wolf, A. (1992). Z. Phys. D, 23, 197-209.
Retardation (Casimir) energy shifts for Rydberg helium-like low-Z ions. An exploratory study

3277 Babb, J.F. and Spruch, L. (1987). Phys. Rev. A, 36, 456-466.
Retardation (Casimir) effects on high and not-so-high Rydberg states of helium

3278 Babb, J.F. and Spruch, L. (1988). Phys. Rev. A, 38, 13-18.
Evaluation of retardation energy shifts in a Rydberg helium atom

3279 Babb, J.F. and Spruch, L. (1989). Phys. Rev. A, 40, 2917-2927.
Retardation (Casimir) effect for a multielectron core system and a Rydberg electron

3280*Bachelet, G.B. and Christensen, N.E. (1985). Phys. Rev. B, 31, 879-887.
Relativistic and core-relaxation effects on the energy bands of gallium arsenide and germanium

3281*Bacry, H. and Richard, J.L. (1967). J. Math. Phys., 8, 2230-2233.
Partial group-theoretical treatment for the relativistic hydrogen atom

3282 Badnell, N.R. and Pindzola, M.S. (1991). Phys. Rev. A, 43, 570-571.
Dielectronic recombination of Bi^{80+} and Bi^{82+}

3283 Badnell, N.R. and Pindzola, M.S. (1992). Phys. Rev. A, 45, 2820-2824.
Unified dielectronic and radiative-recombination cross sections for U^{90+}

3284 Baeck, K.K., Lee, M.S. and Lee, Y.S. (1992). Chem. Phys. Lett., 198, 273-278.
All-electron relativistic self-consistent-field calculations for CO and CO^+

3285 Baeck, K.K. and Lee, Y.S. (1988). Chem. Phys. Lett., 147, 367-372.
All-electron relativistic SCF calculations for B and CH

3286 Baeck, K.K. and Lee, Y.S. (1990). J. Chem. Phys., 93, 5775-5782.
Relativistic self-consistent-field calculations of spin-orbit splittings in diatomic hydrides

3287 Baerends, E.J., Schwarz, W.H.E., Schwerdtfeger, P. and Snijders, J.G. (1990). J. Phys. B: At. Mol. Opt. Phys.,

23, 3225-3240.
Relativistic atomic orbital contractions and expansions: magnitudes and explanations

3288 Bagrov, V.G. and Gitman, D.M. (1990).
Exact Solutions of Relativistic Wave Equations
Kluwer Academic, Dordrecht, pp 330.

3289 Bagrov, V.G., Kopytov, G.F., Tlyachev, V.B. and Fedosov, N.I. (1988). Zh. Eksp. Teor. Fiz., 94, 1-8.
Mechanism of polarization of relativistic photoelectrons (in Russian)

3290 Bagrov, V.G. and Obukhov, V.V. (1989). Eesti NSV Teaduste Akadeemia Füüsika Instituudi Uurimused, 65, 137-143.
A nonstandard example in the separation theory for Dirac-Fock equation (in Russian)

3291 Bagrov, V.G., Shapovalov, A.V. and Evseevich, A.A. (1989). Izv. Vyssh. Uchebn. Zaved., Fiz., 32(10), 30-34; Sov. Phys. J., 32, 781-784.
Yang-Mills fields permitted by Abelian complete sets of first-order symmetry operators of the Dirac equation

3292 Bagus, P.S., Pacchioni, G. and Philpott, M.R. (1989).
J. Chem. Phys. 90, 4287-4295.
Bond ionicity of the halogen-silver interaction

3293 Baik, D.H., Ohr, Y.G., Kim, K.S., Lee, J.M., Indelicato, P. and Kim, Y.-K. (1991). At. Data Nucl. Data Tables, 47, 177-203.
Electric dipole, electric quadrupole, and magnetic dipole transition probabilities of Na-like ions, $56 \leq Z \leq 92$

3294 Bakalov, D. (1988). Muon Catal. Fusion, 3, 321-334.
Relativistic effects in mesic molecules

3295 Baker, J., Hill, R.N. and Morgan III, J.D. (1989).
High precision calculation of helium atom energy levels
In "Relativistic, Quantum Electrodynamic and Weak Interaction Effects in Atoms", ed. W. Johnson, P. Mohr and J. Sucher, AIP Conf. Proc., 189, pp. 123-145.

3296 Balakrishnan, A., Smith, V. and Stoicheff, B.P. (1992).
Phys. Rev. Lett., 68, 2149-2152.
Dissociation energy of the hydrogen molecule

3297 Balasubramanian, K. (1986a). J. Phys. Chem., 90, 1043-1045.
Relativistic configuration interaction calculation of low-lying states of BiH^+

3298 Balasubramanian, K. (1986b). J. Mol. Spectrosc., 115, 258-268.
Relativistic quantum calculations of spectroscopic properties of BiH

3299 Balasubramanian, K. (1986c). Chem. Phys. Lett., 127, 324-329.
On the low-lying electronic states of BiF

3300 Balasubramanian, K. (1986d). J. Chem. Phys., 85, 1443-1446.
Electronic states of PbS

3301 Balasubramanian, K. (1986e). J. Chem. Phys., 85, 3401-3406.
Relativistic calculations of electronic states and potential energy surfaces of Sn_3

3302 Balasubramanian, K. (1986f). J. Phys. Chem., 90, 6786-6790.
Electronic states of Ga_2

3303 Balasubramanian, K. (1986g). Chem. Phys. Lett., 127, 585-589.
CAS SCF/CI calculations of low-lying states of SnH_2

3304 Balasubramanian, K. (1987a). Chem. Phys. Lett., 135, 288-293.
Electronic states and potential energy surfaces of ScH_2

3305 Balasubramanian, K. (1987b). J. Chem. Phys., 86, 3410-3413.
Electronic states of GaAs and $GaAs^+$

3306 Balasubramanian, K. (1987c). J. Mol. Spectrosc., 123, 228-236.
Relativistic quantum calculations of spectroscopic properties of Ge_2

3307 Balasubramanian, K. (1987d). Chem. Phys. Lett., 139, 262-266.
Relativistic calculations of electronic states of SnS

3308 Balasubramanian, K. (1987e). J. Phys. Chem., 91, 5166-5169.
Low-lying electronic states of Se_2

3309 Balasubramanian, K. (1987f). J. Chem. Phys., 87, 2800-2805.
CASSCF/CI calculations of electronic states and potential energy surfaces of PtH_2

3310 Balasubramanian, K. (1987g). J. Chem. Phys., 87, 3518-3521.
Electronic structure of $GaAs_2$

3311 Balasubramanian, K. (1987h). J. Chem. Phys., 87, 6573-6578.
Electronic states of Pt_2

3312 Balasubramanian, K. (1988a). Chem. Phys., 119, 41-50.
CAS SCF/CI calculations on electronic states of Br_2 and Br_2^+

3313 Balasubramanian, K. (1988b). J. Mol. Spectrosc., 127, 97-103.
Theoretical study of spectroscopic properties of PbCl

3314 Balasubramanian, K. (1988c). J. Chem. Phys., 89, 5731-

5738.
Relativistic configuration interaction calculations for polyatomics: Applications to PbH_2, SnH_2, and GeH_2

3315 Balasubramanian, K. (1988d). J. Chem. Phys., 89, 6310-6315.
Spectroscopic properties of 41 electronic states of Pd_2

3316 Balasubramanian, K. (1988e). J. Mol. Spectrosc., 132, 280-283.
Spectroscopic properties and potential energy curves of SnCl: Comparison with PbCl

3317 Balasubramanian, K. (1988f). Int. J. Quantum Chem., S22, 465-476.
Relativistic CASSCF/CI calculations: Applications to transition metal dihydrides

3318 Balasubramanian, K. (1989a). J. Chem. Phys., 91, 307-313.
Ten low-lying electronic states of Pd_3

3319 Balasubramanian, K. (1989b). J. Chem. Phys., 91, 2443-2451.
The single-triplet splittings in AsH_2^+, SbH_2^+, and BiH_2^+ and bond energies and ionization potentials of AsH_2

3320 Balasubramanian, K. (1989c). J. Phys. Chem., 93, 6585-6596.
Relativity and chemical bonding

3321 Balasubramanian, K. (1989d). Chem. Phys. Lett., 164, 231-236.
Geometries and bond energies of GaH_n and GaH_n^+ (n=1-3)

3322 Balasubramanian, K. (1989e). Chem. Rev., 89, 1801-1840.
Spectroscopic properties and potential energy curves for heavy p-block diatomic hydrides, halides, and chalconides

3323 Balasubramanian, K. (1989f). J. Mol. Struct., 202, 291-313.
Electronic structure of coinage metal clusters

3324 Balasubramanian, K. (1990a). J. Mol. Spectrosc., 139, 405-423.
Spectroscopic constants and potential energy curves of GaAs, $GaAs^+$, and $GaAs^-$

3325 Balasubramanian, K. (1990b). Chem. Rev., 90, 93-167.
Spectroscopic constants and potential energy curves of heavy p-block dimers and trimers

3326 Balasubramanian, K. (1990c). J. Chem. Phys., 93, 507-519.
Spectroscopic constants and potential energy curves of 47 electronic states of InSb, $InSb^+$, and $InSb^-$

3327 Balasubramanian, K. (1990d). Chem. Phys. Lett., 171, 58-62.
Electronic structure of $(GaAs)_2$

3328 Balasubramanian, K. (1990e). J. Phys. Chem., 94, 6582-6588.
Spectroscopic constants and potential energy curves for InH and InH^+

3329 Balasubramanian, K. (1990f). J. Chem. Phys., 93, 8061-8072.
The low-lying states of the second-row transition metal hydrides (YH-CdH)

3330 Balasubramanian, K. (1990g). J. Chem. Phys., 93, 8403-8404.
Comment on :Is n=4 a magic number for He_n^+?

3331 Balasubramanian, K. (1991a). J. Chem. Phys., 94, 1253-1263.
Potential energy surfaces for the Pt_2+H_2 reaction

3332 Balasubramanian, K., Christiansen, P.A. and Pitzer, K.S. (1991a). Phys. Rev. A, 43, 2581-2582.
Comment on "Relativistic effects in bonding and dipole moments for the diatomic hydrides of the sixth-row heavy elements"

3333 Balasubramanian, K. and Dai, D.-G. (1990). J. chem. Phys., 93, 7243-7255.
Potential energy surfaces for Ir+H_2 and Ir^++H_2 reactions

3334 Balasubramanian, K. and Das, K.K. (1990). J. Phys. Chem., 94, 8379-8380.
Reply to Comments on "Binding energies and potentials of tetramers of Cu, Ag, and Au"

3335 Balasubramanian, K. and Das, K.K. (1991a). J. Mol. Spectrosc., 145, 142-150.
Spectroscopic constants and potential energy curves of HfH

3336 Balasubramanian, K. and Das, K.K. (1991b). J. Chem. Phys., 94, 2923-2927.
Electronic states of Cu_4^+, Ag_4^+, and Au_4^+: Interpretation of the optical spectra of Cu_4^+

3337 Balasubramanian, K. and Das, K.K. (1991c). Chem. Phys. Lett., 186, 577-582.
Excited electronic states of Au_3

3338 Balasubramanian, K., Das, K.K. and Liao, D.W. (1992a). Chem. Phys. Lett., 195, 487-493.
Spectroscopic constants and potential energy curves for Hg_2

3339 Balasubramanian, K. and Feng, P.Y. (1988). Chem. Phys. Lett., 146, 155-161.
The electronic states of Ga_3

3340 Balasubramanian, K. and Feng, P.Y. (1989). Chem. Phys. Lett., 159, 452-458.
The ionization potentials of Ag_n and Au_n and binding energies of Ag_n, Au_n and Au_n^+ (n=1-4)

3341 Balasubramanian, K. and Feng, P.Y. (1990a). J. Chem. Phys., 92, 541-550.
Potential-energy surfaces for Pt_2+H and Pt+H interactions

3342 Balasubramanian, K. and Feng, P.Y. (1990b). J. Phys. Chem., 94, 1536-1544.
Geometries and energy separations of low-lying electronic states of Ag_4 and Cu_4

3343 Balasubramanian, K. and Feng, P.Y. (1991a). J. Chem. Phys., 94, 6664-6669.
Electronic states of Ga_4 and In_4

3344 Balasubramanian, K. and Feng, P.Y. (1992a). J. Chem. Phys., 96, 2881-2888.
Potential energy surfaces for Pd_2+H and Ni_2+H interactions

3345 Balasubramanian, K., Feng, P.Y., Kaufman, J.J., Hariharan, P.C. and Koski, W.S. (1988a). Phys. Rev. A, 37, 3204-3210.
Energy transfer in Ar-Cl^+ collisions

3346 Balasubramanian, K., Feng, P.Y. and Liao, M.Z. (1987e). J. chem. Phys., 87, 3981-3985.
Relativistic calculations of electronic states of PdH

3347 Balasubramanian, K., Feng, P.Y. and Liao, M.Z. (1988b). J. Chem. Phys., 88, 6955-6961.
Electronic states and potential energy surfaces of PdH_2: Comparison with PtH_2

3348 Balasubramanian, K., Feng, P.Y. and Liao, M.Z. (1989). J. Chem. Phys., 91, 3561-3570.
Geometries and energy separations of 14 electronic states of Au_4

3349 Balasubramanian, K., Han, M. and Liao, M.Z. (1987a). J. Chem. Phys., 86, 4979-4981.
Relativistic calculations of electronic states of TeH

3350 Balasubramanian, K., Kaufman, J.J., Hariharan, P.C. and Koski, W.S. (1986). Chem. Phys. Lett., 129, 165-171.
Energy transfer in Br^+-Kr collisions

3351 Balasubramanian, K. and Li, J.-Q. (1988a). J. Mol. Spectrosc., 128, 413-426.
Spectroscopic properties and potential energy surfaces of GeH

3352 Balasubramanian, K. and Li, J.-Q. (1988b). J. Chem. Phys., 88, 4979-4986.
Spectroscopic properties and potential energy surfaces of In_2

3353 Balasubramanian, K. and Li, J.-Q. (1989). J. Mol. Spectrosc., 135, 169-187.
Spectroscopic properties and potential energy curves of Sb_2

3354 Balasubramanian, K. and Li, J.-Q. (1990). J. Phys. Chem.,

94, 4415-4419.
Spectroscopic properties and potential energy curves of MoH and MoH^+

3355 Balasubramanian, K. and Liao, D.-W. (1988a). J. Phys. Chem., 92, 6259-6264.
Electronic states and potential energy surfaces of RhH_2

3356 Balasubramanian, K. and Liao, D.-W. (1988b). J. Chem. Phys., 88, 317-321.
Electronic states and potential energy surfaces of RhH

3357 Balasubramanian, K. and Liao, D.-W. (1989). J. Phys. Chem., 93, 3989-3992.
Spectroscopic properties of low-lying electronic states of Rh_2

3358 Balasubramanian, K. and Liao, D.-W. (1991a). J. Chem. Phys., 94, 5233-5236.
Is Au_6 a circular ring?

3359 Balasubramanian, K. and Liao, D.-W. (1991b). J. Chem. Phys., 95, 3064-3073.
Spectroscopic constants and potential energy curves of Bi_2 and Bi_2^-

3360 Balasubramanian, K. and Liao, M.Z. (1987). J. Chem. Phys., 86, 5587-5590.
CASSCF/CI calculations of low-lying states and potential energy surfaces of Au_3

3361 Balasubramanian, K. and Liao, M.Z. (1988a). J. Phys. Chem., 92, 361-364.
Electronic states and potential energy surfaces of AuH_2

3362 Balasubramanian, K. and Liao, M.Z. (1988b). J. Phys. Chem., 92, 4595-4599.
Electronic states and potential energy surfaces of SeH_2 and SeH_2^+

3363 Balasubramanian, K. and Liao, M.Z. (1988c). Chem. Phys., 127, 313-324.
Electronic states and potential energy surfaces of gold and silver trimers

3364 Balasubramanian, K. and Liao, M.Z. (1989). J. Phys. Chem., 93, 89-94.
Electronic states and potential energy surfaces of AgH_2: comparison with AuH_2

3365 Balasubramanian, K., Liao, M.Z. and Han, M. (1987c). Chem. Phys. Lett., 139, 551-557.
Theoretical investigation of electronic states of SeH and SeH^+

3366 Balasubramanian, K., Liao, M.Z. and Lin, S.H. (1987b). Chem. Phys. Lett., 138, 49-54.
Relativistic configuration interaction calculations for electronic states of $KrHe^+$

3367 Balasubramanian, K. and Ma, Z. (1991). J. Phys. Chem., 95, 9794-9800.
Potential energy surfaces for the insertion of W into H_2 and W^+ into H_2

3368 Balasubramanian, K. and Nannegari, V. (1989). J. Mol. Spectrosc., 138, 482-496.
Spectroscopic constants and potential energy curves for 16 electronic states of AsH

3369 Balasubramanian, K. and Pitzer, K.S. (1987). Adv. Chem. Phys., 67, 287-319.
Relativistic quantum chemistry

3370 Balasubramanian, K. and Ravimohan, Ch. (1987). J. Mol. Spectrosc., 126, 220-230.
Theoretical investigation of spectroscopic properties of Te_2

3371 Balasubramanian, K. and Ravimohan, Ch. (1988). Chem. Phys. Lett., 145, 39-43.
Electronic states and potential energy surfaces of YH_2

3372 Balasubramanian, K. and Ravimohan, Ch. (1989). J. Phys. Chem., 93, 4490-4494.
Potential energy surfaces for $Nb+H_2$ reaction

3373 Balasubramanian, K. and Ravimohan, Ch. (1990). J. Chem. Phys., 92, 3659-3667.
Spectroscopic properties of 34 electronic states of zirconium dimer (Zr_2)

3374 Balasubramanian, K., Sumathi, K. and Dai, D.-G. (1991b). J. Chem. Phys., 95, 3494-3505.
Group V trimers and their positive ions: The electronic structure and potential energy surfaces

3375 Balasubramanian, K., Tanpipat, N. and Bloor, J.E. (1987d). J. Mol. Spectrosc., 124, 458-463.
Spectroscopic properties of SbH

3376 Balasubramanian, K. and Tao, J.X. (1991). J. Chem. Phys., 94, 3000-3010.
Electronic states, ionization potentials, and bond energies of TlH_n, InH_n, TlH_n^+, and InH_n^+ (n=1-3)

3377 Balasubramanian, K., Tao, J.X. and Liao, D.W. (1991c). J. Chem. Phys., 95, 4905-4913.
Model calculations for halogen-etching of GaAs and InSb. Potential-energy surfaces for GaAs+Cl, GaAs+Br, InSb+Cl, and InSb+Br reactions

3378 Balasubramanian, K. and Wang, J.Z. (1989a). Chem. Phys. Lett., 154, 525-530.
Spectroscopic properties and potential energy curves of thirty-six electronic states of ZrH

3379 Balasubramanian, K. and Wang, J.Z. (1989b). J. Mol. Spectrosc., 133, 82-89.
Spectroscopic properties and potential energy curves

of 29 electronic states of YH

3380 Balasubramanian, K. and Wang, J.Z. (1989c). J. Chem. Phys., 91, 7761-7771.
Potential energy surfaces for $Tc+H_2$ and $Ru+H_2$ reactions

3381 Balasubramanian, K. and Wang, J.Z. (1990). Chem. Phys., 140, 243-253.
Spectroscopic constants and potential energy curves of 21 electronic states of RuH

3382 Balcar, E. and Lovesey, S.W. (1992). J. Phys.: Condens. Matter, 4, 2271-2280.
Theory of magnetic neutron spectroscopy for Pr^{3+} and Tm^{3+}

3383 Balint-Kurti, G.G., Moss, R.E., Sadler, I.A. and Shapiro, M. (1990). Phys. Rev. A, 41, 4913-4921.
Calculations of vibration-rotation energy levels of HD^+

3384 Ball, J.M., Boorman, P.M., Fait, J.F. and Ziegler, T. (1989). J. Chem. Soc., Chem. Commun. 722-723.
Synthesis, x-ray crystal structure, and electronic structure of a trinuclear WCu_2 cluster containing a hexathiolato tungsten(IV) core

3385 Ballone, P. and Galli, G. (1990). Phys. Rev. B, 42, 1112-1123.
Accurate pseudopotential local-density-approximation computations for neutral and ionized dimers of the IB and IIB groups

3386 Baltin, R. and March, N.H. (1988). Phys. Rev. A, 37, 3492-3955.
Charge displaced in a relativistic electron gas by a weak local potential

3387 Baluja, K.L. (1986). J. Phys. B: At. Mol. Phys., 19, L815-L818.
Intercombination transition probabilities for the transitions $1s^2 2s 2p^3\ {}^5S_2^{\circ}$ --> $1s^2 2s^2 2p^2\ {}^3P_{1,2}^{e}$ in the C-sequence ions

3388 Band, I.M., Kikoin, K.A., Trzhaskovskaya, M.B. and Khomskii, D.I. (1988). Zh. Eksp. Teor. Fiz., 94, 79-95; Sov. Phys. JETP, 67, 1561-1569.
Atomic mechanisms for the appearance of mixed valence states in lanthanides and actinides

3389*Band, I.M., Listengarten, M.A. and Trzhaskovskaya, M.B. (1985). Izv. Akad. Nauk SSSR, Ser. Fiz., 49, 2202-2207.
Calculations of hyperfine splittings in highly charged ions (in Russian)

3390 Band, I.M., Listengarten, M.A. and Trzhaskovskaya, M.B. (1986). Izv. Akad. Nauk SSSR, Ser. Fiz., 50, 1290-1295.
X-ray transitions in relativistic Hartree-Fock-Dirac atoms (in Russian)

3391 Band, I.M., Listengarten, M.A. and Trzhaskovskaya, M.B. (1987). Izv. Akad. Nauk SSSR, Ser. Fiz., 51, 1998-2005.

Probability of beta-decay in the Hartree-Fock-Dirac model of an atom and the effect of the chemical environment on beta-decay (in Russian)

3392 Band, I.M., Listengarten, M.A. and Trzhaskovskaya, M.B. (1989a). Zh. Eksp. Theor. Fiz., 96, 525-527.
On the ^{193m}Ir decay

3393 Band, I.M., Listengarten, M.A. and Trzhaskovskaya, M.B. (1989b). Report LIYF-1479, Leningrad Inst. Yad. Fiz., 54p.
RAINE computer programs complex - Relativistic atom. Interaction of electromagnetic radiation and nucleus with atomic electrons, 6. Description of RAINE programs complex

3394 Band, I.M., Listengarten, M.A. and Trzhaskovskaya, M.B. (1990a). Izv. Akad. Nauk SSSR, Ser. Fiz., 54, 15-19; Bull. Acad. Sci. USSR, 54, 14-19.
Internal-conversion coefficients in the Dirac-Fock atomic field

3395 Band, I.M., Listengarten, M.A. and Trzhaskovskaya, M.B. (1991). Opt. Spektrosk., 71, 225-234; Opt. Spectrosc., 71, 130-135.
Relativistic spectrum of 4d excitations in the Ba^{2+} ion

3396 Band, I.M. and Trzhaskovskaya, M.B. (1986). At. Data Nucl. Data Tables, 35, 1-13.
Electron-wave-function expansion amplitudes near the origin calculated in the Dirac-Fock-Slater and Dirac-Fock potentials

3397 Band, I.M. and Trzhaskovskaya, M.B. (1991a). Izv. Akad. Nauk SSSR, Ser. Fiz., 55, 2121-2131.
Tables of internal conversion coefficients for two observable low-energy nuclear transitions (in Russian)

3398 Band, I.M. and Trzhaskovskaya, M.B. (1991b). Izv. Akad. Nauk SSSR, Ser. Fiz., 55, 2141-2146.
Including the hole in calculations of internal conversion coeffients (in Russian)

3399 Band, I.M. and Trzhaskovskaya, M.B. (1992). J. Phys. B: At. Mol. Opt. Phys., 25, L145-L149.
On the 5d photoabsorption spectra in the gaseous and metallic states of uranium and thorium

3400 Band, I.M., Trzhaskovskaya, M.B.,Verner, D.A. and Yakovlev, D.G. (1990b). Astronomy and Astrophysics, 237, 267-269.
K-shell photoionization cross sections: calculations and simple fitting formulae

3401 Banichevich, A., Klotz, R. and Peyerimhoff, S.D. (1992). Mol. Phys., 75, 173-188.
Predissociation of the ions HBr^+ and DBr^+

3402 Banichevich, A., Peyerimhoff, S.D., Hess, B.A. and van Hemert, M.C. (1991). Chem. Phys., 154, 199-209.
Potential curves and predissociation rates for the heavy species HBr^{2+} and DBr^{2+}

3403 Banichevich, A., Peyerimhoff, S.D., van Hemert, M.C. and Fournier, P.G. (1988). Chem. Phys., 121, 351-359.
Potential curves and predissociation rates for the species HCl^{2+} and DCl^{2+}

3404 Bao, M.-Q., Tong, X.-M. and Li, J.-M. (1989). Acta Phys. Sin., 38, 1802-1808.
Excited structure of atomic ions (in Chinese)

3405 Bär, M. and Ahlrichs, R. (1991). Chem. Phys. Lett., 178, 241-245.
Theoretical investigation of the AlCu ground state

3406 Barandiarán, Z. and Seijo, L. (1992). Can. J. Chem., 70, 409-415.
The ab initio model potential method. Cowan-Griffin relativistic core potentials and valence basis sets from Li (Z=3) to La (Z=57)

3407 Barandiarán, Z., Seijo, L. and Huzinaga, S. (1990). J. Chem. Phys., 93, 5843-5850.
The ab initio model potential method. Second series transition metal elements

3408*Barbieri, R. and Sucher, J. (1978). Nucl. Phys., B134, 155-168.
General theory of radiative corrections to atomic decay rates

3409 Bardsley, J.N., Penetrante, B.M. and Mittleman, M.H. (1989). Phys. Rev. A, 40, 3823-3835.
Relativistic dynamics of electrons in intense laser fields

3410 Baretty, R. and Garcia, C. (1988). Int. J. Quantum Chem., S22, 425-430.
Modified Breit-Pauli Hamiltonian suitable for variational calculations

3411 Baretty, R., Ishikawa, Y. and Nieves, J.F. (1986). Int. J. Quantum Chem., S20, 109-117.
Momentum space approach to the relativistic atomic structure calculations

3412 Barnes, L.A., Rosi, M. and Bauschlicher Jr., C.W. (1990). J. Chem. Phys., 93, 609-624.
Theoretical studies of the first- and second-row transition-metal mono- and dicarbonyl positive ions

3413 Barone, V., Minichino, C. Russo, N., Toscano, M., Illas, F. and Rubio, J. (1990). J. Mol. Struct.(Theochem), 204, 325-329.
Non-empirical cluster-model study of the relaxation of (111) surfaces of C, Si, Ge

3414 Barra, A.L., Robert, J.B., and Wiesenfeld, L. (1986). Phys. Lett. A, 115, 443-447.
Parity non-conservation and NMR observables. Calculation of Tl resonance frequency differences in enantiomers

3415 Barra, A.L., Robert, J.B. and Wiesenfeld, L. (1988). Europhys. Lett., 5, 217-222.
Possible observation of parity nonconservation by high-resolution NMR

3416*Barron, L.D. and Buckingham, A.D. (1973). J. Phys. B: At. Mol. Phys., 6, 1295-1300.
Spin-orbit coupling and the interaction of molecules with the radiation field

3417 Bar-Shalom, A., Klapisch, M. and Oreg, J. (1988). Phys. Rev. A, 38, 1773-1784.
Electron collision excitations in complex spectra of ionized heavy atoms

3418 Barthelat, J.C., Chaudret, B., Daudey, J.P., De Loth, Ph. and Poilblanc, R. (1991). J. Am. Chem. Soc., 113, 9896-9898.
Theoretical calculations on Nb and Ta trihydride complexes. Relations with the problem of quantum mechanical exchange coupling

3419*Barthélémy, M.C. (1967). Ann. Inst. H. Poincaré A, 7, 115-143.
Contribution à l'étude de la diffusion par un potentiel central dans la théorie de l'électron de Dirac II

3420 Bartram, R.H., Fockele, M., Lohse, F. and Spaeth, J.-M. (1989). J. Phys.:Condens. Matter, 1, 27-34.
Crystal-field model of the $Pb^0(2)$ centre in SrF_2

3421 Bartschat, K. (1990). J. Phys. B: At. Mol. Opt. Phys., 23, 2341-2348.
Spin-orbit and interference asymmetries in elastic scattering of electrons from heavy one-electron atoms

3422 Bartschat, K. (1991). J. Phys. B: At. Mol. Opt. Phys., 24, 4615-4625.
Spin-dependent effects in elastic and inelastic low-energy electron scattering from sodium atoms

3423 Bartschat, K. and Burke, P.G. (1986). J. Phys. B: At. Mol. Opt. Phys., 19, 1231-1240.
The influence of relativistic effects and electron correlation on the low-energy scattering of electrons from mercury atoms

3424 Bartschat, K., Rudge, M.R.H. and Scott, P. (1986). J. Phys. B: At. Mol. Phys., 19, 2469-2478.
Photoionisation of barium

3425 Bartschat, K., Thumm, U. and Norcross, D.W. (1992). J. Phys. B: At. Mol. Opt. Phys., 25, L641-L646.
Characteristics of light emission after low-energy electron impact excitation of caesium atoms

3426 Barut, A.O. (1986). Phys. Rev. A, 34, 3502-3503.
Quantum electrodynamics based on self-energy versus quantization of fields: Illustration by a simple model

3427 Barut, A.O. (1987). Phys. Scr., 36, 493-496.
On the treatment of Möller and Breit-potentials and the covariant two-body equation for positronium and muonium

3428 Barut, A.O. (1988). Found. Phys., 18, 95-105.
Combining relativity and quantum mechanics: Schrödinger's interpretation of (psi)

3429 Barut, A.O. (1990a). Z. Phys. A, 336, 317-320.
The electron-positron system at short distances

3430 Barut, A.O. (1990b). Phys. Lett. B, 237, 436-439.
Excited states of Zitterbewegung

3431 Barut, A.O. (1991). J. Math. Phys., 32, 1091-1095.
The covariant many-body problem in quantum electrodynamics

3432 Barut, A.O. and Bozic, M. (1988). Z. Phys. A, 330, 319-330.
Passage of relativistic magnetic dipole moments through magnetic fields

3433*Barut, A.O. and Bracken, A.J. (1981). Phys. Rev. D, 23, 2454-2463.
Zitterbewegung and the internal geometry of the electron

3434 Barut, A.O. and Dowling, J.P. (1987). Phys. Rev. A, 36, 2550-2556.
Quantum electrodynamics based on self-energy, without second quantization: The Lamb shift and long-range Casimir-Polder van der Waals forces near boundaries

3435 Barut, A.O. and Dowling, J.P. (1989a). Phys. Rev. A, 39, 2796-2805.
Quantum electrodynamics based on self-fields, without second quantization: Apparatus dependent contributions to g-2

3436 Barut, A.O. and Dowling, J.P. (1989b). Z. Naturforsch. A, 44A, 1051-1056.
QED based on self-fields: A relativistic calculation of g-2

3437 Barut, A.O. and Dowling, J.P. (1990a). Phys. Rev. A, 41, 2277-2283.
Quantum electrodynamics based on self-fields: On the origin of thermal radiation detected by an accelerating observer

3438 Barut, A.O. and Dowling, J.P. (1990b). Phys. Rev: A, 41, 2284-2294.
Self-field quantum electrodynamics: The two-level atom

3439 Barut, A.O., Dowling, J.P. and van Huele, J.F. (1988). Phys. Rev. A, 38, 4405-4412.
Quantum electrodynamics based on self-fields, without second quantization: A nonrelativistic calculation of g-2

3440 Barut, A.O. and Duru, I.H. (1989). Phys. Reports, 172, 1-32.

Path integral formulation of quantum electrodynamics from classical particle trajectories

3441 Barut, A.O., Inomata, A. and Wilson, R. (1989). Fortschr. Phys., 37, 541-597.
Algebraic scattering theory of relativistic composite particles

3442 Barut, A.O., Kraus, J., Salamin, Y. and Ünal, N. (1992). Phys. Rev. A, 45, 7740-7745.
Relativistic theory of the Lamb shift in self-field quantum electrodynamics

3443 Barut, A.O., Onem, C. and Unal, N. (1990). J. Phys. A, 1113-1125.
The classical relativistic two-body problem with spin and self-interactions

3444 Barut, A.O. and Pavsic, M. (1987). Classical Quantum Gravity, 4, L41-L45.
Classical model of the Dirac electron in curved space

3445 Barut, A.O. and Salamin, Y.I. (1988). Phys. Rev. A, 37, 2284-2296.
Relativistic theory of spontaneous emission

3446 Barut, A.O. and Salamin, Y.I. (1991a). Phys. Rev. A, 43, 2524-2527.
QED based on self-energy: The relativistic $2S_{1/2} \rightarrow 1S_{1/2} + 1\gamma$ decay rates of hydrogenlike atoms

3447 Barut, A.O. and Salamin, Y.I. (1991b). Z. Phys. D, 21, 1-5.
The Einstein A-coefficient of spontaneous emission: A relativistic calculation in the Heisenberg representation

3448 Barut, A.O. and Strobel, G.L. (1986). Few-Body Sys., 1, 167-180.
Center-of mass motion of a system of relativistic Dirac particles

3449 Barut, A.O. and Ünal, N. (1986). J. Math. Phys., 27, 3055-3060.
An exactly soluble relativistic quantum two-fermion problem

3450 Barut, A.O. and Ünal, N. (1990). Phys. Rev. D, 41, 3822-3828.
Regularized analytic evaluation of vacuum polarization in a Coulomb field

3451 Basch, H. (1989). Chem. Phys. Lett., 157, 129-134.
The electronic structure of X_3, X=O, S, Se and Te and their anions

3452 Basch, H. (1991). J. Mol. Struct.(Theochem), 234, 185-200.
Geometry and electronic structure studies using computational quantum chemistry

3453 Basch, H., Cohen, D. and Albeck, M. (1988). Chem. Phys.

Lett., 144, 450-454.
The electronic states of TeO_2

3454 Basch, H., Krauss, M. and Stevens, W.J. (1986a). Inorg. Chem., 25, 4777-4779.
Electronic structure of diammine(ascorbato)platinum(II) and the trans influence on the ligand dissociation energy

3455 Basch, H., Krauss, M., Stevens, W.J. and Cohen, D. (1986b). Inorg. Chem., 25, 684-688.
Binding of $Pt(NH_3)_3^{2+}$ to nucleic acid bases

3456 Bassetto, A. and Griguolo, L. (1991). J. Math. Phys., 32, 3195-3198.
Chiral anomalies for vortex potentials in two dimensions and a decompactification limit

3457 Bastug, T., Sepp, W.D., Fricke, B., Heinemann, D. and Kolb, D. (1992). Z. Phys. D, 22, 641-644.
Electronic structure calculations of small Al_n (n=2-8) clusters

3458 Batenburg, P. and Schakel, A.M.J. (1989). Mod. Phys. Lett. A, 4, 1355-1363.
A geometrical interpretation of spin

3459 Bates, C.A., Dunn, J.L., Hallam, L.D., Kirk, P.J. and Polinger, V.Z. (1991). J. Phys.: Cond. Matter, 3, 3441-3453.
Vibronic reduction factors for second-order spin-orbit coupling

3460 Batkin, I.S., Bushahma, K.A., Churakova, T.A. and Demakov, S.L. (1992). J. Phys. G, 18, 1995-2001.
A relativistic study of K-shell auto-ionization in beta decay

3461 Batkin, I.S., Demakov, S.L. and Churakova, T.A. (1988). Izv. Akad. Nauk SSSR, Ser. Fiz., 52, 2109-2114.
Relativistic approach to the self-ionization of internal atomic shells in beta decay

3462 Bauche, J. and Bauche-Arnoult, C. (1989). J. Phys. B: At. Mol. Opt. Phys., 22, 2503-2515.
Mixing of many transition arrays in atomic spectra

3463 Bauche, J., Bauche-Arnoult, C. and Klapisch, M. (1991). J. Phys. B: At. Mol. Opt. Phys., 24, 1-11.
Breakdown of jj coupling in spin-orbit-split atomic transition arrays

3464 Bauche, J., Duong, H.T., Juncar, P., Liberman, S., Pinard, J., Coc, A., Thibault, C., Touchard, F., Lermé, J., Vialle, J.L., Büttgenbach, S., Mueller, A.C., Pesnelle, A. and the ISOLDE Collaboration (1986). J. Phys. B: At. Mol. Phys., 19, L593-L599.
Accurate wavenumbers of the first resonance doublet of francium atoms and relativistic contribution to isotope shifts in the resonance doublets of francium and caesium atoms

3465 Baumann, M. and Braun, M. (1986). Z. Phys. D, 1, 247-251.
Hyperfine structure investigations in excited D-states of ^{171}Yb and ^{173}Yb

3466 Baur, G. (1990). Phys. Rev. A, 42, 5736-5738.
Multiple electron-positron pair production in relativistic heavy-ion collisions: A strong-field effect

3467 Bauschlicher Jr., C.W. (1989). Chem. Phys. Lett., 156, 91-94.
On the electron affinity of Au_3

3468 Bauschlicher Jr., C.W. and Langhoff, S.R. (1986). Chem. Phys. Lett., 126, 163-168.
Theoretical study of the low-lying electronic states of ZnO and ZnS

3469 Bauschlicher Jr., C.W. and Langhoff, S.R. (1989). Chem. Phys. Lett., 161, 383-387.
Theoretical study of the bonding in $LaFe^{2+}$, $LaRu^{2+}$ and YRu^{2+}

3470 Bauschlicher Jr., C.W. and Langhoff, S.R. (1990). Int. Rev. Phys. Chem., 9, 149-185.
Theoretical study of the bonding in molecular transition-metal cations

3471 Bauschlicher Jr., C.W. and Langhoff, S.R. (1991). J. Phys. Chem., 95, 2278-2282.
Theoretical study of the bonding of Sc, Y, and La singly charged and dipositive ions to C_2H_2, C_2H_4, and C_3H_6

3472 Bauschlicher Jr., C.W. and Langhoff, S.R. (1992). Transition-metal hydrides: structure and bonding In "Transition Metal Hydrides", Ed. A. Dedieu, VCH, New York, pp.103-126.

3473 Bauschlicher Jr., C.W., Langhoff, S.R. and Partridge, H. (1986). J. Chem. Phys., 84, 901-909.
Ab initio study of the alkali and alkaline-earth monohydroxides

3474 Bauschlicher Jr., C.W., Langhoff, S.R. and Partridge, H. (1989a). J. Chem. Phys., 91, 2412-2419.
Theoretical study of the structures and electron affinities of the dimers and trimers of the group IB metals (Cu, Ag, and Au)

3475 Bauschlicher Jr., C.W., Langhoff, S.R. and Partridge, H. (1990a). J. Phys. Chem., 94, 8378-8379.
Comments on "Binding energies and ionization potentials of the tetramers of Cu, Ag, and Au"

3476 Bauschlicher Jr., C.W., Langhoff, S.R. and Partridge, H. (1990b). J. Chem. Phys., 93, 8133-8137.
Theoretical study of the homonuclear tetramers and pentamers of the group IB metals (Cu, Ag, and Au)

3477 Bauschlicher Jr., C.W., Langhoff, S.R. and Partridge, H.

(1991a). J. Phys. Chem., 95, 6191-6194.
Theoretical study of the bonding of Nb^{2+} to CH_2, C_2H_2, and C_2H_4

3478 Bauschlicher Jr., C.W., Langhoff, S.R., Partridge, H. and Barnes, L.A. (1989b). J. Chem. Phys., 91, 2399-2411.
Theoretical studies of the first- and second-row transition-metal methyls and their positive ions

3479 Bauschlicher Jr., C.W., Langhoff, S.R., and Taylor, P.R. (1988). J. Chem. Phys., 88, 1041-1045.
Theoretical study of the electron affinities of Cu, Cu_2, and Cu_3

3480 Bauschlicher Jr., C.W., Partridge, H. and Dyall, K.G. (1992c). Chem. Phys. Lett., 199, 225-228.
The $^2\Sigma^+$-$^2\pi$ separation in KO

3481 Bauschlicher Jr., C.W., Partridge, H. and Langhoff, S.R. (1990c). Chem. Phys., 148, 57-68.
An ab initio study of the low-lying doublet states of AgO and AgS

3482 Bauschlicher Jr., C.W., Partridge, H. and Langhoff, S.R. (1992a). Chem. Phys. Lett., 195, 360-364.
Comparative study of the dissociation energies of Ni_2 and Ni_2^+

3483 Bauschlicher Jr., C.W., Partridge, H., Langhoff, S.R. and Rosi, M. (1991b). J. Chem. Phys., 95, 1057-1063.
A theoretical study of the low-lying states of Ti_2 and Zr_2

3484 Bauschlicher Jr., C.W., Partridge, H., Sheehy, J.A., Langhoff, S.R. and Rosi, M. (1992b). J. Phys. Chem., 96, 6969-6973.
Theoretical study of the bonding of the first- and second-row transition-metal positive ions to methylene

3485 Bawin, M. and Cugnon, J. (1991). Europhys. Lett., 16, 617-621.
Charge screening in classical spinor electrodynamics

3486 Baykara, N.A., Andzelm, J., Salahub, D.R. and Baykara, S.Z. (1986). Int. J. Quantum Chem., 29, 1025-1032.
Hydrogen chemisorption on, and diffusion through, palladium clusters

3487 Baylis, W.E. (1992). Phys. Rev. A, 45, 4293-4302.
Classical eigenspinors and the Dirac equation

3488 Baylis, W.E., Huschilt, J. and Wei, J.-S. (1992). Am. J. Phys., 60, 788-797.
Why i?

3489 Baylis, W.E. and Jones, G. (1989a). J. Phys. A, 22, 17-29.
Relativistic dynamics of charges in external fields: the Pauli algebra approach

3490 Baylis, W.E. and Jones, G. (1989b). J. Phys. A, 22, 1-15.
The Pauli algebra approach to special relativity

3491 Bechler, A. and Pratt, R.H. (1989). Phys. Rev. A, 39, 1774-1779.
Higher retardation and multipole corrections to the dipole angular distribution of 1s photoelectrons at low energies

3492 Beck, D.R. (1988). Phys. Rev. A, 37, 1847-1857.
Relativistic and many-body effects in first-row transition-metal negative ions: Three Zn^- bound states

3493 Beck, D.R. (1989). Phys. Rev. A, 40, 2887-2893.
Wavelength, oscillator strength, and fine structure for the ^4P to $^4S°$ transition of Mg^-

3494 Beck, D.R. (1992). Phys. Rev. A, 45, 1399-1402.
Magnetic-dipole hyperfine-structure constants for the nd^2 triplet states of Sc II and Y II

3495 Beck, D.R. and Cai, Z.-Y. (1988). Phys. Rev. A, 37, 4481-4483.
Bound states in Mn^-

3496 Beck, D.R. and Cai, Z.-Y. (1990). Phys. Rev. A, 41, 301-311.
Relativistic many-body methodology of electric dipole oscillator strengths with application to Tl^+ $6s^2$ --> 6s6p

3497 Beck, D.R., Cai, Z.-Y. and Aspromallis, G. (1987). Int. J. Quantum Chem., S21, 457-468.
Electron affinities of states of Sc and Cu

3498 Beck, D.R. and Datta, D. (1991). Phys. Rev. A, 44, 758-759.
Relativistic many-body results for the hyperfine structure of the ground state of the Ag I isoelectronic sequence

3499*Beck, D.R. and Nicolaides, C.A. (1976).
Anomalies in the fine and hyperfine structure of alkali isoelectronic sequences
In "Beam-Foil Spectroscopy", Ed. I.A. Sellin and D.J. Pegg, Plenum, New York, 1, pp. 105-109.

3500*Beck, D.R. and Nicolaides, C.A. (1979). NATO ASI, C46, 329-359.
Theory of one electron binding energies including correlation, relativistic and radiative effects: Application to free atoms and metals

3501*Beck, D.R. and Nicolaides, C.A. (1984). Int. J. Quantum Chem., S18, 467-481.
Fine and hyperfine structure of the two lowest bound states of Be^- and their first two ionization thresholds

3502 Beck, D.R. and Nicolaides, C.A. (1986). Phys. Rev. A, 33, 3885-3890.
Theory of Auger energies in free atoms: Application to the alkaline earths

3503*Beck, D.R., Nicolaides, C.A. and Musher, J.I. (1974).

Phys. Rev. A, 10, 1522-1527.
Calculation of the fine structure of the a $^3\Sigma_u{}^+$ state of molecular helium

3504 Becker, S.R., Butler, K. and Zeippen, C.J. (1989). Astron. Astrophys., 221, 375-384.
Improved M1 and E2 transition probabilities for forbidden lines in ions of the nitrogen isoelectronic sequence

3505 Becker, U., Grün, N., Scheid, W. and Soff, G. (1986). Phys. Rev. Lett., 56, 2016-2019.
Nonperturbative treatment of excitation and ionization in $U^{92+} + U^{91+}$ collisions at 1 GeV/amu

3506 Beckers, J. and Debergh, N. (1990). Phys. Rev. D, 42, 1255-1259.
Supersymmetry, Foldy-Wouthuysen transformations, and relativistic oscillators

3507 Beckers, J., Debergh, N. and Nikitin, A.G. (1992a). Phys. Lett. B, 279, 333-335.
Extended Dirac symmetries and hidden supersymmetry

3508 Beckers, J., Debergh, N. and Nikitin, A.G. (1992b). J. Phys. A, 25, 6145-6154.
Lie extended symmetries and relativistic particles

3509 Beckers, J., Debergh, N. and Nikitin, A.G. (1992c). J. Math. Phys., 33, 3387-3392.
On pararelativistic quantum oscillators

3510*Behncke, H. (1985). J. Math. Phys., 26, 2556-2559.
Spectral properties of the Dirac equation with anomalous magnetic moment

3511 Beier, T. and Soff, G. (1988). Z. Phys. D, 8, 129-134.
Källén-Sabry contribution to the Lamb-shift in hydrogen-like atoms

3512 Bel, L. and Ruiz, E. (1988). J. Math. Phys., 29, 1840-1846.
Relativistic Schrödinger equations

3513 Belinfante, F.J. (1991). Comput. Phys., 5, 319-322.
Computing Dirac's atomic hydrogen wave functions of the continuum, using summation of mathematically divergent series

3514 Bell, F. (1989). J. Phys. B: At. Mol. Opt. Phys., 22, 287-297.
Double and triple differential cross sections for K-shell ionisation by relativistic electron impact

3515 Belloni, L. and Reina, C. (1986). Eur. J. Phys., 7, 55-61.
Sommerfeld's way to the Thomas precession

3516 Belov, V.V. and Maslov, V.P. (1989). Dokl. Akad. Nauk SSSR, 305, 574-580; Sov. Phys. Doklady, 34, 220-223.
Semiclassical path-coherent states of a Dirac operator with an anomalous Pauli interaction

3517 Benavides-Garcia, M. and Balasubramanian, K. (1991).

J. Mol. Spectrosc., 150, 271-279.
Spectroscopic constants and potential energy curves for OsH

3518 Benavides-Garcia, M. and Balasubramanian, K. (1992a). J. Chem. Phys., 97, 7537-7544.
The singlet-triplet energy separations of $GeCl_2$, $GeBr_2$, GeI_2, GeHCl, GeHBr, and GeHI

3519*Bendazzoli, G.L. and Palmieri, P. (1974). Int. J. Quantum Chem., 8, 941-950.
Spin-orbit interaction in polyatomic molecules: Ab initio computations with Gaussian orbitals

3520 Bender, B.R., Norton, J.R., Miller, M.M., Anderson, O.P. and Rappé, A.K. (1992). Organometallics, 11, 3427-3434.
Structure of $Os(CO)_4(C_2H_4)$, an osmacyclopropane

3521 Benítez, J., Martínez y Romero, R.P., Núñez-Yépez, H.N. and Salas-Brito, A.L. (1990). Phys. Rev. Lett., 64, 1643-1645.
Solution and hidden supersymmetry of a Dirac oscillator

3522 Berezin, A.V., Kurochkin, Yu.A. and Tolkachev, E.A. (1989). Kvaterniony v Relyativistskoi Fizike.
Nauka i Tekhnika, Minsk, 198p.

3523 Berghammer, H., Vretenar, D. and Ring, P. (1992). Phys. Lett. B, 296, 290-295.
Relativistic description of Coulomb dissociation. A one-dimensional model

3524*Bergou, J. and Ehlotzky, F. (1983). Phys. Rev. A, 27, 2291-2296.
Relativistic quantum states of a particle in an electromagnetic plane wave and a homogeneous magnetic field

3525 Bergstrom Jr., P.M., Suric, T., Pisk, K. and Pratt, R.H. (1992). Nucl. Instr. Meth. Phys. Res., B71, 1-6.
Some preliminary calculations of whole atom Compton scattering of unpolarized photons

3526 Berksoy, E.M. and Whitehead, M.A. (1988a). J. Chem. Soc. Faraday Trans. 2, 84, 1707-1717.
Molecular electronic structure of aluminium trihalide dimers and their quadrupole couplings by the SCF-MS molecular-orbital method

3527 Berksoy, E.M. and Whitehead, M.A. (1988b). Tetrahedron, 44, 7517-7530.
The electronic structure of diphenyltin(IV) dichloride. Calculation of the Mössbauer isomer shift and the quadrupole splitting of tin and the nuclear quadrupole resonance frequency of chlorine

3528 Berksoy-Bölük, E.M. (1991). Diss. Abstr. Int. B, 51, 4356-4357.
Multiple scattering calculations of large inorganic systems

3529 Bernabeu, J. and Garcia-Molina, R. (1990).

Anales de Fisica, Ser. A, 86, 1-8.
Leptonic mass effects on the relativistic atomic wave functions for non point-like nuclei

3530*Bernabéu, J. and Tarrach, R. (1976). Ann. Phys.(N.Y.) 102, 323-343.
Long-range potentials and the electromagnetic polarizabilities

3531 Bernier, A. and Millié, P. (1987). Chem. Phys. Lett., 134, 245-250.
Potential energy surfaces of HgH_2

3532 Bernier, A., Millié, P. and Pelissier, M. (1986). Chem. Phys., 106, 195-203.
Three-electron approach of HgH using relativistic effective core, core polarization and spin-orbit operators: The low-lying states

3533 Berseth, W.C. and Darewych, J.W. (1991). Phys. Lett. A, 158, 361-366.
Relativistic three-fermion wave equations

3534*Berthier, A.M. and Georgescu, V. (1980). C. R. Ac. Sci., 291, 603-606.
Property of unique continuation for a Dirac operator

3535 Bertoncello, R., Daudey, J.P., Granozzi, G. and Russo, U. (1986). Organometallics, 5, 1866-1873.
He I and He II photoelectron spectra and pseudopotential ab initio calculations of some tetracoordinated tin(IV) derivatives

3536 Bessis, N. and Bessis, G. (1988).
Space-curvature effects in atomic and molecular-structure calculations
In "Molecules in Physics, Chemistry and Biology", Ed. J. Maruani, Vol.2, pp. 145-171.

3537*Bessis, N., Bessis, G. and Roux, D. (1984). Phys. Rev. A, 30, 1094-1097.
Atomic fine-structure calculations in a space of constant negative curvature

3538 Best, C., Greiner, W. and Soff, G. (1992). Phys. Rev. A, 46, 261-269.
Multiplicity distribution of electron-positron pairs created by strong external fields

3539*Bethe, H.A. (1948). Z. Naturforschg., 3a, 470-477.
Bemerkungen über die Wasserstoff-Eigenfunktionen in der Diracschen Theorie

3540*Bhagwat, K.V. and Subramanian, R. (1972). Physica, 62, 614-622.
Relativistic effects on impurity states

3541 Bhattacharya, J., Laha, U. and Talukdar, B. (1987). J. Phys. B: At. Mol. Phys., 20, 1725-1731.
L-shell x-ray energy shifts for an additional vacancy in the M shell

3542*Bhattacharyya, B.K., Bylander, D.M. and Kleinman, L. (1985). Phys. Rev. B, 32, 7973-7978.
Comparison of fully relativistic energy bands and cohesive energies of $MoSi_2$ and WSi_2

3543 Bialynicki-Birula, I. (1986). Phys. Rev. A, 34, 3500-3501.
Comment on "Quantum electrodynamics based on self-energy: Lamb shift and spontaneous emission without field quantization"

3544 Bialynicki-Birula, I., Górnicki, P. and Rafelski, J. (1991). Phys. Rev. D, 44, 1825-1835.
Phase-space structure of the Dirac vacuum

3545 Bice, J.-A.E. (1987). Diss. Abstr. Int. B, 48, 1679-1680.
Gas phase photoelectron studies using He I, He II and synchrotron radiation.

3546*Biedenharn, L.C., Bohm, A., Tarlini, M., van Dam, H. and Mukunda, N. (1984). Phys. Rev. D, 30, 409-420.
Spinorial relativistic rotator: The transformation from quasi-Newtonian to Minkowski coordinates

3547*Biedenharn, L.C. and Horwitz, L.P. (1984). Found. Phys., 14, 953-961.
Chiral two-component spinors and the factorization of Kramers's equation

3548 Biémont, E. (1986). Phys. Scr., 34, 318-327.
Wavelengths and transition probabilities in the sulfur sequence: the $3s^2 3p^4$-$3s3p^5$, $3s^2 3p^4$-$3s^2 3p^3 3d$ and $3s^2 3p^4$-$3s^2 3p^3 4s$ transitions of Ca V, Sc VI and Ti VII

3549 Biémont, E. (1988). At. Data Nucl. Data Tables., 39, 157-181.
Energy levels, wavelengths, and weighted oscillator strengths for n=3-4 and 4-4 transitions in Cu-like ions (Sr X to Nd XXXII)

3550 Biémont, E. (1989). At. Data Nucl. Data Tables., 43, 163-244.
Energy levels, wavelengths, transition probabilities, and oscillator strengths for n=4-4 transitions in zinc-like ions

3551 Biémont, E. (1990a). Bull. Soc. Roya. Sci. Liège, 59, 319-325.
Forbidden lines in the potassium sequence(Cd^{29+}-Ce^{39+})

3552 Biémont, E. (1990b).
Oscillator strength determination for heavy elements In "Atomic Spectra and Oscillator Strengths for Astrophysics and Fusion Research", Ed. J.E. Hansen, North-Holland, Amsterdam (1990), pp. 48-57.

3553 Biémont, E. (1991). At. Data Nucl. Data Tables, 48, 1-23.
Wavelengths and transition probabilities for some Delta n ≥ 0 allowed transitions in aluminum ions (Al V-Al VII)

3554 Biémont, E., El Himdy, A. and Garnir, H.P. (1990). J. Quant. Spectrosc. Radiat. Transfer, 43, 437-443. Oscillator strengths for some n=4 $\Delta n=0$ transitions of germanium-like ions (Rb VI-Ag XVI)

3555 Biémont, E. and Hansen, J.E. (1986a). Phys. Scr., 33, 117-134. Magnetic-dipole and electric-quadrupole transitions in the ground-state configurations of the germanium and arsenic isoelectronic sequences

3556 Biémont, E. and Hansen, J.E. (1986b). Phys. Scr., 34, 116-130. Forbidden transitions in $3p^4$ and $4p^4$ configurations

3557 Biémont, E. and Hansen, J.E. (1987a). At. Data and Nucl. Data Tables., 37, 1-15. Energy levels and weighted oscillator strengths for neon-like ions K X through Se XXV

3558 Biémont, E. and Hansen, J.E. (1987b). Nucl. Instr. Meth., B23, 274-277. Calculations of transition probabilities for forbidden lines in $3p^N$ and $4p^N$ configurations

3559 Biémont, E. and Hansen, J.E. (1989). Phys. Scr., 39, 308-313. Energy levels and transition probabilities in 3d and $3d^9$ configurations

3560 Biémont, E., Hansen, J.E., Quinet, P. and Zeippen, C.J. (1992). J. Phys. B, 25, 5029-5042. Transition rates in the $3d^2$ configuration of the calcium isoelectronic sequence

3561 Biémont, E. and Quinet, P. (1990). J. Quant. Spectrosc. Radiat. Transfer, 44, 233-244. Energy levels and oscillator strengths for gallium-like ions (Br V-In XIX)

3562 Biémont, E., Quinet, P. and Fawcett, B.C. (1989). Phys. Scr., 39, 562-573. Energy levels and transition probabilities along the zinc isoelectronic sequence

3563 Bieron, J.R. (1988). J. Quant. Spectrosc. Radiat. Transfer, 40, I-III. Implementation of the Grant's relativistic multiconfiguration Dirac-Fock program on microcomputers

3564 Bieron, J.R., Marcinek, R. and Migdalek, J. (1991). J. Phys. B: At. Mol. Opt. Phys., 24, 31-43. Relativistic oscillator strengths for the np^2-np(n+1)s transitions

3565 Bieron, J.R. and Migdalek, J. (1992). J. Phys. B: At. Mol. Opt. Phys., 25, 4099-4106. Relativistic lifetimes of the levels belonging to the np^2(n+1)s configuration in As I, Sb I and Bi I spectra

3566 Bieron, J.R. and Migdalek, J. (1988). Acta Phys. Pol. A, A74, 723-728.
Limited MCDF study of $4s^2 4p^2$ configuration in germanium isoelectronic sequence

3567 Bijtebier, J. (1988). Nuovo Cimento A, 100, 91-118.
Relativistic bound states: Transition matrix elements for the scattering by an external field

3568 Bijtebier, J. (1989a). Nuovo Cimento A, 102, 1235-1246.
Relativistic two-body plus potential problem: Compatible equations and double bound-state solutions (spinless case)

3569 Bijtebier, J. (1989b). Nucl. Phys. B, 6, 278-280.
Relativistic bound states: Transition matrix elements for the scattering by an external potential

3570 Bijtebier, J. (1990a). Nucl. Phys. A, 508, 305c-310c.
Separable bound state equations for three relativistic fermions

3571 Bijtebier, J. (1990b). Nuovo Cimento A, 103, 317-348.
The scattering of a separable relativistic bound state by an external perturbative potential

3572 Bijtebier, J. (1990c). Nuovo Cimento A, 103, 639-651.
Relativistic two-body plus potential problem: Compatible equations and double bound-state solutions (two-fermion case)

3573 Bijtebier, J. (1990d). Nuovo Cimento A, 103, 669-682.
Three-fermion relativistic bound states and cluster decomposition

3574 Bismut, J.-M. and Freed, D.S. (1986). Commun. Math. Phys., 107, 103-163.
The analysis of elliptic families II. Dirac operators, êta invariants, and holonomy theorem

3575 Blaha, P. and Schwarz, K. (1992). J. Mol. Struct. (Theochem), 261, 355-365.
Solid state effects in the molecular crystals of Cl_2, Br_2 and I_2

3576 Blaha, P., Schwarz, K., Sorantin, P. and Trickey, S.B. (1990). Comp. Phys. Comm., 59, 399-415.
Full-potential, linearized augmented plane wave programs for crystalline systems

3577 Blaive, B., Barut, A.O. and Boudet, R. (1991). J. Phys. B: At. Mol. Opt. Phys., 24, 3121-3147.
Contribution of the individual discrete levels to the Lamb shift in hydrogen atoms

3578 Blanchard, C. and Girault, P. (1991). Semicond. Sci. Technol., 6, A127-A130.
Energy states of the cation vacancy in II-VI compounds

3579 Blanke, J.H., Fricke, B., Heinemann, D., Sepp, W.-D., Thies,

B., Göbenli, T., Heckmann, P.H., Möller, G. and Träbert, E. (1989). Z. Phys. D, 13, 1-7.
Study of 3d-4f and 3d-5f quartet and doublet transitions of doubly excited three-electron ions. Experiment and MCDF theory along the isoelectronic sequence

3580*Blockley, C.A. and Stedman, G.E. (1985). Eur. J. Phys., 6, 218-224.
Simple supersymmetry: I. Basic examples

3581 Blomberg, M.R.A., Lebrilla, C.B. and Siegbahn, P.E.M. (1988). Chem. Phys. Lett., 150, 522-528.
The binding of carbonyl to a single palladium atom and to a palladium dimer

3582 Blomberg, M.R.A., Schüle, J. and Siegbahn, P.E.M. (1989). J. Am. Chem. Soc., 111, 6156-6163.
Ligand effects on metal-R bonding, where R is hydrogen or alkyl. A quantum chemical study

3583 Blomberg, M.R.A. and Siegbahn, P.E.M. (1992). Theor. Chim. Acta, 81, 365-374.
A comparison between multireference CI and effective medium theories for diatomic FeN

3584 Blomberg, M.R.A., Siegbahn, P.E.M., Nagashima, U. and Wennerberg, J. (1991a). J. Am. Chem. Soc., 113, 424-433.
Theoretical study of the activation of alkane C-H and C-C bonds by different transition metals

3585 Blomberg, M.R.A., Siegbahn, P.E.M. and Svensson, M. (1991b). New. J. Chem., 15, 727-733.
An important ligand effect in the activation of C-H bonds by transition metal complexes

3586 Blomberg, M.R.A., Siegbahn, P.E.M. and Svensson, M. (1992a). J. Am. Chem. Soc., 114, 6095-6102.
Mechanisms for the reactions between methane and the neutral transition metal atoms from yttrium to palladium

3587 Blomberg, M.R.A., Siegbahn, P.E.M. and Svensson, M. (1992b). J. Phys. Chem., 96, 5783-5789.
A theoretical study of the reactivity of Pd clusters with methane

3588 Blomberg, M.R.A., Siegbahn, P.E.M. and Svensson, M. (1992c). J. Phys. Chem., 96, 9794-9800.
Theoretical study of the binding of ethylene to second-row transition-metal atoms

3589 Blomberg, M.R.A. and Wahlgren, U. (1988). Chem. Phys. Lett., 145, 393-398.
On the effect of core orbital relaxation in first-order relativistic calculations

3590 Blügel, S. (1992). Phys. Rev. Lett., 68, 851-854.
Two-dimensional ferromagnetism of 3d, 4d, and 5d transition metal monolayers on noble metal (001) substrates

3591 Blügel, S., Akai, H., Zeller, R. and Dederichs, P.H.

(1987). Phys. Rev. B, 35, 3271-3283.
Hyperfine fields of 3d and 4d impurities in nickel

3592 Blundell, S.A. (1989).
Many-body calculations of parity nonconservation in cesium
In "Relativistic, Quantum Electrodynamic and Weak Interaction Effects in Atoms", ed. W. Johnson, P. Mohr and J. Sucher, AIP Conf. Proc., 189, pp 285-304.

3593 Blundell, S.A. (1992). Phys. Rev. A, 46, 3762-3775.
Accurate screened QED calculations in high-Z many-electron ions

3594 Blundell, S.A., Baird, P.E.G., Palmer, C.W.P., Stacey, D.N. and Woodgate, G.K. (1987a). J. Phys. B: At. Mol. Phys., 20, 3663-3681.
A reformulation of the theory of field isotope shift in atoms

3595 Blundell, S.A., Guo, D.S., Johnson, W.R. and Sapirstein, J. (1987b). At. Data Nucl. Data Tables., 37, 103-119.
Formulas from first-, second-, and third-order perturbation theory for atoms with one valence electron

3596 Blundell, S.A., Hartley, A.C., Liu, Z., Mårtensson-Pendrill, A.-M. and Sapirstein, J. (1991b). Theor. Chim. Acta, 80, 257-288.
Calculations of the parity non-conserving 6s --> 7s transition in caesium

3597 Blundell, S.A., Johnson, W.R., Liu, Z.W. and Sapirstein, J. (1989a). Phys. Rev. A, 40, 2233-2252.
Relativistic all-order calculations of energies and matrix elements for Li and Be^+

3598 Blundell, S.A., Johnson, W.R., Liu, Z.W. and Sapirstein, J. (1989b). Phys. Rev. A, 39, 3768-3775.
Relativistic all-order equations for helium

3599 Blundell, S.A., Johnson, W.R. and Sapirstein, J. (1988). Phys. Rev. A, 38, 4961-4966.
Higher-order many-body perturbation-theory calculations of energy levels in cesium

3600 Blundell, S.A., Johnson, W.R. and Sapirstein, J. (1990a). Phys. Rev. A, 41, 1698-1700.
Improved many-body perturbation-theory calculations of the n=2 states of lithiumlike uranium

3601 Blundell, S.A., Johnson, W.R. and Sapirstein, J. (1990b). Phys. Rev. A, 42, 3751-3762.
Third-order many-body perturbation theory calculations of the ground-state energies of cesium and thallium

3602 Blundell, S.A., Johnson, W.R. and Sapirstein, J. (1990c). Phys. Rev. Lett., 65, 1411-1414.
High-accuracy calculation of the $6s_{1/2}$ --> $7s_{1/2}$ parity-nonconserving transition in atomic cesium and implications for the standard model

3603 Blundell, S.A., Johnson, W.R. and Sapirstein, J. (1991a). Phys. Rev. A, 43, 3407-3418.
Relativistic all-order calculations of energies and matrix elements in cesium

3604 Blundell, S.A. and Snyderman, N.J. (1991). Phys. Rev. A, 44, R1427-R1430.
Basis-set approach to calculating the radiative self-energy in highly ionized atoms

3605 Blunden, P.G. (1990). Phys. Rev. C, 41, 1851-1854.
Evaluation of Dirac sea effects in a finite system

3606 Boatz, J.A., Bak, K.L. and Simons, J. (1992). Theor. Chim. Acta, 83, 209-225.
Ab initio studies of ground and excited electronic states of MgAr, CdAr, and BeAr

3607 Boča, R. (1987). Int. J. Quantum Chem., 31, 941-950.
Inclusion of relativistic effects into ZDO methods. I. A quasi-relativistic CNDO/1

3608 Boča, R. (1988a). Int. J. Quantum Chem., 33, 159-167.
Inclusion of relativistic effects into ZDO methods. II. Solvation of metal complexes

3609 Boča, R. (1988b). Int. J. Quantum Chem., 34, 385-399.
Inclusion of relativistic effects into ZDO methods. III. A quasi-relativistic INDO/1 version

3610 Boča, R. (1989). Int. J. Quantum Chem., 36, 727-739.
Inclusion of relativistic effects into ZDO methods. V. Effect of core and polarization functions in bonding of first-group elements

3611 Boča, R. (1990a). Int. J. Quantum Chem., 37, 209-220.
Inclusion of relativistic effects into ZDO methods. IV. Relativistic CNDO/1

3612 Boča, R. (1990b). Czech. J. Phys., 40, 629-645.
Inclusion of relativistic effects into ZDO methods. VI. Stability or Jahn-Teller instability of octahedral gold and copper complexes

3613 Boča, R., Benkovsky, I. and Benco, L. (1992a). Czech. J. Phys., 42, 1005-1013.
Thin layers of gallium-arsenide: a molecular orbital study

3614 Boča, R. and Breza, M. (1990). Czech. J. Phys., 40, 778-789.
Electronic structure of high-temperature superconductors. I. Y-Ba-Cu-O systems

3615 Boča, R., Hajko, P. and Benco, L. (1992b). Czech. J. Phys., 42, 685-694.
Molecular orbital study of germanium, germanium-gallium and germanium-arsenic clusters

3616 Boča, R. and Pelikán, P. (1992). Coord. Chem. Rev., 118,

1-405.
Quantum chemistry of coordination compounds

3617 Bock, H., Kremer, M., Dolg, M. and Preuss, H.-W. (1991).
Angew. Chem., 103, 1200-1202.
SiI_2, ein neues dreiatomiges Molekül mit relativistischem Touch

3618 Boehm, R.C. and Banerjee, A. (1992). Chem. Phys., 167, 291-297.
Ab initio study of the oxides of rhenium and their anions

3619 Boerrigter, P.M. (1987).
Spectroscopy and Bonding of Heavy Element Compounds
Thesis, Vrije Universiteit, Amsterdam.

3620 Boerrigter, P.M., Baerends, E.J. and Snijders, J.G. (1988a).
Chem. Phys., 122, 357-374.
A relativistic LCAO Hartree-Fock-Slater investigation of the electronic structure of the actinocenes $M(COT)_2$, M=Th, Pa, U, Np and Pu

3621 Boerrigter, P.M., Buijse, M.A. and Snijders, J.G. (1987).
Chem. Phys., 111, 47-53.
Spin-orbit interaction in the excited states of the dihalogen ions F_2^+, Cl_2^+ and Br_2^+

3622 Boerrigter, P.M., Snijders, J.G. and Dyke, J.M. (1988b).
J. Electron Spectr. Rel. Phen., 46, 43-53.
A reassignment of the gas-phase photoelectron spectra of the actinide tetrahalides UF_4, UCl_4, ThF_4 and $ThCl_4$ by relativistic Hartree-Fock-Slater calculations

3623 Böhm, M.C., Elsässer, C., Fähnle, M. and Brandt, E.-H. (1989). Chem. Phys., 130, 65-87.
Local magnetic anisotropy in amorphous transition metals: A problem of metal physics treated with methods of quantum chemistry

3624*Bohr, N. and Coster, D. (1923). Z. Physik, 12, 342-374.
Röntgenspektren und periodisches System der Elemente

3625 Boland, Ø. (1986).
Ikke-Relativistiske og Kvasi-Relativistiske Modell Potensial Beregninger på AgH og AuH.
Cand. Real. Thesis, University of Tromsø, Norway.

3626 Boldyrev, A.I. and Simons, J. (1992). J. Chem. Phys., 97, 2826-2827.
Is TeF_8^{2-} the MX_n^{2-} dianion with the largest electron detachment energy (5 eV)

3627 Boldyrev, A.I., Zhdankin, V.V., Simons, J. and Stang, P.J. (1992). J. Am. Chem. Soc., 114, 10569-10572.
Macrocyclic, square planar, tetraalkynyl tetraiodonium salts: structures, stabilities, and vibrational frequencies via ab initio calculations

3628 Bonacic-Koutecky, V., Fantucci, P. and Koutecky, J. (1991).
Chem. Rev., 91, 1035-1108.

Quantum chemistry of small clusters of elements of Group Ia, Ib, and IIa: Fundamental concepts, predictions, and interpretation of experiments

3629 Bonesteel, N.E., Rice, T.M. and Zhang, F.C. (1992). Phys. Rev. Lett., 68, 2684-2687.
Spin-orbit coupling and spirals in doped La_2CuO_4

3630*Bongaarts, P.J.M. and Ruijsenaars, S.N.M. (1976). Ann. Phys.(NY), 101, 289-318.
The Klein paradox as a many particle problem

3631*Bongaarts, P.J.M. and Ruijsenaars, S.N.M. (1977). Ann. Inst. H. Poincaré A, 26, 1-17.
Scattering theory for one-dimensional step potentials

3632 Bordovitsyn, V.A. and Ternov, I.M. (1991). Izv. VUZov, Fizika (2), 5-9.
The Zitterbewegung in quantum theory with definite-parity operators (in Russian)

3633*Borie, E. (1984).
Vacuum polarization corrections and fine structure in antiprotonic atoms
In "Physics at LEAR with Low-Energy Cooled Antiprotons", Ed. U. Gastaldi and R. Klapisch, Plenum, pp. 561-566.

3634 Borie, E. (1986). Metrologia, 22, 140-145.
The role of the fine-structure constant in QED

3635 Borie, E. and Jödicke, B. (1988). Comp. Phys., 2(6), 61-63.
PBAR: A code for calculating the electromagnetic contribution to the energy levels of antiprotonic atoms

3636*Borie, E. and Rinker, G.A. (1982). Revs. Mod. Phys., 54, 67-118.
The energy levels of muonic atoms

3637 Boring, A.M., Albers, R.C., Schadler, G.H., Lawson, A.C., Weinberger, P. and Christensen, N.E. (1987). Phys. Rev. B, 36, 5507-5511.
Electronic structure and magnetic behavior of UMn_2 and UFe_2

3638 Boring, M., Schadler, G., Albers, R.C. and Weinberger, P. (1988). J. Less-Com. Met., 144, 71.
Relativistic electronic structure of Np

3639*Borisoglebskii, L.A. and Yakimovich, A.D. (1985). Vestn. Belorus. Un-ta, Ser. 1, 3, 10-13.
Relativistic probabilities of forbidden transitions induced by the quadrupole interaction of a nucleus with the electron shell of an atom (in Russian)

3640 Borkovskii, I.V. and Lyudchik, A.M. (1989). Koord. Khim., 15, 1147-1152.
Influence of spin-orbit interactions on the electronic structure of uranyl complexes (in Russian)

3641 Borkovskii, N.B., Svetashev, A.G. and Tsvirko, M.P.

(1990). Opt. Spektrosk., 69, 1064-1068; Opt. Spectrosc., 69, 630-633.
Absorption spectra and electronic structure of Ce^{+3} aquo centers in crystals and solutions

3642 Bosanac, S.D. (1991). Fizika, 23, 375-384.
Scattering of relativistic wave packet

3643 Bosch, F. (1986).
Superheavy atoms - electrons in strong fields
In " Atoms in Unusual Situations", ed. J.P. Briand, Plenum, New York, pp. 313-341.

3644 Boschi Filho, H. and Vaidya, A.N. (1989). J. Phys. A, 22, 3223-3228.
SO(2,1) coherent-state Green function for the Klein-Gordon Coulomb problem

3645 Boston, E.R. and Sandars, P.G.H. (1990). J. Phys. B: At. Mol. Opt. Phys., 23, 2663-2672.
The anapole moments of hydrogenic atoms

3646 Botham, C.P., Blundell, S.A., Mårtensson-Pendrill, A.-M. and Sandars, P.G.H. (1987). Phys. Scr., 36, 481-484.
PNC equations of motion method

3647 Bottcher, C. and Strayer, M.R. (1987). Ann. Phys.(N.Y.), 175, 64-111.
Relativistic theory of fermions and classical fields on a collocation lattice

3648 Bottcher, C. and Strayer, M.R. (1988). Nucl. Instrum. Meth. B, 31, 122-130.
Probing the vacuum with highly charged ions

3649 Bottcher, C., Strayer, M.R., Umar, A.S. and Reinhard, P.-G. (1989). Phys. Rev. A, 40, 4182-4189.
Damped relaxation techniques to calculate relativistic bound states

3650 Bouazza, S. and Bauche, J. (1988). Z. Phys. D, 10, 1-11.
Hyperfine structure in low-lying levels of Pt II, Au II and Bi II

3651 Bouazza, S., Guern, Y. and Bauche, J. (1986). J. Phys. B: At. Mol. Phys., 19, 1881-1889.
Isotope shift and hyperfine structure in low-lying levels of Pb II

3652 Bouchaud, J.P. and Sornette, D. (1992). Europhys. Lett., 17, 721-726.
Magnetoconductance of Anderson insulators in the presence of strong spin-orbit scattering: a simple approach

3653 Bouchiat, C. and Piketty, C.A. (1988). J. Phys.(France), 49, 1851-1856.
Magnetic dipole and electric quadrupole amplitudes induced by the hyperfine interaction in the cesium 6S-7S transition and the parity violation calibration

3654 Boudjedaa, T., Chetouani, L., Guechi, L. and Hammann, T.F. (1992). Phys. Scr., 46, 289-294.
Path integral for particles of spin zero and 1/2 in the field of an electromagnetic plane wave

3655 Boudreaux, E.A., Doussa, S.P. and Klobukowski, M. (1986). Int. J. Quantum Chem., S20, 239-252.
Nonempirical self-consistent modified extended Huckel calculations on heavy-metal systems.
II. Electronic structure, bonding, and spectra of the binuclear $Pt_2(P_2O_5H_2)_4^{4-}$ ion

3656 Boulanger, D., Curie, D. and Parrot, R. (1991). J. Luminescence, 48 & 49, 680-684.
Strong and very strong covalency effects on the lifetime of Mn^{2+} in II-VI compounds

3657 Bouyssy, A., Mathiot, J.-F., Nguyen, V.G. and Marcos, S. (1987). Phys. Rev. C, 36, 380-401.
Relativistic description of nuclear systems in the Hartree-Fock approximation

3658 Bowmaker, G.A., Görling, A., Häberlen, O., Rösch, N., Goodman, G.L. and Ellis, D.E. (1992). Inorg. Chem., 31, 577-581.
A molecular orbital study of bonding and ionization energies in pentavalent uranium imide/amide complexes

3659 Bowmaker, G.A. and Schwerdtfeger, P. (1990). J. Mol. Struct.(Theochem), 205, 295-300.
Ab-initio calculations of the electronic structure and properties of the diatomic zinc monohalides ZnX (X=F, Cl, Br, I)

3660 Bowmaker, G.A., Schwerdtfeger, P. and von Szentpály, L. (1989). J. Mol. Struct.(Theochem), 184, 87-101.
Pseudopotential and multiple scattering Xα calculations of nuclear quadrupole coupling constants and other properties of diatomic halogen molecules and their monoanions and monocations

3661 Brage, T. (1988).
Theoretical Studies of Electron Correlation and Relativistic Effects in Complex Atomic Systems.
Thesis, Lund.

3662 Brage, T. and Litzén, U. (1987). Phys. Scr., 35, 662-667.
Nickel-like 4s-4p transitions and energy levels in Zr XIII, Nb XIV and Mo XV

3663 Bransden, B.H. and McDowell, M.R.C. (1992).
Charge Exchange and the Theory of Ion-Atom Collisions
Clarendon Press, Oxford

3664 Braun, M.A. (1988). Zh. Eksp. Teor. Fiz., 94, 145-158; Sov. Phys. JETP, 67, 2039-2046.
Decay law of unstable levels and spectral lineshape in the theory of a relativistic atom

3665*Braun, M.A. and Parera, J.J. (1985). Rev. Cubana Fis., 5(2-3), 63-76.
Electron pair in a relativistic atom. II. Bethe-Salpeter equation for the pair

3666 Braun, M.A. and Parera, J. (1986). Izv. Akad. Nauk SSSR, Ser. Fiz., 50, 1303-1308.
Transition probabilities between degenerate levels of the relativistic many-electron atom (in Russian)

3667 Bravo, G., Flores-Riveros, A. and Novaro, O. (1990). Int. J. Quantum Chem., 37, 785-796.
A CI pseudopotential-based description of the low-lying states of AgO_2

3668 Breidung, J. and Thiel, W. (1992). J. Comp. Chem., 13, 165-176.
A systematic ab initio study of the group V trihalides MX_3 and pentahalides MX_5 (M = P - Bi, X = F - I)

3669 Breidung, J., Thiel, W. and Komornicki, A. (1991). Inorg. Chem., 30, 1067-1073.
Ab initio calculation of harmonic force fields and vibrational spectra of the fluoroarsines AsH_nF_{3-n} (n=0-3) and the fluoroarsoranes AsH_nF_{5-n} (n=0-5)

3670*Breit, G. (1931c). Proc. Nat. Acad. Sci. USA, 17, 70-73.
On the interpretation of Dirac's alpha matrices

3671*Breit, G. (1937). Phys. Rev., 51, 248-262.
Approximately relativistic equations for nuclear particles

3672 Breza, M. (1992). Czech. J. Phys., 42, 439-446.
On electronic structure of H_2 in palladium

3673 Breza, M. and Boca, R. (1990a). Solid State Comm., 73, 783-785.
Electronic structure of $YBa_2Cu_3O_y$

3674 Breza, M. and Boca, R. (1990b). Solid State Comm., 74, 801-804.
Electronic structure of Tl-Ba-Ca-Cu-O systems

3675 Briand, J.P., Indelicato, P., Simionovici, A., San Vicente, V., Liesen, D. and Dietrich, D. (1989). Europhys. Lett., 9, 225-230.
Spectroscopic study of hydrogenlike and heliumlike xenon ions

3676*Brockmann, R. (1978). Phys. Rev. C, 18, 1510-1524.
Relativistic Hartree-Fock description of nuclei

3677 Brockmann, R. and Machleidt, R. (1990). Phys. Rev. C, 42, 1965-1980.
Relativistic nuclear structure. I Nuclear matter

3678 Brockmann, R. and Toki, H. (1992). Phys. Rev. Lett., 68, 3408-3411.
Relativistic density-dependent Hartree approach for finite nuclei

3679*Brockmann, R. and Weise, W. (1981). Nucl. Phys., A355, 365-382.
Relativistic single particle motion and spin-orbit coupling in nuclei and hypernuclei

3680*Brodsky, S.J. and Primack, J.R. (1968). Phys. Rev., 174, 2071-2073.
Electromagnetic interactions of loosely-bound composite systems

3681 Brooks, M.S.S. (1987). J. Magn. Magn. Mater., 63-64, 649-651.
Plutonium sulfide (PuS), plutonium selenide (PuSe), plutonium telluride (PuTe): relativistic semiconductors

3682 Brooks, M.S.S., Johansson, B., Eriksson, O. and Skriver, H.L. (1986). Physica, 144B, 1-13.
Relativistic effects in heavy elements

3683 Brouder, C. and Hikam, M. (1991). Phys. Rev. B, 43, 3809-3820.
Multiple-scattering theory of magnetic x-ray circular dichroism

3684 Brown, G.E. (1986). AIP Conf. Proc., 142, 2-26.
Relativistic effects in nuclear physics

3685 Brown, G.E. (1987a). Phys. Scr., 36, 71-76.
The relativistic atomic many-body problem

3686 Brown, G.E. (1987b). Phys. Scr., 36, 209-213.
Relativistic effects in nuclear physics

3687*Brown, G.E. and Schaefer, G.W. (1956). Proc. Roy. Soc. (London), A233, 527-536.
Expansion in angular momenta in bound-state perturbation theory

3688 Brown, G.E., Weise, W., Baym, G. and Speth, J. (1987). Comments Nucl. Part. Phys., 17, 39-62.
Relativistic effects in nuclear physics

3689 Broyles, A.A. (1988). Phys. Rev. A, 38, 1137-1148.
Relativistic equation for the multielectron atom

3690 Broyles, A.A. (1989). Phys. Rev. A, 39, 4939-4945.
Dissolution of relativistic atoms into negative energy states

3691 Bruce, S.A. (1986). Am. J. Phys., 54, 446-447.
A simple solution of the two-dimensional relativistic Kepler problem

3692*Bruch, R., Chung, K.T., Träbert, E., Heckmann, P.H., Raith, B. and Müller, H.R. (1984). J. Phys. B: At. Mol. Phys., 17, 333-344.
High-resolution EUV spectra of core-excited ^2P, $^2D^\circ$ and ^2F states of doubly ionised boron

3693 Bruch, R., Schneider, D., Chen, M.H., Chung, K.T. and Davis, B.F. (1992). Phys. Rev. A, 45, 4476-4480.
Atomic K-shell binding energies of multiply charged neon ions studied by zero-degree Auger-electron spectroscopy

3694 Bruneau, J. (1987). J. Phys. B: At. Mol. Opt. Phys., 20, 713-722.
A relativistic close-coupling approach to the Auger process

3695 Bruno, G., Ciliberto, E., Di Bella, S., Fragalà, I. and Müller, G. (1990). Organometallics, 9, 1629-1643.
Electronic structure of bis(dimethylmethylenephosphoranyl) dihydroborato(1-) main-group and transition-metal complexes. A case study involving theoretical pseudopotential ab initio calculations and UV photoelectron spectroscopy

3696 Brusich, M.J. (1989). Diss. Abstr. Int. B, 49, 4832.
Theoretical insights into the bonding in thorium organometallic complexes: A comparison with group IV transition metal chemistry

3697 Bueckert, H., Rothstein, S.M. and Vrbik, J. (1992). Chem. Phys. Lett., 190, 413-416.
A method for relativistic variational Monte Carlo calculations

3698 Buenker, R.J. (1992a). Mol. Phys., 76, 277-291.
Evidence from the special relativity and blackbody radiation theories for the existence of photons possessing zero kinetic energy

3699 Buenker, R.J. (1992b). Mol. Phys., 77, 1095-1122.
Exponentially damped Breit-Pauli Hamiltonian for the description of positronium decay and other high energy processes

3700 Buenker, R.J. and Chandra, P. (1988). Pure & Appl. Chem., 60, 167-173.
Application of configuration interaction for the study of relativistic effects in atoms and molecules

3701 Bukowski, R., Jeziorski, B., Moszynski, R. and Kolos, W. (1992). Int. J. Quantum Chem., 42, 287-319.
Bethe logarithm and Lamb shift for the hydrogen molecular ion

3702 Bulla, W., Gesztesy, F. and Unterkofler, K. (1988). Lett. Math. Phys., 15, 313-324.
On relativistic energy band corrections in the presence of periodic potentials

3703 Bulla, W., Gesztesy, F. and Unterkofler, K. (1992). Commun. Math. Phys., 144, 391-416.
Holomorphy of the scattering matrix with respect to c^{-2} for Dirac operators and an explicit treatment of relativistic corrections

3704 Burns, C.J. and Bursten, B.E. (1989). Comm. Inorg. Chem., 9, 61-93.
Covalency in f-element organometallic complexes: theory and experiment

3705 Bursten, B.E. and Clark, D.L. (1987). Polyhedron, 6, 695-704.
Theoretical description of metal-metal multiple bonds in $M_2(O_2CH)_4$ compounds using the Xα-SW MO method

3706 Bursten, B.E. and Novo-Gradac, K.J. (1987). J. Am. Chem. Soc., 109, 904-905.
Metal-metal bonds involving the f-elements. 2. Nature of the bonding in $(\eta^5\text{-}C_5H_5)_2(I)M\text{-}Ru(\eta^5\text{-}C_5H_5)(CO)_2$ (M=Zr, Th) complexes

3707 Bursten, B.E., Rhodes, L.F. and Strittmatter, R.J. (1989a). J. Am. Chem. Soc., 111, 2756-2758.
Bonding in tris(η^5-cyclopentadienyl) actinide complexes. 2. On the ground electronic configurations of "base-free" Cp_3An complexes (An=Th, Pa, U, Np, Pu)

3708 Bursten, B.E., Rhodes, L.F. and Strittmatter, R.J. (1989b). J. Am. Chem. Soc., 111, 2758-2766.
Bonding in tris(η^5-cyclopentadienyl) actinide complexes. 3. Interaction of π-neutral, π-acidic, and π-basic ligand with $(\eta^5\text{-}C_5H_5)_3U$

3709 Bursten, B.E., Rhodes, L.F. and Strittmatter, R.J. (1989c). J. Less-Comm. Met., 149, 207-211.
The bonding in tris(η^5-cyclopentadienyl) actinide complexes. 4. Electronic structural effects in $AnCl_3$ and $(\eta^5\text{-}C_5H_5)_3An$ (An=Th-Cf) complexes

3710 Bursten, B.E. and Strittmatter, R.J. (1987). J. Am. Chem. Soc., 109, 6606-6608.
f-orbital to carbonyl 2π back-bonding: The electronic structures of $(\eta^5\text{-}C_5H_5)_3UCO$ and $(\eta^5\text{-}C_5H_5)_3UOC$

3711 Bursten, B.E. and Strittmatter, R.J. (1991). Angew. Chem., 103, 1085-1103.
Cyclopentadienylkomplexe der actinoide: Bindungsverhältnisse und elektronenstruktur

3712 Burton, N.A., Hillier, I.H., Guest, M.F. and Kendrick, J. (1989). Chem. Phys. Lett., 155, 195-198.
Pseudopotential calculations of the geometry and ionization energies of palladium (II) acetylacetonate

3713 Butler, M.N., Chu, M.-C., Koonin, S.E. and Piekarewicz, J. (1988). Phys. Rev. A, 38, 2274-2283.
Positron interactions with high-Z atoms at relativistic energies

3714 Bychkov, Yu.A. (1988). Fiz. Tverd. Tela, 30, 3672-3676.
Energy spectra of two-dimensional systems with the spin-orbit interaction in an inclined magnetic field

3715 Bychkov, Yu.A., Mel'nikov, V.I. and Rashba, E.I. (1991). Ukr. Fiz. Zh., 36, 761-765.
Two-dimensional electrons with spin-orbit interactions in a tilted field (in Russian)

3716 Bylander, D.M. and Kleinman, L. (1990). Phys. Rev. B, 41, 907-912.
4f resonances with norm-conserving pseudopotentials

3717 Caballol, R., Català, J.A. and Problet, J.M. (1986). Chem. Phys. Lett., 130, 278-284.
Jahn-Teller distortions in XH_4^+ radical cations (X=Si, Ge, Sn). An ab initio CI study

3718 Caballol, R., Sanchez Marcos, E. and Barthelat, J.-C. (1987). J. Phys. Chem., 91, 1328-1333.
Theoretical study of the different coordination modes of copper-carbon dioxide complex

3719 Cabot, P.Ll., Illas, F., Ricart, J.M. and Rubio, J. (1986). J. Phys. Chem., 90, 30-33.
Ab initio study of the superoxides of group 13 metals

3720 Cai, Z.-Y. (1991). Diss. Abstr. Int. B, 52, 2097.
An ab initio relativistic many-body calculation of Hg $6p^2$ resonance states, including continuum effects

3721 Cai, Z.-Y. and Beck, D.R. (1989). Phys. Rev. A, 40, 1657-1659.
Bound excited states of Cr^-, Fe^-, Co^-, and Ni^-

3722 Cai, Z.-Y., Beck, D.R. and Perger, W.F. (1991). Phys. Rev. A, 43, 4660-4667.
Ab initio relativistic many-body calculations for the Hg $6p^2$ resonances

3723 Cai, Z.-Y., Umar, V.M. and Froese Fischer, C. (1992). Phys. Rev. Lett., 68, 297-300.
Large-scale relativistic correlation calculations: Levels of Pr^{+3}

3724 Caldas, M.J., Fazzio, A., Dabrowski, J. and Scheffler, M. (1990). Int. J. Quantum Chem., S24, 563-567.
Anion-antisite defects in GaAs: As and Sb

3725 Calkin, M.G., Kiang, D. and Nogami, Y. (1987). Am. J. Phys., 55, 737-739.
Proper treatment of the delta function potential in the one-dimensional Dirac equation

3726 Calkin, M.G., Kiang, D. and Nogami, Y. (1988). Phys. Rev. C, 38, 1076-1077.
Nonlocal separable potential in the one-dimensional Dirac equation

3727 Campbell, J.L. and Wang, J.-X. (1989). At. Data Nucl. Data Tables, 43, 281-291.
Interpolated Dirac-Fock values of L-subshell X-ray emission rates including overlap and exchange effects

3728 Cannata, F. and Ferrari, L. (1988). Am. J. Phys., 56, 721-725.
Dimensions of relativistic quantum mechanical paths

3729*Canuto, V. and Chiuderi, C. (1969). Lett. Nuovo Cim., 2,

223-227.
Solution of the Dirac equation in orthogonal electric and magnetic fields

3730 Capri, A.Z. and Roy, S.M. (1992). Mod. Phys. Lett. A, 7, 793-800.
A causal, Lorentz covariant and gauge-invariant vacuum for a Dirac field in the presence of an external electromagnetic field

3731 Cardona, M., Christensen, N.E. and Fasol, G. (1988). Phys. Rev. B, 38, 1806-1827.
Relativistic band structure and spin-orbit splitting of zinc-blende-type semiconductors

3732 Cardoso, J.G. (1992). J. Math, Phys., 33, 1109-1113.
Bessel function expressions for Dirac fields

3733 Carlacci, L. (1989). Diss. Abstr. Int. B, 49, 3777.
Molecular electronic structure theory: Application to spin-orbit interactions and hyperpolarizability.

3734 Carlsson, J. (1988). Phys. Rev. A, 38, 1702-1710.
Configuration interaction in the neutral copper atom

3735 Carnell, M., Peyerimhoff, S.D. and Hess, B.A. (1989). Z. Phys. D, 13, 317-333.
Ab initio MRD CI calculations on the cesium hydride (CsH) molecule

3736 Carroll, T.X., Ji, D., MacLaren, D.C., Thomas, T.D. and Saethre L.J. (1986). J. Electron Spectr. Rel. Phen., 42, 281-284.
Relativistic corrections to reported sulfur 1s ionization energies

3737 Carson, T.R. (1988). Astron. Astrophys. Suppl. Ser., 72, 531-541.
Relativistic radiative transitions.
Dirac-Coulomb oscillator strengths

3738 Casarin, M., Ciliberto, E., Di Bella, S., Gulino, A., Fragalà, I. and Marks, T.J. (1992). Inorg. Chem., 31, 2835-2842.
Electronic structure of tetracoordinate transition-metal complexes. 5. Comparative theoretical ab initio/Hartree-Fock-Slater and ultraviolet-photoelectron spectroscopic studies of building blocks for low-dimensional conductors. Dibenzo[b,i][1,4,8,11]tetraazacyclotetradecine complexes of nickel(II) and palladium(II)

3739 Case, D.A. (1986).
Spin-orbit and spin-polarization effects in metalloporphyrins
In "Porphyrins: Excited States & Dynamics", Ed. M. Gouterman, P.M. Rentzepis and K.D. Straub, ACS Symp. 321, pp. 59-71.

3740 Castaños, O., Frank, A., López, R. and Urrutia, L.F. (1991). Phys. Rev. D, 43, 544-547.
Soluble extensions of the Dirac oscillator with exact and broken supersymmetry

3741 Castillo, S., Poulain, E. and Novaro, O. (1989).
Int. J. Quantum Chem., S23, 509-520.
A theoretical study of the photochemistry of methylcopper hydride. Activation of the carbon-hydrogen bond

3742 Castillo, S., Poulain, E. and Novaro, O. (1991).
Int. J. Quantum Chem., S25, 577-585.
A theoretical study of the photochemistry of methylcopper hydride: II. Formation and stability of the $HCuCH_3$ intermediate complex

3743*Castorina, P., Cea, P., Nardulli, G. and Paiano, G. (1984). Phys. Rev. D, 29, 2660-2663.
Partial-wave analysis of a relativistic Coulomb problem

3744 Caswell, W.E. and Lepage, G.P. (1986). Phys. Lett., 167B, 437-442.
Effective lagrangians for bound state problems in QED, QCD, and other field theories

3745 Cauletti, C., de Simone, M. and Stranges, S. (1991).
J. El. Sp. Rel. Phen., 57, R1-.
Electronic structure of $SnCl_2$ by He I and He II gas-phase photoelectron spectroscopy and pseudopotential calculations

3746 Cauletti, C., Grandinetti, F., Granozzi, G., Sebald, A. and Wrackmeyer, B. (1988). Organometallics, 7, 262-266.
Sn-S and Sn-Se bonding in some Tin(IV) compounds studied by UV photoelectron and NMR spectroscopy and pseudopotential ab initio calculations

3747 Cave, R.J., Davidson, E.R., Sautet, P., Canadell, E. and Eisenstein, O. (1989). J. Am. Chem. Soc., 111, 8105-8111.
A theoretical study of models for X_2Y_2 Zintl ions

3748 Cavell, R.G. and Sodhi, R.N.S. (1986). J. El. Spectr. Rel. Phen., 41, 25-35.
Absolute total static relaxation and extra-atomic relaxation terms for volatile phosphorus compounds obtained from the Auger parameter and relativistic atomic calculations: comparisons with empirical relaxation shifts

3749 Cayton, R.H. (1988). Diss. Abstr. Int. B, 49, 1161.
Relationships between electronic structure and the reactivity patterns and conformational preferences in organometallic piano-stool systems

3750 Cayton, R.H., Novo-Gradac, K.J. and Bursten, B.E. (1991).
Inorg. Chem., 30, 2265-2272.
Metal-metal bonds involving the f elements. 4. Molecular orbital studies of metal-metal and metal-ligand interactions in dinuclear uranium(V) systems

3751 Centelles, M., Viñas, X., Barranco, M. and Schuck, P. (1990). Nucl. Phys. A, 519, 73c-82c.
On the relativistic extended Thomas-Fermi method

3752 Chaix, P. and Iracane, D. (1989). J. Phys. B: At. Mol. Opt. Phys., 22, 3791-3814.
From quantum electrodynamics to mean-field theory: I. The Bogoliubov-Dirac-Fock formalism

3753 Chaix, P., Iracane, D. and Lions, P.L. (1989). J. Phys. B: At. Mol. Opt. Phys., 22, 3815-3828.
From quantum electrodynamics to mean-field theory: II. Variational stability of the vacuum of quantum electrodynamics in the mean-field approximation

3754 Chang, A.H.H. (1991). Diss. Abstr. Int. B, 52, 2571.
Electronic structure of C_{60}, C_{60} complexes, and actinocenes

3755 Chang, A.H.H., Ermler, W.C. and Pitzer, R.M. (1991). J. Chem. Phys., 94, 5004-5010.
The ground and excited states of $C_{60}M$ and $C_{60}M^+$ (M=O, F, K, Ca, Mn, Cs, Ba, La, Eu, U)

3756 Chang, A.H.H. and Pitzer, R.M. (1989). J. Am. Chem. Soc., 111, 2500-2507.
Electronic structure and spectra of uranocene

3757 Chang, Ch., Pelissier, M. and Durand, Ph. (1986). Phys. Scr., 34, 394-404.
Regular two-component Pauli-like effective Hamiltonians in Dirac theory

3758 Chang, T.N. (1989). Phys. Rev. A, 39, 6129-6133.
Combined perturbation and configuration-interaction calculation for the fine-structure level splittings of the 1s4f states of helium

3759 Chang, T.N. and Bryan, E.T. (1988). Phys. Rev. A, 38, 645-653.
Effect of the configuration interaction on the fine-structure-level splittings for quasi-two-electron atoms

3760 Chang, T.N. and Chung, K.T. (1989). Chi. J. Phys., 27 425-434.
Electron correlation due to intrashell core excitation and intershell core-valence interaction

3761*Chang, T.Y. (1967). Rev. Mod. Phys., 39, 911-942.
Moderately long-range interatomic forces

3762 Chang, Y.-C. and James, R.B. (1992). Phys. Rev. B, 46, 15040-15045.
Electronic and optical properties of HgI_2

3763 Chao, C.Y.-P. and Chuang, S.L. (1992). Phys. Rev. B, 46, 4110-4122.
Spin-orbit-coupling effects on the valence-band structure of strained semiconductor quantum wells

3764 Chapman, D.A. (1989). Diss. Abstr. Int. B, 50, 191.
One-electron properties in the relativistic CI approximation: Calculation of electric dipole and transition moments

3765 Chapman, D.A., Balasubramanian, K. and Lin, S.H. (1987a). J. Chem. Phys., 87, 5325-5328.
Electronic dipole and transition moments from the relativistic CI wave function: Application to HI

3766 Chapman, D.A., Balasubramanian, K. and Lin, S.H. (1987b). Chem. Phys., 118, 333-343.
A theoretical study of spectroscopic properties and transition moments of HBr

3767 Chapman, D.A., Balasubramanian, K. and Lin, S.H. (1988a). Phys. Rev. A, 38, 6098-6106.
Theoretical study of the negative ions of HBr and HI

3768 Chapman, D.A., Balasubramanian, K., Lin, S.H., Kaufman, J.J., Hariharan, P.C. and Koski, W.S. (1989). Phys. Rev. A, 39, 4428-4431.
Transition moments for energy transfers in $Ar-Cl^+$ collisions

3769 Chapman, D.A., Li, J.-Q., Balasubramanian, K. and Lin, S.H. (1988b). J. Chem. Phys., 88, 3826-3833.
Theoretical study of electric dipole and transition moments of GeH, SnH, and PbH

3770 Chatterjee, A. (1986). J. Math. Phys., 27, 2331-2335.
Large-N solution of the Klein-Gordon equation

3771 Chatterjee, A. (1990). Phys. Rep., 186, 249-370.
Large N expansions in quantum mechanics, atomic physics and some O(N) invariant systems

3772 Chatterjee, R. and Buckmaster, H.A. (1989). Can. J. Phys., 67, 632-633.
New relativistic crystalline electric field effective operators

3773 Chatterjee, R. and Buckmaster, H.A. (1991). J. Phys.: Condens. Matter, 3, 7079-7084.
A relativistic effective Hamiltonian for S-state ions

3774 Chatterjee, R., Lulek, T. and Biel, J.K. (1992). Acta Phys. Pol. A, 81, 467-474.
Electronic spaces and classification of multielectron states in multicentre systems

3775 Chen, C.-T. and Huang, K.N. (1990). Phys. Rev. A, 42, 6934-6937.
Anomalous doublet splittings of binding energies in Dirac-Fock calculation

3776 Chen, M.H. (1986a). Phys. Rev. A, 33, 994-999.
Multiconfiguration Dirac-Fock calculations of dielectronic recombination coefficients for the He isoelectronic sequence

3777 Chen, M.H. (1986b). Phys. Rev. A, 34, 1073-1078.
Relativistic calculations of dielectronic recombination coefficients for the Ne isoelectronic sequence

3778 Chen, M.H. (1986c). Phys. Rev. A, 34, 1079-1083.
Dielectronic recombination coefficient for F-like selenium

3779 Chen, M.H. (1988a). Phys. Rev. A, 38, 2332-2337.
Dielectronic recombination coefficients for fluorinelike ions: The multiconfiguration Dirac-Fock calculation

3780 Chen, M.H. (1988b). Phys. Rev. A, 38, 3280-3285.
Effects of relativity and configuration interaction on dielectronic recombination of hydrogenlike ions

3781 Chen, M.H. (1988c). Phys. Rev. A, 38, 6430-6432.
Effect of Coster-Kronig transition on dielectronic recombination of the He-like ions

3782 Chen, M.H. (1989a). Phys. Rev. A, 40, 2365-2372.
Effects of relativity and configuration interaction on L-shell Auger and radiative decays of the doubly excited 3l3l' states of sodiumlike ions

3783 Chen, M.H. (1989b). Phys. Rev. A, 40, 2758-2761.
Effective L-shell fluorescence yields for sodiumlike and neonlike low-lying autoionizing states

3784 Chen, M.H. (1989c). Nucl. Instrum. Meth. B, 43, 366-371.
Relativity: X-ray and Auger transitions of highly charged ions

3785 Chen, M.H. (1990a). Phys. Rev. A, 42, 5228-5231.
Resonant transfer and excitation in collisions of Ca^{q+} with H_2 and He targets

3786 Chen, M.H. (1990b). AIP Conf. Proc., 215, 391-407.
X-ray and Auger transitions in atoms and ions

3787 Chen, M.H. (1991a). Nucl. Instrum. Meth. Phys. Res. B, B56-57(1), 149-153.
Effects of relativity on RTEX in collisions of U^{q+} with light targets

3788 Chen, M.H. (1991b). Phys. Rev. A, 44, 239-242.
Auger transistion rates and fluorescence yields for the double-K-hole state

3789 Chen, M.H. (1991c). Phys. Rev. A, 44, 4215-4223.
Relativistic intermediate-coupling calculations of dielectronic-recombination coefficients for the Li isoelectronic sequence

3790 Chen, M.H. (1992). Phys. Rev. A, 45, 1684-1689.
Effect of intermediate coupling on angular distribution of Auger electron

3791 Chen, M.H. and Crasemann, B. (1986). Phys. Rev. A, 34, 87-92.
Relativistic calculation of atomic N-shell ionization by protons

3792 Chen, M.H. and Crasemann, B. (1987a). Phys. Rev. A, 35, 4839-4841.
Lifetime of the 1s2p3d $^4F_{9/2}$ metastable state of Li-like ions

3793 Chen, M.H. and Crasemann, B. (1987b). Phys. Rev. A, 35, 4579-4585.
Auger and radiative decay rates of 1s vacancy states in the boron isoelectronic sequence: Effects of relativity and configuration interaction

3794 Chen, M.H. and Crasemann, B. (1987c). At. Data Nucl. Data Tables, 37, 419-439.
Dielectronic satellite spectra of Li-like ions

3795 Chen, M.H. and Crasemann, B. (1988a). At. Data Nucl. Data Tables, 38, 381-424.
K-shell Auger and raditive transitions in the boron isoelectronic sequence

3796 Chen, M.H. and Crasemann, B. (1988b). Phys. Rev. A, 37, 2886-2892.
Effects of autoionizing resonances on electron-impact excitation rates for Be-like ions

3797 Chen, M.H. and Crasemann, B. (1988c). Phys. Rev. A, 38, 5595-5601.
Relativistic calculations of dielectronic recombination rate coefficients for Be-like ions

3798 Chen, M.H. and Crasemann, B. (1989a). At. Data Nucl. Data Tables, 41, 59-69.
Contributions of autoionizing resonances to the n=2-2 electron collisional excitation rates for Be-like ions

3799 Chen, M.H. and Crasemann, B. (1989b). At. Data Nucl. Data Tables, 41, 257-285.
Atomic K-, and M-shell cross sections for ionization by protons: A relativistic Hartree-Slater calculation

3800 Chen, M.H. and Crasemann, B. (1989c). Phys. Rev. A, 40, 4330-4337.
Systematic trends of the oscillator strengths for n=3 and 2 electric dipole transitions in oxygenlike ions

3801 Chen, M.H. and Crasemann, B. (1991). Phys. Rev. A, 43, 5886-5890.
Resonance enhancement of electron-impact excitation rates for argonlike ions

3802 Chen, M.H., Larkins, F.P. and Crasemann, B. (1990a). At. Data Nucl. Data Tables, 45, 1-205.
Auger and Coster-Kronig radial matrix elements for atomic numbers 6 ≤ Z ≤ 92

3803 Chen, M.H. and Reed, K.J. (1989). Phys. Rev. A, 40, 2292-2300.
Direct and resonance contributions to electron-impact excitation of n=2 to n=3 transitions in neonlike ions

3804 Chen, M.H. and Reed, K.J. (1990). Phys. Rev. A, 41, 4759-4765.
Effects of autoionizing resonances on electron-impact excitation rates for transitions amond n=3 levels in neonlike selenium

3805 Chen, M.H. and Reed, K.J. (1992). Phys. Rev. A, 45, 4525-4529.
Electron-impact ionization of highly charged lithiumlike ions

3806 Chen, M.H., Reed, K.J. and Moores, D.L. (1990b). Phys. Rev. Lett., 64, 1350-1353.
Contributions of resonant excitation double autoionization to the electron-impact ionization of Fe^{15+}

3807 Chen, M.H., Weber, W. and Mehlhorn, W. (1989). J. Electron Spectr. Rel. Phen., 49, 77-85.
Very strong correlation effects in Auger spectra through the collapse of D orbitals: The 2p-3p4s Auger spectrum of free Ca atoms

3808 Chen, M.-K. (1990). J. Phys. B: At. Mol. Opt. Phys., 23, 4041-4049.
Hyperfine splitting in muonic ^{4}He and ^{3}He atoms: a global-operator approach

3809 Chen, M.-K. (1992). Phys. Rev. A, 45, 1479-1492.
Correlated wave functions and hyperfine splittings of the 2s state of muonic ^{3,4}He atoms

3810 Chen, W.Q. and Cook, A.H. (1987). J. Phys. B: At. Mol. Phys., 20, 3587-3598.
A quasi-relativistic treatment of nuclear motion in atoms and molecules

3811 Chen, Z.-H. and Goldman, S.P. (1991a). Phys. Rev. A, 44, 4459-4463.
Relativistic calculations for hydrogenic atoms in strong magnetic fields with Slater-type basis functions

3812 Chen, Z.-H. and Goldman, S.P. (1991b). AIP Conf. Proc., 233, 525-527.
Relativistic variational calculations for hydrogenic atoms in strong magnetic fields

3813 Chen, Z.-H. and Goldman, S.P. (1992). Phys. Rev. A, 45, 1722-1731.
Relativistic and nonrelativistic finite-basis-set calculations of low-lying levels of hydrogenic atoms in intense magnetic fields

3814 Chen, Z.-H., Rosenberg, L. and Spruch, L. (1987). Phys. Rev. A, 35, 1981-1990.
Hartree-Fock calculations of the ratio of bound-state to continuum β decay of ^{187}Re, the chronometric partner of ^{187}Os

3815 Cheng, H.-P. and Ellis, D.E. (1991). J. Chem. Phys., 94,

3735-3747.
Electronic structure, binding energies, and interaction potentials of transition metal clusters

3816 Cheng, K.-T., Johnson, W.R. and Sapirstein, J. (1991). Phys. Rev. Lett., 66, 2960-2963.
Screened Lamb shift calculations for lithiumlike uranium, sodiumlike platinum and copperlike gold

3817 Cheng, K.-T. and Wagner, R.A. (1987). Phys. Rev. A, 36, 5435-5438.
Quantum electrodynamics effects in the 4s-4p transitions in Cu-like and Zn-like ions

3818 Cheng, T.-C. and Huang, K.-N. (1992). Phys. Rev. A, 45, 4367-4373.
Resonance transitions of Zn-like ions from the multiconfiguration relativistic random-phase approximation

3819 Cheng, W. and Balasubramanian, K. (1991). J. Mol. Spectrosc., 149, 99-108.
Spectroscopic constants and potential energy curves for TaH

3820 Cheon, I.-T. (1988). Phys. Rev. A, 37, 2785-2796.
Possible new frequency shifts for the $2S_{1/2}$ --> $2P_{1/2}$ transition in a hydrogen atom within a specific boundary

3821 Chermette, H. (1992). New. J. Chem., 16, 1081-1088.
20 years of the MS-Xα method

3822 Chermette, H., Bellafrouh, K., Goursot, A. and Waltz, W.L. (1991). Chem. Phys. Lett., 184, 282-288.
Electronic structure and photobehavior of hexaamminechromium (III) ion

3823 Cheung, C.Y. and Li, S.P. (1990). Phys. Lett. B, 234, 444-448.
Solving a phenomenological two-body Dirac equation

3824 Cheung, C.Y. and Li, S.P. (1992). Phys. Lett. B, 293, 265-269.
A quasi-relativistic equation for a fermion-boson system

3825 Chi, H.-C. and Huang, K.-N. (1991). Phys. Rev. A, 43, 4742-4745.
Photoionization of beryllium in the multiconfiguration relativistic random-phase approximation

3826 Chi, H.-C., Huang, K.-N. and Cheng, K.T. (1991). Phys. Rev. A, 43, 2542-2545.
Autoionizing levels of beryllium from the multiconfiguration relativistic random-phase approximation

3827 Childs, W.J., Goodman, L.S. and Cheng, K.T. (1986). Phys. Rev. A, 33, 1469-1479.
Hyperfine structure of excited $4f^{11}5d6s^2$ levels in ^{167}Er I: Measurements and multiconfiguration Dirac-Fock calculations

3828 Childs, W.J. and Nielsen, U. (1988). Phys. Rev. A, 37,

6-15.
Hyperfine structure of the $(5d+6s)^3$ configuration of ^{139}La I: New measurements and ab initio multiconfigurational Dirac-Fock calculations

3829 Chiu, Y.-N. (1968). J. Chem. Phys., 48, 3476-3492.
Relativistic effect and nonconservative spin-orbital forces in the radiative transition of molecules

3830 Cho, H.T., Diek, A. and Kantowski, R. (1990). J. Math. Phys., 31, 2192-2200.
A Clifford algebra quantization of Dirac's electron positron field

3831*Chodos, A., Jaffe, R.L., Johnson, K., Thorn, C.B. and Weisskopf, V.F. (1974). Phys. Rev. D, 9, 3471-3495.
New extended model of hadrons

3832 Choe, A.S., Yoo, B.-D. and Lee, J.-M. (1992). New Phys. (Korea), 32, 472-477.
The calculation of the photoionization cross sections for the $^1S(6s^2)$ and $^{1,3}P(6s6p)$ states of Yb using the local spin density functional theory

3833 Chong, D.P., Langhoff, S.R., Bauschlicher Jr., C.W., Walch, S.P. and Partridge, H. (1986). J. Chem. Phys., 85, 2850-2860.
Theoretical dipole moments for the first-row transition metal hydrides

3834 Chou, H.-S. and Huang, K.-N. (1991).
Multiconfiguration Dirac-Fock theory including parity and time-reversal nonconserving interactions
In "Prog. Nucl. Phys., Proc. 2nd Int. Conf. and Spring School on Medium and High En. Nucl. Phys., Part B", ed. W.-Y. Hwang et al., North-Holland, Amsterdam, pp. 315-328.

3835 Chou, H.-S. and Huang, K.-N. (1992a). Phys. Rev. A, 45, 1403-1406.
Relativistic excitation energies and oscillator strengths for the $6s^2\ ^1S_0$ --> $6s6p\ ^1P_1$, $^3P_1$ transitions in Hg-like ions

3836 Chou, H.-S. and Huang, K.-N. (1992b). Phys. Rev. A, 46, 3725-3729.
Relativistic excitation energies and oscillator strengths for the $(5s^2)\,^1S_0$ -> $(5s5p)\,^1P_1$, $^3P_1$ transitions in Cd-like ions

3837 Chou, H.-S., Huang, K.-N. and Johnson, W.R. (1991). Phys. Rev. A, 44, R2769-R2770.
Magnetic dipole transitions of Pb-like ions from the multiconfiguration relativistic random-phase approximation

3838 Christensen, N.E., Satpathy, S. and Pawlowska, Z. (1986). Phys. Rev. B, 34, 5977-5980.
First-principles theory of tetrahedral bonding and crystal structure of lead

3839 Christiansen, P.A. (1986). Chem. Phys. Lett., 127,

50-54.
Core-valence correlation in relativistic effective potentials

3840 Christiansen, P.A. (1989).
Relativistic effective potentials in quantum Monte Carlo studies
In "The Challenge of d and f Electrons", Ed. D.R. Salahub and M.C. Zerner, ACS Symposium Series No.394, pp. 309-321.

3841 Christiansen, P.A. (1990). J. Phys. Chem., 94, 7865-7867.
Core polarization in aluminum

3842 Christiansen, P.A. (1991). J. Chem. Phys., 95, 361-363.
Relativistic effective potentials in transition metal quantum Monte Carlo simulations

3843 Chulkov, E.V., Koroteev, Yu.M. and Silkin, V.M. (1991). Surface Science, 247, 115-119.
Relativistic electronic structure of metal surfaces: first-principles calculation

3844 Chung, K.T. (1989a). Phys. Rev. A, 40, 4203-4208.
$1s2s2p^2\ ^5P$-$1s2p^3\ ^5S^o$ transition and fine structure of Be-like systems

3845 Chung, K.T. (1989b). Chi. J. Phys., 27, 90-102.
Spin-induced autoionization and differential metastability of lithium-like systems

3846 Chung, K.T. (1990a). Phys. Scr., 42, 530-536.
Identification of 1s core-excited four-electron resonances in high resolution spectra of carbon

3847 Chung, K.T. (1990b). Phys. Scr., 42, 537-539.
The 1s2l2l'2l" states of oxygen

3848 Chung, K.T. (1990c). J. Phys. B: At. Mol. Opt. Phys., 23, 2929-2943.
Identification of high resolution Auger spectra of boron and beryllium

3849 Chung, K.T. (1991a). Phys. Rev. A, 44, 5421-5433.
Ionization potential of the lithiumlike $1s^2 2s$ states from lithium to neon

3850 Chung, K.T. (1991b). J. Phys. B: At. Mol. Opt. Phys., 24, 3923-3932.
2s and 2p energy inversion in a doubly-core-excited four-electron system

3851 Chung, K.T. (1992). Phys. Rev. A, 45, 7766-7773.
Ionization potential for the lithiumlike $1s^2 2s$ states from Na IX to Ca XVIII

3852 Churilov, S.S., Ryabtsev, A.N. and Wyart, J.-F. (1988). Phys. Scr., 38, 326-335.
Identification of n=4, $\Delta n=0$ transitions in the spectra of nickel-like and zinc-like ions through tin

3853 Cinal, M. and Englert, B.-G. (1992). Phys. Rev. A, 45, 135-139.
Thomas-Fermi-Scott model in momentum space

3854 Ciullo, G., Rosi, M., Sgamellotti, A. and Floriani, C. (1991). Chem. Phys. Lett., 185, 522-528.
On the relative stability of side-on and end-on coordination of dinitrogen to nickel(0): "ab initio" calculations on the model compounds [NiN_2] and [$Ni(PH_3)_2N_2$]

3855*Cízek, J. and Paldus, J. (1977). Int. J. Quantum Chem., 12, 875-896.
An algebraic approach to bound states of simple one-electron systems

3856 Clemence, D.P. (1990). Inverse Problems, 6, 175-184.
Levinson's theorem for a Dirac system

3857 Clerk, G.J. and Davies, A.J. (1992). Am. J. Phys., 60, 537-539.
Inversion of transmission for the one-dimensional Dirac equation

3858 Clerk, G.J. and McKellar, B.H.J. (1991). Phys. Lett. A, 158, 261-264.
On the occurrence of relativistic band gaps in one-dimensional disordered systems

3859 Clerk, G.J. and McKellar, B.H.J. (1992). J. Phys. A, 25, L861-L866.
Scalar symmetry of the massless Dirac equation

3860 Coffey, D., Rice, T.M. and Zhang, F.C. (1991). Phys. Rev. B, 44, 10112-10116.
Dzyaloshinskii-Moriya interaction in the cuprates

3861 Cogordan, J.A. and Lunell, S. (1986). Phys. Scr., 33, 406-411.
Energies of $2p^5 3s$, 3p and 3d levels of neon-like ions from relativistic MCDF calculations, $20 \le Z \le 54$

3862*Cohen, D. and Basch, H. (1983). J. Am. Chem. Soc., 105, 6980-6982.
Ethylene and acetylene complexes of the silver atom

3863 Cohen, J.M. and Mustafa, E. (1986). Phys. Lett. A, 115, 265-267.
Can classical physics give g-factors > 1 ?

3864*Cole, L.A. and Harmon, B.N. (1981). J. Appl. Phys., 52, 2149-2151.
Calculations for 4f atoms using new exchange-correlation schemes

3865 Colmenares-Landin, F., Castillo, S., Martinez-Magadán, J.M., Novaro, O. and Poulain, E. (1992). Chem. Phys. Lett., 189, 378-382.
Role of excited states of maximal d-shell occupancy in the $Ru+H_2$ reaction

3866 Combet Farnoux, F. (1987).
Giant resonances in heavy elements: Open-shell effects, multiplet structure and spin-orbit interaction
In "Giant Resonances in Atoms, Molecules, and Solids", Ed. J.P. Connerade, J.M. Esteva and R.C. Karnatak, New York, Plenum Press, pp. 153-170.

3867 Connerade, J.P. (1991). Comments At. Mol. Phys., 26, 155-177.
Today and tomorrow in photoionization

3868 Connerade, J.P. (1992). Adv. At. Mol. Phys., 29, 325-367.
Cooperative effects in atomic physics

3869 Connerade, J.P. and Karnatak, R.C. (1990). Comments At. Mol. Phys., 24, 1-35.
Centrifugal barrier effects in atoms, clusters and solids: Do they control valence changes?

3870 Connerade, J.P. and Sarpal, B.K. (1992). J. Phys. B: At. Mol. Opt. Phys., 25, 2245-2256.
A fully relativistic analysis of the 3p spectrum of Ca I

3871 Cooper, F., Khare, A., Musto, R. and Wipf, A. (1988). Ann. Phys.(NY), 187, 1-28.
Supersymmetry and the Dirac equation

3872 Cordova, N.J. (1991). Mod. Phys. Lett. A, 6, 3071-3077.
Fractional charge in 1+1, 2+1 and 3+1 dimensions

3873 Cortona, P. (1989). Phys. Rev. B, 40, 12105-12110.
Simplified method for fully relativistic spin-polarized density-functional calculations

3874 Cortona, P. (1991).
Relativistic spin-polarized density-functional theory: Simplified method for fully relativistic calculations
In "The Effects of Relativity in Atoms, Molecules, and the Solid State", Ed. S. Wilson, I.P. Grant and B.L. Gyorffy, Plenum, New York and London, pp. 285-293.

3875 Contona, P. (1992). Phys. Rev. B, 46, 2008-2014.
Direct determination of self-consistent total energies and charge densities of solids: A study of the cohesive properties of the alkali halides

3876 Cortona, P. and Sommers, C. (1987). J. Magn. Magn. Mater., 63-64, 658-660.
Properties of rare earth systems by the relativistic self-interaction-corrected local-density-approximation

3877 Costa Cabral, B.J., Cordeiro, M.N.D.S. and Telo da Gama, M.M. (1991). J. Phys.:Condens. Matter, 3, 5615-5620.
The structure of molten CsAu: ab initio and Monte Carlo study

3878 Cotton, F.A. and Feng, X.-J. (1989). Inorg. Chem., 28, 1180-1183.
Electronic structures of the $Ru_2(RNNNR)_4$, $Rh_2(RNCHNR)_4$, and $Co_2(RNNNR)_4$ molecules: Ordering and separation of M-M δ^{*2} and π^* orbitals

3879 Cotton, F.A. and Feng, X.-J. (1991). Inorg. Chem., 30, 3666-3670.
Electronic structure and bonding in trinuclear molybdenum and tungsten cluster compounds of M_3X_{13} type

3880 Couch, R. and Scott, B.L. (1991). Phys. Rev. A, 44, 3016-3022.
Relativistic calculations of Kα satellite properties for medium-Z elements

3881 Coutinho, F.A.B., Glöckle, W., Nogami, Y. and Toyama, F.M. (1988b). Can. J. Phys., 66, 769-775.
Two-body Dirac equation: illustration in one space dimension

3882 Coutinho, F.A.B. and Nogami, Y. (1987a). Phys. Rev. A, 36, 1008-1012.
Exactly solvable relativistic Hartree-Fock equation

3883 Coutinho, F.A.B. and Nogami, Y. (1987b). Phys. Lett. A, 124, 211-214.
Conditions for the existence of bound states of a Dirac particle in one dimension

3884 Coutinho, F.A.B. and Nogami, Y. (1990). Phys. Rev. A, 42, 5716-5719.
Zero-range potential for the Dirac equation in two and three dimensions: Elementary proof of Svendsen's theorem

3885 Coutinho, F.A.B., Nogami, Y., Parent, N.P. and Suzuki, A. (1991). Phys. Lett. A, 161, 26-29.
Conditions for the existence of bound states of Dirac particle in two and three dimensions

3886 Coutinho, F.A.B., Nogami, Y. and Toyama, F.M. (1988a). Am. J. Phys., 56, 904-907.
General aspects of the bound-state solutions of the one-dimensional Dirac equation

3887 Cowan, R.D., Martin, N.L.S. and Wilson, M. (1988). J. Phys. B: At. Mol. Opt. Phys., 21, L1-L3.
Non-metastability of the $5d^9(^2D_{3/2})6s^26p_{1/2}$ J=2 level in Hg

3888 Cowan, R.D. and Wilson, M. (1991a). J. Phys. B: At. Mol. Opt. Phys., 24, 111-114.
Theoretical autoionization widths for the levels of the $4d^95s^25p$ configuration of Cd I

3889 Cowan, R.D. and Wilson, M. (1991b). Phys. Scr., 43, 244-247.
Simple estimates of atomic negative ion structures

3890 Crasemann, B. (1989). Acta Phys. Hung., 65, 171-181.
Relativistic phenomena in atomic and chemical physics: Opportunities for studies with synchrotron radiation

3891 Crasemann, B., Karim, K.R. and Chen, M.H. (1987).

At. Data Nucl. Data Tables, 36, 355-374.
Resource note: Theoretical atomic-electron bonding energies

3892 Crater, H.W. (1991). Phys. Rev. A, 44, 7065-7070.
Singlet-positronium decay using a relativistic wave function

3893 Crater, H.W., Becker, R.L., Wong, C.Y. and Van Alstine, P. (1992). Phys. Rev. D, 46, 5117-5155.
Nonperturbative solution of two-body Dirac equations for quantum electrodynamics and related field theories

3894 Crater, H.W. and van Alstine, P. (1987). Phys. Rev. D, 36, 3007-3036.
Two-body Dirac equations for particles interacting through world scalar and vector potentials

3895 Crater, H.W. and van Alstine, P. (1992). Phys. Rev. D. 46, 766-776.
Restrictions imposed on relativistic two-body interactions by classical relativistic field theory

3896 Crawford, J.P. (1991). J. Math. Phys., 32, 576-583.
Clifford algebra: Notes on the spinor metric and Lorentz, Poincaré, and conformal groups

3897 Crépin, C. and Millié, P. (1989). Chem. Phys., 133, 377-393.
Mechanism of Hg(^{3}P) relaxation in nitrogen matrices. I. Theoretical study of HgN_2

3898 Cruz Barrios, S. and Nemes, M.C. (1992). Mod. Phys. Lett. A, 7, 1915-1921.
Anatomy of relativistic mean-field approximations

3899 Cuellar, L.E., Compton, R.N., Carman Jr, H.S. and Feigerle, C.S. (1990). Phys. Rev. Lett., 65, 163-166.
Photoelectron angular distributions for ns (n=8-12) subshells of cesium: Relativistic effects

3900 Culberson, J. C., Knappe, P., Rösch, N. and Zerner, M.C. (1987). Theor. Chim. Acta, 71, 21-39.
An intermediate neglect of differential overlap (INDO) technique for lanthanide complexes: Studies on lanthanide halides

3901 Cundari, T.R. (1992a). J. Am. Chem. Soc., 114, 7879-7888.
Transition metal imido complexes

3902 Cundari, T.R. (1992b). Int. J. Quantum Chem., S26, 793-806.
The activation and elimination of H_2 by Zr complexes

3903 Cundari, T.R. (1992c). J. Am. Chem. Soc., 114, 10557-10563.
Methane activation by Group IVB imido complexes

3904 Cundari, T.R. and Gordon, M.S. (1992a). Organometallics, 11, 55-63.
Theoretical investigations of olefin metathesis catalysts

3905 Cundari, T.R. and Gordon, M.S. (1992b). J. Am. Chem. Soc., 114, 539-548.
High-valent transition-metal alkylidene complexes: Effect of ligand and substituent modification

3906 Curtis, L.J. (1987). Phys. Rev. A, 35, 2089-2094.
Fine-structure intervals for the lowest P terms in the Cu, Zn, Ga, and Br isoelectronic sequences for Z≤92

3907 Curtis, L.J. (1989). J. Phys. B: At. Mol. Opt. Phys., 22, L267-L271.
Computation of M1 decay rates in intermediate coupling using empirical spectroscopic data

3908 Curtis, L.J. (1990).
Semiempirical formulations of line strengths using single-triplet mixing angles
In "Atomic Spectra and Oscillator Strengths for Astrophysics and Fusion Research", Ed. J.E. Hansen, North-Holland, Amsterdam, pp. 134-135.

3909 Curtis, L.J., Haar, R.R. and Martinson, I. (1988). Nucl. Instrum. Meth. B, 31, 280-283.
Multiconfiguration Dirac-Fock calculations of energy levels and transition probabilities in Ne-like S VII, Cl VIII and Ar IX

3910 Curtis, L.J. and Theodosiou, C.E. (1989). Phys. Rev. A, 39, 605-615.
Comprehensive calculations of 4p and 4d lifetimes for the Cu sequence

3911 Czuchaj, E., Rebentrost, F., Stoll, H. and Preuss, H. (1989). Chem. Phys., 138, 303-314.
Semi-local pseudopotential calculations for the potential energies of the CaHe and CaNe systems

3912 Czuchaj, E., Rebentrost, F., Stoll, H. and Preuss, H. (1991). Chem. Phys. Lett., 178, 246-254.
Pseudopotential calculations for the potential energy curves and transition dipole moments of the NaHg system

3913 Czuchaj, E., Rebentrost, F., Stoll, H. and Preuss, H. (1992a). Chem. Phys. Lett., 197, 187-194.
Calculation of the potential energies and transition dipole moments of the CdHg pair

3914 Czuchaj, E., Rebentrost, F., Stoll, H. and Preuss, H. (1992b). Chem. Phys. Lett., 199, 47-54.
Calculation of the potential energies and transition dipole moments of the KHg pair

3915 Czuchaj, E., Stoll, H. and Preuss, H. (1987). J. Phys. B: At. Mol. Phys., 20, 1487-1507.
Pseudopotential SCF/CI calculations for the potential energies of the MHe and MNe (M=Mg, Cd, Hg) systems

3916 Czyzewski, Z., O'Neill MacCallum, D., Romig, A., and Joy, D.C. (1990). J. Appl. Phys., 68, 3066-3072.

Calculations of Mott scattering cross section

3917 Dagdigian, P.J. and Campbell, M.L. (1987). Chem. Rev., 87, 1-18.
Spin-orbit effects in gas-phase chemical reactions

3918 Dai, D.-G. and Balasubramanian, K. (1990). J. Chem. Phys., 93, 1837-1846.
Geometries and energies of electronic states of AsH_3, SbH_3, and BiH_3 and their positive ions

3919 Dai, D.-G. and Balasubramanian, K. (1991a). Chem. Phys. Lett., 185, 165-171.
Potential energy surfaces for the insertion of Os^+ into H_2

3920 Dai, D.-G. and Balasubramanian, K. (1991b). J. Chem. Phys., 95, 4284-4295.
Potential energy surfaces for the insertion of Re and Re^+ into H_2

3921 Dai, D.-G. and Balasubramanian, K. (1991c). New. J. Chem., 15, 721-726.
Spectroscopic constants and potential energy curves of IrH

3922 Dai, D.-G. and Balasubramanian, K. (1992a). J. Phys. Chem., 96, 3279-3282.
Potential energy surfaces for Pt_3 + H and Pd_3 + H interactions

3923 Dai, D.-G. and Balasubramanian, K. (1992b). Chem. Phys. Lett., 193, 565-572.
Electronic states of Zr_4

3924 Dai, D.-G. and Balasubramanian, K. (1992c). Chem. Phys. Lett., 195, 207-213.
High-spin electronic states of the rhodium trimer (Rh_3)

3925 Dai, D.-G. and Balasubramanian, K. (1992d). J. Chem. Phys., 96, 8345-8353.
Electronic structure of group IV tetramers (Si_4-Pb_4)

3926 Dai, D.-G. and Balasubramanian, K. (1992e). Theor. Chim. Acta, 83, 141-154.
Potential energy surfaces for OsH_2

3927 Dai, D.-G. and Balasubramanian, K, (1992f). J. Phys. Chem., 96, 9236-9243.
Electronic structure and potential energy surfaces of positive ions of group IV tetramers (Ge_4^+ - Pb_4^+)

3928 Dai, D.-G., Cheng, W. and Balasubramanian, K. (1991a). J. Chem. Phys., 95, 9094-9105.
Potential energy surfaces for the insertion of Ta and Ta^+ into H_2

3929 Dai, D.-G., Das, K.K. and Balasubramanian, K. (1991b). Chem. Phys. Lett., 184, 589-595.
Electronic states of Pd_3^+

3930 Dai, D.-G., Huang, A. and Balasubramanian, K. (1992b). J. Mol. Spectrosc., 154, 345-354.
Spectroscopic constants and potential energy curves for 18 electronic states of IrH^+

3931 Dai, D.-G., Sumathi, K. and Balasubramanian, K. (1992a). Chem. Phys. Lett., 193, 251-257.
Eight electronic states and potential energy surfaces of Ge_3

3932 Dai, S.-S. and Ma, Z.-X. (1988a). Chin. J. Chem. Phys., 1(5), 339-346.
Ab initio studies of the electronic structure of $MoCl_2^{2+}$, and $MoCl_4$ using relativistic effective core potentials (in Chinese)

3933 Dai, S.-S. and Ma, Z.-X. (1988b). Huaxue Tongbao, (9), 1-5.
Development of the relativistic quantum chemistry (in Chinese)

3934 Dai, S.-S. and Ma, Z.-X. (1990). Acta Chimica Sinica, 48, 315-319.
The study of non-empirically parametrized relativistic extended Hückel method (in Chinese)

3935*Dakshinamoorthy, M. and Iyakutti, K. (1984). Phys. Rev. B, 30, 6943-6950.
Band structure, Fermi surface, superconductivity, and resistivity of actinium under high pressure

3936 Daniel, C., Koga, N., Han, J., Fu, X.Y. and Morokuma, K. (1988). J. Am. Chem. Soc., 110, 3773-3787.
Ab initio MO study of the full catalytic cycle of olefin hydrogenation by the Wilkinson catalyst $RhCl(PR_3)_3$

3937 Danilov, L.I. (1990). Teor. Mat. Fiz. 85(1), 41-53; Theor. Math. Phys., 85, 1039-1048.
Spectrum of the Dirac operator in R^n with periodic potential

3938 Daoudi, A., Suard, M., Barthelat, J.C. and Berthier, G. (1989). J. Mol. Struct., 198, 215-221.
Electronic structure and bonding of transition metal complexes MCO (M=Ru, Os)

3939 Darewych, J.W. (1990). Phys. Lett. A, 147, 403-405.
Relativistic three-fermion wave equation with Coulomb interactions

3940 Darewych, J.W. and Berseth, W. (1992). Phys. Rev. A, 45, 1325-1332.
Gauge invariance and approximate variational calculations of relativistic few-particle states in the Hamiltonian QED

3941 Darewych, J.W. and Horbatsch, M. (1989). J. Phys. B: At. Mol. Opt. Phys., 22, 973-986.
The relativistic two-body Coulomb system

3942 Darewych, J.W. and Horbatsch, M. (1990). J. Phys. B: At. Mol. Opt. Phys., 23, 337-354.
A variational treatment of the relativistic two-fermion bound-state system in quantum electrodynamics

3943 Das, B.P. (1987). Phys. Scr., 36, 497-499.
Configuration interaction approach to parity non-conservation in atoms

3944 Das, B.P. (1989).
Relativistic many-body theory of electric dipole moment of atoms due to parity and time reversal violation
In "Aspects of Many-Body Effects in Molecules and Extended Systems", Ed. D. Mukherjee, Lecture Notes in Chemistry, 50, Springer, Berlin, pp. 411-416.

3945 Das, B.P. and Grant, I.P. (1986a). J. Phys. B, 19, L7-L11.
Multiconfiguration Dirac-Fock approach to the fine-structure splitting in the 3s3p configuration of the magnesium sequence

3946 Das, B.P. and Idrees, M. (1990). Phys. Rev. A, 42, 6900-6902.
Some theoretical aspects of the group-IIIA-ion atomic clocks: Intercombination transition probabilities

3947 Das, B.P., Venugopal, E.P. and Idrees, M. (1990). Phys. Rev. A, 42, 6903-6904.
First-principles relativistic calculations of the fine-structure interval and magnetic dipole transition probabilities in the $1s^2 2p$ configuration of the lithium isoelectric sequence

3948 Das, K.C., Nayak, K. and Misra, S.K. (1988). Phys. Lett. A, 133, 517-520.
Spin and spin-orbit contribution to magnetic susceptibility of divalent HCP metals

3949 Das, K.C., Nayak, K. and Misra, S.K. (1991). J. Phys.: Cond. Matter, 3, 1307-1321.
Theory of total magnetic susceptibility of interacting electrons in metals

3950 Das, K.K. and Balasubramanian, K. (1989a). J. Chem. Phys., 91, 2433-2442.
Potential energy surfaces for YH_2^+ and ZrH_2^+

3951 Das, K.K. and Balasubramanian, K. (1989b). J. Chem. Phys., 91, 6254-6267.
Potential energy surfaces for NbH_2^+ and MoH_2^+

3952 Das, K.K. and Balasubramanian, K. (1990a). J. Chem. Phys., 93, 625-632.
Potential energy surfaces of eight low-lying electronic states of Rh_3

3953 Das, K.K. and Balasubramanian, K. (1990b). J. Mol. Spectrosc., 140, 280-294.

Spectroscopic properties of low-lying electronic states of Au_2

3954 Das, K.K. and Balasubramanian, K. (1990c). J. Chem. Phys., 92, 6697-6709.
Potential-energy surfaces for Tc^++H_2 and Ru^++H_2 reactions

3955 Das, K.K. and Balasubramanian, K. (1990d). Chem. Phys. Lett., 172, 372-378.
Spectroscopic constants and potential energy curves of electronic states of LaH

3956 Das, K.K. and Balasubramanian, K. (1990e). J. Chem. Phys., 93, 5883-5889.
Geometries and energies of GeH_n and GeH_n^+ (n=1-4)

3957 Das, K.K. and Balasubramanian, K. (1990f). J. Chem. Phys., 93, 6671-6675.
Geometries and energy separations of low-lying states of YNH and NYH

3958 Das, K.K. and Balasubramanian, K. (1990g). J. Mol. Spectrosc., 144, 245-256.
Spectroscopic properties and potential energy curves of 28 electronic states of NbH

3959 Das, K.K. and Balasubramanian, K. (1990h). J. Chem. Phys., 93, 625-632.
Potential energy surfaces of eight low-lying electronic states of Rh_3

3960 Das, K.K. and Balasubramanian, K. (1991a). Chem. Phys. Lett., 176, 571-574.
Electron affinities of Ag_4 and Au_4

3961 Das, K.K. and Balasubramanian, K. (1991b). J. Phys. Chem., 95, 42-46.
Electronic states and potential energy surfaces of LaH_2

3962 Das, K.K. and Balasubramanian, K. (1991c). J. Chem. Phys., 94, 3722-3729.
Potential energy surfaces of LaH^+ and LaH_2^+

3963 Das, K.K. and Balasubramanian, K. (1991d). J. Mol. Spectrosc., 147, 114-119.
Spectroscopic constants and potential energy curves of RhH^+

3964 Das, K.K. and Balasubramanian, K. (1991e). J. Phys. Chem., 95, 3979-3984.
Potential energy surfaces for insertion of Hf atoms into H_2

3965 Das, K.K. and Balasubramanian, K. (1991f). J. Chem. Phys., 94, 6620-6631.
Geometries and potential energy curves of $InSb_2$, $SbIn_2$, $GaAs_2$,$AsGa_2$, and their ions

3966 Das, K.K. and Balasubramanian, K. (1991g). J. Chem. Phys., 95, 2568-2571.
Electronic states of Ru_2

3967 Das, K.K. and Balasubramanian, K. (1991h). J. Phys. Chem., 95, 6880-6883.
Potential energy surfaces for RhH_2^+

3968 Dasgupta, A. and Whitney, K.G. (1990). Phys. Rev. A, 42, 2640-2654.
Scaling of dielectronic-recombination data in fluorinelike ions

3969 Daskalopoulou, M., Bohmer, H.-U. and Peyerimhoff, S.D. (1990). Z. Phys. D, 15, 161-169.
Multi-reference configuration interaction calculations on the systems Xe_2^+ and Xe_3^+

3970 Datta, S. and Datta, S.N. (1992). Pramana, 38, 521-530.
Examples of the minimax technique in relativistic atomic one-electron calculations

3971 Datta, S.N. (1987). Pramana, 28, 633-639.
Relativistic extension of the Hohenberg-Kohn theorem

3972 Datta, S.N. (1989).
Relativistic equations of motion method
In "Aspects of Many-Body Effects in Molecules and Extended Systems", Ed. D. Mukherjee, Lecture Notes in Chemistry, 50, Springer, Berlin, pp. 393-410.

3973 Datta, S.N. (1992). Pramana, 38, 51-75.
Bound state solutions of the two-electron Dirac-Coulomb equation

3974 Datta, S.N. and Devaiah, G. (1988). Pramana-J. Phys., 30, 387-405.
The minimax technique in relativistic Hartree-Fock calculations

3975 Datta, S.N. and Madhusudana Rao, R. (1989). Pramana, 32, L171-L176.
Bound state solutions of the two-electron Dirac-Coulomb equation

3976 Daul, C.A. (1989). J. Chim. Phys., 86, 703-713.
The calculation of multiplet states from MS-Xα wave functions in metal complexes

3977 Davenport, J.W., Watson, R.E. and Weinert, M. (1988). Phys. Rev. B, 37, 9985-9992.
Linear augmented-Slater-type-orbital method for electronic-structure calculations. V. Spin-orbit splitting in Cu_3Au

3978 Davidson, E.R., Hagstrom, S.A., Chakravorty, S.J., Umar, V.M. and Froese Fischer, C. (1991). Phys. Rev. A, 44, 7071-7083.
Ground-state correlation energies for two- to ten-electron atomic ions

3979 Davies, A.J. (1990). Phys. Rev. D, 41, 2628-2630.
Quaternionic Dirac equation

3980 Davis, B.F. and Chung, K.T. (1989). Phys. Rev. A, 39, 3942-3955.
Spin-induced autoionization and radiative transition rates for the (1s2s2p) $^4P_J{}^0$ states in lithiumlike ions

3981 Davis, B.F. and Chung, K.T. (1990). Phys. Rev. A, 42, 5121-5131.
Radiative decay of triply excited $2p^2np$ $^{4,2}S^\circ$ states to the 1s2pmp $^{4,2}P$ states in lithiumlike ions

3982 Dawson, J.F. and Furnstahl, R.J. (1990). Phys. Rev. C, 42, 2009-2022.
Relativistic spectral random-phase approximation in finite nuclei

3983 De Angelis, G.F. and Serva, M. (1992). J. Phys. A, 25, 6539-6550.
Brownian path integral from Dirac equation: a probabilistic approach to the Foldy-Wouthuysen transformation

3984 Decker, F. (1990). Phys. Rev. A, 41, 6552-6554.
Second Born approximation for relativistic electron capture: Exact Monte Carlo calculations for C^{6+}-Au and Ar^{18+}-Ag collisions

3985 Decker, F. (1991). Phys. Rev. A, 44, 2883-2889.
Electron-positron pair production in relativistic heavy-ion collisions

3986 Decker, F. and Eichler, J. (1991a). Phys. Rev. A, 44, 377-387.
Exact second-order Born calculations for relativistic electron capture

3987 Decker, F. and Eichler, J. (1991b). Phys. Rev. A, 44, 2195-2197.
Second-order Born calculations for electron capture in relativistic U + U collisions

3988 Deco, G. and Grün, N. (1989). J. Phys. B: At. Mol. Opt. Phys., 22, 1357-1364.
Asymptotic behaviour of distorted-wave models for ionisation at relativistic energies

3989 Deco, G.R. and Rivarola, R.D. (1987a). J. Phys. B: At. Mol. Opt. Phys., 20, 317-324.
A theoretical model for charge transfer at relativistic energies

3990 Deco, G.R. and Rivarola, R.D. (1987b). J. Phys. B: At. Mol. Opt. Phys., 20, 3853-3865.
Relativistic electron capture in ion-atom collisions

3991 Deco, G.R. and Rivarola, R.D. (1987c). J. Phys. B: At. Mol. Opt. Phys., 20, 5117-5125.
Relativistic continuum distorted-wave model for arbitrary initial and final shells -- electron capture in ion-atom collisions

3992 Deco, G.R. and Rivarola, R.D. (1989a). J. Phys. B: At. Mol. Opt. Phys., 22, 1043-1050.
Pair production with electron capture in relativistic heavy-ion collisions

3993 Deco, G.R. and Rivarola, R.D. (1989b). Phys. Rev. A, 39, 5451-5454.
Two-center effects in relativistic radiative electron capture

3994 De Frutos, J. (1989). Rev. Int. Mét. Num. Cálc. Dis. Ing., 5, 185-202.
Estabilidad y convergencia de esquemas numericos para sistemas de Dirac no lineales

3995 De Frutos, J. and Sanz-Serna, J.M. (1989). J. Comp. Phys., 83, 407-423.
Split-step spectral scheme for nonlinear Dirac systems

3996*Dehmer, J.L. (1973). Phys. Rev. A, 7, 4-9.
Phase-amplitude method in atomic physics. II. Z dependence of spin-orbit coupling

3997 Dehnen, H. and Shahin, M. (1991). Int. J. Mod. Phys. A, 6, 1031-1049.
Magnetic resonances of the e^+e^--system

3998 Deiseroth, H.J., Stupperich, A., Pankaluoto, R. and Christensen, N.E. (1991). Z. Anorg. Allg. Chem., 597, 41-50.
NaHg: A variant of the cesiumchloride structure. Structural relations and electronic structure

3999 DeKock, R.L., Peterson, M.A., Timmer, L.K., Baerends, E.J. and Vernooijs, P. (1990). Polyhedron, 9, 1919-1934.
A theoretical study of the linear versus bent geometry for several MX_2 molecules: MgF_2, CaH_2, CaF_2, CeO_2 and $YbCl_2$

4000 De La Jara F., J. and Tabensky R., R. (1992). Int. J. Mod. Phys. A, 7, 6537-6553.
Spinor operators

4001 de Lange, O.L. (1989a). J. Math. Phys., 30, 858-866.
An operator analysis for the Schrödinger, Klein-Gordon, and Dirac equations with a Coulomb potential

4002 de Lange, O.L. (1989b). Am. J. Phys., 57, 883-886.
A simple solution for the Dirac hydrogenlike atom

4003 de Lange, O.L. (1991a). J. Phys. A, 24, 667-677.
Algebraic properties of the Dirac oscillator

4004 de Lange, O.L. (1991b). J. Math. Phys., 32, 1296-1300.
Shift operators for a Dirac oscillator

4005 de Lange, O.L. and Raab, R.E. (1991).
Operator Methods in Quantum Mechanics
Clarendon Press, Oxford, 388p.

4006 de Lange, O.L. and Welter, A. (1992a). J. Phys. A, 25, 5753-5760.
Shape invariance with application to the momentum representation

4007 de Lange, O.L. and Welter, A. (1992b). Am. J. Phys., 60, 254-257.
Shape invariance of Coulomb problems

4008 Delgado-Barrio, G., Fernandez-Rico, J. and Varade, A. (1988). J. Mol. Struct. (Theochem), 166, 159-162.
Approximate relativistic effect on Hartree-Fock type methods

4009 De Loth, P., Karafiloglou, P., Daudey, J.-P. and Kahn, O. (1988). J. Am. Chem. Soc., 110, 5676-5680.
Ab initio calculation of the ferromagnetic interaction in a $Cu^{II}V^{II}O$ heterodinuclear system

4010 Demkov, Yu.N. and Rebane, T.K. (1991). Opt. Spectrosc., 71, 414-415; Opt. Spektrosk. 71, 714-716.
Atomic and molecular magnetic moments related to the anomalous electron magnetic moment

4011 Denisenko, K.G. and Mrówczynski, S. (1987). Phys. Rev. D, 36, 1529-1537.
Interaction of relativistic elementary atoms with matter. II. Numerical results

4012 De Sanctis, M. (1989). Nucl. Phys., A503, 779-800.
Relativistic corrections to photon scattering polarizabilities

4013 De Sanctis, M. (1991). Nuovo Cimento A, 104A, 661-681.
Polarizability corrections in exotic atoms

4014 De Sanctis, M. and Prosperi, D. (1987). Nuovo Cim. A, 98, 621-.
Relativistic corrections for one- and two-photon matrix elements of weakly bound systems

4015 De Sanctis, M. and Prosperi, D. (1991). Nuovo Cim. A, 104, 921-924.
A two-body relativistic equation for interacting nucleons

4016*Desclaux, J.-P. (1971).
Contribution des effets relativistes à la structure fine et hyperfine des spectres de quelques atomes
Thesis, Université de Paris, 78p.

4017 Desclaux, J.P. (1989).
The multiconfiguration Dirac-Fock method
In "Relativistic, Quantum Electrodynamic and Weak Interaction Effects in Atoms", Ed. W. Johnson, P. Mohr and J. Sucher, AIP Conf. Proc., 189, pp. 265-271.

4018 Desclaux, J.P. and Indelicato, P. (1987).
Relativistic methods for few-electron ions
In "Atoms in Unusual Situations", Ed. J.P. Briand, New York, Plenum, pp. 309.

4019*Desclaux, J.-P. and Pyykkö, P. (1980b). La Recherche, 11, 592-594.
Quand la chimie rejoint la relativité

4020 Deshmukh, P.C. and Manson, S.T. (1986). Phys. Rev. A, 34, 4757-4761.
Extensive relativistic-random-phase-approximation study of photoionization from atomic ytterbium

4021 Deshmukh, P.C., Nasreen, G. and Manson, S.T. (1988). Phys. Rev. A, 38, 504-505.
Photoionization of Al^+ ions: A relativistic random-phase-approximation study

4022 Deshmukh, P.C., Radojevic, V. and Manson, S.T. (1986b). Phys. Rev. A, 34, 5162-5163.
Multiple minima in ground-state photoionization: Interchannel coupling at high Z

4023 Deshmukh, P.C., Radojevic, V. and Manson, S.T. (1986a). Phys. Lett. A, 117, 293-296.
Relativistic splittings of Cooper minima in radon: An RRPA study

4024 Deshmukh, P.C., Radojevic, V. and Manson, S.T. (1992). Phys. Rev. A, 45, 6339-6348.
Photoionization of the outer shells of radon and radium: Relativistic random-phase approximation for high-Z atoms

4025 Deshmukh, P.C., Tambe, B.R. and Manson, S.T. (1986c). Aust. J. Phys., 39, 679-685.
Relativistic effects in the photoionization of heavy atoms: Cooper minima

4026 Detcheva, V., Metodieva, M., Kandilarov, B.D. and Primatarowa, M.T. (1990). Helv. Phys. Acta, 63, 908-921.
Dirac equation approach to scattering from interface potential steps and existence of interface states

4027 Detcheva, V., Metodieva, M., Khristov, T. and Kandilarov, B.D. (1989a). Phys. Stat. Sol. B, 152, 163-169.
One-dimensional Dirac equation approach to the effective mass concept in diatomic crystals

4028 Detcheva, V., Metodieva, M., Lungarova, M. and Kandilarov, B.D. (1989b). Czech. J. Phys. B, 39, 696-701.
Effective masses in a simple relativistic energy-band scheme

4029 Deutsch, M. and Kizler, P. (1992). Phys. Rev. A, 45, 2112-2115.
X-ray multielectronic photoexcitations near the K edge of xenon

4030*Dewar, M.J.S., Grady, G.L., Merz Jr., K.M. and Stewart, J.J.P. (1985a). Organometallics, 4, 1964-1966.
MNDO calculations for compounds containing mercury

4031*Dewar, M.J.S., Holloway, M.K., Grady, G.L. and Stewart,

J.J.P. (1985b). Organometallics, 4, 1973-1980.
MNDO calculations for compounds containing lead

4032 Dewar, M.J.S. and Jie, C. (1989). Organometallics, 8, 1547-1549.
AM1 calculations for compounds containing mercury

4033 Di Bella, S., Casarin, M., Fragalà, I., Granozzi, G. and Marks, T.J. (1988). Inorg. Chem., 27, 3993-4002.
Electronic structure of tetracoordinate transition-metal complexes. 2. Comparative theoretical ab initio/Hartree-Fock-Slater and UV-photoelectron spectroscopic studies of building blocks for low-dimensional conductors: Glyoximate complexes of palladium(II) and platinum(II)

4034 Di Bella, S., Fragalà, I. and Granozzi, G. (1986). Inorg. Chem., 25, 3997-4003.
Electronic structure of transition-metal tetracoordinated complexes. 1. Theoretical ab initio and UV-photoelectron spectroscopy study of palladium(II) and platinum(II) square-planar acetylacetonate complexes

4035 Dieks, D. and Nienhuis, G. (1990). Am. J. Phys., 58, 650-655.
Relativistic aspects of nonrelativistic quantum mechanics

4036 Diener, G. and Grafenstein, J. (1989). J. Phys. Condensed Matter, 1, 8445-8451.
Relativistic Kohn-Sham formalism and the microscopic stress tensor

4037 Dietz, K. (1986). Z. Phys. D, 3, 359-364.
Vacuum-induced Friedel-type oscillations in relativistic atoms

4038 Dietz, K. (1987a).
Relativistic correlations in atoms
In "Giant Resonances in Atoms, Molecules and Solids", Ed. J.P. Connerade, J.M. Esteva, R.C. Karnatak, New York, Plenum Press, pp. 49-70.

4039 Dietz, K. (1987b). Phys. Scr., T17, 22-27.
Relativistic quantum electrodynamics of atoms

4040 Dietz, K. and Hess, B.A. (1989). Phys. Scr., 39, 682-688.
Single particle orbitals for configuration interaction derived from quantum electrodynamics

4041 Dietz, K. and Hess, B.A. (1990). Phys. Rev. A, 41, 2329-2338.
Functional representation of correlations in inhomogeneous many-electron systems

4042 Dietz, K. and Hess, B.A. (1991). J. Phys. B: At. Mol. Opt. Phys., 24, 1129-1142.
Hartree-Fock-type equations in relativistic quantum electrodynamics with non-linear gauge fixing

4043 Dietz, K. and Roemer, H. (1989). Lett. Math. Phys., 18, 133-141.

Avoided level crossings and pair production in strong Coulomb fields

4044 Dietz, K., Schmidt, C., Warken, M. and Hess, B. (1992). J. Phys. B: At. Mol. Opt. Phys., 25, 1705-1718.
A comparative study of standard and non-standard mean-field theories for the energy, the first and the second moments of Be and LiH

4045 Dittrich, J., Exner, P. and Seba, P. (1989). J. Math. Phys., 30, 2875-2882.
Dirac operators with a spherically symmetric δ-shell interaction

4046 Dittrich, J., Exner, P. and Seba, P. (1990). Oper. Theor., 46, 209-219.
Dirac hamiltonian with Coulomb potential and contact interaction on a sphere

4047 Dittrich, J., Exner, P. and Seba, P. (1992). J. Math. Phys., 33, 2207-2214.
Dirac hamiltonian with Coulomb potential and spherically symmetric shell contact interaction

4048*Dittrich, W. (1970). Phys. Rev. D, 1, 3345-3348.
Equivalence of the Dirac equation to a subclass of Feynman diagrams

4049*Dittrich, W. (1974). J. Phys. A, 7, 2273-2283.
High-energy limit versus infinite-mass limit in the Coulomb problem

4050 Dixon, D.A., Matsuzawa, N. and Walker, S.C. (1992). J. Phys. Chem., 96, 10740-10746.
Conformational analysis of 1,2-dihaloethanes: a comparison of theoretical methods

4051 Dixon, R.N. and Tuckett, R.P. (1987). Chem. Phys. Lett., 140, 553-557.
The sign of the spin-orbit coupling constant in $(t_2)^5\ {}^2T_2$ states of MX_4^+ ions

4052 Dmitriev, Yu.Yu., Khait, Yu.G., Kozlov, M.G., Labzovsky, L.N., Mitrushenkov, A.O., Shtoff, A.V. and Titov, A.V. (1992). Phys. Lett. A, 167, 280-286.
Calculation of the spin-rotational Hamiltonian including P- and P,T-odd weak interaction terms for HgF and PbF molecules

4053 Dodelson, S. (1986). J. Phys. B: At. Mol. Phys., 19, 2871-2879.
Relativistic treatment of ortho-para-H_2 transitions

4054 Dolg, F.M. (1989).
Energiejustierte quasirelativistische Pseudopotentiale für die 4f-Elemente
Ph. D. Thesis, Universität Stuttgart.

4055 Dolg, M., Fulde, P., Küchle, W., Neumann, C.-S. and Stoll, H. (1991a). J. Chem. Phys., 94, 3011-3017.

Ground state calculations of di-π-cyclooctatetraene cerium

4056 Dolg, M., Küchle, W., Stoll, H., Preuss, H. and Schwerdtfeger, P. (1991b). Mol. Phys., 74, 1265-1285.
Ab initio pseudopotentials for Hg to Rn. II. Molecular calculations on the hydrides of Hg to At and the fluorides of Rn

4057 Dolg, M. and Stoll, H. (1989). Theor. Chim. Acta, 75, 369-387.
Pseudopotential study of the rare earth monohydrides, monoxides and monofluorides

4058 Dolg, M., Stoll, H., Flad, H.-J. and Preuss, H. (1992a). J. Chem. Phys., 97, 1162-1173.
Ab initio pseudopotential study of Yb and YbO

4059 Dolg, M., Stoll, H. and Preuss, H. (1989a). J. Chem. Phys., 90, 1730-1734.
Energy-adjusted ab initio pseudopotentials for the rare earth elements

4060 Dolg, M., Stoll, H. and Preuss, H. (1990a). Chem. Phys. Lett., 174, 208-212.
Ab initio pseudopotential study of the $^9\Sigma^-$ and $^7\Sigma^-$ states of GdO

4061 Dolg, M., Stoll, H. and Preuss, H. (1990b). Chem. Phys., 148, 219-227.
Ab initio pseudopotential study of europium monoxide EuO: $^8\Sigma^-$ ground state and $^8\Sigma^-$ first excited state

4062 Dolg, M., Stoll, H. and Preuss, H. (1991c). J. Mol. Struct. (Theochem), 231, 243-255.
The low-lying electronic states of cerium monoxide CeO: Ab initio calculations using energy-adjusted pseudopotentials and spin-orbit operators

4063 Dolg, M., Stoll, H. and Preuss, H. (1991d). J. Mol. Struct. (Theochem), 235, 67-79.
Pseudopotential study on rare earth dihalides and trihalides

4064 Dolg, M., Stoll, H. and Preuss, H. (1991e). J. Mol. Struct. (Theochem), 251, 327-334.
Helium chemistry of rare earth elements: pseudopotential study of the cations $LnHe^{3+}$

4065 Dolg, M., Stoll, H. and Preuss, H. (1992b). Chem. Phys., 165, 21-30.
Ab initio pseudopotential study of YbH and YbF

4066 Dolg, M., Stoll, H. and Preuss, H. (1992c). J. Mol. Struct.(Theochem), 277, 239-249.
Homonuclear diatomic lanthanoid compounds: a pseudopotential configuration interaction and correlation energy density functional study

4067 Dolg, M., Stoll, H., Savin, A. and Preuss, H. (1989b). Pseudopotential study of rare earth compounds
In "Quantum Chemistry -- Basic Aspects, Actual Trends",

Studies in Phys. Theor. Chem., Elsevier, 62, Amsterdam, pp.265-273.

4068 Dolg, M., Stoll, H., Savin, A. and Preuss, H. (1989c). Theor. Chim. Acta, 75, 173-194.
Energy-adjusted pseudopotentials for the rare earth elements

4069 Dolg, M., Wedig, U., Stoll, H. and Preuss, H. (1987a). J. Chem. Phys., 86, 866-872.
Energy-adjusted ab initio pseudopotentials for the first row transition elements

4070 Dolg, M., Wedig, U., Stoll, H. and Preuss, H. (1987b). J. Chem. Phys., 86, 2123-2131.
Ab initio pseudopotential study of the first row transition metal monoxides and iron monohydride

4071 Dollard, J. and Velo, G. (1966). Nuovo Cim. A, 45, 801-812.
Asymptotic behaviour of a Dirac particle in a Coulomb field

4072 Domingos, M. and Caldeira, M.H. (1991). Nuovo Cim. B, 106, 623-626.
Localization and noncausal behaviour of relativistic quantum particles

4073 Dominguez-Adame, F. (1987). Am. J. Phys., 55, 1003-1006.
Relativistic and nonrelativistic Kronig-Penney models

4074 Dominguez-Adame, F. (1989). J. Phys. Condensed Matter, 1, 109-112.
A generalised Dirac-Kronig-Penney model

4075 Dominguez-Adame, F. (1990a). Am. J. Phys., 58, 886-888.
Dirac particles in the potential $-1/|x|$

4076 Dominguez-Adame, F. (1990b). Eur. J. Phys., 11, 149-151.
Relativistic one-dimensional hydrogen atom in momentum representation

4077 Dominguez-Adame, F. (1990c). J. Phys. A, 23, 1993-1999.
Exact solutions of the Dirac equation with surface delta interactions

4078 Dominguez-Adame, F. (1991). Europhys. Lett., 15, 569-574.
Spectroscopy of a perturbed Dirac oscillator

4079 Dominguez-Adame, F. (1992). Phys. Lett. A, 162, 18-20.
A relativistic interaction without Klein paradox

4080 Dominguez-Adame, F. and González, M.A. (1990). Europhys. Lett., 13, 193-198.
Solvable linear potentials in the Dirac equation

4081 Dominguez-Adame, F. and González, M.A. (1992). Physica B, 176, 180-188.
A generalized Dirac-Kronig-Penney model with nonlocal separable potentials

4082 Dominguez-Adame, F. and Maciá, E. (1989a). Europhys. Lett., 8, 711-715.
On relativistic singular harmonic-oscillator potentials

4083 Dominguez-Adame, F. and Maciá, E. (1989b). J. Phys. A, 22, L419-L423.
Bound states and confining properties of relativistic point interaction potentials

4084 Dominguez-Adame, F. and Méndez, B. (1991). Can. J. Phys., 69, 780-785.
A solvable two-body Dirac equation in one space dimension

4085 Dominguez-Adame, F. and Sanchez, A. (1991). Phys. Lett. A, 159, 153-157.
Relativistic effects in Kronig-Penney models on quasiperiodic lattices

4086 Doncheski, M., Grotch, H. and Erickson, G.W. (1991). Phys. Rev. A, 43, 2152-2170.
Pure recoil corrections to the Lamb shift in hydrogenic atoms

4087 Doncheski, M., Grotch, H. and Owen, D.A. (1990). Phys. Rev. A, 41, 2851-2853.
Lamb shift: Additional recoil corrections and estimates of radiative proton effects

4088 Dong, Q., Kuang, Y. and Pratt, R.H. (1989). Nucl. Instrum. Meth. A, 280, 189-194.
Photoionization of atoms and ions

4089 Doolen, G. and Liberman, D.A. (1987). Phys. Scr., 36, 77-79.
Calculations of photoabsorption by atoms using a linear response method

4090 Drake, G.W.F. (1987). Phys. Rev. Lett., 59, 1549-1552.
New variational techniques for the 1snd states of helium

4091 Drake, G.W.F. (1988a). Nucl. Instrum. Meth. B, 31, 7-13.
High precision variational calculations for the $1s^2\ {}^1S$ state of H^- and the $1s^2\ {}^1S$, $1s2s\ {}^1S$ and $1s2s\ {}^3S$ states of helium

4092 Drake, G.W.F. (1988b). Can. J. Phys., 66, 586-611.
Theoretical energies for the n=1 and 2 states of the helium isoelectronic sequence up to Z=100

4093 Drake, G.W.F. (1989).
New variational techniques and high precision eigenvalues for helium
In "Relativistic, Quantum Electrodynamic and Weak Interaction Effects in Atoms", Ed. W. Johnson, P. Mohr and J. Sucher, AIP Conf. Proc., 189, pp.146-164.

4094 Drake, G.W.F. (1990). Phys. Rev. Lett., 65, 2769-2772.
Variational eigenvalues for the Rydberg states of helium:

Comparison with experiment and with asymptotic expansions

4095*Drake, G.W.F. and Dalgarno, A. (1968). Astrophys. J., 152, L121-L123.
The two-photon decay of metastable triplet helium

4096 Drake, G.W.F. and Goldman, S.P. (1988). Adv. At. Mol. Phys., 25, 393-416.
Relativistic Sturmian and finite basis set methods in atomic physics

4097 Drake, G.W.F. and Makowski, A.J. (1988). J. Opt. Soc. Am. B, 5, 2207-2214.
High-precision eigenvalues for the 1s2p ^1P and ^3P states of helium

4098 Drake, G.W.F., Patel, J. and van Wijngaarden, A. (1988). Phys. Rev. Lett., 60, 1002-1005.
High-precision anisotropy measurement of the Lamb shift in He^+

4099 Drake, G.W.F. and Swainson, R.A. (1990). Phys. Rev. A, 41, 1243-1246.
Bethe logarithms for hydrogen up to n=20, and approximations for two-electron atoms

4100 Drake, G.W.F. and van Wijngaarden, A. (1989).
Radiative transitions in one- and two-electron ions
In "Physics of Highly-Ionised Atoms. Proceedings of a NATO Advanced Study Institute", Ed. R. Marrus, Plenum, New York, pp. 143-188.

4101*Drake, G.W.F., Victor, G.A. and Dalgarno, A. (1969). Phys. Rev., 180, 25-32.
Two-photon decay of the singlet and triplet metastable states of helium-like ions

4102 Drake, G.W.F. and Yan, Z.-C. (1992). Phys. Rev. A, 46, 2378-2409.
Energies and relativistic corrections for the Rydberg states of helium: Variational results and asymptotic analysis

4103 Dreitlein, J.F. and Sauer, J.R. (1990). Comp. Phys., 4, 64-72.
Spinor software tools in C++

4104 Dreizler, R.M., Engel, E. and Malzacher, P. (1987). NATO ASI Ser. B, 153, 565-583.
Relativistic density functional theory

4105 Dreizler, R.M. and Gross, E.K.U. (1990).
Density Functional Theory
Springer, Berlin, Ch.8.

4106 Druzbick, J. and Williams, H.T. (1987). Phys. Rev. A, 35, 417-420.
Hyperfine structure of ^3He

4107*Dubler, Th. (1978). Helv. Phys. Acta, 51, 743-749.

Methode zur schnellen und genauen Berechnung des Vakuum-Polarisations-Potentials der Ordnung $\alpha(Z\alpha)$

4108*Dunning Jr., T.H., Valley, M. and Taylor, H.S. (1978). J. Chem. Phys., 69, 2672-2681.
Theoretical studies of the low-lying electronic states of GaKr, including extrapolation to InKr and TlKr

4109*Durand, B. and Durand, L. (1983). Phys. Rev. D, 28, 396-406.
Analytic solution of the relativistic Coulomb problem for a spinless Salpeter equation

4110 Durand, G., Spiegelmann, F. and Bernier, A. (1987). J. Phys. B: At. Mol. Phys., 20, 1161-1174.
Electronic structure of doubly charged dimers Mg_2^{2+} and Hg_2^{2+}

4111 Durand, L. and Gara, A. (1990). J. Mat. Phys., 31, 2237-2243.
Matrix methods for the numerical solution of relativistic wave equations

4112 Durand, Ph. (1986a). Int. J. Quantum Chem., 29, 915-935.
Core-valence correlations and relativistic effects in heavy atoms for molecular applications

4113 Durand, Ph. (1986b). C. R. Acad. Sc. Paris, 303, 119-124.
Transformation du hamiltonien de Dirac en hamiltoniens variationnels de type Pauli. Application à des atomes hydrogénoïdes

4114 Durand, Ph. and Malrieu, J.-P. (1987). Adv. Chem. Phys., 67, 321-412.
Effective hamiltonians and pseudo-operators as tools for rigorous modelling

4115 Dutta, J., Laha, U., Mukhopadhyaya, S. and Talukdar, B. (1986). Chem. Phys. Lett., 128, 305-309.
Relativistic Compton profiles for rare gases

4116 Dyall, K.G. (1986). Aust. J. Phys., 39, 667-678.
Inclusion of relativistic effects in electronic structure calculations

4117 Dyall, K.G. (1992a). J. Chem. Phys., 96, 1210-1217.
All-electron molecular Dirac-Hartree-Fock calculations: Properties of the XH_4 and XH_2 molecules and the reaction energy XH_4 --> XH_2 + H_2, X=Si, Ge, Sn, Pb

4118 Dyall, K.G. (1992b). Chem. Phys. Lett., 196, 178-182.
Negative-energy states in the Dirac-Hartree-Fock problem: the effect of omission of two-electron integrals involving the small component

4119 Dyall, K.G. and Fægri Jr., K. (1990). Chem. Phys. Lett., 174, 25-32.
Kinetic balance and variational bounds failure in the solution of the Dirac equation in a finite Gaussian basis set

4120 Dyall, K.G., Fægri Jr., K. and Taylor, P.R. (1991a). Polyatomic molecular Dirac-Hartree-Fock calculations with Gaussian basis sets
In "The Effects of Relativity in Atoms, Molecules and the Solid State", Ed. S. Wilson, I.P. Grant and B. Gyorffy, Plenum, pp. 167-184.

4121*Dyall, K.G. and Grant, I.P. (1984). J. Phys. B: At. Mol. Phys., 17, 1281-1300.
The K_β x-ray emission spectrum of argon: a theoretical account

4122 Dyall, K.G., Grant, I.P., Johnson, C.T., Parpia, F.A. and Plummer, E.P. (1989). Comp. Phys. Comm., 55, 425-456.
GRASP: A general-purpose relativistic atomic structure program

4123 Dyall, K.G., Taylor, P.R., Faegri Jr., K. and Partridge, H. (1991b). J. Chem. Phys., 95, 2583-2594.
All-electron molecular Dirac-Hartree-Fock calculations: The group IV tetrahydrides CH_4, SiH_4, SnH_4, and PbH_4

4124 Dyke, J.M., Elbel, S., Morris, A. and Stevens, J.C.H. (1986a). J. Chem. Soc., Faraday Trans. 2, 82, 637-645.
High-temperature photoelectron spectroscopy. A study of atomic and molecular arsenic

4125 Dyke, J.M., Fehér, M., Gravenor, B.W.J. and Morris, A. (1987). J. Phys. Chem., 91, 4476-4481.
High-temperature photoelectron spectroscopy: A study of the alkaline earth oxides SrO and BaO

4126 Dyke, J.M., Morris, A. and Stevens, J.C.H. (1986b). Chem. Phys., 102, 29-36.
High-temperature photoelectron spectroscopy: A study of atomic and molecular antimony

4127 Dykshoorn, W. and Koniuk, R. (1990). Phys. Rev. A, 41, 64-67.
Spin effects in highly relativistic systems

4128 Dykshoorn, W., Koniuk, R. and Muñoz-Tapia, R. (1990). Phys. Rev. A, 41, 60-63.
Ultrarelativistic bound states in spinor and scalar QED

4129 Dyubo, Zh. and Safronova, U.I. (1989). Opt. Spectrosc., 66, 562-565.
Probabilities of nonradiative decays of Be-like ions

4130 Dzuba, V.A. (1989). Report IYaF 89-92, Novosibirsk, 25p.
Complex of programs for atomic physics problems

4131 Dzuba, V.A., Flambaum, V.V., Gribakin, G.F. and Sushkov, O.P. (1991a). Phys. Rev. A, 44, 2823-2827.
Fine structure of negative ions of alkaline-earth-metal atoms

4132 Dzuba, V.A., Flambaum, V.V. and Khriplovich, I.B. (1986a).

Z. Phys. D, 1, 243-245.
Enhancement of P- and T-nonconserving effects in rare-earth atoms

4133 Dzuba, V.A., Flambaum, V.V., Kraftmakher, A.Ya. and Sushkov, O.P. (1989b). Phys. Lett. A, 142, 373-377.
Summation of the high orders of perturbation theory in the correlation correction to the hyperfine structure and to the amplitudes of E1-transitions in the caesium atom

4134 Dzuba, V.A., Flambaum, V.V., Silvestrov, P.G. and Sushkov, O.P. (1986b). Phys. Lett. A, 118, 177-180.
Shielding of an external electric field in atoms

4135 Dzuba, V.A., Flambaum, V.V., Silvestrov, P.G. and Sushkov, O.P. (1987a). J. Phys. B: At. Mol. Phys., 20, 1399-1412.
Correlation potential method for the calculation of energy levels, hyperfine structure and E1 transition amplitudes in atoms with one unpaired electron

4136 Dzuba, V.A., Flambaum, V.V., Silvestrov, P.G. and Sushkov, O.P. (1987b). J. Phys. B: At. Mol. Phys., 20, 3297-3311.
Calculation of parity non-conservation in thallium

4137 Dzuba, V.A., Flambaum, V.V., Silvestrov, P.G. and Sushkov, O.P. (1987c). Phys. Scr., 36, 69-70.
Parity non-conservation in thallium and caesium

4138 Dzuba, V.A., Flambaum, V.V., Silvestrov, P.G. and Sushkov, O.P. (1988). Europhys. Lett., 7, 413-418.
Relativistic many-body calculations of parity nonconservation in lead and bismuth atoms

4139 Dzuba, V.A., Flambaum, V.V., Silvestrov, P.G. and Sushkov, O.P. (1989a).
Screening of residual Coulomb interaction and structure of many-body perturbation theory in heavy atoms
In "Relativistic, Quantum Electrodynamic and Weak Interaction Effects in Atoms", Ed. W. Johnson, P. Mohr and J. Sucher, AIP Conf. Proc., 189, 254-264.

4140 Dzuba, V.A., Flambaum, V.V., Silvestrov, P.G. and Sushkov, O.P. (1991b). Phys. Rev. A, 44, 2828-2831.
Many-body perturbation-theory calculations in atoms with open shells

4141 Dzuba, V.A., Flambaum, V.V. and Sushkov, O.P. (1989c). Phys. Lett. A, 141, 147-153.
Summation of the high orders of perturbation theory for the parity nonconserving E1-amplitude of the 6s-7s transition in the caesium atom

4142 Dzuba, V.A., Flambaum, V.V. and Sushkov, O.P. (1991c). Phys. Rev. A, 44, 4224-4227.
Resonance enhancement of relativistic effects for the scattering of very slow electrons by heavy atoms

4143 Dzyuba, V.A., Sushkov, O.P. and Flambaum, V.V. (1990).

Izv. Vyss. Uche. Zave., Fiz., 33, 108-119; Sov. Phys. J., 33, 713-721.
Multiparticle perturbation theory and an accurate calculation of parity nonconservation in cesium

4144 Eberlein, C. (1991). J. Phys. B: At. Mol. Opt. Phys., 24, 849-859.
Electrostatic level shift in a hydrogen atom inserted between perfectly conducting plates

4145 Ebert, H. (1988). Phys. Rev. B, 38, 9390-9397.
Two ways to perform spin-polarized relativistic linear muffin-tin-orbital calculations

4146 Ebert, H. (1989). J. Phys. Condensed Matter, 1, 9111-9116.
Fully relativistic treatment of core states for spin-dependent potential

4147 Ebert, H. (1992). Phys. Blätter, 48, 795-798.
Relativistische Effekte in magnetischen Festkörpern

4148 Ebert, H. and Akai, H. (1992). Mat. Res. Soc. Symp. Proc., 253, 329-340.
Spin-polarized relativistic band structure calculations for dilute and concentrated disordered alloys

4149 Ebert, H., Drittler, B. and Akai, H. (1992). J. Magn. Magn. Materials, 104-107, 733-734.
Spin-polarized relativistic electronic structure calculations for disordered alloys using the CPA: application to Fe_xCo_{1-x} and Co_xPt_{1-x}

4150 Ebert, H., Drittler, B., Strange, P., Zeller, R. and Gyorffy, B.L. (1991b).
Theory of magnetic X-ray dichroism
In "The Effects of Relativity in Atoms, Molecules, and the Solid State", Ed. S. Wilson, I.P. Grant and B.L. Gyorffy, Plenum, New York and London, pp. 333-348.

4151 Ebert, H., Drittler, B., Zeller, R. and Dederichs, P.H. (1991a).
Influence of relativistic effects on the magnetic moments and hyperfine fields of 5d-impurity atoms dissolved in ferromagnetic Fe
In "The Effects of Relativity in Atoms, Molecules, and the Solid State", Ed. S. Wilson, I.P. Grant and B.L. Gyorffy, Plenum, New York and London, pp. 275-283.

4152 Ebert, H. and Gyorffy, B.L. (1988). J. Phys. F: Met. Phys., 18, 451-455.
On the scattering solutions to the Dirac equation for nonspherically-symmetric targets

4153 Ebert, H., Strange, P. and Gyorffy, B.L. (1988a). J. Appl. Phys., 63, 3052-3054.
Spin-polarized relativistic LMTO method

4154 Ebert, H., Strange, P. and Gyorffy, B.L. (1988b). J. Appl. Phys., 63, 3055-3057.
Theory of circularly polarized x-ray absorption by ferromagnetic Fe

4155 Ebert, H., Strange, P. and Gyorffy, B.L. (1988c). J. Phys. F: Met. Phys., 18, L135-L139.
The influence of relativistic effects on the magnetic moments and hyperfine fields of Fe, Co and Ni

4156 Ebert, H., Strange, P. and Gyorffy, B.L. (1989). Hyperfine Interactions, 51, 929-932.
A fully relativistic description of the hyperfine interaction in magnetic systems

4157*Ebert, H., Weinberger, P. and Voitländer, J. (1985a). Phys. Rev. B, 31, 7557-7565.
Electronic structure of Ag_xPt_{1-x} alloys

4158*Ebert, H., Weinberger, P. and Voitländer, J. (1985b). Phys. Rev. B, 31, 7566-7571.
Relativistic formulation for the nuclear-spin-lattice relaxation rate in metallic systems: Applications to Ag_xPt_{1-x}

4159 Eby, P.B. (1991). Phys. Rev. A, 43, 2258-2269.
Cross section for production of low-energy electron-positron pairs by relativistic heavy ions

4160 Eckert, J., Kubas, G.J., Hall, J.H., Hay, P.J. and Boyle, C.M. (1990). J. Am. Chem. Soc, 112, 2324-2332.
Molecular hydrogen complexes. 6. The barrier to rotation of η^2-H_2 in $M(CO)_3(PR_3)_2(\eta^2$-$H_2)$ (M=W, Mo; R=Cy, i-Pr): Inelastic neutron scattering, theoretical, and molecular mechanics studies

4161 Edelstein, V.M. (1992). Solid State Commun., 81, 351-355.
A novel anomalous contribution to the Hall effect in some high-T_c compounds

4162 Edlund, U., Lejon, T., Pyykkö, P., Venkatachalam, T.K. and Buncel, E. (1987). J. Am. Chem. Soc., 109, 5982-5985.
^7Li, ^{29}Si, ^{119}Sn, and ^{207}Pb NMR studies of phenyl-substituted group 4 anions

4163 Ehlotzky, F. (1988). Opt. Comm., 66, 265-269.
Sum rules in relativistic potential scattering in a strong laser field

4164 Ehlotzky, F. (1989). J. Phys. B: At. Mol. Opt. Phys., 22, 601-616.
Scattering of x-rays by relativistic electrons in a strong laser field

4165 Eichler, J. (1986). Paper presented at ICPEAC, 14, Stanford University.
Relativistic eikonal theory of electron capture

4166 Eichler, J. (1987). Phys. Rev. A, 35, 3248-3255.
Theory of relativistic charge exchange with Coulomb boundary conditions; Erratum, ibid. 37(1988), 287.

4167*Eisenberg, J.M. (1981). Nucl. Phys., A355, 269-276.
Observational tests of models for a relativistic nucleon bound in scalar and vector potentials

4168 Eissner, W. (1991).
Atomic structure calculations in Breit-Pauli approximation In "The Effects of Relativity in Atoms, Molecules, and the Solid State", Ed. S. Wilson, I.P. Grant and B.L. Gyorffy, Plenum, New York and London, pp. 55-66.

4169 Eissner, W.B. and Tully, J.A. (1992). Astron. Astrophys., 253, 625-631.
Spectral lines in the beryllium sequence

4170 El-Ashry, F.M., Darwish, N.Z., Omar, K.M. and Saad, S.S. (1989). Proc. Indian Nat. Sci. Acad. A, 55, 66-74.
Binding, Coulomb and relativistic effects in K-shell ionization

4171 El-Issa, B.D., Ali, A.A.M. and Zanati, H. (1989). Inorg. Chem., 28, 3297-3305.
Theoretical study of the bonding in sulfur- and oxygen-containing molybdenum species of the general formula $MoS_nO_{4-n}^{2-}$ (n=0-4)

4172 El-Issa, B.D., Katrib, A., Ghodsian, R., Salsa, B.A. and Addassi, S.H. (1988). Int. J. Quantum Chem., 33, 195-216.
A comparative study of the bonding in different halides of iridium

4173 El-Issa, B.D., Makhyoun, M.A. and Salsa, B.A. (1987). Int. J. Quantum Chem., 31, 295-307.
A molecular orbital study of cis and trans diaminodichloroplatinum(II)

4174 El-Issa, B.D., Pyykkö, P. and Zanati, H.M. (1991). Inorg. Chem., 30, 2781-2787.
MS Xα studies on the colors of $BiPh_5$, $PbCl_6^{2-}$, and WS_4^{2-}: Are relativistic effects on the LUMO important?

4175 El-Issa, B.D. and Zeedan, M.M. (1991). Inorg. Chem., 30, 2594-2605.
Quasirelativistic effects in the electronic structure of the thiomolybdate and thiotungstate complexes of nickel, palladium, and platinum

4176 Elkhovskii, A.S. and Mil'shtein, A.I. (1991). Zh. Eksp. Teor. Fiz., 99, 1068-1076; Sov. Phys. JETP, 72, 592-597.
Vacuum polarization and nuclear multipole moments

4177 Elkhovsky, A.S., Milstein, A.I. and Khriplovich, I.B. (1992). Zh. Eksp. Teor. Fiz, 102, 1768-1780; Sov. Phys. JETP, 75, 954-959.
Logarithmic corrections in the two-body problem in QED

4178*Elliott, R.J. (1954). Phys. Rev., 96, 280-287.
Spin orbit coupling in band theory - Character tables for some "double" space groups

4179*Ellis, D.E. (1985).
Molecular clusters
In "Handbook on the Physics and Chemistry of the Actinides", Ed. A.J. Freeman and G.H. Lander, Elsevier Science Publishers B.V., pp. 1-27.

4180 Ellis, D.E., Guo, J., Cheng, H.-P. and Low, J.J. (1991). Adv. Quantum Chem., 22, 125-165.
Transition metal clusters: Electronic structure and interaction with hydrogen and oxides

4181 Ellis, D.E. and Holland, G.F. (1986). Chem. Scr., 26, 441-448.
Molecular cluster studies of charge and magnetization densities in transition and heavy metal compounds

4182 Elsässer, C., Takeuchi, N., Ho, K.M., Chan, C.T., Braun, P. and Fahnle, M. (1990). J. Phys.: Condens. Matter, 2, 4371-4394.
Relativistic effects on ground state properties of 4d and 5d transition metals

4183*Elyashar, N. and Koelling, D.D. (1977). Phys. Rev. B, 15, 3620-3632.
Self-consistent relativistic APW calculation of the electronic structure of niobium with a non-muffin-tin potential

4184 Engel, E. and Dreizler, R.M. (1987). Phys. Rev. A, 35, 3607-3618.
Field-theoretical approach to a relativistic Thomas-Fermi-Dirac-Weizsäcker model

4185 Engel, E. and Dreizler, R.M. (1988). Phys. Rev. A, 38, 3909-3917.
Solution of the relativistic Thomas-Fermi-Dirac-Weizsäcker model for the case of neutral atoms and positive ions

4186 Engel, E. and Dreizler, R.M. (1989). J. Chim. Phys., 86, 831-840.
The relativistic Thomas-Fermi-Dirac-Weizsäcker model: First applations to atomic physics

4187 Engel, E., Dreizler, R.M. and Malzacher, P. (1987). J. Phys.(Paris), C2, 321-328.
Density functionals on the basis of the relativistic field theory

4188 Engel, E., Müller, H. and Dreizler, R.M. (1989). Phys. Rev. A, 39, 4873-4876.
Alternative approach to the gradient expansion of Green's functions of noninteracting particles

4189 Englert, B.-G. (1988).
Semiclassical Theory of Atoms
Springer-Verlag, Berlin.

4190*Englman, R. (1966). J. Chem. Phys., 45, 2669-2672.
Operator for spin-forbidden transitions

4191 Enss, V. and Thaller, B. (1986). Ann. Inst. H. Poincaré, 45, 147-171.
Asymptotic observables and Coulomb scattering for the Dirac equation

4192 Eppenga, R., Schuurmans, M.F.H. and Rompa, H.W.A.M. (1988). J. Phys. Chem. Solids, 49, 1119-1124.
Ab-initio determinations of semiconductor spin-orbit splittings from ASW

4193*Epstein, J.H. and Epstein, S.T. (1962). Am. J. Phys., 30, 266-268.
Some applications of hypervirial theorems to the calculation of average values

4194 Epstein, S.T. (1989). Phys. Lett. A, 141, 14.
On Zitterbewegung

4195*Eriksen, E. (1958). Phys. Rev., 111, 1011-1016.
Foldy-Wouthuysen approximation. Exact solution with generalization to the two-particle problem

4196*Eriksen, E. and Kolsrud, M. (1960). Nuovo Cim. Suppl., 18, 1-39.
Canonical transformations of Dirac's equation to even forms. Expansion in terms of the external fields

4197 Eriksson, O., Albers, R.C. and Boring, A.M. (1991). Phys. Rev. B, 43, 5649-5652.
Calculated magnetic properties of the antiferromagnetic UNiSn compound

4198 Eriksson, O., Brooks, M.S.S. and Johansson, B. (1990). J. Less Common Metals, 158, 207-220.
Calculated cohesive properties of lanthanide and rare-earth like actinide elements

4199 Eriksson, O., Johansson, B., Brooks, M.S.S. and Skriver, H.L. (1989a). Phys. Rev. B, 39, 5647-5654.
Electronic structure, cohesive, and magnetic properties of the actinide-iridium Laves phases

4200 Eriksson, O., Johansson, B., Brooks, M.S.S. and Skriver, H.L. (1989b). Phys. Rev. B, 40, 9508-9518.
Electronic structure of the actinide-Rh_3 systems and the 5f localization in UPd_3

4201 Eriksson, O., Söderlind, P. and Wills, J.M. (1992a). Phys. Rev. B, 45, 12588-12591.
fcc -> bct phase transition in Th at extreme compressions: Theory

4202 Eriksson, O. and Wills, J.M. (1992). Phys. Rev. B, 45, 3198-3203.
Theoretical aspects of the 5f delocalization of americium under pressure

4203 Eriksson, O., Wills, J.M. and Boring, A.M. (1992b). Phys. Rev. B, 46, 12981-12989.

Calculations of crystal-structure stabilities of Ce under pressure

4204 Ermler, W.C., Ross, R.B. and Christiansen, P.A. (1988a). Adv. Quantum Chem., 19, 139-182.
Spin-orbit coupling and other relativistic effects in atoms and molecules

4205 Ermler, W.C., Ross, R.B. and Christiansen, P.A. (1991). Int. J. Quantum Chem., 40, 829-846.
Ab initio relativistic effective potentials with spin-orbit operators. VI. Fr through Pu

4206 Ermler, W.C., Ross, R.B., Sawamura, M. and Sandroff, C.J. (1988b). Supercomputing Symposium 88 Proc. Edmonton, Canada, 19-21 June 1988
Ab initio calculations and experimental studies of anisotropic semiconductor clusters

4207 Eschrig, H. and Richter, M. (1986). Solid State Comm., 59, 861-864.
Kramers' theorem in the relativistic electronic structure calculation

4208 Estiu, G.L. and Zerner, M.C. (1992). Int. J. Quantum Chem., S26, 587-604.
Studies on CO bonding to Rh clusters using an intermediate neglect of differential overlap theory to model heterogereous catalytic reactions

4209 Ettema, A.R.H.F., de Groot, R.A., Haas, C. and Turner, T.S. (1992). Phys. Rev. B, 46, 7363-7373.
Electronic structure of SnS deduced from photoelectron spectra and band-structure calculations

4210*Euwema, R.N. and Stukel, D.J. (1971). Int. J. Quantum Chem., Symp., 4, 55-75.
Relativistic exchange approximation study for He, Ne, Ar, Kr and Xe

4211 Evans, A.B. (1991). Found. Phys., 21, 633-647.
Klein's paradox in a four-space formulation of Dirac's equation

4212 Faegri Jr., K. and Bauschlicher Jr., C.W. (1991). Chem. Phys., 153, 399-408.
Heteronuclear transiticn metal diatomics: the bonding and electronic structure of ScNi, YNi, ScPd, and YPd

4213 Faiez, R. and Ruiz Lopez, M.F. (1988). J. Mol. Struct. (Theochem), 166, 103-108.
Spin-orbit coupling calculations using finite perturbation theory

4214 Fainshtein, A.G., Manakov, N.L. and Nekipelov, A.A. (1990). J. Phys. B: At. Mol. Opt. Phys., 23, 559-569.
Vacuum polarization by a Coulomb field. Analytical approximation of the polarization potential

4215 Falkensteiner, P. and Grosse, H. (1987). Lett. Math.

Phys., 14, 139-148.
Quantization of fermions interacting with point-like external fields

4216 Fanchi, J.R. (1986). Phys. Rev. A, 34, 1677-1681.
Parametrizing relativistic quantum mechanics

4217 Fanizzi, F.P., Maresca, L., Natile, G., Lanfranchi, M., Tiripicchio, A. and Pacchioni, G. (1992). J. Chem. Soc., Chem. Commun., 333-335.
Five-coordination in platinum(II) species: When and why

4218 Farazdel, A. and Smith Jr., V.H. (1986). Int. J. Quantum Chem., 29, 311-314.
Invalidity of the ubiquitous mass-velocity operator in quasirelativistic theories

4219 Fassari, S. (1990). Helv. Phys. Acta, 63, 884-907.
On the bound states of relativistic Krönig-Penney hamiltonians with short range impurities

4220 Fawcett, B.C. (1986). Phys. Scr., 34, 331-336.
Calculated oscillator strength, wavelengths and energy levels for allowed 3-3 and 3-4 transitions for iron (Fe XV) with isoelectronic comparisons between Dirac-Fock and Hartree-Fock

4221 Fawcett, B.C. (1987a). At. Data Nucl. Data Tables, 37, 367-409.
Calculated wavelengths, oscillator strengths, and energy levels for allowed 2-2 and 2-3 transitions for ions in the C-like isoelectronic sequence between F IV and Ni XXIII

4222 Fawcett, B.C. (1987b). At. Data Nucl. Data Tables, 37, 411-418.
Oscillator strengths of allowed transitions for C I, N II, and O III

4223 Fawcett, B.C. (1991).
On the accuracy of oscillator strengths
In "The Effects of Relativity in Atoms, Molecules, and the Solid State", Ed. S. Wilson, I.P. Grant and B.L. Gyorffy, Plenum, New York and London, pp. 45-54.

4224 Fawcett, B.C. and Cowan, R.D. (1988). J. Quant. Spectrosc. Radiat. Transfer, 40, 15-20.
A study of theoretical computations of oscillator strengths for Fe I

4225 Fawcett, B.C. and Wilson, M. (1991). At. Data Nucl. Data Tables, 47, 241-317.
Computed oscillator strengths, Lande g values, and lifetimes in YbII

4226*Fazzio, A., Caldas, M.J. and Zunger, A. (1985). Phys. Rev. B, 32, 934-954.
Electronic structure of copper, silver, and gold impurities in silicon

4227 Fechner, B., Rinkleff, R.-H. and Steudel, A. (1987).

Z. Phys. D, 6, 31-34.
Radiative lifetime of levels of the configuration $4f^7(^8S)6s6p$ in Eu I

4228*Feder, R., Rosicky, F. and Ackermann, B. (1983). Z. Phys. B, 52, 31-36.
Relativistic multiple scattering theory of electrons by ferromagnets

4229*Fedorov, M.V. and Kazakov, A.E. (1973). Z. Phys., 261, 191-202.
An electron in a quantized plane wave and in a constant magnetic field

4230 Fehér, M. (1988). Diss. Abstr. Int. B, 49, 1193.
High temperature photoelectron and chemielectron studies of some small molecules in the gas phase.

4231 Feinberg, G. and Sucher, J. (1992). Phys. Rev. D, 45, 2493-2517.
Spin-dependent two-photon-exchange forces: spin-0 particle and charged spin-1/2 particle

4232 Feldman, G. and Fulton, T. (1986). Ann. Phys.(N.Y.), 172, 40-66.
The relativistic Hartree-Fock and random phase approximations for many-electron atoms

4233 Feldman, G. and Fulton, T. (1987). Ann. Phys.(N.Y.), 179, 20-51.
Relativistic many-electron atoms in perturbation and Dirac-Fock theory

4234 Feldman, G. and Fulton, T. (1990). Ann. Phys.(N.Y.), 201, 193-220.
Log terms for leading radiative corrections in atoms in the Dirac-Fock approximation

4235 Feldman, G., Fulton, T. and Ingham, J. (1992). Ann. Phys. (N.Y.), 219, 1-41.
Leading radiative corrections for atoms in the Dirac-Fock approximation

4236 Feldman, U., Indelicato, P. and Sugar, J. (1991). J. Opt. Soc. Am. B, 8, 3-5.
Magnetic dipole line from U LXXI ground-term levels predicted at 3200 Å

4237 Felser-Wenz, C., Hohlneicher, G., Andersen, O.K. and Jepsen, O. (1992). Ber. Bunsenges. Phys. Chem., 96, 1580-1585.
Investigation of sub-shell occupation in some Ba compounds by LMTO-ASA calculations

4238 Feng, P.Y. and Balasubramanian, K. (1989). Chem. Phys., 138, 89-98.
The low-lying electronic states of indium trimer

4239 Feng, P.Y., Balasubramanian, K., Kaufman, J.J. and Koski, W.S. (1990). Phys. Rev. A, 42, 6428-6435.

Energy tranfer in collisions of I^+ with Xe

4240 Feng, S.-C. (1989). Phys. Rev. B, 39, 8722-8724.
Mesoscopic conductance fluctuations in the presence of spin-orbit coupling and Zeeman splitting

4241 Fernandez, J., Lespes, G. and Dargelos, A. (1986a).
Chem. Phys., 103, 85-91.
The vacuum ultraviolet spectrum of stannane

4242 Fernandez, J., Lespes, G. and Dargelos, A. (1986b).
Chem. Phys., 111, 97-103.
Theoretical and experimental study of the vacuum ultraviolet spectrum of tetrasubstituted tin derivatives $SnCl_4$ and $Sn(CH_3)_4$

4243 Fernandez, J., Teichteil, C. and Dargelos, A. (1987).
Chem. Phys., 114, 201-207.
Ab initio calculations of spin-orbit coupling in the SnH_4^+ ion

4244 Fernandez Pacios, L. (1990). Chem. Phys. Lett., 169, 281-284.
Modification of Ar-core effective potentials for first-row transition metals

4245 Fernandez Pacios, L. (1992). J. Mol. Struct. (Theochem), 254, 13-19.
Modified relativistic effective potentials for transition elements

4246 Fernandez Pacios, L. and Botella Olcina, V. (1991a).
Int. J. Quantum Chem., 39, 197-209.
A correction to effective core potentials for transition metals

4247 Fernandez Pacios, L. and Botella Olcina, V. (1991b).
J. Chem. Phys., 95, 441-450.
Modified Ar core ab initio relativistic effective potentials for transition metals Sc through Cu

4248 Fernandez Pacios, L., Fernandez Nuñez, M. and Gomez Calzada, P.C. (1988). J. Mol. Struct., 166, 163-168.
Core polarization corrections on the relativistic core effective potentials formalism

4249 Fernandez Pacios, L. and Gomez Calzada, P. (1988).
Int. J. Quantum Chem., 34, 267-277.
Core/valence partition and relativistic effects in effective potentials for transition metals

4250 Fernández Sanz, J. and Márquez, A. (1989a). J. Phys. Chem., 93, 7328-7333.
Molecular structure and vibrational analysis of distannane from ab initio second-order perturbation calculation.
A theoretical approach to the tin-X bond (X=C, Si, Ge, Sn)

4251 Fernández Sanz, J., Márquez, A. and Pouchan, C. (1989b).
Chem. Phys., 130, 451-456.
Vibrational spectra of stannane: Harmonic force field,

Raman and IR intensities from ab initio correlated wavefunctions

4252 Fernando, G.W., Davenport, J.W., Watson, R.E. and Weinert, M. (1989). Phys. Rev. B, 40, 2757-2766.
Full-potential linear augmented-Slater-type-orbital method

4253 Fernando, G.W., Watson, R.E., Weinert, M., Wang, Y.J. and Davenport, J.W. (1990). Phys. Rev. B, 41, 11813-11821.
Cohesion and lattice stabilities in the 5d transition metals: Full versus muffin-tin potentials

4254 Ferrarese, G., Stazi, L., Cattani, C. and Bini, D. (1990). Nuovo Cim. B, 105, 1131-1140.
Kinematics of relativistic deformations and Thomas precession

4255 Ferrari, L. and Russo, G. (1990). Phys. Rev. B, 42, 7454-7461.
Nonrelativistic zitterbewegung in two-band systems

4256 Fink, M.G.J. and Johnson, W.R. (1986). Phys. Rev. A, 34, 3754-3759.
Low-energy photoionization of alkali-metal atoms: Relativistic random-phase-approximation calculations

4257 Fink, M.G.J. and Johnson, W.R. (1990). Phys. Rev. A, 42, 3801-3818.
Electron correlations and spin-orbit interaction in two-photon ionization of closed-shell atoms: A relativistic time-dependent Dirac-Fock approach

4258 Firth, N.C. and Grice, R. (1987). J. Chem. Soc. Faraday Trans. 2, 83, 1011-1022.
Topography of potential-energy surfaces
Spin-orbit interaction for $H+F_2$, Cl_2

4259 Fisher, J.J. and Michl, J. (1987). J. Am. Chem. Soc., 109, 583-584.
External and internal heavy-atom effects on the rate of spin-forbidden proton tunneling in the triplet ground state biradical, 1,3-perinaphthadiyl

4260 Flad, H.-J., Savin, A. and Preuss, H. (1992). J. Chem. Phys., 97, 459-463.
Reduction of the computational effort in quantum Monte Carlo calculations with pseudopotentials through a change of the projection operators

4261*Flad, J., Igel-Mann, G., Dolg, M., Preuss, H. and Stoll, H. (1985a). Surface Sci., 156, 930-932.
Model calculations for the adsorption of H on Cu and Ag

4262*Flad, J., Igel-Mann, G., Dolg, M., Preuss, H. and Stoll, H. (1985b). Surface Sci., 163, 285-302.
Quantum chemical studies of the chemisorption of atomic hydrogen on copper and silver clusters

4263*Flad, J., Igel-Mann, G., Preuss, H. and Stoll, H. (1985c). Surface Sci., 156, 379-385.

Comparison of group IA and group IB homonuclear clusters

4264 Flad, J., Stoll, H., Nicklass, A. and Preuss, H. (1990). Z. Phys. D, 15, 79-86.
Quantum chemical investigations of the latent image formation. III. Adsorption of Ag_n^+, Ag_n, and Ag_n^- ($n \leq 2$) om a (100) silver bromide surface

4265 Flad, J., Stoll, H. and Preuss, H. (1987). Z. Phys. D, 6, 193-198.
Quantum chemical investigations of the latent image formation. I. Modelling the silver halide surface

4266 Flambaum, V.V. (1992). Phys. Rev. A, 45, 6174-6179.
Long-range parity-nonconserving interaction

4267*Fleischer, W. and Soff, G. (1984). Z. Naturforsch., 39a, 703-719.
Bound state soluations of the Klein-Gordon equation for strong potentials

4268 Florescu, V., Marinescu, M. and Pratt, R.H. (1990). Phys. Rev. A, 42, 3844-3851.
Relativistic effects on low-frequency Rayleigh scattering including spin flip for hydrogenic atoms.
Erratum: ibid. 43(1991), 6432.

4269 Florio, G.M., Ginatempo, B., Giuliano, E.S. and Staunton, J.B. (1989). J. Phys.: Condensed Matter, 1, 8385-8396.
The electronic structure of Pd-Pt random alloys: A relativistic KKR-CPA calculation

4270*Flügge, S. (1974).
The relativistic Dirac equation
In "Practical Quantum Mechanics", 2, Springer, Berlin, pp. 165-225.

4271 Flurer, R.A. and Busch, K.L. (1991). J. Am. Chem. Soc., 113, 3656-3663.
Structures of Ag_2X^+ and Cu_2X^+ ions: Comparison of theoretical predictions with experimental results from mass spectrometry/mass spectrometry

4272 Focher, P., Lastri, A., Covi, M. and Bachelet, G.B. (1991). Phys. Rev. B, 44, 8486-8495.
Pseudopotentials and physical ions

4273*Fock, V. (1933). Z. Physik, 81, 195-208.
Über Austauschenergie

4274*Fock, V.A. (1934). Zh. Eksp. Teor. Fiz., 4, 5-16.
On the energy of quantum exchange (in Russian)

4275*Foldy, L.L. (1956). Phys. Rev., 102, 568-581.
Synthesis of covariant particle equations

4276 Folga, E. and Ziegler, T. (1992). Can. J. Chem., 70, 333-342.
A theoretical study on the activation of hydrogen-hydrogen and hydrogen-alkyl bonds by electron-poor early transition metals

4277 Folga, E., Ziegler, T. and Fan, L.-Y. (1991). New. J. Chem., 15, 741-748.
A theoretical study on the hydrogen exchange reaction between Cl_2Lu-H and H_2

4278 Ford, M.J., Pejcev, V., Smith, D., Ross, K.J. and Wilson, M. (1990). J. Phys. B: At. Mol. Opt. Phys., 23, 4247-4262.
The ejected-electron spectra of manganese and samarium vapour atoms arising from autoionizing and Auger transitions following electron impact excitation

4279*Forman, P. (1968). Isis, 59, 156-174.
The doublet riddle and atomic physics circa 1924

4280 Foucrault, M., Millie, Ph. and Daudey, J.P. (1992). J. Chem. Phys., 96, 1257-1264.
Nonperturbative method for core-valence correlation in pseudopotential calculations: Application to the Rb_2 and Cs_2 molecules

4281 Fox, A.G., O'Keefe, M.A. and Tabbernor, M.A. (1989). Acta Cryst., A45, 786-793.
Relativistic Hartree-Fock X-ray and electron atomic scattering factors at high angles

4282 Fraga, S., Klobukowski, M., Muszynska, J., Saxena, K.M.S. and Sordo, J.A. (1986). Phys. Rev. A, 34, 23-28.
Matrix elements of the Breit Hamiltonian

4283 Fraga, S., Klobukowski, M., Muszynska, J., Saxena, K.M.S., Sordo, J.A., Climenhaga, J.D. and Clark, P. (1987). Comp. Phys. Comm., 47, 159-172.
Research in atomic structure: A configuration interaction program with relativistic corrections

4284 Fraga, S., San Fabian, E., Sordo, J.A., Campillo, M., Climenhaga, J.D. and Klobukowski, M. (1989). Int. J. Quantum Chem., 35, 325-330.
Atomic energy levels from configuration interaction calculations with relativistic corrections

4285 Francisco, E. and Pueyo, L. (1987). Phys. Rev. A, 36, 1978-1982.
Accurate calculation of spin-orbit coupling constants for 3d atoms and ions with effective core potentials and reduced basis sets

4286 Francisco, E. and Pueyo, L. (1988). Phys. Rev. B, 37, 5278-5288.
Theoretical spin-orbit coupling constants for 3d ions in crystals

4287 Franke, R. and Kutzelnigg, W. (1992). Chem. Phys. Lett., 199, 561-567.
Perturbative relativistic calculations for one-electron systems in a Gaussian basis

4288 Franosch, T. and Soff, G. (1991). Z. Phys. D, 18, 219-222.

The influence of the nuclear shape and of the muonic vacuum polarization on strongly bound electrons

4289 Frey, R.F. and Davidson, E.R. (1989). J. Chem. Phys., 90, 5541-5554.
Analysis of the bond energy of ScCO

4290*Friar, J.L. (1979b). Ann. Phys. (New York), 122, 151-196.
Nuclear finite-size effects in light muonic atoms

4291*Friar, J.L. and Tomusiak, E.L. (1984). Phys. Rev. C, 29, 1537-1539.
Relativistically corrected Schrödinger equation with Coulomb interaction

4292*Fricke, B. and Greiner, W. (1969). Phys. Lett., 30B, 317-319.
On the chemistry of superheavy elements around Z = 164

4293*Fricke, B. and Sepp, W.-D. (1981). J. Phys. B: At. Mol. Phys., 14, L549-L551.
Kr-Kr correlation diagram and its use in heavy-ion collisions

4294 Fritsche, L., Kroner, C. and Reinert, Th. (1992). J. Phys. B: At. Mol. Opt. Phys., 25, 4287-4312.
A consistent relativistic theory of electron scattering by atoms

4295 Fritsche, L., Noffke, J. and Eckardt, H. (1987). J. Phys. F, 17, 943-965.
A relativistic treatment of interacting spin-aligned electron systems: Application to ferromagnetic iron, nickel and palladium metal

4296 Fritzsche, S. and Fricke, B. (1992). Phys. Scr., T41, 45-50.
Interchannel interactions and relaxation in the 2p Auger spectra of Mg-like ions

4297 Fritzsche, S., Fricke, B. and Sepp, W.-D. (1992). Phys. Rev. A, 45, 1465-1470.
Reduced L_1 level width and Coster-Kronig yields by relaxation and continuum interactions in atomic zinc

4298 Fritzsche, S. and Zschornack, G. (1991). Z. Phys. D, 21, S155-156.
Effects of final-state interactions on the L-MM Auger spectra of krypton

4299 Fritzsche, S., Zschornack, G., Musiol, G. and Soff, G. (1991). Phys. Rev. A, 44, 388-391.
Interchannel interactions in highly energetic radiationless transitions of neonlike ions

4300 Froese Fischer, C. (1986). Phys. Rev. A, 34, 1667-1676.
Energy levels and lifetimes for some core-excited quartet states in Na I and Mg II

4301 Froese Fischer, C. (1987). J. Phys. B, 20, 1193-1202.

The 2p-3s transitions in atomic oxygen

4302 Froese Fischer, C. (1988). Nucl. Instrum. Meth. B, 31, 265-272.
Accurate oscillator strengths

4303 Froese Fischer, C. (1989). Phys. Rev. A, 39, 963-970.
Binding energies for p electrons in negative alkaline-earth element

4304 Froese Fischer, C. (1991a). Comp. Phys. Comm., 64, 369-398.
The MCHF atomic-structure package

4305 Froese Fischer, C. (1991b). Comp. Phys. Comm., 64, 473-485.
A configuration interaction program

4306 Froese Fischer, C. and Chen, D. (1989). J. Mol. Struct. (Theochem), 199, 61-73.
Numerical multiconfiguration Hartree-Fock calculations for nsnp ^3P states of Rb and Cs negative ions

4307 Froese Fischer, C. and Godefroid, M. (1986). J. Phys. B, 19, 137-148.
Multi-configuration Hartree-Fock plus Breit-Pauli results for some forbidden transitions

4308 Froese Fischer, C. and Godefroid, M.R. (1991). Comp. Phys. Comm., 64, 501-519.
Programs for computing LS and LSJ transitions from MCHF wave function

4309 Froese Fischer, C. and Idrees, M. (1989). Phys. Scr., 39, 70-72.
Autoionization rates for core excited ^5P states in Na^-

4310 Fröhlich, J., Lieb, E.H. and Loss, M. (1986). Commun. Math. Phys., 104, 251-270.
Stability of Coulomb systems with magnetic fields. I. The one-electron atom

4311 Frost, L., Mitroy, J. and Weigold, E. (1986). J. Phys. B: At. Mol. Phys., 19, 4063-4074.
The electron momentum spectroscopy of lead

4312 Fuentealba, P. and Reyes, O. (1987). Mol. Phys., 62, 1291-1296.
Pseudopotential calculations on the ground state of the alkaline-earth monohydride ions

4313 Fuentealba, P., Reyes, O., Stoll, H. and Preuss, H. (1987). J. Chem. Phys., 87, 5338-5345.
Ground state properties of alkali and alkaline-earth hydrides

4314 Fuentealba, P., Savin, A., Stoll, H. and Preuss, H. (1989). Phys. Rev. A, 40, 2163-2165.
Electron affinities of alkaline-earth atoms by means of different density functionals

4315 Fuentealba, P., Savin, A., Stoll, H. and Preuss, H. (1990). Phys. Rev. A, 41, 1238-1242.
Electron affinities of alkaline-earth-metal atoms

4316 Fuentealba, P., Stoll, H. and Savin, A. (1988). Phys. Rev. A, 38, 483-486.
Atomic correlation energy differences by means of a polarization potential

4317 Fujimura, N. and Matsuoka, O. (1992). Int. J. Quantum Chem., 42, 751-759.
Molecular integrals of Breit interaction over Laguerre Gaussian-type functions

4318 Fulton, T. (1987). Phys. Scr., 36, 441-443.
Approximation methods for many-electron atoms from a field theory point of view

4319 Fulton, T. (1989).
An alternative approach to the application of QED to many-electron atoms
In "Relativistic, Quantum Electrodynamic and Weak Interaction Effects in Atoms", Ed. W. Johnson, P. Mohr and J. Sucher, AIP Conf. Proc., 189, pp. 429-441.

4320 Fulton, T. and Johnson, W.R. (1986). Phys. Rev. A, 34, 1686-1695.
Numerical test of the equality of the "length" and "velocity" forms of oscillator strengths for Li I in the Dirac-Fock approximation

4321 Gaida, R.P., Tretjak, V.I. and Yaremko, Yu.G. (1991). Ukr. Fiz. Zh., 36, 1807-1818.
Changing centers-of-mass in the relativistic Lagrangean mechanics of a system of particles (in Ukrainian)

4322 Gajek, Z. and Mulak, J. (1992). J. Phys.: Condens. Matter, 4, 427-444.
The ab initio computational basis of the angular overlap model and its phenomenological effectiveness in the 5f ion compounds

4323*Gajzágó, E. and Marx, G. (1974). Atomki Közlemények Suppl., 16(2), 177-184.
Energy difference of mirror molecules

4324 Galeao, A.P. and Ferreira, P.L. (1992). J. Math. Phys., 33, 2618-2625.
General method for reducing the two-body Dirac equation

4325 Galic, H. (1988). Am. J. Phys., 56, 312-317.
Fun and frustration with hydrogen in a 1+1 dimension

4326 Gallego-Planas, N. and Whitehead, M.A. (1992). J. Mol. Struct.(Theochem), 260, 419-432.
The molecular structure of pentastanna[1.1.1]propellane and bicyclo[1.1.1]pentastannane from the quasi-relativistic SIC-MS-$X\alpha$ theory

4327 Garcia, A., Elsässer, C., Zhu, J., Louie, S.G. and Cohen, M.L. (1992). Phys. Rev. B, 46, 9829-9832.
Use of gradient-corrected functionals in total-energy calculations for solids

4328 Garcia, J.D. (1986). Phys. Rev. A, 34, 4396-4398.
Quantum solutions and classical limits for strong Coulomb fields

4329 Garcia-Cuesta, I., Miralles-Sabater, J., Sanchez-De-Merás, A.M., Merchán, M. and Nebot-Gil, I. (1989). Mol. Phys., 66, 659-668.
A MO-CI study on the π-coordinated interaction between palladium atom and acetylene

4330 Gavezzotti, A., Tantardini, G.F. and Miessner, H. (1988). J. Phys. Chem., 92, 872-878.
SCF-MO-pseudopotential and extended Hückel calculations on the interaction of CO with Pt clusters

4331 Gavezzotti, A., Tantardini, G.F. and Simonetta, M. (1986). Chem. Phys. Lett., 129, 577-581.
Molecular-orbital calculations on Pt_2, PtH and PtCO with an optimized relativistic pseudopotential for Pt

4332 Gay, J.G. and Richter, R. (1991). J. Magnetism Magnetic Mat., 93, 315-318.
Spin-orbit induced orbital angular momentum in monolayers

4333 Gayen, S. and Ermler, W.C. (1992). Int. J. Quantum Chem., 42, 1091-1099.
Ab initio study of K adsorption on a Ag(100) surface

4334 Gayen, S., Ermler, W.C. and Sandroff, C.J. (1991). J. Phys. Chem., 95, 7357-7361.
Ab initio study of GaAs surface passivation by sulfur

4335 Gaylord, R.H. and Kevan, S.D. (1987). Phys. Rev. B, 36, 9337-9340.
Spin-orbit-interaction-induced surface resonance on W(011)

4336 Geiger, K., Reinhardt, J. and Mueller, B. (1987). Verh. Dtsch. Phys. Ges., 22, F-7.4
A relativistic two-body equation for positronium

4337 Gemein, B. and Peyerimhoff, S.D. (1990). Chem. Phys. Lett., 173, 7-14.
Ab initio study of the $HeAr^+$ ion: Potential energy curves for the lowest states, spin-orbit coupling and predissociation rates

4338 Gerasyuk, A.K., Repkin, O.M. and Starostin, N.V. (1986). Opt. Spectrosc., 61, 785-789.
Relativistic calculation of cerium impurity centers in a fluorite crystal

4339*Gesztesy, F., Grosse, H. and Thaller, B. (1985a). Adv. Appl. Math., 6, 159-176.
First-order relativistic corrections and spectral concentration

4340 Gesztesy, F. and Seba, P. (1987). Lett. Math. Phys., 13, 345-358.
New analytically solvable models of relativistic point interactions

4341*Gesztesy, F., Simon, B. and Thaller, B. (1985b). Proc. Am. Math. Soc., 94, 115-118.
On the selfadjointness of Dirac operators with anomalous magnetic moment

4342 Gilbert, J. and Murray, M. (1991).
Clifford Algebras and Dirac Operators in Harmonic Analysis
Cambridge University Press, 350pp.

4343 Ginatempo, B., Durham, P.J. and Gyorffy, B.L. (1989). J. Phys.: Condens. Matter, 1, 6483-6501.
A relativistic theory of photoemission from crystalline metals and alloys

4344 Ginatempo, B. and Gyorffy, B.L. (1991).
The spin polarized photoemission from non-magnetic metals
In "The Effects of Relativity in Atoms, Molecules, and the Solid State", Ed. S. Wilson, I.P. Grant and B.L. Gyorffy, Plenum, New York and London, pp. 319-331.

4345 Ginatempo, B. and Staunton, J.B. (1988). J. Phys. F: Met. Phys., 18, 1827-1837.
The electronic structure of disordered alloys containing heavy elements - an improved calculational method illustrated by a study of a copper-gold alloy

4346*Ginsburg, J.L. and Goodman, L. (1968). Mol. Phys., 15, 441-451.
Effect of two-electron spin-orbit interactions on singlet-triplet transition probabilities

4347 Gjaja, I. and Bhattacharjee, A. (1988). Phys. Rev. A, 37, 1009-1012.
Relativistic quantum dynamics and quantum noise in short-wavelength free-electron lasers

4348 Glass, J.T., McCann, J.F. and Crothers, D.S.F. (1992). J. Phys. B: At. Mol. Opt. Phys., 25, L541-L544.
Electron capture at semirelativistic energies: distorted wave models

4349 Glass, R. (1987a). Z. Phys. D, 5, 133-141.
The atomic form factor for vanadium: A treatment including correlation and LS-dependent relativistic effects

4350 Glass, R. (1987b). J. Phys. C, 20, 4941-4951.
Spin magnetic form factors for cobalt and nickel: including atomic correlation and LS-dependent relativistic effects

4351 Glass, S.J. and Mendlowitz, H. (1989). Am. J. Phys., 57, 466-467.
Spin-flips in Dirac scattering by a one-dimensional barrier

4352 Glebov, V.A., Kasztura, L., Nefedov, V.S. and Zhuikov, B.L. (1989). Radiochimica Acta, 46, 117-121.
Is element 104 (kurchatovium) a p-element?
II. Relativistic calculations of the electronic atomic structure

4353 Glebov, V.A., Nefedov, V.S. and Chebotarev, N.T. (1990). J. Radioanalytical Nucl. Chem., 143, 143-148.
Quantum-chemical investigation of the nature of interatomic bonding in molybdenum and uranium clusters

4354 Gleghorn, J. (1986). C. R. Acad. Sc. Paris, 303, 1425-1428.
A theoretical study of the mercuration of benzene

4355 Gleghorn, J.T. and Hammond, N.D.A. (1990). Chem. Phys. Lett., 168, 400.
Erratum to "Charge iterative relativistic extended Hückel theory and its application to the digermene, distannene and diplumbene systems, Chem. Phys. Lett., 105(1984)621."

4356 Glöckle, W., Lee, T.-S.H. and Coester, F. (1986). Phys. Rev. C, 33, 709-716.
Relativistic effects in three-body bound states

4357 Glöckle, W., Nogami, Y. and Fukui, I. (1987). Phys. Rev. D, 35, 584-590.
Structure of a composite system in motion in relativistic quantum mechanics

4358 Glotov, G.E., German, M.M., Farberovich, O.V. and Remeta, E. Yu. (1990). Opt. Spectrosc., 68, 287-289.
Photoabsorption of ytterbium in the nonstationary theory of the local-density functional

4359 Glushkov, A.V. (1988). Opt. Spectrosc., 64, 267-268.
Model of a multielectron atom

4360 Glushkov, A.V. (1989). Opt. Spectrosc., 66, 18-20.
Universal quasiparticle energy functional in density-functional theory for a relativistic atom

4361 Glushkov, A.V. (1990). Izv. VUZov, Fiz., 33(1), 5-9; Sov. Phys. J., 33, 1-4.
Relativistic polarization potential of a many-electron atom

4362 Glushkov, A.V. (1991a). Opt. Spektrosk., 70, 952-955; Opt. Spectrosc., 70, 555-557.
Relativistic calculation and extrapolation of oscillator strengths in franciumlike ions by a model-potential method

4363 Glushkov, A.V. (1991b). Izv. VUZov, Fiz. (4), 34-39.
Calculation and extrapolation of spectroscopic properties for Ga-like ions (in Russian)

4364 Glushkov, A.V. (1991c). Izv. VUZov, Fiz. (10), 29-34.
Relativistic multiconfiguration "time-dependent" SCF theory for molecules (in Russian)

4365 Glushkov, A.V. (1992a). Opt. Spectrosc., 72, 293-296;
Opt. Spektrosk., 72, 542-547.
Semiempirical calculations of oscillator strengths in multiply charged ions with an electron above closed core

4366 Glushkov, A.V. (1992b). JETP Lett., 55, 97-100; Pis'ma Zh. Eksp. Teor. Fiz. 55, 104-107.
Negative ions of inert gases

4367 Glushkov, A.V. (1992c). J. Appl. Sp., 56, 5-9; Zh. Prikl. Sp., 56, 13-17.
Oscillator strengths of Cs and Rb-like ions

4368 Glushkov, A.V., Malinovskaya, S.V. and Filatov, V.V. (1987). Opt. Spectrosc., 62, 716-718.
Calculation and extrapolation of the spectroscopic characteristics of Al-like ions by the model-potential method

4369 Goalwin, P.W. and Kunz, A.B. (1987). Phys. Rev. B, 35, 5795-5801.
Ab initio calculations of the excitonic spectra of the alkali iodides including relativistic effects

4370 Godwal, B.K. and Jeanloz, R. (1989). Phys. Rev. B, 40, 7501-7507.
First-principles equation of state of gold

4371*Goeppert Mayer, M. (1941). Phys. Rev., 60, 184-187.
Rare-earth and transuranic elements

4372 Gogava, A.L., Ivanov, L.N. and Ivanova, E.P. (1986). Izv. Akad. Nauk, Ser. Fiz., 50, 1349-1355.
Nuclear and electrodynamic effects in the spectra of Li-like ions (in Russian)

4373 Gogava, A.L., Ivanova, E.P. and Tsirekidze, M.A. (1988). Opt. Spectrosc., 64, 435-438.
Highly excited energy levels in the Ne-like ions Ar IX-Mo XXXIII

4374 Goldberg, I.B. and Pratt, R.H. (1987a). J. Math. Phys., 28, 1351-1359.
Schrödinger-like equation for relativistic particles

4375 Goldberg, I.B. and Pratt, R.H. (1987b). Phys. Rev. A, 36, 2108-2111.
Possible narrow and deep minima in the photoionization cross sections of excited states

4376 Goldberg, I.B., Stein, J., Ron, A. and Pratt, R.H. (1989). Phys. Rev. A, 39, 506-513.
Relativistic continuum normalizations and phase shifts in a screened Coulomb potential from a relativistic generalized WKB approximation

4377 Goldberg, I.B., Stein, J., Ron, A. and Pratt, R.H. (1990). Phys. Rev. A, 42, 2501-2512.
Relativistic Fermi-Segre formula

4378 Goldman, S.P. (1987). Phys. Rev. A, 36, 3054-3058.
Variational Dirac-Hartree-Fock energy optimization by the virial-theorem method

4379 Goldman, S.P. (1988a). Phys. Rev. A, 37, 16-30.
Variational Dirac-Hartree-Fock method: Results for the He, Be, C, and Ne isoelectronic sequences

4380 Goldman, S.P. (1988b). Nucl. Instrum. Meth. B, 31, 131-133.
Variational Dirac-Hartree-Fock calculation of the Breit interaction

4381 Goldman, S.P. (1989a). Phys. Rev. A, 39, 976-980.
Gauge-invariance method for accurate atomic-physics calculations: Application to relativistic polarizabilities

4382 Goldman, S.P. (1989b). Phys. Rev. A, 40, 1185-1193.
Generalized Laguerre representation: Application to relativistic two-photon decay rates

4383 Goldman, S.P. (1990). Phys. Rev. A, 41, 6526-6528.
Comment on "Finite-basis-set variational solution of the Dirac equation"

4384 Goldman, S.P. (1992). J. Phys. B: At. Mol. Opt. Phys., 25, 629-648.
Extended relativistic representation of spin-1/2 particles

4385 Goldman, S.P. and Chen, Z.-H. (1991). Phys. Rev. Lett., 67, 1403-1406.
Generalized relativistic variational calculations for hydrogenic atoms in arbitrary magnetic fields

4386 Goldman, S.P. and Dalgarno, A. (1986). Phys. Rev. Lett., 57, 408-411.
Finite-basis-set approach to the Dirac-Hartree-Fock equations

4387 Goldman, S.P. and Drake, G.W.F. (1992). Phys. Rev. Lett., 68, 1683-1686.
Asymptotic Lamb shifts for helium Rydberg states

4388 Golshan, S.M. and Kobe, D.H. (1986). Phys. Rev. A, 34, 4449-4452.
Retarded and nonretarded interactions between a pair of nonidentical two-level atoms

4389 González, J., Guinea, F. and Vozmediano, M.A.H. (1992). Phys. Rev. Lett., 69, 172-175.
Continuum approximation to fullerene molecules

4390 Gonze, X., Michenaud, J.-P. and Vigneron, J.-P. (1988). Phys. Scr., 37, 785-789.
Ab initio calculations of bismuth properties, including spin-orbit coupling

4391 Gonze, X., Michenaud, J.-P. and Vigneron, J.-P. (1990). Phys. Rev. B, 41, 11827-11836.

First-principles study of As, Sb, and Bi electronic properties

4392 Good Jr., R.H. (1989). Ann. Phys.(N.Y.), 196, 1-11.
Relativistic wave equations for particles in electromagnetic fields

4393 Goodman, G.L., Loong, C.-K. and Soderholm, L. (1991). J. Phys.: Condens. Matter, 3, 49-67.
Crystal-field properties of f-electron states in $RBa_2Cu_3O_7$ for R=Ho, Nd and Pr

4394 Goodman, G.L. and Soderholm, L. (1990). Physica C, 171, 528-536.
Molecular-orbital cluster calculations on $MBa_2Cu_3O_7$ for M=Y, Pr, Nd, Ho and Cm

4395 Goodson, D.Z., Morgan III, J.D. and Herschbach, D.R. (1991). Phys. Rev. A, 43, 4617-4624.
Dimensional singularity analysis of relativistic equations

4396 Gorceix, O. and Indelicato, P. (1988). Phys. Rev. A, 37, 1087-1094.
Effect of the complete Breit interaction on two-electron ion energy levels

4397 Gorceix, O., Indelicato, P. and Desclaux, J.P. (1987). J. Phys. B: At. Mol. Phys., 20, 639-649.
Multiconfiguration Dirac-Fock studies of two-electron ions: I. Electron-electron interaction

4398 Gordienko, A.B., Zhuravlev, Yu.N. and Poplavnoi, A.S. (1992). Izv. VUZov, Fiz., 35 (2), 38-40.
Energy band structure of silver azide (AgN_3) (in Russian)

4399 Görling, A., Rösch, N., Ellis, D.E. and Schmidbaur, H. (1991). Inorg. Chem., 30, 3986-3994.
Electronic structure of main-group-element-centered octahedral gold clusters

4400 Gorshkov, V.G., Karasev, V.V., Labzovsky, L.N., Nefedov, A.V. and Sultanaev, A.A. (1992). Opt. Spectrosc., 72, 16-23; Opt. Spektrosk., 72, 31-43.
Quantum electrodynamic theory of spectral line profile: application to the U^{90+} spectrum

4401 Gough, W. (1986). Eur. J. Phys., 7, 35-42.
The analysis of spin and spin-orbit coupling in quantum and classical physics by quaternions

4402 Gough, W. (1989). Eur. J. Phys., 10, 318-319.
On the probability density of a relativistic free electron

4403 Goursot, A., Chermette, H., Waltz, W.L. and Lilie, J. (1989). Inorg. Chem., 28, 2241-2247.
Theoretical study by the $X\alpha$ method of platinum(III) complex ions containing aquo and chloro ligands

4404 Goursot, A., Kirk, A.D., Waltz, W.L., Porter, G.B. and Sharma, D.K. (1987a). Inorg. Chem., 26, 14-18.

Experimental and theoretical study of the nascent photoredox behavior of the aqueous hexachloroplatinate(IV) ion

4405 Goursot, A., Motais, B.L., Daul, C., Penigault, E. and Koulkes-Pujo, A.M. (1987b). Chem. Phys. Lett., 138, 339-345.
Electronic structure and absorption spectra of a dimeric tantalum(III) compound and its transient anion

4406 Goursot, A., Papai, I. and Salahub, D.R. (1992). J. Am. Chem. Soc., 114, 7452-7458.
Density functional study of CO chemisorption on model clusters of Rh and Pd: A comparative analysis of the site selection

4407 Grabandt, O., De Lange, C.A. and Mooyman, R. (1989). Chem. Phys. Lett., 160, 359-365.
He(I) photoelectron spectroscopy of InF

4408 Graf, S., Muller, B., Stein, E., Reinhardt, J., Soff, G. and Greiner, W. (1991).
QED in strong Coulomb fields: charged vacuum, atomic clock, and correlated, narrow e^+e^--lines
In "International Symposium on Heavy Ion Physics and Its Application", Ed. W.Q. Shen, Y.X. Luo and J.Y. Liu, World Scientific, Singapore, pp. 106-139.

4409 Graff, M.M. and Wagner, A.F. (1990). Chem. Phys. Lett., 174, 287-293.
Theoretical studies of fine-structure effects and long-range forces: approximating the reactive surface of $O(^3P) + OH(^2\Pi)$

4410 Grant, I.P. (1986a). J. Phys. B: At. Mol. Phys., 19, 3187-3205.
Variational methods for Dirac wave equations

4411 Grant, I.P. (1986b). Aust. J. Phys., 39, 649-665.
Current developments in the relativistic quantum mechanics of atoms and molecules

4412 Grant, I.P. (1987). J. Phys. B: At. Mol. Phys., 20, L735-L740.
On the relativistic electron-electron interaction

4413 Grant, I.P. (1988a). AIP Conf. Proc., 168, 78-93.
Theory of high-Z atoms: Principles and practice

4414 Grant, I.P. (1988b).
Relativistic atomic structure calculations
In "Methods in Computational Chemistry", Ed. S. Wilson, Plenum, 2, pp. 1-71.

4415 Grant, I.P. (1989).
Notes on basis sets for relativistic atomic structure and QED
In "Relativistic, Quantum Electrodynamic and Weak Interaction Effects in Atoms", Ed. W. Johnson, P. Mohr and J. Sucher, AIP Conf. Proc., 189, pp. 235-253.

4416 Grant, I.P. (1990). AIP Conf. Proc., 215, 46-71.
Relativistic atomic structure and electron-atom collisions

4417 Grant, I.P. (1991).
Relativistic atomic structure and electron-atom collisions
In "The Effects of Relativity in Atoms, Molecules, and the Solid State", Ed. S. Wilson, I.P. Grant and B.L. Gyorffy, Plenum, New York and London, pp. 17-43.

4418 Grant, I.P. (1992).
Relativistic electronic structure theory
In "Methods in Computational Molecular Physics", Ed. S. Wilson and G.H.F. Diercksen, Plenum.

4419 Grant, I.P. and Quiney, H.M. (1988). Adv. At. Mol. Phys., 23, 37-86.
Foundations of the relativistic theory of atomic and molecular structure

4420 Grave de Peralta, L. and Boada, R.L. (1986). Rev. Cubana Fis., 6(2), 65-69.
Sobre la formulacion y aplicacion del metodo de los seudopotenciales en atomos alcalinos

4421 Gray, L.G., Sun, X. and MacAdam, K.B. (1988). Phys. Rev. A, 38, 4985-4993.
Resonance measurements of d-f-g-h intervals in Rydberg states of sodium and a redetermination of the core polarizabilities

4422*Green, H.S. (1965).
Matrix Mechanics
Noordhoff, Groningen, pp. 110-115.

4423 Green, J.C., Hohl, D. and Rösch, N. (1987). Organometallics, 6, 712-720.
(Cyclopentadienyl)lanthanide compounds $Ln(C_5Me_5)_2$ (Ln=Sm, Eu, Yb): Photoelectron spectra and molecular orbital calculations

4424*Green, T.A. and Rose, M.E. (1951). Phys. Rev., 110, 105-122.
Nuclear structure effects in internal conversion

4425 Greene, C.H. (1990). Phys. Rev. A, 42, 1405-1415.
Photoabsorption spectra of the heavy alkali-metal negative ions

4426 Greene, C.H. and Aymar, M. (1991). Phys. Rev. A, 44, 1773-1790.
Spin-orbit effects in the heavy alkaline-earth atoms

4427 Greene, C.H. and Theodosiou, C.E. (1990). Phys. Rev. A, 42, 5773-5775.
Photoionization of the Ba 6s6p $^1P_1$ state

4428*Greenside, H.S. and Schlüter, M. (1983). Phys. Rev. B, 28, 535-543.

Pseudopotentials for the 3d transition-metal elements

4429 Grein, F. and Peyerimhoff, S.D. (1987). J. Chem. Phys., 87, 4684-4692.
Theoretical studies on excited states of Ne_2. II. Potential curves for states dissociating to Ne+Ne*(3s) with semiempirical spin-orbit interaction, and comparison with spectroscopic results

4430 Grein, F., Peyerimhoff, S.D. and Klotz, R. (1987). Theor. Chim. Acta, 72, 403-409.
Theoretical studies on excited states of Ne_2

4431 Greiner, W. (1987).
The Physics of Strong Fields
Pergamon - Vieweg.

4432 Greiner, W. (1990).
Relativistic Quantum Mechanics. Wave Equations
Springer-Verlag, Berlin, 345p.

4433 Greiner, W. and Reinhardt, J. (1992).
Quantum Electrodynamics
Springer-Verlag, 308 pp.

4434 Grexa, M., Hermann, G., Lasnitschka, G. and Fricke, B. (1988). Phys. Rev. A, 38, 1263-1269.
Hyperfine structure and isotopic shift of the n ^2P_J levels (n=7-10) of $^{203,205}Tl$ measured by Doppler-free two-photon spectroscopy

4435*Gribov, L.A., Baranov, V.I. and Novosadov, B.K. (1984).
Calculation Methods for Electron-Vibration Spectra of Polyatomic Molecules (in Russian), Nauka, Moscow.
(Ch. 11. Spin-orbit interactions and relativistic corrections)

4436 Grigis, A. and Mohamed, A. (1992). Compt. Rend. Acad. Sci., Ser. I, 315, 1249-1252.
Finitude des lacunes dans le spectre de l'opérateur de Schrödinger et celui de Dirac avec des potentiels électrique et magnétique périodiques

4437 Grigore, D.R., Nenciu, G. and Purice, R. (1989).
Ann. l'Institut Henri Poincare Phys. Theor., 51, 231-263.
On the nonrelativistic limit of the Dirac Hamiltonian

4438 Grochmalicki, J., Lewenstein, M. and Rzazewski, K. (1991).
Phys. Rev. Lett., 66, 1038-1041.
Stabilization of atoms in superintense laser fields: is it real?

4439 Gropen, O. (1988).
The relativistic effective core potential method
In "Methods in Computational Chemistry", Ed. S.Wilson, Plenum, 2, pp. 109-135.

4440 Gropen, O., Almlöf, J. and Wahlgren, U. (1992). J. Chem. Phys., 96, 8363-8366.
Relativistic calculations on platinum hydride using

effective core potentials and first-order perturbation theory

4441 Gross, F. (1986). Few-Body Syst., Suppl., 1, 432-443.
Relativistic equations

4442*Grosse, H. (1984). Lett. Math. Phys., 8, 313-319.
Solitons of the modified KdV equation

4443 Grosse, H. (1986). Phys. Reports, 134, 297-304.
New solitons connected to the Dirac equation

4444 Grosse, H. (1987). Phys. Lett. B, 197, 413-417.
On the level order for Dirac operators

4445 Grosse, H., Martin, A. and Stubbe, J. (1991). Phys. Lett. B, 255, 563-566.
Order of energy levels for the Klein-Gordon equation

4446 Grosse, H., Martin, A. and Stubbe, J. (1992). Phys. Lett. B, 284, 347-350.
Order of energy levels in relativistic one-electron models

4447 Grotch, H. (1989).
Status report on the theory of the Lamb shift in hydrogen
In "Relativistic, Quantum Electrodynamic and Weak Interaction Effects in Atoms", Ed. W. Johnson, P. Mohr and J. Sucher, AIP Conf. Proc., 189, pp. 108-119.

4448 Grotch, H., Zhang, X.-G. and Sebastian, K.J. (1992). Phys. Rev, D, 45, 4337-4339.
Electric dipole transition rates of the bound states of the $b\bar{b}$ system

4449 Grozdanov, T.P., Rakovic, M.J. and Solov'ev, E.A. (1992). J. Phys. B: At. Mol. Opt. Phys., 25, 4455-4466.
Hydrogen atoms in circularly polarized microwave fields: photoexcitation of Rydberg states

4450 Gudder, S. (1992). Int. J. Theor. Phys., 31, 1973-1991.
Schrödinger and Dirac quantum random walks

4451 Guet, C., Blundell, S.A. and Johnson, W.R. (1990). Phys. Lett. A, 143, 384-386.
Relativistic many-body calculations of oscillator strengths for sodium-like ions

4452 Guet, C. and Johnson, W.R. (1991). Phys. Rev. A, 44, 1531-1535.
Relativistic many-body calculations of transition rates for Ca^+, Sr^+, and Ba^+

4453 Gulmaro Corona, C. (1989). J. Phys. A, 22, 2341-2353.
A one-parameter family of approximations of the solution of the initial Cauchy problem for the Dirac equation

4454 Gulov, A.V. and Ivanova, E.P. (1989a). Phys. Lett. A, 140, 39-43.
Radiative lifetimes of $2p^53p$, $2p^53d$ states in Ne-like ions

4455 Gulov, A.V. and Ivanova, E.P. (1989b). Sov. Phys.-Lebedev Inst. Rep., 11, 24-27.
Radiational lifetimes of levels in Ne-like ions

4456 Guo, C.-X., Chang, S. and Xia, S.-D. (1991). J. Lumin., 48 & 49, 469-472.
Density of states and energy levels of $BaF_2:Sm^{2+}$ using nonrelativistic and relativistic cluster models

4457 Guo, D.-S. and Åberg, T. (1988). J. Phys. A, 21, 4577-4591.
Quantum electrodynamical approach to multiphoton ionization in the high-intensity H field

4458 Guo, D.-S. and Åberg, T. (1991). J.Phys. B: At. Mol. Opt. Phys., 24, 349-359.
Orthogonality and completeness of the quantum electrodynamical Volkov solutions

4459 Guo, D.-S., Åberg, T. and Crasemann, B. (1989). Phys. Rev. A, 40, 4997-5005.
Scattering theory of multiphoton ionization in strong fields

4460 Guo, G.Y., Ebert, H., Temmerman, W.M., Schwarz, K. and Blaha, P. (1991a). Solid State Comm., 79, 121-124.
Relativistic effects on the structural and magnetic properties of iron

4461 Guo, G.Y., Temmerman, W.M. and Ebert, H. (1991b). J. Phys.: Condens. Matter, 3, 8205-8212.
First-principles determination of the magnetization direction of Fe monolayer in noble metals

4462 Guo, J., Ellis, D.E., Alp, E., Soderholm, L. and Shenoy, G.K. (1989). Phys. Rev. B, 39, 6125-6139.
Multiple-scattering approach to the M-edge x-ray-absorption spectra of UO_2 and UCl_4

4463 Guo, S.H., Yang, X.L., Chan, F.T., Wong, K.W. and Ching, W.Y. (1991). Phys. Rev. A, 43, 1197-1205.
Analytic solution of a two-dimensional hydrogen atom. II. Relativistic theory

4464 Guo, Y.-F. and Whitehead, M.A. (1988). Phys. Rev. A, 38, 3166-3174.
Calculations with the quasirelativistic local-spin-density-functional theory for high-Z atoms

4465 Guo, Y.-F. and Whitehead, M.A. (1989). Phys. Rev. A, 40, 28-34.
Electron affinities of alkaline-earth and actinide elements calculated with the local-spin-density-functional theory

4466 Guo, Y.-F. and Whitehead, M.A. (1992). J. Mol. Struct. (Theochem), 258, 243-250.
Application of the relativistic theory to the spin-polarized local-density functional theory

4467 Guo, Y.-F., Wrinn, M.C. and Whitehead, M.A. (1989). Phys. Rev. A, 40, 6685-6688.
Electron affinities for rare gases and some actinides from local-spin-density-functional theory

4468 Gurchumeliya, A.D., Ivanova, E.P., Tsirekidze, M.A. and Tsirekidze, T.A. (1986). Izv. VUZov, Fiz., 28(10), 99-105; Sov. Phys. J., 28, 852-858.
Relativistic calculation of the energy spectrum of Zn-like highly charged ions

4469*Gürsey, F. (1957). Nuovo Cim., 5, 784-809.
Relativistic kinematics of a classical point particle in spinor form

4470 Gyorffy, B.L., Staunton, J.B., Ebert, H., Strange, P. and Ginatempo, B. (1991).
Relativistic density-functional theory for electrons in solids
In "The Effects of Relativity in Atoms, Molecules, and the Solid State", Ed. S. Wilson, I.P. Grant and B.L. Gyorffy, Plenum, New York and London, pp. 255-273.

4471*Gyulassy, M. (1975). Nucl. Phys., A244, 497-525.
Higher order vacuum polarization for finite radius nuclei

4472 Haar, R.R. and Curtis, L.J. (1987). Am. J. Phys., 55, 1044-1045.
The Thomas precession gives g_e - 1, not $g_e/2$

4473 Haberland, R. and Fritsche, L. (1986). Int. J. Quantum Chem., S20, 289-297.
A generalized Kohn-Sham theory for low-energy electron scattering by atoms including relativistic effects

4474 Haberland, R. and Fritsche, L. (1987). J. Phys. B: At. Mol. Phys., 20, 121-126.
On the elastic scattering of low-energy electrons by Hg, Tl, Pb and Bi atoms

4475 Haberlandt, H. (1990). Theochem, 64, 25-34.
Pseudopotential MRD-CI calculations of nickel-containing molecules. I. The electronic ground state of NiH

4476 Häberlen, O.D. and Rösch, N. (1992). Chem. Phys. Lett., 199, 491-496.
A scalar-relativistic extension of the linear combination of Gaussian-type orbitals local density functional method: application to AuH, AuCl and Au_2

4477 Häberlen, O.D., Rösch, N. and Dunlap, B.I. (1992). Chem. Phys. Lett., 200, 418-423.
Are endohedral metal(IV)C_{28} compounds hypervalent?

4478 Hafner, J. (1989).
Quantum theory of structure: sp-bonded systems
In "The Structures of Binary Compounds", Vol. 3 in the series "Cohesion and Structure", Ed. F.R. De Boer and D.G. Pettifor, North-Holland, Amsterdam.

4479 Hafner, J. and Jank, W. (1990). Phys. Rev. B, 42, 11530-11539.
Structural and electronic properties of the liquid polyvalent elements. III. The trivalent elements

4480 Hafner, J. and Jank, W. (1992). Phys. Rev. B, 45, 2739-2749.
Structural and electronic properties of the liquid polyvalent elements. IV. The pentavalent semimetals and trends across the periodic table

4481 Hagelberg, F., Sahoo, N., Das, T.P., Weil, K.G. and Speidel, K.-H. (1992). Phys. Rev. A, 46, 6087-6090.
Hartree-Fock investigations of the structure of Sb_4 clusters

4482 Hagelstein, P.L. (1986a). Phys. Rev. A, 34, 874-884.
Relativistic distorted-wave results for nickel-like gadolinium

4483 Hagelstein, P.L. (1986b). Phys. Rev. A, 34, 924-942.
Electron collisional excitation in F-like selenium

4484 Hagelstein, P.L. (1987). J. Phys. B, 20, 5785-5806.
Dielectronic capture into excited levels of Ne-like selenium

4485 Hagelstein, P.L. (1991). Rad. Effects Defects in Solids, 122-123, 163-231.
On the partial wave method for self-energy calculations for non-hydrogenic electrons

4486 Hagelstein, P.L. and Jung, R.K. (1987). At. Data Mol. Data Tables, 37, 121-188.
Relativistic distorted-wave calculations of electron collision cross sections and rate coefficients for Ne-like ions

4487 Hall, R.L. (1986). J. Phys. A, 19, 2079-2086.
Dirac eigenvalues for screened Coulomb potentials

4488 Hall, R.L. (1987). J. Math. Phys., 28, 457-462.
Dirac spinor orbits

4489 Hamacher, P. and Hinze, J. (1991). Phys. Rev. A, 44, 1705-1711.
Finite-volume variational method for the Dirac equation

4490 Hamann, D.R. (1989). Phys. Rev. B, 40, 2980-2987.
Generalized norm-conserving pseudopotentials

4491*Hameka, H.F. and Goodman, L. (1965). J. Chem. Phys., 42, 2305-2306.
On the effect of the ground-state dipole moment on optical transition probability

4492 Hansen, J.E. (Ed.) (1990).
Atomic Spectra and Oscillator Strengths for Astrophysics and Fusion Reseach, North-Holland, Amsterdam.

4493 Hardekopf, E.E. (1987). Diss. Abstr. Int. B, 47, 3817.
Relativistic momentum-space equations with applications to atomic and elementary particle physics

4494 Harrison, W.A. and Straub, G.K. (1987). Phys. Rev. B, 36, 2695-2706.
Electronic structure and properties of d- and f-shell-metal compounds

4495 Harston, M.R. and Pyper, N.C. (1992). Phys. Rev. A, 45, 6282-6295.
Exchange effects in ß decays of many-electron atoms

4496 Hartley, A.C., Lindroth, E. and Mårtensson-Pendrill, A.-M. (1990). J. Phys. B: At. Mol. Opt. Phys., 23, 3417-3436.
Parity non-conservation and electric dipole moments in caesium and thallium

4497 Hartley, A.C. and Mårtensson-Pendrill, A.-M. (1990). Z. Phys. D, 15, 309-319.
Relativistic correlation effects on the hyperfine structure and electric dipole transitions in Cs and Tl

4498 Hartley, A.C. and Mårtensson-Pendrill, A.-M. (1991). J. Phys. B: At. Mol. Opt. Phys., 24, 1193-1207.
Calculations of isotope shifts in caesium and thallium using many-body perturbation theory

4499 Hartley, A.C. and Sandars, P.G.H. (1991).
Relativistic calculations of parity non-conserving effects in atoms
In "The Effects of Relativity in Atoms, Molecules, and the Solid State", Ed. S. Wilson, I.P. Grant and B.L. Gyorffy, Plenum, New York and London, pp. 67-81.

4500 Hartley, D. and Tucker, R.W. (1991). Phys. Lett. A, 159, 193-197.
on the motion of charged relativistic particles with spin

4501 Hartung, H., Fricke, B., Sepp, W.-D., Thies, B., Kolb, D., Heinemann, D. and Loftager, P. (1987). Phys. Lett. A, 119, 457-461.
Total differential scattering cross section of argon ion(1+)-argon atom at 15 to 400 keV

4502 Hasegawa, A. and Yamagami, H. (1992). J. Magn. Magn. Materials, 104-107, 65-66.
Electronic structures of plutonium monochalcogenides

4503 Hasegawa, T., Fujimura, N. and Matsuoka, O. (1991). Int. J. Quantum Chem., 39, 805-812.
Hydrogenlike atoms in uniformly charged sphere model of atomic nucleus. I. Reference calculations of energy levels

4504 Hasenburg, K., Madison, D.H., Bartschat, K. and Blum, K. (1986). J. Phys. B: At. Mol. Phys., 19, 1803-1816.

Relativistic effects in elastic scattering of electrons from heavy atoms

4505 Hashizume, K. (1987). Phys. Stat. Sol. B, 144, K23-K28.
Band structure of solid argon within relativistic APW approximation

4506*Hautot, A. (1973). J. Math. Phys., 14, 2011-2017.
Motion in radial magnetic fields

4507 Hay, P.J. (1987). J. Am. Chem. Soc., 109, 705-710.
Ab initio theoretical studies of dihydrogen coordination vs. oxidative addition of H_2 to five-coordinate tungsten complexes

4508 Hay, P.J. (1991). New. J. Chem., 15, 735-740.
The model reaction $Pt(PH_3)_2+H_2$ revisited. Influence of outer core orbitals on electron correlation effects using relativistic effective core potentials

4509 Hay, P.J. (1992).
Theoretical studies of oxidative addition and reductive elimination of hydrogen and alkanes
In "Transition Metal Hydrides", Ed. A. Dedieu, VCH, New York, pp. 127-148.

4510 Hay, P.J., Ryan, R.R., Salazar, K.V., Wrobleski, D.A. and Sattelberger, A.P. (1986). J. Am. Chem. Soc., 108, 313-315.
Synthesis and X-ray structure of $(C_5Me_5)_2Th(\mu\text{-}PPh_2)_2Pt(PMe_3)$: A complex with a thorium-platinum bond

4511 Hayaishi, T., Murakami, E., Morioka, Y., Aksela, H., Aksela, S., Shigemasa, E. and Yagishita, A. (1991). Phys. Rev. A, 44, R2771-R2774.
Manifestation of Kr 3d and Xe 4d conjugate shake-up satellites in threshold-electron spectra

4512 Hayaishi, T., Murakami, E., Morioka, Y., Aksela, H., Aksela, S., Shigemasa, E. and Yagishita, A. (1992). J. Phys. B: At. Mol. Opt. Phys., 25, 4119-4124.
Ar 2p shake-up structures studied by threshold electron spectroscopy

4513 Haynes, G.R., Martin, R.L. and Hay, P.J. (1992). J. Am. Chem. Soc., 114, 28-36.
Theoretical investigations of classical and nonclassical structures of MH_7L_2 polyhydride complexes of Re and Tc

4514 He, M.-K., Rao, J.-X. and Lin, D.-H. (1986). Phys. Energ. Fortis Phys. Nucl., 10, 146-152.
The relativistic corrections for ground state energy of the positronium negative ion (in Chinese)

4515 Hegarty, D. (1986). Theor. Chim. Acta, 70, 351-356.
On the Dirac equation in the algebraic approximation

4516 Hegarty, D. and Aerts, P.J.C. (1987). Phys. Scr., 36, 432-435.

Variational solutions of the Dirac equation for atoms and molecules

4517 Hegerfeldt, G.C., Hinders, M.K., Rhodes, B.A. and Sandri, G.v.H. (1992). Phys. Rev. D, 45, 713-716.
Novel quantization phenomenon for relativistic wave equations

4518*Hegstrom, R.A., Rein, D.W. and Sandars, P.G.H. (1980). J. Chem. Phys., 73, 2329-2341.
Calculation of parity nonconserving energy difference between mirror-image molecules

4519 Heifets, E.N. (1992). Chem. Phys., 160, 363-373.
The formulae for calculation of g-tensor shifts in third-order perturbation theory

4520 Herath Banda, M.A., Koenig, W., Martin, B., Güttner, F., Skapa, H., Soltani, J., Dworschak, K., Bosch, F., Kozhuharov, Ch. and Ramayya, A.V. (1986). Phys. Rev. A, 33, 861-870.
δ-electron emission in superheavy quasiatomic systems with total charge Z_u=110-171

4521 Herbst, J.F. and Wilkins, J.W. (1988). Phys. Rev. B, 38, 1027-1039.
Relativistic calculations of 2p and 3d excitation energies in the actinide metals

4522 Hernandez, P. and Sanquer, M. (1992). Phys. Rev. Lett., 68, 1402-1405.
Negative magnetoconductance in an Anderson insulator with strong spin-orbit scattering

4523 Hersbach, H. (1992). Phys. Rev. A, 46, 3657-3670.
Relativistic quantum theory in momentum space for weak and strong coupling

4524 Hersbach, H. and Ruijgrok, T.W. (1991). Z. Phys. D, 18, 209-217.
A quasipotential theory for the relativistic hydrogen atom

4525 Hess, B.A. (1986). Phys. Rev. A, 33, 3742-3748.
Relativistic electronic-structure calculations employing a two-component no-pair formalism with external-field projection operators

4526 Hess, B.A., Buenker, R.J. and Chandra, P. (1989). Int. J. Quantum Chem., 29, 737-753.
Toward the variational treatment of spin-orbit and other relativistic effects for heavy atoms and molecules

4527 Hess, B.A. and Chandra, P. (1987). Phys. Scr., 36, 412-415.
Relativistic ab initio CI study of the $X^1\Sigma^+$ and $A^1\Sigma^+$ states of the AgH molecule

4528 Hess, B.A. and Ohno, M. (1989). Phys. Rev. A, 39, 5637-5641.
Ab initio g-Hartree calculations on the atoms Be and Ne

that employ fully numerical and basis-set techniques

4529 Hessels, E.A. (1992). Phys. Rev. A, 46, 5389-5396.
Higher-order relativistic corrections to the polarization energies of helium Rydberg states

4530*Hestenes, D. (1966).
Space-Time Algebra
Gordon & Breach, New York.

4531 Hestenes, D. (1990). Found. Phys., 20, 1213-1232.
The zitterbewegung interpretation of quantum mechanics

4532 Hestenes, D. (1991).
The electron, new theory and experiment
In "Fundamental Theories of Physics", Ed. A. Weingartshofer, 45, 410p.
Dordrecht, Kluwer.

4533*Heully, J.-L. (1985).
Atomic many-body calculation with a relativistic formalism
Ph. D. Thesis, University of Gothenburg.

4534 Heully, J.-L., Lindgren, I., Lindroth, E., Lundqvist, S. and Mårtensson-Pendrill, A.-M. (1986a). J. Phys. B: At. Mol. Phys., 19, 2799-2815.
Diagonalisation of the Dirac Hamiltonian as a basis for a relativistic many-body procedure

4535 Heully, J.-L., Lindgren, I., Lindroth, E. and Mårtensson-Pendrill, A.-M. (1986b). Phys. Rev. A, 33, 4426-4429.
Comment on relativistic wave equations and negative-energy states

4536 Hibbert, A. (1989). Phys. Scr., 39, 574-580.
The use of model potentials in oscillator strength calculations

4537 Hibbert, A., Glass, R. and Froese Fischer, C. (1991).
Comp. Phys. Comm., 64, 455-472.
A general program for computing angular integrals of the Breit-Pauli Hamiltonian

4538 Hibbert, A., Mohan, M. and Le Dourneuf, M. (1992). J. Phys. B: At. Mol. Opt. Phys., 25, 1107-1117.
Lifetimes of $2p^5 3l$ states in S VII, Cl VIII and Ar IX

4539 Hickman, A.P., Huestis, D.L. and Saxon, R.P. (1992).
J. Chem. Phys., 96, 2099-2113.
Interatomic potentials for excited states of XeHe and XeAr

4540*Hilgevoord, J. and Wouthuysen, S.A. (1963). Nucl. Phys., 40, 1-12.
On the spin angular momentum of the Dirac particle

4541 Hill, S.H., Grout, P.J. and March, N.H. (1987). J. Phys. B: At. Mol. Phys., 20, 11-22.
Total binding energy of heavy positive ions including

density treatment of Darwin and Breit corrections

4542 Hill III, W.T., Sugar, J., Lucatorto, T.B. and Cheng, K.T. (1987). Phys. Rev. A, 36, 1200-1206.
Analysis of the $5p^6$ --> $5p^5nl$ (J=1) Rydberg series in Ba^{2+}

4543 Hillion, P. (1989). Phys. Rev. A, 40, 1194-1197.
Front form of electromagnetism

4544 Hillion, P. (1992a). J. Math. Phys., 33, 1817-1821.
Nondispersive solutions of the Klein-Gordon equation

4545 Hillion, P. (1992b). J. Math. Phys., 33, 1822-1830.
Nondispersive solutions of the Dirac equation

4546*Hinkley, R.K., Walker, T.E.H. and Richards, W.G. (1972). Mol. Phys., 24, 1095-1102.
Spin-orbit coupling constants from semi-empirical wavefunctions

4547 Hino, K.-I. and Watanabe, T. (1987). Phys. Rev. A, 36, 581-590.
Cross sections of relativistic radiative electron capture by use of the strong-potential Born calculation

4548 Hino, K.-I. and Watanabe, T. (1989). Phys. Rev. A, 39, 3373-3387.
Theory of the relativistic radiative electron capture incorporating effects of the internal conversion process

4549 Hinton, D.B., Jordan, A.K., Klaus, M. and Shaw, J.K. (1991). J. Math. Phys., 32, 3015-3030.
Inverse scattering on the line for a Dirac system

4550 Hinton, D.B., Klaus, M. and Shaw, J.K. (1988). Proc. R. Soc. Edinburgh A, 109, 173-186.
Levinson's theorem and Titchmarsh-Weyl theory for Dirac systems

4551 Hippe, D. and Peyerimhoff, S.D. (1992a). J. Chem. Phys., 96, 3503-3509.
Spin-orbit coupling in the copper monoxide molecule

4552 Hippe, D. and Peyerimhoff, S.D. (1992b). Mol. Phys., 76, 293-318.
Theoretical spectroscopy of copper monoxide

4553 Hjelm, A. and Calais, J.-L. (1991). Phys. Rev. Lett., 67, 2064-2067.
Zeeman splitting in Pd and Pt calculated from self-consistent band structure including an external magnetic field

4554 Hjelm, A. and Calais, J.-L. (1992). J. Magn. Magn. Materials, 104-107, 727-728.
Self-consistent band structure calculations including external magnetic fields

4555 Hliwa, M., Barthelat, J.C., Pélissier, M. and Spiegelmann, F. (1986). Chem. Phys. Lett., 132, 205-208.

Calculation and analysis of the low-lying excited $^{3,1}\Sigma^+$ states of copper hydride

4556 Hoang, L.V., Hai, L.X., Komarov, L.I. and Romanova, T.S. (1992). J. Phys. A, 25, 6461-6469.
Relativistic analogy of the Aharonov-Bohm effect in the presence of Coulomb field and magnetic charge

4557 Hofer, D. and Stocker, W. (1989). Phys. Lett. A, 138, 463-464.
A Dirac particle in a scalar potential depending on one spatial coordinate

4558 Hoffman, D.C. (1990).
Chemistry of the transactinide elements
In "The Robert A. Welch Foundation Conference on Chemical Research XXXIV Fifty Years With Transuranium elements", Houston, Texas, pp. 255-276.

4559 Hofmann, P. and Rösch, N. (1986). J. Chem. Soc., Chem. Commun., 843-844.
Theoretical evidence for a paramagnetic ground state of the cluster ion $[(\eta^5\text{-}C_5Me_5)_3Re_3(\mu_2\text{-}O)_6]^{2+}$

4560 Hofmann, P., Rösch, N. and Schmidt, H.R. (1986). Inorg. Chem., 25, 4470-4478.
Electronic structure of $[(\eta^5\text{-}C_5Me_5)_3Re_3(\mu\text{-}O)_6]^{2+}$ and of related early-transition-metal arene clusters

4561 Hofstetter, S., Hofmann, C. and Soff, G. (1992). Z. Phys. D, 23, 227-232.
Ionization of hydrogen by a relativistic heavy projectile

4562 Hohl, D., Ellis, D. and Rösch, N. (1987). Inorg. Chim. Acta, 127, 195-202.
A relativistic muffin-tin free X molecular orbital study of metal tetrakis tetrahydroborates: $M(BH_4)_4$, M=Zr, U

4563 Hohl, D. and Rösch, N. (1986). Inorg. Chem., 25, 2711-2718.
On the electronic structure of metal tetrahydroborates: Quasi-relativistic Xα-SW study of $M(BH_4)_4$ (M=Zr, Hf, Th, U)

4564*Hojman, S. (1971). Nuovo Cim. A, 4, 676-682.
Algebraic method for solving the Dirac equation with a Coulomb potential

4565 Hokkyo, N. (1987). Nuovo Cim. A, 98, 787-795.
Radiationless Zitterbewegung (oscillation) of Dirac particles and mass formula

4566 Holladay, W.G. (1992). Am. J. Phys., 60, 281.
The derivation of relativistic energy from the Lorentz γ

4567 Holland, P.R. (1992). Found. Phys., 22, 1287-1301.
The Dirac equation in the de Broglie-Bohm theory of motion

4568 Holm, P. (1988). Phys. Rev. A, 37, 3706-3719.
Relativistic Compton cross section for general central-field Hartree-Fock wave functions

4569 Holz, J. (1987). Z. Phys. D, 4, 211-225.
Self-energy of electrons in a Coulomb field: Momentum-space method

4570*Hönl, H. and Papapetrou, A. (1940). Z. Physik, 116, 153-183.
Über die innere Bewegung des Elektrons. III.

4571 Horbatsch, M. (1992). Phys. Lett. A, 161, 360-364.
Are there resonances in relativistic charged particle scattering due to spin-spin interactions?

4572 Hörmandinger, G. and Weinberger, P. (1992). J. Phys.: Condens. Matter, 4, 2185-2199.
Symmetry aspects in spin-polarized relativistic electronic structure calculations based on the Green function method

4573 Hörmandinger, G., Weinberger, P., Marksteiner, P. and Badralexe, E. (1991). Physica B, 172, 237-243.
Spin-orbit effects in the band structure of $Bi_2Sr_2Ca_1Cu_2O_8$

4574*Hossein Partovi, M. and Jabbarian-Lotfabadi, A. (1978). Phys. Rev. A, 17, 1575-1582.
Application of the new fermion-antifermion equation to positronium and the numerical solution of its static-interaction limit, the Breit equation

4575*Hostler, L.C. (1982). J. Math. Phys., 23, 1179-1184.
An SL(2,C)-invariant representation of the Dirac equation

4576*Hostler, L.C. (1985b). J. Math. Phys., 26, 1348-1354.
Scalar formalism for quantum electrodynamics
Erratum, J. Math. Phys., 27, 2208.

4577 Hostler, L.C. (1986). J. Math. Phys., 27, 2423-2428.
Scalar formalism for non-Abelian gauge theory

4578 Hostler, L.C. (1987a). J. Math. Phys., 28, 720-727.
Self-energy operator for an electron in an external Coulomb potential

4579 Hostler, L.C. (1987b). J. Math. Phys., 28, 2984-2989.
Relativistic Coulomb Sturmian matrix elements and the Coulomb Green's function of the second-order Dirac equation

4580 Hostler, L.C. (1989). J. Math. Phys., 30, 1621-1630.
Foldy-Wouthuysen-type reduction of the second-order Dirac equation

4581 Hota, R.L. and Tripathi, G.S. (1991a). Phys. Rev. B, 44, 1918-1921.
Theory of spin, orbital, and spin-orbit Knight shifts of ^{207}Pb in n-type PbTe

4582 Hota, R.L. and Tripathi, G.S. (1991b). J. Phys.: Condens. Matter, 3, 6299-6311.
Theory of effective g-factors in ternary semiconductors: application to $Pb_{1-x}Sn_xTe$

4583 Howells, M.H. and Kennedy, R.A. (1990). J. Chem. Soc. Faraday Trans., 86, 3495-3503.
Relativistic corrections for the ground and first excited states of H_2^+, HD^+ and D_2^+

4584 Hsu, J.-J., Chung, K.T. and Huang, K.-N. (1991). Phys. Rev. A, 44, 5485-5491.
Term values of 1s2s2p $^4P°$ for lithiumlike ions

4585 Hsu, S.-W., Kuo, T.-Y., Chen, C.-M.J. and Huang, K.-N. (1992). Phys. Lett. A, 167, 277-279.
Discrepancy in total cross sections of positron-impact ionization of hydrogen

4586 Hsuan, H., Bitter, M., Hill, K.W., von Goeler, S., Grek, B., Johnson, D., Johnson, L.C., Sesnic, S., Bhalla, C.P., Karim, K.R., Bely-Dubau, F. and Faucher, P. (1987). Phys. Rev. A, 35, 4280-4285.
Satellite spectra of heliumlike nickel

4587 Hu, S.-Z. and Su, R.-K. (1991). Acta. Phys. Sin., 40, 1201-1206.
Bound states of the Dirac equations with Hulthen-type potentials (in Chinese)

4588*Huang, K.-N. and Hughes, V.W. (1979). Phys. Rev. A, 20, 706-717.
Theoretical hyperfine structure of muonic helium

4589 Hurley, M.M. and Christiansen, P.A. (1987). J. Chem. Phys., 86, 1069-1070.
Relativistic effective potentials in quantum Monte Carlo calculations

4590 Hurley, M.M., Fernandez Pacios, L., Christiansen, P.A., Ross, R.B. and Ermler, W.C. (1986). J. Chem. Phys., 84, 6840-6853.
Ab initio relativistic effective potentials with spin-orbit operators: II. K through Kr

4591 Hussein, M.S., Spehler, D. and Marques, G.C. (1991). Ann. Phys. (NY), 207, 343-373.
Symmetrical representation of the transition matrix. II. Relativistic extensions

4592 Hutton, R., Prior, M.H., Chantrenne, S., Chen, M.H. and Schneider, D. (1989). Phys. Rev. A, 39, 4902-4905.
Double and single electron capture in slow collisions of $Ar^{9+,8+}$ ions with He atoms

4593 Huzinaga, S. (1991). J. Mol. Struct.(Theochem), 234, 51-73.
Effective hamiltonian method for molecules

4594 Hwang, C.-Y., Chi, H.-C. and Huang, K.-N. (1991). Phys. Rev. A, 44, 7189-7194.
Autoionizing levels of zinc from the multiconfiguration relativistic random-phase approximation

4595 Hybertsen, M.S. and Louie, S.G. (1986). Phys. Rev. B, 34, 2920-2922.
Spin-orbit splitting in semiconductors and insulators from the ab initio pseudopotential

4596 Hylton, D.J. (1988).
The Nuclear Size Correction to the Lamb Shift in Hydrogen
New Haven, CT (USA), Yale Univ., 119p.

4597 Ichinose, T. and Tamura, H. (1988). J. Math. Phys., 29, 103-109.
The Zitterbewegung of a Dirac particle in two-dimensional space-time

4598 Igel-Mann, G., Dolg, M., Wedig, U., Preuss, H. and Stoll, H. (1987). J. Chem. Phys., 86, 6348-6351.
Comparison of ab initio and semiempirical pseudopotenticals for Ca in calculations for CaO

4599 Igel-Mann, G., Feller, C., Flad, H.-J., Savin, A., Stoll, H. and Preuss, H. (1989). Mol. Phys., 68, 209-223.
Comparative study of spectroscopic properties of some indium, tin and antimony compounds

4600 Igel-Mann, G., Flad, H.-J., Feller, C. and Preuss, H. (1990). J. Mol. Struct.(Theochem), 209, 313-331.
Pseudopotential investigations on indium, tin and antimony compounds

4601 Igel-Mann, G., Savin, A. and Preuss, H. (1992).
Theoretische Untersuchungen an Verbindungen von Elementen der Gruppen IIIA, IVA und VA
In "Unkonventionelle Wechselwirkungen in der Chemie metallischer Elemente", Ed. B. Krebs, VCH, Weinheim, pp. 3-16.

4602 Igel-Mann, G., Stoll, H. and Preuss, H. (1988a). Mol. Phys., 65, 1321-1328.
Pseudopotentials for main group elements (IIIa through VIIa)

4603 Igel-Mann, G., Stoll, H. and Preuss, H. (1988b). Mol. Phys., 65, 1329-1336.
Pseudopotential study of monohydrides and monoxides of main group elements K through Br

4604 Igel-Mann, G., Wedig, U., Fuentealba, P. and Stoll, H. (1986). J. Chem. Phys., 84, 5007-5012.
Ground-state properties of alkali dimers XY(X,Y=Li to Cs)

4605 Iglesias, L., Cabeza, M.I., Rico, F. and Kaufman, V. (1988). Phys. Scr., 37, 855-870.
Fundamental configurations of doubly-ionized molybdenum (Mo III)

4606 Illas, F., Rubio, J., Bachs, M. and Ricart, J.M. (1989). J. Mol. Struct.(Theochem), 202, 315-324.
Taking into account non-dynamical correlation effects in ab initio chemisorption cluster model calculations

4607 Illas, F., Rubio, J., Cañellas, J. and Ricart, J.M. (1990).

J. Chem. Phys., 93, 2603-2610.
Electronic structure of Rh, RhH, and Rh_2 as derived from ab initio configuration interaction calculations

4608 Illas, F., Rubio, J. and Ricart, J.M. (1988). J. Chem. Phys., 88, 260-271.
Dynamical and nondynamical correlation effects in ab initio chemisorption cluster model calculations. Ground and low lying states of H on Cu(100) and Ag(100)

4609 Illas, F., Rubio, J. and Ricart, J.M. (1991). J. Chem. Phys., 95, 4225-4229.
An analysis of 3d correlation effects in the bonding of atomic oxygen to Cu(100)

4610 Illas, F., Rubio, J., Ricart, J.M. and Cabot, P.L. (1986). J. Am. Chem. Soc., 108, 7893-7897.
Theoretical evidence for two geometrical isomers of AgO_2

4611 Ilyabaev, E. and Kaldor, U. (1992a). Chem. Phys. Lett., 194, 95-98.
Relativistic coupled-cluster calculations for closed-shell atoms

4612 Ilyabaev, E. and Kaldor, U. (1992b). J. Chem. Phys., 97, 8455-8458.
The relativistic open shell coupled cluster method: Direct calculation of excitation energies in the Ne atom

4613 Indelicato, P. (1986). J. Phys. B: At. Mol. Phys., 19, 1719-1730.
Effects of the Breit interaction on the 1s binding energy of superheavy elements

4614 Indelicato, P. (1987). J. Phys.(Paris), Colloq., 1, C9-297 - C9-300.
MCDF calculations of relativistic effects in lithium-like ions

4615 Indelicato, P. (1988). Nucl. Instrum. Meth. B, 31, 14-20.
Multiconfiguration Dirac-Fock calculations of transition energies in two electron ions with $10 \leq Z \leq 92$

4616 Indelicato, P. (1989). J. Phys.(Paris), Colloq., 50, C1-239 - C1-254.
Recent progress in the theory of relativistic effects and QED corrections in two and three-electron ions

4617 Indelicato, P. (1990). AIP Conf. Proc., 215, 591-602.
K alpha transitions in few-electron ions and in atoms

4618 Indelicato, P., Birkett, B.B., Briand, J.-P., Charles, P., Dietrich, D.D., Marrus, R. and Simionovici, A. (1992). Phys. Rev. Lett., 68, 1307-1310.
Hyperfine quenching and precision measurement of the $2^3P_0-2^3P_1$ fine-structure splitting in heliumlike gadolinium (Gd^{62+})

4619 Indelicato, P., Briand, J.P., Tavernier, M. and Liesen, D. (1986a). Z. Phys. D, 2, 249-250.
Experimental study of relativistic correlations and QED effects in heliumlike krypton ions

4620 Indelicato, P. and Desclaux, J.P. (1990). Phys. Rev. A, 42, 5139-5149.
Multiconfiguration Dirac-Fock calculations of transition energies with QED corrections in three-electron ions

4621 Indelicato, P., Gorceix, O. and Desclaux, J.P. (1987a). J. Phys. B: At. Mol. Phys., 20, 651-663.
Multiconfigurational Dirac-Fock studies of two-electron ions: II. Radiative corrections and comparison with experiment

4622 Indelicato, P., Gorceix, O. and Desclaux, J.P. (1987b). J. Phys.(Paris), Colloq., 1, C9-293 - C9-296.
MCDF calculations of transition energies in uranium helium-like ions

4623 Indelicato, P. and Lindroth, E. (1992). Phys. Rev. A, 46, 2426-2436.
Relativistic effects, correlation, and QED corrections on Kα transitions in medium to very heavy atoms

4624 Indelicato, P. and Mohr, J. (1991a). Theor. Chim. Acta, 80, 207-214.
Quantum electrodynamic effects in atomic structure

4625 Indelicato, P. and Mohr, P.J. (1991b). AIP Conf. Proc., 233, 501-503.
Electron screening correction to the self energy in high-Z atoms

4626 Indelicato, P. and Mohr, P.J. (1992). Phys. Rev. A, 46, 172-185.
Coordinate-space approach to the bound-electron self-energy

4627*Infeld, L. and Hull, T.E. (1951). Rev. Mod. Phys., 23, 21-68.
The factorization method

4628 Ionescu, D.C., Reinhardt, J., Müller, B., Greiner, W. and Soff, G. (1988). Phys. Rev. A, 38, 616-620.
Nonlinear extensions of the Dirac equation and their implications in QED

4629 Ionova, G.V. and Gerasimova, G.A. (1992). Uspekhi Khim., 61, 1656-1686; Russ. Chem. Rev., 61, 907-924.
The influence of relativistic effects and the crystal field on the properties of actinyl ions

4630 Ionova, G.V., Pershina, V., Johnson, E., Fricke, B. and Schädel, M. (1992). J. Phys. Chem., 96, 11096-11101.
Redox reactions for Group 5 elements, including element 105, in aqueous solutions

4631 Ishii, M., Morihashi, K. and Kikuchi, O. (1991). J. Mol.

Struct.(Theochem), 235, 39-49.
Ab initio calculations of g values of free radicals by finite perturbation theory

4632 Ishikawa, Y. (1990a). Chem. Phys. Lett., 166, 321-325.
Dirac-Fock Gaussian basis calculations: Inclusion of the Breit interaction in the self-consistent field procedure

4633 Ishikawa, Y. (1990b). Phys. Rev. A, 42, 1142-1150.
Relativistic Dirac-Fock and many-body perturbation calculations on He, He-like ions, Ne, and Ar

4634 Ishikawa, Y. (1990c). Int. J. Quantum Chem., S24, 383-391.
Atomic Dirac-Fock-Breit self-consistent field calculations

4635 Ishikawa, Y. (1991a). Chem. Phys. Lett., 179, 291-296.
Relativistic many-body perturbation-theory calculations on Kr: Accurate representation of second- and third-order energies with contracted Gaussian basis set

4636 Ishikawa, Y. (1991b).
Basis set expansion Dirac-Fock SCF calculations and MBPT refinement
In "The Effects of Relativity in Atoms, Molecules, and the Solid State", Ed. S. Wilson, I.P. Grant and B.L. Gyorffy, Plenum, New York and London, pp. 149-162.

4637 Ishikawa, Y. (1991c). Int. J. Quantum Chem., S25, 113-120.
Effects of basis set contraction on relativistic and Breit interaction energies

4638 Ishikawa, Y. (1992a). Can. J. Chem., 70, 399-403.
Relativistic many-body perturbation theory using discrete basis expansion method: accurate representation of second-order energy of Xe atom with contracted well-tempered Gaussian basis set

4639 Ishikawa, Y. (1992b). Int. J. Quantum Chem., S26, 127-135.
Relativistic many-body perturbation theory using the discrete basis expansion method: analysis of relativistic pair correlation energies of the Xe atom

4640 Ishikawa, Y. and Binning Jr., R.C. (1991).
Accurate relativistic Dirac-Fock and MBPT calculations on argon with basis sets of contracted Gaussian functions
In "The Effects of Relativity in Atoms, Molecules, and the Solid State", Ed. S. Wilson, I.P. Grant and B.L. Gyorffy, Plenum, New York and London, pp. 207-214.

4641 Ishikawa, Y., Malli, G.L. and Pyper, N.C. (1992a).
Chem. Phys. Lett., 194, 481-484.
Ab initio Dirac-Fock self-consistent field calculations for open-shell heavy-atom systems: bonding in AuH^+ ion

4642 Ishikawa, Y., Malli, G.L. and Stacey, A.J. (1992b).
Chem. Phys. Lett., 188, 145-148.
Matrix Dirac-Fock-Breit SCF calculations on heavy atoms using geometric basis sets of Gaussian functions

4643 Ishikawa, Y. and Quiney, H.M. (1987). Int. J. Quantum Chem., S21, 523-532.
On the use of an extended nucleus in Dirac-Fock Gaussian basis set calculations

4644 Ishikawa, Y., Quiney, H.M. and Malli, G.L. (1991). Phys. Rev. A, 43, 3270-3278.
Dirac-Fock-Breit self-consistent-field method: Gaussian basis-set calculations on many-electron atoms

4645 Ishikawa, Y., Rodriguez, W. and Alexander, S.A. (1987). Int. J. Quantum Chem., S21, 417-423.
Solution of the integral Dirac equation in momentum space

4646 Ishikawa, Y., Rodriguez, W., Torres, S. and Alexander, S.A. (1988). Chem. Phys. Lett., 143, 289-292.
Solving the Dirac equation in momentum space a numerical study of H_2^+

4647 Ishikawa, Y. and Sekino, H. (1988). Int. J. Quantum Chem., S22, 457-463.
On the use of Gaussian-type functions in Dirac-Fock basis set expansion calculations

4648 Ishikawa, Y. and Sekino, H. (1990). Chem. Phys. Lett., 165, 243-249.
Variational bounds in Dirac-Fock basis set expansion calculations

4649 Ishikawa, Y., Sekino, H. and Binning Jr., R.C. (1989). Chem. Phys. Lett., 160, 206-211.
Relativistic many-body perturbation theory calculations on Be, Ne^{6+}, Ar^{14+} and Ne

4650 Ishikawa, Y., Sekino, H. and Binning Jr., R.C. (1990). Chem. Phys. Lett., 165, 237-242.
Effects of basis set contraction in relativistic calculations on neon, argon, and germanium

4651 Ito, H. (1988). Progr. Theor. Phys., 80, 874-880.
Nonrelativistic limit of the renormalized BS wave function at the origin

4652 Ivanov, I.A., Nilsen, J., Safronova, M.S. and Safronova, U.I. (1990). Izv. VUZov, Fiz., 33(8), 55-72; Sov. Phys. J., 33, 670-684.
Z-dependences of atomic parameters of autoionization states of two-electron system

4653 Ivanov, I.A. and Safronova, U.I. (1987a). Opt. Spektrosk., 62, 14-19.
Use of screening for the calculation of relativistic corrections (in Russian)

4654 Ivanov, I.A. and Safronova, U.I. (1987b). Opt. Spektrosk., 63, 237-243.
Dirac-based perturbation theory of many-electron systems (in Russian)

4655 Ivanov, I.A. and Safronova, U.I. (1989a). Opt. Spektrosk., 65, 5-9; Opt. Spectrosc., 65, 2-5.
Satellite structure of spectra

4656 Ivanov, I.A. and Safronova, U.I. (1989b). Opt. Spektrosk., 66, 489-494. Opt. Spectrosc., 66, 284-287.
Some aspects of the identification of levels of highly excited states of two-electron systems (the 2l4l' states)

4657 Ivanov, I.A. and Safronova, U.I. (1991). Z. Phys. D, 21, S151-152.
Autoionization states 1s2l4l'-energy and autoionization rates

4658 Ivanov, L.N., Ivanova, E.P. and Aglitsky, E.V. (1988). Phys. Reports, 164, 316-375.
Modern trends in the spectroscopy of multicharged ions

4659 Ivanov, L.N., Ivanova, E.P. and Zaridze, G.G. (1990). Izv. VUZov, Fiz., 33(8), 34-42; Sov. Phys. J., 33, 653-660.
Green's function for the Dirac equation with nonsingular central potential and complex energy. Estimation algorithm

4660 Ivanova, E.P. and Gulov, A.V. (1991). At. Data Nucl. Data Tables, 49, 1-64.
Theoretical investigation of the neon isoelectronic sequence

4661 Ivanova, E.P., Ivanov, L.N., Gogava, A.L. and Sretenskaya, Zh.L. (1989). Opt. Spektrosk., 67, 505-509; Opt. Spectrosc., 67, 295-297.
Forbidden 2-3 radiative transitions in ions of the neon isoelectronic series

4662 Ivanova, E.P., Ivanov, L.N. and Tsirekidze, M.A. (1986). At. Data Nucl. Data Tables, 35, 419-428.
Energy levels of magnesium-like ions calculated in the model-potential relativistic perturbation theory: Z = 25-84

4663 Ivrii, V.Ya. (1990). Sov. Phys. Dokl., 35, 340-341.
Asymptotic eigenvalues for two-dimensional Dirac operator with strong magnetic field

4664 Iwazaki, A. and Kumano, S. (1988). Phys. Lett. B, 212, 99-104.
Oscillations of the polarized vacuum around a large-Z "nucleus"

4665 Ixaru, L.Gr. and Pantea, D. (1988). Rev. Roum. Phys., 33, 1079-1092.
Periodicity effect in the two centre Dirac equation

4666 Iyer B.R. and Kamran, N. (1991). J. Math. Phys., 32, 2497-2502.
Separation of variables for the Dirac equation in an extended class of Lorentzian metrics with local rotational symmetry

4667 Jablonski, A. (1991). Phys. Rev. B, 43, 7546-7554.
Elastic electron backscattering from gold

4668 Jackiw, R. and Nair, V.P. (1991). Phys. Rev. D, 43, 1933-1942.
Relativistic wave equation for anyons

4669 Jaffe, M.D. and Singh, J. (1987). Solid State Comm., 62, 399-402.
Inclusion of spin-orbit coupling into tight binding bandstructure calculations for bulk and superlattice semiconductors

4670 Jaffe, R.L. and Manohar, A. (1989). Ann. Phys. (NY), 192, 321-330.
Bound states of the Dirac equation outside a hard sphere

4671 Jagannathan, A. (1991). Europhys. Lett., 16, 217-222.
On multiple spin orbit scattering and anisotropic spin interactions in metals

4672 Jagannathan, A. (1992). J. Magn. Magn. Materials, 104-107, 1673-1675.
The nature of middle-range spin-spin interactions in metallic alloys

4673 Jagannathan, R. (1990). Phys. Rev. A, 42, 6674-6689.
Quantum theory of electron lenses based on the Dirac equation

4674 Jagannathan, R., Simon, R., Sudarshan, E.C.G. and Mukunda, N. (1989). Phys. Lett. A, 134, 457-464.
Quantum theory of magnetic electron lenses based on Dirac equation

4675 Jahns, V., Köstlmeier, S., Kotzian, M., Rösch, N. and Watson, P.L. (1992). Int. J. Quantum Chem., 44, 853-867.
On the agostic interaction in lanthanide phenylene complexes: An INDO study of $[(Cp^*_2M)_2C_6H_4]$ (M=Sc,Lu) employing an energy-partitioning analysis

4676 Jakubassa-Amundsen, D.H. (1989). Z. Phys. D, 11, 305-313.
Relativistic theory for K-shell ionization by fast electrons

4677 Jakubassa-Amundsen, D.H. (1990). Phys. Rev. A, 42, 653-654.
Relativistic second-order Born theory for electron capture

4678 Jakubassa-Amundsen, D.H. (1991). Z. Phys. D, 21, S291-292.
Electron transfer in nuclear reactions

4679 Jakubassa-Amundsen, D.H. (1992). J. Phys. B: At. Mol. Opt. Phys., 25, 1297-1305.
A systematic study of relativistic (e, 2e) collisions in comparison with experiment

4680 Janiak, C. and Hoffmann, R. (1989a). Angew. Chem., 101, 1706-1708.
Tl^{I}-Tl^{I}-Wechselwirkung in Molekülen - eine MO-Analyse

4681 Janiak, C. and Hoffmann, R. (1989b). Inorg. Chem., 28, 2743-2747.
Electronic structure of one-dimensional linear halogen-bridged gold chains

4682 Janiak, C. and Hoffmann, R. (1990). J. Am. Chem. Soc., 112, 5924-5946.
Tl^{I}-Tl^{I} and In^{I}-In^{I} interactions: From the molecular to the solid state

4683 Janiak, C. and Hoffmann, R. (1992).
Tl^{I}-Tl^{I}-und In^{I}-In^{I}-Wechselwirkungen und die Ligandengeometrie. Eine theoretische Studie an Molekülen und Festkörpern In "Unkonventionelle Wechselwirkungen in der Chemie metallischer Elemente", Ed. B. Krebs, VCH, Weinheim, pp. 45-80.

4684 Jank, W. and Hafner, J. (1990a). Phys. Rev. B, 41, 1497-1515.
Structural and electronic properties of the liquid polyvalent elements: The group-IV elements Si, Ge, Sn, and Pb

4685 Jank, W. and Hafner, J. (1990b). Phys. Rev. B, 42, 6926-6938.
Structural and electronic properties of the liquid polyvalent elements. II. The divalent elements

4686 Jank, W. and Hafner, J. (1990c). J. Non-Cryst. Sol., 117/118, 304-307.
Trends in the electronic structure of the polyvalent liquid metals

4687 Jankowski, K. and Polasik, M. (1989). J. Phys. B: At. Mol. Opt. Phys., 22, 2369-2376.
On the calculation of Kß/Kα x-ray intensity ratios

4688 Jankowski, K. and Rutkowski, A. (1987). Phys. Scr., 36, 464-467.
Correlation and relativistic effects for many-electron systems

4689 Jankowski, K. and Rutkowski, A. (1989). J. Phys. B: At. Mol. Opt. Phys., 22, 2669-2678.
Correlation and relativistic effects for many-electron systems: II. Second-order energies for closed-shell atoms

4690*Jannussis, A. (1966). Z. Phys. 190, 129-134.
Die relativistische Bewegung eines Elektrons in äusseren homogenen Magnetfeld

4691 Jansen, G. and Hess, B.A. (1989a). Phys. Rev. A, 39, 6061-6062.
Revision of the Douglas-Kroll transformation

4692 Jansen, G. and Hess, B.A. (1989b). Chem. Phys. Lett., 160, 507-513.

Relativistic all-electron configuration interaction calculations on the gold atom

4693 Jansen, G. and Hess, B.A. (1989c). Z. Phys. D, 13, 363-375.
Relativistic all electron configuration interaction calculation of ground and excited states of the gold hydride molecule

4694 Jansen, G., Pusch, M. and Soff, G. (1988). Z. Phys. D, 8, 315-327.
Continuum solutions of the Klein-Gordon equation

4695 Jansen, H.J.F. (1988a). Phys. Rev. B, 38, 8022-8029.
Magnetic anisotropy in density-functional theory

4696 Jansen, H.J.F. (1988b). J. Appl. Phys., 64, 5604-5606.
Model calculations of the magnetic anisotropy in 3d transition metals

4697 Jansen, H.J.F. (1989). J. Appl. Phys., 67, 4555-4557.
Origin of orbital momentum and magnetic anisotropy in transition metals

4698 Jarid, A., Aaid, M., Legoux, Y., Merini, J., Loudet, M., Gonbeau, D. and Pfister-Guillouzo, G. (1991). Chem. Phys., 150, 353-360.
Fluorides of ruthenium and chromium: theoretical studies

4699 Jaroszewicz, T. and Kurzepa, P.S. (1991). Ann. Phys.(NY), 210, 255-322.
Spin, statistics, and geometry of random walks

4700 Jarque, C., Novaro, O. and Ruiz, M.E. (1987). Mol. Phys., 62, 129-139.
Variational and perturbational CI study of the Pd-C_2H_4 molecule

4701 Jarque, C., Novaro, O., Ruiz, M.E. and Garcia-Prieto, J. (1986). J. Am. Chem. Soc., 108, 3507-3509.
On the stability of the PdH_2 molecule

4702 Jarrett-Sprague, S.A., Hillier, I.H., Burton, N.A. and Guest, M.F. (1990). Chem. Phys. Lett., 170, 35-38.
Pseudopotential calculations of the structure and harmonic frequencies of $HSbO_3$ and $HAsO_3$

4703 Jarvis, P.D. and Stedman, G.E. (1986). J. Phys. A, 19, 1373-1385.
Supersymmetry in second-order relativistic equations for the hydrogen atom

4704 Jasien, P.G. and Stevens, W.J. (1988). Chem. Phys. Lett., 147, 72-78.
Theoretical calculations of diatomic metal oxides. CrO and CrO^+

4705 Jaskólski, W. (1989). J. Phys. B: At. Mol. Opt. Phys., 22, 3275-3287.
Quasirelativistic approach to many-channel electron scattering from atoms

4706 Jaskólski, W. and Karwowski, J. (1986). J. Phys. B: At. Mol. Phys., 19, 1093-1105.
Quasi-relativistic approach to low-energy electron scattering from closed-shell atoms

4707 Jaskólski, W., Karwowski, J. and Kobus, J. (1987). Phys. Scr., 36, 436-440.
Quasirelativistic calculations of the elastic scattering of slow electrons from Xe atoms

4708 Jaskólski, W., Karwowski, J. and Kobus, J. (1990). Acta Phys. Pol. A, A78, 693-695.
A remark on the outward integration procedure for quasirelativistic radial equations

4709*Jena, S.N. (1983). Pramana, 21, 247-255.
Mass spectra of light and heavy mesons in the Dirac equation with power-law potential

4710 Jeong, K., Gaylord, R.H. and Kevan, S.D. (1988). Phys. Rev. B, 38, 10302-10312.
Relativistic effects on the surface electronic structure of Mo(011)

4711 Jeong, K., Gaylord, R.H. and Kevan, S.D. (1989). J. Vacuum Sci. Tech. A, 7, 2199-2202.
Symmetry effects in hydrogen chemisorption electronic structure

4712 Jeung, G.H. and Koutecky, J. (1988). J. Chem. Phys., 88, 3747-3760.
Molecular bonding with scandium: Diatomics ScH, ScO, ScC, and ScN

4713 Jiang, C.-G., Fletcher, G., Fry, J.L. and Papaconstantopoulos, D.A. (1991). Phys. Rev. B, 44, 2268-2275.
Calculation of the superconducting parameter $\langle I^2 \rangle$ for hcp transition metals

4714 Jin, Y.-H., Onley, D.S. and Wright, L.E. (1992). Phys. Rev. C, 45, 1311-1320.
Electron Coulomb effects in quasielastic (e,e'p) reactions

4715 Joc, K. and Bojara, A. (1987). J. Quant. Spectrosc. Radiat. Transfer, 37, I.
Adaptation of the relativistic multiconfiguration Dirac-Fock/MCDF/program to microcomputers

4716 Johansson, B., Eriksson, O. and Brooks, M.S.S. (1990). J. Radioanalytical Nucl. Chem., 143, 21-34.
Actinide metals and intermetallic compounds: Electronic structure and magnetic properties

4717 Johansson, B., Eriksson, O., Brooks, M.S.S. and Skriver, H.L. (1987). Inorg. Chim. Acta, 140, 59-66.
Theoretical studies of actinide intermetallic compounds

4718 Johnson, E. and Fricke, B. (1991). J. Phys. Chem., 95, 7082-7084.
Prediction of some thermodynamic properties of selected compounds of element 104

4719 Johnson, E., Fricke, B., Keller Jr., O.L., Nestor Jr., C.W. and Tucker, T.C. (1990). J. Chem. Phys., 93, 8041-8050.
Ionization potentials and radii of atoms and ions of element 104 (unnilquadium) and of hafnium (2+) derived from multiconfiguration Dirac-Fock calculations

4720*Johnson, W.R. (1972). Phys. Rev. Lett., 29, 1123-1126.
Radiative decay rates of metastable one-electron atoms

4721 Johnson, W.R. (1987). Phys. Scr., 36, 765-775.
Calculations of P- and CP-violating matrix elements for heavy atoms

4722 Johnson, W.R. (1988). Adv. At. Mol. Phys., 25, 375-391.
Relativistic random-phase approximation

4723 Johnson, W.R. (1989).
Applications of relativistic MBPT to lithium-like atoms
In "Relativistic, Quantum Electrodynamic and Weak Interaction Effects in Atoms", Ed. W. Johnson, P. Mohr and J. Sucher, AIP Conf. Proc., 189, 209-216.

4724 Johnson, W.R. (1991). J. Phys.(Paris) IV, 1, C1/15-21.
Relativistic many-body calculations

4725 Johnson, W.R. (1992). APS News, 1(3), 17-18.
Relativistic many-body calculations

4726 Johnson, W.R., Blundell, S.A., Liu, Z.W. and Sapirstein, J. (1988d). Phys. Rev. A, 37, 1395-1400.
Correlation effects in the parity-nonconserving 6s --> 7s transition in cesium

4727 Johnson, W.R., Blundell, S.A. and Sapirstein, J. (1988a). Phys. Rev. A, 37, 307-315.
Finite basis sets for the Dirac equation constructed from B splines

4728 Johnson, W.R., Blundell, S.A. and Sapirstein, J. (1988b). Phys. Rev. A, 37, 2764-2777.
Many-body perturbation-theory calculations of energy levels along the lithium isoelectronic sequence

4729 Johnson, W.R., Blundell, S.A. and Sapirstein, J. (1988c). Phys. Rev. A, 38, 2699-2706.
Many-body perturbation-theory calculations of energy levels along the sodium isoelectronic sequence

4730 Johnson, W.R., Blundell, S.A. and Sapirstein, J. (1990). Phys. Rev. A, 42, 1087-1095.
Many-body perturbation-theory calculations of energy levels along the copper isoelectronic sequence

4731 Johnson, W.R. and Cheng, K.T. (1992). Phys. Rev. A,

46, 2952-2954.
Relaxed relativistic random-phase-approximation calculations of photoionization amplitudes and phases for the 4d subshell of xenon

4732 Johnson, W.R., Guo, D.S., Idrees, M. and Sapirstein, J. (1986). Phys. Rev. A, 34, 1043-1057.
Weak-interaction effects in heavy atomic system. II

4733 Johnson, W.R., Idrees, M. and Sapirstein, J. (1987). Phys. Rev. A, 35, 3218-3226.
Second-order energies and third-order matrix elements of alkali-metal atoms

4734 Johnson, W., Mohr, P. and Sucher, J. (ed.) (1989b).
Relativistic, Quantum Electrodynamic, and Weak Interaction Effects in Atoms
AIP Conf. Proc., 189, 513p.

4735 Johnson, W.R. and Sapirstein, J. (1986). Phys. Rev. Lett., 57, 1126-1129.
Computation of second-order many-body corrections in relativistic atomic systems

4736 Johnson, W.R. and Sapirstein, J. (1992). Phys. Rev. A, 46, R2197-R2200.
Relativistic many-body perturbation theory applied to n=2 triplet states of heliumlike ions

4737 Johnson, W.R., Sapirstein, J. and Blundell, S.A. (1989a). J. Phys. B: At. Mol. Opt. Phys., 22, 2341-2347.
Many-body theory applied to negative ions

4738*Johnson Jr., M.H. and Breit, G. (1933). Phys. Rev., 44, 77-83.
The magnetic interaction of a valence electron with inner shells

4739 Jokisaari, J., Lazzeretti, P. and Pyykkö, P. (1988). Chem. Phys., 123, 339-350.
NMR chemical shifts of CX and XCY molecules (X,Y=O, S, Se, Te): A comparison of coupled Hartree-Fock, semi-empirical REX and experimental results

4740 Jolly, C.A. and Marynick, D.S. (1992). J. Mol. Struct. (Theochem), 253, 9-16.
Electronic structure of the polyatomic Zintl anion Sb_7^{3-}. A scattered-wave Xα study

4741 Jolly, W.L. (1991).
Modern Inorganic Chemistry, 2nd Ed.
McGraw-Hill, New York.

4742 Jonas, V., Frenking, G. and Gauss, J. (1992). Chem. Phys. Lett., 194, 109-117.
The geometry of TiH_6^{2-} and VH_6^-

4743 Jones, T. (1992).
Relativistic quantum mechanics, QED and QCD
In "Report RAL--92-018", Ed. R.J. Barlow, Rutherford

Appleton Lab., Chilton, p. 113-190.

4744*Jortner, J. (1962). Proc. Int. Symp. Mol. Struct. Spectr., Tokyo, B, 212, 1-4.
The doublet structure of halide ions spectra

4745 Jostes, R. (1988). Theor. Chim. Acta, 74, 229-235.
Extended-Hückel parameters for third-row transition metals

4746 Jostes, R., Müller, A. and Diemann, E. (1986). J. Mol. Struct.(Theochem), 137, 311-328.
Electronic structure and spectra of oxothiometalates $MO_{4-n}S_n^{2-}$ (M=Mo, W; n=0-4)

4747 Jové, J., He, L., Proust, J., Pagès, M. and Pyykkö, P. (1991). J. Alloys Compounds, 177, 285-310.
Mössbauer spectroscopy as a nuclear probe for solid state transuranium chemistry

4748*Kaburagi, M., Kawaguchi, M., Morii, T., Kitazoe, T. and Morishita, J. (1980). Phys. Lett. B, 97, 143-146.
Spectroscopy of atomlike mesons Qanti q in the Dirac equation with logarithmic confining potential

4749*Kaburagi, M., Kawaguchi, M., Morii, T., Kitazoe, T. and Morishita, J. (1981). Z. Phys. C, 9, 213-220.
Spectroscopy of atomlike mesons Qanti q in the Dirac equation with scalar confining potential

4750 Kagawa, T. (1992). Comp. Phys. Commun., 72, 165-174.
A program to calculate fractional parentage coefficients for jj-coupling states with equivalent particles

4751 Kagawa, T., Honda, Y. and Kiyokawa, S. (1991). Phys. Rev. A, 44, 7092-7107.
Relativistic configuration-interaction theory for atomic systems

4752 Kagawa, T. and Safronova, U.I. (1992). Phys. Scr., 45, 569-577.
Relativistic perturbation method for two-electron systems

4753 Kahane, S., Shahal, O. and Moreh, R. (1989). Phys. Rev. A, 40, 6950-6957.
Rayleigh scattering of neutron capture γ rays from U

4754*Kalf, H. (1976). J. Funct. Anal., 21, 389-396.
The virial theorem in relativistic quantum mechanics

4755 Kalinkin, A.N., Skorikov, V.M. and Soldatov, A.A. (1992). Inorg. Mater., 28, 337-530; Neorg. Mater., 28, 558-562.
Electron structure of $Bi_{12}GeO_{20}$

4756 Kalinkin, A.N., Skorikov, V.M., Soldatov, A.A. and Dobretsov, V.Yu. (1991). Phys. Stat. Sol., 166, 185-190.
Electronic structure of $Bi_2Sr_2CaCu_2O_8$ doped by sulphur in the cluster approximation

4757 Kalnins, E.G. and Miller Jr., W. (1992). J. Math. Phys.,

33, 286-296.
Series solutions for the Dirac equation in Kerr-Newman space-time

4758 Kamath, S.G. (1989). Phys. Rev. A, 40, 6791-6799.
Classical solutions of Maxwell's equations in (2+1)-dimensional electrodynamics with the Chern-Simons term

4759 Kane, P.P., Kissel, L., Pratt, R.H. and Roy, S.C. (1986).
Phys. Rep., 140, 75-159.
Elastic scattering of γ-rays and X-rays by atoms

4760 Kao, H.-C., Kuo, T.-Y., Yen, H.-P., Wei, C.-M. and Huang, K.-N. (1992). Phys. Rev. A, 45, 4646-4652.
Relativistic cross sections of electron-impact ionization of hydrogenic ions

4761 Karasiov, V.V., Labzowsky, L.N. and Nefiodov, A.V. (1992b).
Phys. Lett. A, 172, 62-65.
Parity violation in heliumlike uranium

4762 Karasiov, V.V., Labzowsky, L.N., Nefiodov, A.V. and Shabaev, V.M. (1992a). Phys. Lett. A, 161, 453-457.
Overlapping resonances in the process of recombination of an electron with hydrogenlike uranium

4763 Karazija, R. (1987).
The Theory of X-Ray and Electronic Spectra of Free Atoms. An Introduction (in Russian)
Vilnius, Mokslas.

4764 Karazija, R. and Rudzikaite, L. (1988). Lit. Fiz. Sbornik, 28, 3-14; Sov. Phys.-Coll., 28, 1-10.
Mean energy of the Auger spectrum and its spin-doublet groups of lines

4765*Karaziya, R.I. (1981). Usp. Fiz. Nauk, 135, 79-115; Sov. Phys. Usp., 24, 775-794.
Excited electron orbit collapse and atomic spectra

4766 Kargabaev, S.A., Shabaeva, M.B. and Shabaev, V.M. (1990).
Vestn. Leningr. Univ., Fiz. Khim., (2), 71-76.
The Coulomb Green function of the linear Dirac equation taking into account the finite size of the nucleus

4767 Karim, K.R. and Bhalla, C.P. (1986). Phys. Rev. A, 34, 4743-4750.
Dielectronic satellite spectra of hydrogenlike chromium

4768 Karna, S.P., Dupuis, M., Perrin, E. and Prasad, P.N. (1990). J. Chem. Phys., 92, 7418-7425.
Theoretical and experimental studies of optical nonlinearities of haloforms CHX_3, X=F, Cl, Br, I

4769 Karpenko, I.K. (1987). Izv. Vyssh. Uchebn. Zaved., Fiz., 30, 126.
Relativistic fermion in a square-well potential (in Russian)

4770 Karwowski, J., Jaskólski, W. and Kobus, J. (1988).

Phys. Scr., 38, 554-556.
Comment on "A comparison of relativistic and quasirelativistic line strengths" by A.K. Mohanty and D.H. Sampson

4771 Karwowski, J. and Kobus, J. (1986). Int. J. Quantum Chem., 30, 809-819.
The Dirac second-order equation and an improved quasirelativistic theory of atoms

4772 Karwowski, J. and Martin, I. (1991). Phys. Rev. A, 43, 4832-4836.
Quasirelativistic formulation of the quantum-defect-orbital method

4773 Karwowski, J., Styszynski, J. and Schwarz, W.H.E. (1991). J. Phys. B: At. Mol. Opt. Phys., 24, 4877-4886.
A numerical study on the validity of the Breit-Pauli approximation

4774 Kato, T. and Safronova, U.I. (1991). Opt. Spektrosk., 70, 497-501; Opt. Spectrosc., 70, 291-293.
$1s^2 2s^2$-$1s^2 2s2p$-$1s^2 2p^2$ excitation cross sections for O V, Si XI, Fe XIII, Mo XXXIX

4775 Kato, T. and Yajima, K. (1991). Ann. Inst. Henri Poincaré, 54, 209-221.
Dirac equations with moving nuclei

4776 Katrib, A., El-Issa, B.D. and Ghodsian, R. (1986). Int. J. Quantum Chem., 29, 993-999.
Multiple bonds and Re(4f) binding energies of some rhenium halides

4777 Katsnel'son, M.I., Solov'ev, I.V. and Trefilov, A.V. (1992). JETP Lett., 56, 272-276; Pis'ma Zh. Eksp. Teor. Fiz., 56, 276-279.
α-δ transition in plutonium as a Mott transition in an f subsystem

4778 Katsuki, S. (1990). Chem. Phys. Lett., 165, 535-539.
An ab initio effective hamiltonian method applied to a relativistic molecule, AgH

4779 Katsuki, S. (1992). Can. J. Chem., 70, 285-289.
Molecular orbital calculation using spectral representation technique

4780 Katsuki, S. and Huzinaga, S. (1988). Chem. Phys. Lett., 147, 597-602.
Inclusion of relativistic effects in variational molecular calculations

4781 Katz, A.R. (1988). Lett. Math. Phys., 15, 39-46.
Semiclassical limit of the Dirac equation and the g minus 2 experiment

4782 Kaupp, M., Charkin, O.P. and Schleyer, P.v.R. (1992a). Organometallics, 11, 2765-2767.
The ions MCp_2^+ (M=Sc, La): Significantly bent sandwich species

4783 Kaupp, M. and Schleyer, P.v.R. (1992a). J. Am. Chem. Soc., 114, 491-497.
The structural variations of monomeric alkaline earth MX_2 compounds (M=Ca, Sr, Ba; X=Li, BeH, BH_2, CH_3, NH_2, OH, F). An ab initio pseudopotential study

4784 Kaupp, M. and Schleyer, P.v.R. (1992b). J. Phys. Chem., 96, 7316-7323.
Do low-coordinated group 1-3 cations $M^{n+}L_m$ (M^{n+}= K^+, Rb^+, Cs^+, Ca^{2+}, Sr^{2+}, Ba^{2+}, Sc^{3+}, Y^{3+}, La^{3+}; L=NH_3, H_2O, HF; m=1-3) with a formal noble-gas electron configuration favor regular or "abnormal" shapes?

4785 Kaupp, M. and Schleyer, P.v.R. (1992c). Angew. Chem., 104, 1240-1242.
Struktur- und Stabilitätstendenzen bei Fluor(methyl)plumbanen

4786 Kaupp, M., Schleyer, P.v.R., Dolg, M. and Stoll, H. (1992b). J. Am. Chem. Soc., 114, 8202-8208.
The equilibrium structures of monomeric Group 2 and lanthanide(II) metallocenes MCp_2 (M=Ca, Sr, Ba, Sm, Eu, Yb) studied by ab initio calculations

4787 Kaupp, M., Schleyer, P.v.R. and Stoll, H. (1992c). J. Phys. Chem., 96, 9801-9805.
Model CaH_2(L) and CaF_2(L) complexes (L=Ne, Ar, Kr, Xe, CO, N_2): Consequences of interactions between "inert-gas" ligands and floppy molecules

4788 Kaupp, M., Schleyer, P.v.R., Stoll, H. and Preuss, H. (1991a). J. Chem. Phys., 94, 1360-1366.
Pseudopotential approaches to Ca, Sr, and Ba hydrides. Why are some alkaline earth MX_2 compounds bent?

4789 Kaupp, M., Schleyer, P.v.R., Stoll, H. and Preuss, H. (1991b). J. Am. Chem. Soc., 113, 6012-6020.
The question of bending of the alkaline earth dihalides MX_2 (M= Be, Mg, Ca, Sr, Ba; X= F, Cl, Br, I). An ab initio pseudopotential study

4790 Kaupp, M., Stoll, H. and Preuss, H. (1990a). J. Comp. Chem., 11, 1029-1037.
Pseudopotential calculations for methyl compounds of zinc and magnesium

4791 Kaveh, M. and Mott, N.F. (1987). Phil. Mag. Lett., 56, 97-102.
Effect of spin-orbit interaction on the metal-insulator transition in doped silicon

4792 Kawamura-Kuribayashi, H., Koga, N. and Morokuma, K. (1992). J. Am. Chem. Soc., 114, 8687-8694.
An ab initio MO and MM study of homogeneous olefin polymerization with silylene-bridged zirconocene catalyst and its regio- and stereoselectivity

4793 Kefi, M., Andre, J.-M., Heno, Y., Giorgi, G. and Bonnelle, C. (1992). Phys. Rev. A, 45, 2859-2867.

X-ray attenuation and anomalous scattering factors of medium-Z elements around their K edge

4794 Keister, B.D. and Polyzou, W.N. (1991). Adv. Nucl. Phys., 20.
Relativistic hamiltonian dynamics in nuclear and particle physics

4795 Kelleher, D.E. and Saloman, E.B. (1987). Phys. Rev. A, 35, 3327-3338.
Rydberg states with anisotropic ion cores: Stark effect

4796 Keller, J. (1986). Int. J. Theor. Phys., 25, 779-806.
Generalization of the Dirac equation admitting isospin and color symmetries

4797 Keller, J. and Rodriguez, A. (1992). J. Math. Phys., 33, 161-170.
Geometric superalgebra and the Dirac equation

4798 Keller, J. and Rodriguez-Romo, S. (1990). J. Math. Phys., 31, 2501-2510.
A multivectorial Dirac equation

4799*Keller Jr., O.L. (1976).
Prediction of the chemical and physical properties of superheavy elements (in Russian)
In "Predictions in the Study of Periodicity", Ed. B.M. Kedrov and D.N. Trifonov, Nauka, Moscow., pp. 203-223.

4800 Kellö, V. and Sadlej, A.J. (1990a). J. Chem. Phys., 93, 8122-8132.
Estimates of relativistic contributions to molecular properties

4801 Kellö, V. and Sadlej, A.J. (1990b). Chem. Phys. Lett., 174, 641-648.
Relativistic contributions to molecular electric-field gradients in hydrogen halides

4802 Kellö, V. and Sadlej, A.J. (1991). J. Chem. Phys., 95, 8248-8253.
Quadrupole moments of CuH, AgH, and AuH. A study of the electron correlation and relativistic effects

4803 Kellö, V. and Sadlej, A.J. (1992a). Mol. Phys., 75, 209-220.
Electron correlation and relativistic contributions to molecular electric properties: dipole and quadrupole moments of cyanogen halides

4804 Kellö, V. and Sadlej, A.J. (1992b). Theor. Chim. Acta, 83, 351-366.
Medium-size polarized basis sets for high-level-correlated calculations of molecular electric properties. VI. Fifth-row atoms: Pb through At

4805 Kelly, H.P. (1990). AIP Conf. Proc., 215, 292-311.
Review of our present understanding of the photoionization process for atoms

4806 Kelly, P.J. and Brooks, M.S.S. (1987). J. Chem. Soc., Faraday Trans. 2, 83, 1189-1203.
Electronic structure and ground-state properties of the actinide dioxides

4807*Kemper, F., Rosicky, F. and Feder, R. (1985). J. Phys. B: At. Mol. Phys., 18, 1223-1228.
Relativistic two-channel calculations of the elastic scattering of slow electrons from Ar, Kr and Xe atoms

4808 Kertesz, M. and Guloy, A.M. (1987). Inorg. Chem., 26, 2852-2857.
Bonding and electronic structure of conducting mercury networks: $KHgC_{4n}$ graphite amalgams and Hg_3MF_6 layers and chains

4809 Kessler, J. (1991). Adv. At. Mol. Opt. Phys., 27, 81-163.
Electron-polarization phenomena in electron-atom collisions

4810 Ketov, S.V. and Prager, Ya.S. (1990ab). Acta Phys. Pol. B, 21, 463-467; Izv. Akad. Nauk SSSR, Ser. Fiz.(3), 5-9.
On the "square root" of the Dirac equation within extended supersymmetry

4811 Kheifets, A.S. and Amusia, M.Ya. (1992). Phys. Rev. A, 46, 1261-1269.
Relativistic ab initio calculation of the xenon 5s ionization spectra for the (γ,e) and (e,2e) reactions

4812 Khriplovich, I. (1988).
Parity Non-Conservation in Atomic Phenomena (in Russian)
Nauka, Moscow.

4813 Khriplovich, I.B. (1989). Comm. At. Mol. Phys., 22, 295-306.
T-Violation in atoms

4814 Khriplovich, I.B. (1991).
Parity Nonconservation in Atomic Phenomena
Gordon and Breach, Philadelphia, PA, 314 p.

4815 Khriplovich, I.B. and Pospelov, M.E. (1992). Phys. Lett. A, 171, 349-354.
High-frequency asymptotics of optical activity and energy difference of chiral molecules

4816*Kiefer, H.M. and Fradkin, D.M. (1969). Phys. Rev., 180, 1282-1288.
Noninvariance group for the Dirac Coulomb problem

4817 Kijowski, J., Thielmann, A., Hennig, J.D., Luecke, W. and Tolar, J. (1991).
A lattice approximation of the Dirac equation: Differential geometry, group representations, and quantization
In "Lecture Notes in Physics", 379, pp. 85-97.
Berlin, Springer.

4818 Kim, B.-S., Harmon, B.N. and Lynch, D.W. (1989a). Phys. Rev. B, 39, 5754-5757.
Optical properties of Mo

4819 Kim, B.-S., Harmon, B.N. and Lynch, D.W. (1989b). Phys. Rev. B, 39, 13146-13148.
Interpretation of the optical properties of Nb

4820 Kim, G.-B. and Balasubramanian, K. (1989). J. Mol. Spectrosc., 134, 412-420.
Spectroscopic properties and potential energy curves of GaH

4821 Kim, G.-B. and Balasubramanian, K. (1992a). J. Mol. Spectrosc., 152, 192-198.
Spectroscopic constants and potential energy curves of GaBr

4822 Kim. G.-B. and Balasubramanian, K. (1992b). Chem. Phys. Lett., 193, 109-114.
Spectroscopic properties and potential energy curves of GaCl

4823 Kim, K.J., Harmon, B.N., Chen, L.-Y. and Lynch, D.W. (1990). Phys. Rev. B, 42, 8813-8819.
Optical properties and electronic structures of the intermetallic compounds $AuGa_2$ and $PtGa_2$

4824 Kim, L.-H. and Greene, C.H. (1987). Phys. Rev. A, 36, 4272-4279.
Two-electron excitations in atomic calcium. II. Fine-structure effects

4825 Kim, L.-H. and Pratt, R.H. (1987). Phys. Rev. A, 36, 45-58.
Numerical calculation of classical bremsstrahlung

4826 Kim, L.-H., Pratt, R.H., Seltzer, S.M. and Berger, M.J. (1986). Phys. Rev. A, 33, 3002-3009.
Ratio of positron to electron bremsstrahlung energy loss: An approximate scaling law

4827 Kim, Y.-K. (1989).
Comments on relativistic correlation
In "Relativistic, Quantum Electrodynamic and Weak Interaction Effects in Atoms", Ed. W. Johnson, P. Mohr and J. Sucher, AIP Conf. Proc., 189, 272-282.

4828 Kim, Y.-K. (1990). AIP Conf. Proc., 206, 19-39.
Relativistic and quantum electrodynamic effects in highly-charged ions

4829 Kim, Y.-K., Baik, D.H., Indelicato, P. and Desclaux, J.P. (1991). Phys. Rev. A, 44, 148-166.
Resonance transition energies of Li-, Na-, and Cu-like ions

4830 Kim, Y.-K. and Desclaux, J.P. (1987). Phys. Scr., 36, 796-800.
Relativistic effects in electron-atom collisions

4831 Kim, Y.-K. and Desclaux, J.P. (1988). Phys. Rev. A, 38, 1805-1808.
Incident-energy dependence of electron-ion collision cross sections

4832 Kim, Y.-K., Martin, W.C. and Weiss, A.W. (1988a). J. Opt. Soc. Am. B, 5, 2215-2224.
Relativistic and correlation effects in the 2s3p configuration of berylliumlike ions

4833 Kim, Y.-K., Sugar, J., Kaufman, V. and Ali, M.A. (1988b). J. Opt. Soc. Am. B, 5, 2225-2229.
Quantum-electrodynamic contributions to spin-orbit splitting in the ground state of aluminumlike ions

4834 Kim, Y.-K., Weiss, A.W. (1987).
Interplay between correlation and relativistic effects: $2s^2$ - 2s3p transitions in Be-like ions
In "Atoms in Unusual Situations", Ed. J.P. Briand, New York, Plenum Press, p.310.

4835 Kim, Y.S., Goldberg, I.B. and Pratt, R.H. (1992). Phys. Rev. A, 45, 4542-4552.
Photon-electron polarization correlations in high-Z ns-subshell photoionization

4836 Kimura, H., Kaizu, Y. and Kobayashi, H. (1990). Mol. Phys., 70, 891-902.
Magnetic circular dichroism in the charge-transfer-to-solvent excited states of iodide ion in aqueous solution

4837 King, C., Wang, J.-C., Khan, M.N.I. and Fackler Jr., J.P. (1989). Inorg. Chem., 28, 2145-2149.
Luminescence and metal-metal interactions in binuclear gold(I) compounds

4838 King, H.F. and Furlani, T.R. (1988). J. Comp. Chem., 9, 771-778.
Computation of one and two electron spin-orbit integrals

4839 King, R.B. (1992). Inorg. Chem., 31, 1978-1980.
Chemical applications of topology and group theory. 27. Covalent bonding in actinide derivatives

4840 Kinoshita, T. (Ed.) (1990).
Quantum Electrodynamics
World Scientific, River Edge, N.J. 997 pp.

4841 Kippels, C., Thiel, W., McKean, D.C. and Coats, A.M. (1992). Spectrochim. Acta, 48A, 1067-1082.
The harmonic force fields of dimethyl zinc, cadmium and mercury: a joint theoretical and experimental study

4842 Kirchner, E.J.J., Baerends, E.J., van Slooten, U. and Kleyn, A.W. (1992). J. Chem. Phys., 97, 3821-3830.
Relativistic effect on the interatomic platinum-oxygen potential and its consequences in high energy O_2/Pt(111) scattering

4843 Kiselev, A.A. and Lyaptsev, A.V. (1989).
Kvantovomekhanicheskaya Teoriya Vozmushchenii (Diagrammnyi metod)
Izd. LGU, Leningrad, 358 p.

4844 Kissel, L. and Pratt, R.H. (1990). Acta Cryst., A46, 170-175.
Corrections to tabulated anomalous-scattering factors

4845 Kissel-Phillip, M. and Schwarz, W.H.E. (1988). Phys. Rev. A, 38, 6027-6033.
Relativistically corrected Schrödinger equation

4846*Klaus, M. (1980a). Helv. Phys. Acta, 53, 453-462.
On the point spectrum of Dirac operators

4847*Klaus, M. (1980b). Helv. Phys. Acta, 53, 463-482.
Dirac operators with several Coulomb singularites

4848 Klaus, M. (1990). J. Math. Phys., 31, 182-190.
On the Levinson theorem for Dirac operators

4849 Klein, A. and Dreizler, R.M. (1992). Phys. Rev. A, 45, 4340-4345.
Alternative derivation of relativistic two-body equations

4850*Klein, A. and Rafelski, J. (1975). Phys. Rev. D, 11, 300-311.
Bose condensation in supercritical external fields

4851 Klimenko, Yu.I. and Klimenko, E.Yu. (1991). Teor. Mat. Fiz., 87, 235-240; Theor. Math. Phys., 87, 499-502.
Elastic scattering amplitude for an electron in a plane-wave electromagnetic field

4852 Klobukowski, M. (1991). Chem. Phys. Lett., 183, 417-422.
Comparison of the effective-core-potential and model-potential methods in the studies on spin-orbit effects: zero-field splitting of the X $^3\Sigma^-$ state of pnicogen hydrides

4853 Klobukowski, M. (1992a). Can. J. Chem., 70, 589-595.
Computational studies on hydrates of alkaline-earth dications

4854 Klobukowski, M. (1992b). Theor. Chim. Acta, 83, 239-248.
Comparison of the effective core potential and model potential methods in studies of electron correlation energy in molecules: Dihalides and halogen hydrides

4855 Klobukowski, M. and Fraga, S. (1988). Phys. Rev. A, 38, 1593-1594.
Accurate theoretical prediction of the experimental ground-state total atomic energies

4856 Kluger, Y., Eisenberg, J.M., Svetitsky, B., Cooper, F. and Mottola, E. (1992). Phys. Rev. D, 45, 4659-4671.
Fermion pair production in a strong electric field

4857 Knappe, P. (1990).

Weiterentwicklung und Anwendung der LCGTO-LDF-Methode: Cluster-Modelle für Oberflächen- und Festkörperprobleme und eine quasirelativistische Verallgemeinerung Dissertation, TU München.

4858 Knappe, P. and Rösch, N. (1990). J. Chem. Phys., 92, 1153-1162.
Relativistic linear combination of Gaussian-type orbitals density functional method based on a two-component formalism with external field projectors

4859 Knight Jr., L.B., Cobranchi, S.T., Herlong, J., Kirk, T., Balasubramanian, K. and Das, K.K. (1990). J. Chem. Phys., 92, 2721-2732.
Laser vaporization/rare gas matrix-isolation electron spin resonance studies of PdH_2^+, PdD_2^+, PdH, and PdD: Theoretical investigation of PdH_2^+

4860*Knipp, J.K. (1938). Phys. Rev., 53, 734-745.
Quadrupole-quadrupole interatomic forces

4861 Kober, E.M. and Hay, P.J. (1989).
Ab initio studies of transition metal dihydrogen chemistry In "The Challenge of d and f Electrons", Ed. D.R. Salahub and M.C. Zerner, ACS Symposium Series, 394, pp. 92-105.

4862 Kobus, J. (1986). Acta Phys. Pol., B17, 771-779.
Rationale of quasi-relativistic methods

4863 Kobus, J. and Jaskólski, W. (1987). J. Phys. B: At. Mol. Phys., 20, 4949-4961.
Numerical comparison between DHF and RHF methods

4864 Kobus, J., Karwowski, J. and Jaskolski, W. (1987). J. Phys. A, 20, 3347-3352.
Matrix elements of r^q for quasirelativistic and Dirac hydrogenic wavefunctions

4865 Koc, K. and Migdalek, J. (1990). J. Phys. B: At. Mol. Opt. Phys., 23, L5-L8.
Relativistic many-body calculations of ionisation energies and fine-structure intervals in Ag and Au atoms

4866 Koc, K. and Migdalek, J. (1992). J. Phys. B: At. Mol. Opt. Phys., 25, 907-911.
Relativistic many-body calculations of the $4f^{14}nl$ levels in Yb II and Lu III spectra

4867 Koelling, D.D. (1986). Int. J. Quantum Chem., S20, 377-392.
The band description of materials with localizing orbitals

4868*Koelling, D.D., Boring, A.M. and Wood, J.H. (1983). Solid State Comm., 47, 227-232.
The electronic structure of CeO_2 and PrO_2

4869*Koelling, D.D. and Freeman, A.J. (1976).
Electronic energy band structure of actinide metals and intermetallics
In "Plutonium and Other Actinides", Ed. H. Blank and R.

Lindner, North-Holland, Amsterdam, pp. 291-309.

4870 Koga, N., Jin, S.Q. and Morokuma, K. (1988). J. Am. Chem. Soc., 110, 3417-3425.
Rearrangement through Berry pseudorotation and olefin insertion of d^8 five-coordinate $Rh(H)(C_2H_4)(CO)_2(PH_3)$. An ab initio MO study

4871 Koga, N. and Morokuma, K. (1986). J. Am. Chem. Soc., 108, 6136-6144.
Mechanism of carbonyl insertion reaction of Pd and Pt complexes. An ab initio MO study

4872 Koga, N. and Morokuma, K. (1989).
Potential energy surface of olefin hydrogenation by Wilkinson catalyst
In "The Challenge of d and f Electrons", Ed D.R. Salahub and M.C. Zerner, ACS Symposium Series, 394, pp. 77-91.

4873 Koga, N. and Morokuma, K. (1990). J. Phys. Chem., 94, 5454-5462.
Ab initio potential energy surface and electron correlation effect in CH activation of CH_4 by coordinatively unsaturated $RhCl(PH_3)_2$

4874 Koga, N. and Morokuma, K. (1991a). Chem. Rev., 91, 823-842.
Ab initio molecular orbital studies of catalytic elementary reactions and catalytic cycles of transition-metal complexes

4875 Koga, N. and Morokuma, K. (1991b). New J. Chem., 15, 749-755.
Carbonyl insertion into Pd-H bond of $HPd(R)(CO)(PH_3)$ ($R=CH_3$, H) and comparison with that into Pd-CH_3 bond. An ab initio MO study

4876 Koga, N. and Morokuma, K. (1992).
Hydrogen transfer from transition metal hydrides - Theoretical aspects
In "Transition Metal Hydrides", Ed. A. Dedieu, VCH, New York, pp.185-234.

4877*Kogut, J.B. (1983). Rev. Mod. Phys., 55, 775-836.
The lattice gauge theory approach to quantum chromodynamics

4878 Kohout, M., Savin, A. and Preuss, H. (1991). J. Chem. Phys., 95, 1928-1942.
Contribution to the electron distribution analysis. I. Shell structure of atoms

4879 Kokko, K. (1991).
Investigation into Electronic Properties of Monovalent Metals and Alloys Using Band Theoretical Supercell LMTO Calculation
Thesis, Turku

4880 Koller, D., Malvetti, M. and Pilkuhn, H. (1988). Phys. Lett. A, 132, 259-261.
Positronium and muonium in strong magnetic fields

4881 Kölmel, C. and Ahlrichs, R. (1990). J. Phys. Chem., 94, 5536-5542.
An ab initio investigation of copper complexes with supershort Cu-Cu distances

4882 Kolomietz, V.M. and Kondratjev, V.N. (1990). Z. Phys. A, 335, 379-385.
Nuclear gamma-transitions via the accompanying atomic excitations

4883 Kolsrud, M. (1989). Phys. Scr., 39, 13-29.
The external field approximation of quantum electrodynamics

4884 Komachiya, M. and Fukuda, R. (1992). Int. J. Mod. Phys. A, 7, 5585-5606.
Effective action and the atomic energy levels

4885 Kondratenko, A.V., Nemanova, V.I., Ruzankin, S.F., Bausk, N.V., Zhidomirov, G.M. and Mazalov, L.N. (1987). Zh. Strukt. Khim., 28(5), 16-19.
SCF MS-Xα calculations of XANES and EXAFS of Pt in chloride complexes

4886 Körling, M. and Häglund, J. (1992). Phys. Rev. B, 45, 13293-13297.
Cohesive and electronic properties of transition metals: The generalized gradient approximation

4887 Kosaka, K. and Yonei, K. (1991a). J. Phys. Soc. Japan, 60, 850-861.
A relativistic WKB method for bound states of an electron in screened Coulomb potentials

4888 Kosaka, K. and Yonei, K. (1991b). J. Phys. Soc. Japan, 60, 4081-4088.
A relativistic WKB method for bound states of an electron in screened Coulomb potentials. II. Examination of the wavefunction

4889 Koseki, S., Schmidt, M.W. and Gordon, M.S. (1992). J. Phys. Chem., 96, 10768-10772.
MCSCF/6-31G(d,p) calculations of one-electron spin-orbit coupling constants in diatomic molecules

4890 Kosik, D.W. (1991). J. Phys. A, 24, 2259-2272.
An approximate partial differential matrix equation in the lepton energy for evaluation of Dirac-Coulomb radial integrals

4891 Kostin, M.D. (1991). Nuovo Cim. B, 106, 91-94.
Linearization of the hamiltonian

4892 Kotochigova, S.A. (1986). Opt. Spektrosk., 60, 1116-1121.
Relativistic calculation of radiationless-transition probabilities in barium

4893 Kotochigova, S.A. and Tupitsyn, I.I. (1986a). Opt. Spektrosk., 60, 8-13; Opt. Spectrosc., 60, 4-7.
Calculations of the energy structure of complex atoms by

the Hartree-Fock-Dirac method allowing for superposition of configurations. Barium

4894 Kotochigova, S.A. and Tupitsyn, I.I. (1986b). Opt. Spektrosk., 61, 1161-1166.
Hartree-Fock-Dirac method with consideration for superposition of configurations for the calculation of transition oscillator strengths in heavy atoms

4895 Kotochigova, S.A. and Tupizin, I.I. (1987). J. Phys. B: At. Mol. Phys., 20, 4759-4772.
Theoretical investigation of rare-earth and barium spectra by the Hartree-Fock-Dirac method

4896 Kotzian, M. and Rösch, N. (1991). J. Mol. Spectrosc., 147, 346-358.
Intermediate neglect of differential overlap spectroscopic studies of lanthanide complexes: Diatomic lanthanide oxides PrO and TmO

4897 Kotzian, M. and Rösch, N. (1992). J. Phys. Chem., 96, 7288-7293.
Electronic structure of hydrated cerium(III). An INDO/S-CI molecular orbital study including spin-orbit interaction

4898 Kotzian, M., Rösch, N., Pitzer, R.M. and Zerner, M.C. (1989). Chem. Phys. Lett., 160, 168-174.
A spin-orbit interaction enhanced INDO/S-CI technique: Applications to main group and transition metal heteronuclear diatomic molecules

4899 Kotzian, M., Rösch, N. and Zerner, M.C. (1991). Int. J. Quantum Chem., S25, 545-555.
An INDO/S-CI treatment including spin-orbit interaction based on Rumer spin functions. Application to the hydrated cerium ion

4900 Kotzian, M., Rösch, N. and Zerner, M.C. (1992). Theor. Chim. Acta, 81, 201-222.
Intermediate neglect of differential overlap spectroscopic studies on lanthanide complexes. I. Spectroscopic parametrization and application to diatomic lanthanide oxides LnO (Ln=La, Ce, Gd, and Lu)

4901 Koutroulos, C.G. (1990). Z. Naturforsch., 45a, 14-16.
Relativistic energy states of the Lambda-particle in hypernuclei using Gaussian potentials

4902 Koutroulos, C.G. (1991). J. Phys. G, 17, 1069-1076.
Relativistic versus non-relativistic treatment of hypernuclei using the potential $U(r)=-D(\cosh^2(r/R))^{-1}$

4903 Kovba, V.M. and Topol', I.A. (1986a). Teor. Eksp. Khim., 22, 11-19.
SCF MS-Xα calculations of low-lying electronic states of GaO and InO

4904 Kovba, V.M. and Topol, I.A. (1986b). J. Mol. Struct. (Theochem), 137, 65-79.
Calculations on the electronic structure and spectra

of the AlO, GaO and InO molecules by the X_α-SW method

4905 Kozlov, M.G. (1987). Opt. Spectrosc., 63, 441-444.
Calculation of the degree of parity nonconservation in the spin-rotational spectrum of the PbF molecule

4906 Kozlov, M.G. (1988). Sov. J. Quantum Electron., 18, 713-715.
Calculation sof the degree of parity nonconservation in the spin-rotational spectrum of the PbF molecule

4907 Kozlov, M.G., Fomichev, V.I., Dmitriev, Yu.Yu., Labzovsky, L.N. and Titov, A.V. (1987). J. Phys. B: At. Mol. Phys., 20, 4939-4948.
Calculation of the P- and T-odd spin-rotational Hamiltonian of the PbF molecule

4908 Kraftmakher, A.Ya. (1988a). J. Phys. B: At. Mol. Opt. Phys., 21, 2803-2813.
On the Hartree-Fock calculation of the electron electric dipole moment enhancement factor for the thallium atom

4909 Kraftmakher, A.Ya. (1988b). Phys. Lett. A, 132, 167-170.
Hartree-Fock calculation of parity violation in atomic caesium, induced by the anapole moment of the nucleus

4910 Krainov, V.P. and Roshupkin, S.P. (1992). J. Opt. Soc. Am. B, 9, 1231-1233.
Relativistic effects in the angular distribution of ejected electrons in tunneling ionization of atoms by strong electromagnetic fields

4911*Kramers, H.A. (1957).
The Foundations of Quantum Theory
North-Holland, Amsterdam.

4912 Kratz, J.V., Zimmermann, H.P., Scherer, U.W., Schädel, M., Brüchle, W., Gregorich, K.E., Gannett, C.M., Hall, H.L., Henderson, R.A., Lee, D.M., Leyba, J.D., Nurmia, M.J., Hoffman, D.C., Gäggeler, H., Jost, D., Baltensperger, U., Ya, N.-Q., Türler, A. and Lienert, Ch. (1989). Radiochim. Acta, 48, 121-133.
Chemical properties of element 105 in aqueous solution: Halide complex formation and anion exchange into triisoctyl amine

4913 Krause, J. (1986). Phys. Rev. A, 34, 3692-3699.
Relativistic variational theory for two-electron atoms

4914 Krauss, M., Basch, H. and Miller, K.J. (1988a). J. Am. Chem. Soc., 110, 4517-4521.
Comparison of direct and through-water binding of platinum ammines to the phosphate anion

4915 Krauss, M., Basch, H. and Miller, K.J. (1988b). Chem. Phys. Lett., 148, 577-580.
Hydrogen bonding in Pt ammine complexes

4916 Krauss, M. and Stevens, W.J. (1990). J. Chem. Phys.,

93, 4236-4242.
Effective core potentials and accurate energy curves for Cs_2 and other alkali diatomics

4917 Krewer, J.W., Dürr, W., Pescia, D. and Feder, R. (1990). Solid State Comm., 74, 137-142.
Spin-polarized electron diffraction from Pb(110): Surface barrier and non-spherical potential effects

4918 Krewer, J.W. and Feder, R. (1989). Solid State Comm., 69, 87-90.
Non-spherical potentials in relativistic multiple scattering theory

4919 Krogh-Jespersen, K., Zhang, X.-H., Ding, Y.-B., Westbrook, J.D., Potenza, J.A. and Schugar, H.J. (1992). J. Am. Chem. Soc., 114, 4345-4353.
Molecular and electronic structures of pentaammineruthenium(II)-thioether complexes. The nature of Ru(II)-S back-bonding elucidated by structural, electronic spectral, and molecular orbital studies

4920 Krogh-Jespersen, K., Zhang, X.-H., Westbrook, J.D., Fikar, R., Nayak, K., Kwik, W.-L., Potenza, J.A. and Schugar, H.J. (1989). J. Am. Chem. Soc., 111, 4082-4091.
Crystal, molecular, and electronic structures of pentaammineruthenium(III)-thioether complexes

4921*Krogh-Jespersen, M.-B. (1985). J. Comp. Chem., 6, 614-624.
Theoretically derived absorption and MCD spectral data for the cellular species produced in the hydrolysis of cis-diamminedichloroplatinum(II)

4922 Krogh-Jespersen, M.-B. and Altonen, A. (1987). Inorg. Chem., 26, 2084-2090.
Theoretical studies of the potential hydrolysis products from cis-$Pt(NH_3)_2Cl_2$ and acetamide

4923 Królikowski, W. (1989). Acta Phys. Pol., B20, 119-132.
Asymptotic solution to the Breit relativistic equation for Coulomb parafermionium

4924 Królikowski, W. (1992a). Acta Phys. Pol., B23, 43-52.
Various forms of radial equations for the Dirac two-body problem

4925 Królikowski, W. (1992b). Nuovo Cim. A, 105, 655-662.
Useful radial equations for the Dirac two-body problem

4926 Krstic, P.S. and Milosevic, D.B. (1989). Phys. Rev. A, 39, 1783-1790.
Relativistic effects in potential scattering of electrons in an ultrastrong laser field

4927 Krstic, P.S. and Mittleman, M.H. (1990). Phys. Rev. A, 42, 4037-4044.
Bound states of atomic hydrogen in a laser field beyond the nonrelativistic dipole approximation

4928 Krutzen, B.C.H. and Inglesfield, J.E. (1990). J. Phys.: Condens. Matter, 2, 4829-4847.
First-principles electronic structure calculations for incommensurately modulated calaverite

4929 Krutzen, B.C.H. and Springelkamp, F. (1989). J. Phys.: Cond. Matt., 1, 8369-8383.
Spin-polarised relativistic electronic structure calculations

4930 Kryutchkov, S.V., Mironov, V.S., Antipov, B.G. and Plekhanov, Yu.V. (1991). Metalloorg. Khim., 4, 52-62.
The role of the outer ns-AO and the influence of the change in the effective charge of the metal atoms on the M-M bonding formation in transition metals acidoclusters

4931 Kübler, J. and Harmon, B. (Ed.) (1991). Physica, 172(1-2).
Electronic structure in the 1990s. On the occasion of the 60th birthday of A.J. Freeman

4932 Kubo, Y. (1987). J. Phys. F, 17, 383-396.
Self-consistent relativistic band structures of strontium and ytterbium under normal and high pressure

4933 Kubo, Y. and Yamashita, J. (1986). J. Phys. F, 16, 2017-2025.
Self-consistent relativistic band structure for lead at normal and high pressure

4934 Kuchiev, M.Y., Solov'yov, A.V. and Yakhontov, V.L. (1992). Z. Phys. D, 22, 679-682.
Quadrupole moment of the $NS_{1/2}$-state of the hydrogen atom

4935 Küchle, W., Dolg, M., Stoll, H. and Preuss, H. (1991). Mol. Phys., 74, 1245-1263.
Ab initio pseudopotentials for Hg through Rn. I. Parameter sets and atomic calculations

4936*Kudar, J. (1926). Ann. Physik, 81, 632-636.
Zur vierdimensionalen Formulierung der undulatorischen Mechanik

4937 Kudo, T. and Nagase, S. (1989). Chem. Phys. Lett., 156, 289-295.
Jahn-Teller distortions in SnH_4^+ and PbH_4^+

4938 Kudo, T. and Nagase, S. (1992). J. Phys. Chem., 96, 9189-9194.
The heavier Group 14 analogues of the bicyclo[1.1.0] butane radical cation. A theoretical study

4939 Kukulin, V.I., Loyola, G. and Moshinsky, M. (1991). Phys. Lett. A, 158, 19-22.
A Dirac equation with an oscillator potential and spin-orbit coupling

4940 Kukulin, V.I., Loyola, G. and Moshinsky, M. (1992).
A Dirac equation with an oscillator potential and spin orbit coupling

In "Hadron 91", Ed. S. Oneda and D.S. Peaslee, World Scientific, Singapore, pp. 293-301.

4941 Kulagin, N.A. (1987). Opt. Spektrosk., 63, 964-967. Opt. Spectrosc., 63, 566-568.
Change in the order of occupation of one-electron states of multiply charged ions of the lanthanum and actinium groups

4942 Kulakowski, K. and Lacroix, C. (1991). Phys. Status Solidi B, 165, K17-K21.
The spin-orbit contribution to the magnetic susceptibility in degenerate narrow bands

4943 Kumar, V., Chatterjee, S.N. and Roy, B.N. (1990). Pramana, 34, 447-460.
The effect of the relativistic Hartree-Fock-Roothaan momentum distribution in the binary encounter approximation for inner shell ionization

4944 Kuplyauskene, A.V. and Zhukauskas, G. (1991). Opt. Spectrosc., 71, 8-11; Opt. Spektrosk., 71, 15-20.
Spectra of radiative decay of the $1s^2 2s^2 2p^5 3l3l'$ LSJ states of sodiumlike ions of chlorine and argon

4945 Kurasawa, H. and Suzuki, T. (1986). Phys. Lett. B, 173, 377-381.
The Coulomb sum for electron scattering in the relativistic random phase approximation

4946 Kurasawa, H. and Suzuki, T. (1988a). Nucl. Phys., A490, 571-584.
Effects of the Dirac sea on the Coulomb sum value for electron scattering

4947 Kurasawa, H. and Suzuki, T. (1988b).
The missing strength of the Coulomb sum and the relativistic RPA in the σ-ω model
In "Workshop on Relativistic Nuclear Many-Body Physics", Columbus. Ohio.

4948 Kurdak, Ç., Chang, A.M., Chin, A. and Chang, T.Y. (1992). Phys. Rev. B, 46, 6846-6856.
Quantum interference effects and spin-orbit interaction in quasi-one-dimensional wires and rings

4949 Kürpick, P., Sepp, W.-D. and Fricke, B. (1991a). Z. Phys. D, 21, S293-294.
Inclusive probabilities for the scattering system 16MeV-S^{16+} on Ar

4950 Kürpick, P., Sepp, W.-D. and Fricke, B. (1992). J. Phys. B: At. Mol. Opt. Phys., 25, 5431-5437.
Inclusive probability calculations for the K-vacancy transfer in collisions of S^{15+} on Ar

4951 Kürpick, P., Thies, B., Sepp, W.-D. and Fricke, B. (1991b). J. Phys. B: At. Mol. Opt. Phys., 24, L139-L144.
Multiple vacancy inclusive probabilities for the scattering system 16 MeV S^{16+} on Ar

4952 Kuten', S.A. and Rapoport, V.I. (1989). Opt. Spektrosk., 67, 756-760. Opt. Spectrosc., 67, 443-445.
Existence of magnetic octupole moment of hydrogenlike atoms in the $1S_{1/2}$ and $2P_{1/2}$ states

4953 Kutzelnigg, W. (1987). Phys. Scr., 36, 416-431.
The relativistic many body problem in molecular theory

4954 Kutzelnigg, W. (1988). Theor. Chim. Acta, 73, 173-200.
Origin and meaning of the Fermi contact interaction

4955 Kutzelnigg, W. (1989a). Z. Phys. D, 11, 15-28.
Perturbation theory of relativistic corrections
1. The non-relativistic limit of the Dirac equation and a direct perturbation expansion

4956 Kutzelnigg, W. (1989b). Studies Phys. Theor. Chem., 62, 275-289.
Perturbation theory of relativistic effects

4957 Kutzelnigg, W. (1989c).
Generalization of Kato's cusp conditions to the relativistic case
In "Aspects of Many-Body Effects in Molecules and Extended Systems", Ed. D. Mukherjee, Lecture Notes in Chem., 50, Springer, Berlin, pp. 353-366.

4958 Kutzelnigg, W. (1990). Z. Phys. D, 15, 27-50.
Perturbation theory of relativistic corrections. Pt. 2, Analysis and classification of known and other possible methods

4959 Kutzner, M., Felton, M. and Winn, D. (1992). Phys. Rev. A, 45, 7761-7765.
Frequency-dependent polarizabilities of halide anions and H^-

4960 Kutzner, M., Radojevic, V., Kelly, H.P. and Altun, Z. (1990). Phys. Scr., 41, 823-826.
Photoionization of the 4d subshell of barium

4961 Kyle, K.R., Palke, W.E. and Ford, P.C. (1990). Coord. Chem. Rev., 97, 35-46.
The photoluminescence properties of the copper(I) clusters $Cu_4I_4A_4$ (A= aromatic amine) in solution

4962 Laaksonen, L., Grant, I.P. and Wilson, S. (1988). J. Phys. B: At. Mol. Opt. Phys., 21, 1969-1985.
The Dirac equation in the algebraic approximation: VI. Molecular self-consistent field studies using basis sets of Gaussian-type functions

4963 Laasonen, K., Andreoni, W. and Parrinello, M. (1992). Science, 28, 1916-1918.
Structural and electronic properties of $La@C_{82}$

4964 Ladányi, K. (1987). Acta Phys. Hung., 62, 161-175.
Scalar Bethe-Salpeter equation and the relativistic bound state problem

4965 Lagowski, J.B. and Vosko, S.H. (1989). Phys. Rev. A, 39, 4972-4977.
J-independent relativistic contributions to the s-d interconfiguration energies of the iron series: Some consequences for density-functional theory

4966 LaJohn, L. and Talman, J.D. (1992). Chem. Phys. Lett., 189, 383-389.
Minimax variational solution of the Dirac equation in molecular geometries

4967 LaJohn, L.A. (1990). Diss. Abstr. Int. B, 51, 229.
Inclusion of core valence correlation corrections in frozen core calculations using relativistic effective potentials

4968 LaJohn, L.A., Christiansen, P.A., Ross, R.B., Atashroo, T. and Ermler, W.C. (1987). J. Chem. Phys., 87, 2812-2824.
Ab initio relativistic effective potentials with spin-orbit operators. III. Rb through Xe

4969*Lam, L. (1970a). Lett. Nuovo Cim., 3, 292-294.
Dirac electron in orthogonal electric and magnetic fields

4970*Lam, L. (1970b). Can. J. Phys., 48, 1935-1937.
New exact solutions of the Dirac equation

4971 Lamoreaux, S.K. and Fortson, E.N. (1992). Phys. Rev. A, 46, 7053-7059.
Calculation of a linear Stark effect on the 254-nm line of Hg

4972 Landau, R.H. (1990).
Quantum Mechanics II
Wiley, New York, 483p.

4973 Lander, G.H. (1991). Phys. Scr., 44, 33-37.
Spin and orbital magnetic moments in actinide compounds

4974 Landro, M., Mork, K.J. and Olsen, H.A. (1987a). Ark. Fys. Semin. Trondheim, 1, 35p.
The radiative correction to the equivalent photon spectrum of a relativistic electron and the two-photon process

4975 Landro, M., Mork, K.J. and Olsen, H.A. (1987b). Phys. Rev. D, 36, 44-54.
Radiative correction to the equivalent-photon spectrum of a relativistic electron and the two-photon process

4976 Langhoff, S.R. and Bauschlicher Jr., C.W. (1986). Chem. Phys. Lett., 124, 241-247.
Theoretical study of the $X^2\Pi$ and $A^2\Sigma^+$ states of CuO and CuS

4977 Langhoff, S.R. and Bauschlicher Jr., C.W. (1988a). J. Chem. Phys., 89, 2160-2169.
Theoretical studies of the monoxides and monosulfides of Y, Zr, and Nb

4978 Langhoff, S.R. and Bauschlicher Jr., C.W. (1988b). Ann. Rev. Phys. Chem., 39, 181-212.
Ab initio studies of transition metal systems

4979 Langhoff, S.R. and Bauschlicher Jr., C.W. (1990). J. Mol. Spectrosc., 143, 169-179.
Theoretical study of the spectroscopy of niobium nitride

4980 Langhoff, S.R., Bauschlicher Jr., C.W., Pettersson, L.G.M. and Siegbahn, P.E.M. (1989). Chem. Phys., 132, 49-58.
Theoretical spectroscopic constants for the low-lying states of the oxides and sulfides of Mo and Te

4981 Langhoff, S.R., Bauschlicher Jr., C.W., Walch, S.P. and Laskowski, B.C. (1986). J. Chem. Phys., 85, 7211-7215.
Ab initio study of the ground state surface of Cu_3

4982 Langhoff, S.R., Bauschlicher Jr., C.W. and Partridge, H. (1988). J. Chem. Phys., 89, 396-407.
Theoretical study of the scandium and yttrium halides

4983 Langhoff, S.R., Pettersson, L.G.M., Bauschlicher Jr., C.W. and Partridge, H. (1987). J. Chem. Phys., 86, 268-278.
Theoretical spectroscopic parameters for the low-lying states of the second-row transition metal hydrides

4984 Lao, M.-Z. and Christiansen, P.A. (1992). J. Chem. Phys., 96, 2161-2163.
Relativistic effective potential quantum Monte Carlo simulations for Ne

4985 Larsson, S. and Pyykkö, P. (1986). Chem. Phys., 101, 355-369.
Relativistically parameterized extended Hückel calculations. IX. An iterative version with applications to some xenon, thorium and uranium compounds

4986 Laughlin, C. and Victor, G.A. (1988). Adv. At. Mol. Phys., 25, 163-194.
Model-potential methods

4987 Lauwers, P.G., Ben-Av, R. and Solomon, S. (1992). Nucl. Phys. B, 374, 249-259.
Inverting the Dirac matrix for SU(2) lattice gauge theory by means of parallel transported multigrid

4988 Lavín, C., Martín, I. and Karwowski, J. (1992). J. Mol. Struct.(Theochem), 254, 161-170.
Relativistic effects on the computation of oscillator strengths for the principal series in Na-like systems

4989 Layton, E. and Chu, S.-I. (1991). Chem. Phys. Lett., 186, 100-106.
Generalized Fourier-grid Hamiltonian approach to the Dirac equation: variational solution without basis set

4990 Leal Ferreira, P. (1988). Phys. Rev. D, 38, 2548-2650.
Two-body Dirac equation with a scalar linear potential

4991 Leal Ferreira, P. and Galeao, A.P. (1990). Mod. Phys. Lett. A, 5, 2523-2530.
Solutions for the two-body Dirac equation with phenomenological interactions

4992 Leal Ferreira, P. and Tomio, L. (1992). Mod. Phys. Lett. A, 7, 2781-2788.
Potential bags

4993 Lee, C.-Y. (1987). Phys. Rev. B, 36, 2720-2729.
Relativistic corrections to the intrinsic concentration, the effective density of states, and the conductivity in s-dimensional semiconductors

4994 Lee, S. (1991). J. Am. Chem. Soc., 113, 8611-8614.
Elemental structures of the heavy main group atoms and the second moment scaling hypothesis

4995 Lee, S., Bylander, D.M. and Kleinman, L. (1988). Phys. Rev. B, 37, 10035-10038.
Dissociation energy of Mo_2 and Mo_2^+

4996 Lee, S., Bylander, D.M. and Kleinman, L. (1989). Phys. Rev. B, 39, 4916-4920.
Pd_2: a dimer with two Kohn-Sham triplet ground states

4997 Lee, S.Y. and Lee, Y.S. (1991). Chem. Phys. Lett., 187, 302-308.
Second-order M ller-Plesset perturbation theory calculations with relativistic effective core potentials including spin-orbit operators

4998 Lee, S.Y. and Lee, Y.S. (1992). J. Comp. Chem., 13, 595-601.
Kramers restricted Hartree-Fock method for polyatomic molecules using ab initio relativistic effective core potentials with spin-orbit operators

4999 Lee, Y.S. (1986). Bull. Korean Chem. Soc., 7, 480-481.
An efficient computational approximation to the relativistic self-consistent-field method

5000 Lee, Y.S. and Baeck, K.K. (1986). Bull. Korean Chem. Soc., 7, 428-433.
Basis set requirement for small components besides kinetic balance in relativistic self-consistent-field calculations of many electron systems

5001 Lee, Y.S., Baeck, K.K. and McLean, A.D. (1989). J. Comp. Chem., 10(7), 112-117.
Basis set selections for relativistic self-consistent field calculations

5002 Lee, Y.S. and McLean, A.D. (1987). Bull. Korean Chem. Soc., 8, 122-126.
Relativistic effects on orbital energies in AgH and AuH; A clue to the origin of relativistic correlation effects

5003 Lefebvre, I., Lannoo, M., Allan, G., Ibanez, A., Fourcade, J., Jumas, J.C. and Beaurepaire, E. (1987). Phys. Rev.

Lett., 59, 2471-2474.
Electronic properties of antimony chalcogenides

5004 Lefebvre-Brion, H., Dehmer, P.M. and Chupka, W.A. (1988). J. Chem. Phys., 88, 811-817.
Spin-orbit and electronic autoionization in hydrogen chloride

5005*Lefebvre-Brion, H., Giusti-Suzor, A. and Raseev, G. (1985). J. Chem. Phys., 83, 1557-1566.
Theoretical study of the spin-orbit autoionization in molecules application to the hydrogen iodide photoionization spectrum

5006 Legrand, I.C., Dörner, R., Schmidt-Böcking, H. and Zoran, V. (1991).
Influence of rotating electron wave functions on L-shell ionization and alignment
In "High-Energy Ion-Atom Collisions", Ed. D. Berenyi and G.Hock, Springer, Berlin, pp. 330-337.

5007 Legrand, I.C., Zoran, V., Dörner, R., Schmidt-Böcking, H., Berinde, A., Fluerasu, D. and Ciortea, C. (1992). J. Phys. B: At. Mol. Opt. Phys., 25, 189-197.
Coupled channel treatment of the L-shell ionization in ion-atom collisions

5008 Leiberich, R., Schmidt, P.C., Sahoo, N. and Das, T.P. (1990). Z. Naturforsch., 45a, 368-374.
Augmented plane wave study of electric field gradients in non-cubic metals

5009 Leininger, T., Jeung, G.-H., Rohmer, M.-M. and Pélissier, M. (1992). Chem. Phys. Lett., 190, 342-348.
Comparison of the four widely used HF pseudopotentials: the group 1, 2 and 8 atoms

5010 Lelj, F. and Rosa, A. (1990). Chem. Phys. Lett., 168, 417-422.
Relativistic effects on bonding in CuIn, AgIn, and AuIn

5011 Lespes, G., Fernandez, J. and Dargelos, A. (1987). Chem. Phys., 115, 453-459.
VUV and photoelectronic spectra of methylstannane. Theoretical and experimental approach

5012 Leung, P.T. (1989). Phys. Rev. A, 40, 5417-5419.
Bethe stopping-power theory for heavy-target atoms

5013 Leung, P.T., Rustgi, M.L. and Long, S.A.T. (1986). Phys. Rev. A, 33, 2827-2829.
Relativistic corrections to the Bethe rule

5014 Leung, Y.C. and Pei, S.-Y. (1989). Phys. Rev. A, 40, 2731-2737.
High-density expansions of the relativistic Thomas-Fermi equation

5015 Leutwyler, H. and Smilga, A. (1992). Phys. Rev. D, 46, 5607-5632.

Spectrum of Dirac operator and role of winding number in QCD

5016 Lev, F.M. (1988). Yad. Fiz., 47, 1571-1576; Sov. J. Nucl. Phys., 47, 995-998.
Relativistic correction to the binding energy of four or more nucleons

5017 Leväsalmi, M., Aksela, H. and Aksela, S. (1992). Phys. Scr., T41, 119-121.
Satellite structure in the KLL spectrum of neon

5018 Levin, J.C., Soresen, S.L., Crasemann, B., Chen, M.H. and Brown, G.S. (1986). Phys. Rev. A, 33, 968-976.
Krypton L-MM Auger spectra: New measurements and analysis

5019 Levitan, B.M. and Sargsyan, I.S. (1988).
Operatory Shturma-Liuvillya i Diraka
Nauka, Moscow

5020 Levitan, B.M. and Sargsyan, I.S. (1991).
Sturm-Liouville and Dirac Operators
Kluwer, Dordrecht.

5021*Lévy-Leblond, J.-M. (1974). Riv. Nuovo Cim., 4, 99-143.
The pedagogical role and epistemological significance of group theory in quantum mechanics

5022 Ley-Koo, E. and Wang, R.C. (1988). Rev. Mexicana Fis., 34, 296-313.
Dirac equation in orthogonal curvilinear coordinates

5023 Li, H.-S., Li, Y.-P. and Coey, J.M.D. (1991). J. Phys.: Cond. Matter, 3, 7277-7290.
R-T and R-R exchange interactions in the rare-earth (R)-transition-metal(T) intermetallics: an evaluation from relativistic atomic calculations

5024 Li, J. and Pyykkö, P. (1992). Chem. Phys. Lett., 197, 586-590.
Relativistic pseudo-potential analysis of the weak Au(I) ...Au(I) attraction

5025 Li, J.-Q. and Balasubramanian, K. (1989). J. Mol. Spectrosc., 138, 162-180.
Spectroscopic properties and potential energy curves of I_2 and I_2^+

5026 Li, J.-Q. and Balasubramanian, K. (1990). J. Phys. Chem., 94, 545-552.
Potential energy surfaces for the Mo + H_2 reaction: Collinear versus perpendicular collisions

5027 Li, P. and Gou, Q.-Q. (1991). Chi. J. At. Mol. Phys., 8, 2075-2094.
Relativistic effects in the elastic scattering of slow electrons by mercury

5028 Li, X.-Z. and Zhang, J.-Z. (1992). Phys. Rev. A, 46, 2250-2260.

Spinor structures of the Bethe-Salpeter irreducible kernels

5029 Li, Y., Krieger, J.B., Norman, M.R. and Iafrate, G.J. (1991). Phys. Rev. B, 44, 10437-10443.
Band-structure calculations of noble-gas and alkali halide solids using accurate Kohn-Sham potentials with self-interaction correction

5030 Liao, D.-W. and Balasubramanian, K. (1992a). J. Chem. Phys., 96, 8938-8947.
Electronic structure of the III-V tetramer clusters and their positive ions

5031 Liao, D.-W. and Balasubramanian, K. (1992b). J. Chem. Phys., 97, 2548-2552.
Electronic structure of Cu_6, Ag_6, Au_6, and their positive ions

5032 Liaw, S.-S. (1992). Can. J. Phys., 70, 644-651.
Transition amplitudes in the Dirac-Fock approximation

5033 Liaw, S.S., Feldman, G. and Fulton, T. (1988a). Phys. Rev. A, 38, 1126-1136.
Numerical study of the role of shielding and exchange effects and of correlation and relativistic corrections in the Dirac-Fock approximation

5034 Liaw, S.S., Feldman, G. and Fulton, T. (1988b). Phys. Rev. A, 38, 5985-5994.
Energies and transition amplitudes for atoms with nondegenerate cores, plus or minus two electrons

5035 Liberman, D. and Zangwill, A. (1989). Phys. Rev. A, 39, 415-416.
Quadrupole resonances in the rare-earth metals

5036*Liberman, D.A. (1968). Phys. Rev., 171, 1-3.
Exchange potential for electrons in atoms and solids

5037 Lieb, E.H. and Yau, H.-T. (1988). Phys. Rev. Lett., 61, 1695-1697.
Many-body stability implies a bound on the fine-structure constant

5038 Lieber, M. (1989).
The Coulomb Green's function
In "Relativistic, Quantum Electrodynamic and Weak Interaction Effects in Atoms", Ed. W. Johnson, P. Mohr and J. Sucher, AIP Conf. Proc., 189, 445-459.

5039 Lin, C.D. (1988).
Theoretical studies of heavy particle collisions
In "Fundamental Processes of Atomic Dynamics", Ed. J.S. Briggs, H. Kleinpoppen and H.O. Lutz, NATO ASI Ser. B, 181, Plenum, New York.

5040*Lin, D.L. and Feinberg, G. (1974). Phys. Rev. A, 10, 1425-1429.
Radiative correction to relativistic magnetic-dipole decays of hydrogenlike ions

5041 Lin, J.-S. and Ortiz, J.V. (1990). Chem. Phys. Lett., 171, 197-200.
Electron propagator calculations on the ionization energies of CrH^-, MnH^- and FeH^-

5042 Lin, Z.-Y. and Hall, M.B. (1992a). J. Am. Chem. Soc., 114, 2928-2932.
Transition metal polyhydride complexes. 2. Theoretical methods for the determination of stabilities of classical and nonclassical isomers

5043 Lin, Z.-Y. and Hall, M.B. (1992b). J. Am. Chem. Soc., 114, 6102-6108.
Transition-metal polyhydride complexes. 3. Relative stabilities of classical and nonclassical isomers

5044 Lin, Z.-Y. and Hall, M.B. (1992c). J. Am. Chem. soc., 114, 6574-6575.
Electron-density analysis of the transition states of substitution reactions of 17- and 18-electron hexacarbonyl complexes

5045 Lin, Z.-Y. and Hall, M.B. (1992d). Inorg. Chem., 31, 2791-2797.
Theoretical studies of inorganic and organometallic reaction mechanisms. 5. Substitution reactions of 17- and 18-electron transition-metal hexacarbonyl complexes

5046 Lin, Z.-Y. and Hall, M.B. (1992e). Inorg. Chem., 31, 4262-4265.
Transition metal polyhydride complexes. 4. Highly stable nonclassical isomers with octahedral structures

5047 Lin, Z.-Y. and Hall, M.B. (1992f). J. Am. Chem. Soc., 114, 10054-10055.
Structure and bonding in M_8C_{12} clusters

5048 Lindgren, B. (1989). Hyperfine Interactions, 49, 357-366.
Calculations of hyperfine fields

5049 Lindgren, I.(Ed.) (1987a).
Relativistic Many-Body Problems
Phys. Scr., RS 7.

5050 Lindgren, I. (1987b). Phys. Scr., 36, 591-601.
The coupled-cluster approach to non-relativistic and relativistic many-body calculations

5051 Lindgren, I. (1988). Nucl. Instrum. Meth. B, 31, 102-114.
Towards a relativistic many-body procedure

5052 Lindgren, I. (1989a).
Relativistic and non-relativistic many-body procedure, applied to atomic systems
In "Aspects of Many-Body Effects in Molecules and Extended Systems", Ed. Mukherjee, Lecture Notes in Chem., 50, 367-392, Springer-Verlag.

5053 Lindgren, I. (1989b).
A relativistic coupled-cluster approach with radiative corrections
In "Many-Body Methods in Quantum Chemistry", Ed. U. Kaldor, Lecture Notes in Chem., 52, Springer, Berlin, pp. 293-306.

5054 Lindgren, I. (1989c).
Effective potentials in relativistic many-body theory
In "Relativistic, Quantum Electrodynamic and Weak Interaction Effects in Atoms", Ed. W. Johnson, P. Mohr, and J. Sucher, AIP Conf. Proc., 189, pp. 371-392.

5055 Lindgren, I. (1990). J. Phys. B: At. Mol. Opt. Phys., 23, 1085-1093.
Gauge dependence of interelectronic potentials

5056 Lindgren, I. (1991). Phys. Scr., T34, 36-46.
Relativistic many-body calculations

5057 Lindgren, I. and Lindroth, E. (1987).
Non-relativistic many-body problem perturbation theory calculation: Toward relativistic many-body problem perturbation theory
In "Atoms in Unusual Situations", Ed. J.P. Briand, New York, Plenum, p. 310.

5058 Lindhard, J. (1991). Phys. Rev. A, 43, 6032-6037.
Quantum-radiation spectra of relativistic particles derived by the correspondence principle

5059 Lindroth, E. (1987a). Phys. Scr., 36, 485-492.
The relativistic pair equation and related fundamental problems

5060 Lindroth, E. (1987b). Nucl. Instrum. Meth. B, 27, 543-554.
Correlation in a relativistic framework

5061 Lindroth, E. (1987c).
Pair correlation in atomic systems - towards a relativistic treatment
Ph.D. Thesis, University of Gothenburg.

5062 Lindroth, E. (1988). Phys. Rev. A, 37, 316-328.
Numerical solution of the relativistic pair equation

5063 Lindroth, E., Heully, J.-L., Lindgren, I. and Mårtensson-Pendrill, A.-M. (1987). J. Phys. B: At. Mol. Phys., 20, 1679-1696.
A relativistic pair equation projected onto positive energy states

5064 Lindroth, E. and Hvarfner, J. (1992). Phys. Rev. A, 45, 2771-2776.
Relativistic calculation of the 2 $^1S_0$ - 2 $^{1,3}P_1$ transitions in berylliumlike molybdenum and berylliumlike iron

5065 Lindroth, E., Lynn, B.W. and Sandars, P.G.H. (1989a). J. Phys. B: At. Mol. Opt. Phys., 22, 559-576.

Order α^2 theory of the atomic electric dipole moment due to an electric dipole moment on the electron

5066 Lindroth, E. and Mårtensson-Pendrill, A.-M. (1989). Phys. Rev. A, 39, 3794-3802.
Further analysis of the complete Breit interaction

5067 Lindroth, E., Mårtensson-Pendrill, A.-M., Ynnerman, A. and Öster, P. (1989b). J. Phys. B: At. Mol. Opt. Phys., 22, 2447-2464.
Self-consistent treatment of the Breit interaction, with application to the electric dipole moment in thallium

5068 Lindroth, E., Persson, H., Salomonson, S. and Mårtensson-Pendrill, A.-M. (1992). Phys. Rev. A, 45, 1493-1496.
Corrections to the beryllium ground-state energy

5069 Lindroth, E. and Salomonson, S. (1990). Phys. Rev. A, 41, 4659-4669.
Relativistic calculation of the 2 $^3S_1$ - 1 $^1S_0$ magnetic dipole transition energy for heliumlike argon

5070*Lipas, P.O. (1970). Am. J. Phys., 38, 85-87.
Relativistic quantal oscillator

5071 Lipkin, H.J. (1987). Phys. Rev. Lett., 58, 425.
Anomalous positron peaks and the Dirac equation

5072 Lipnitskii, A.G., Nikiforov, A.V., Silkin, V.M. and Chulkov, E.V. (1991). Izv. VUZov, Fiz. (10), 47-53.
Relativistic electronic structure of the Pb(001) surface (in Russian)

5073 Lisini, A. and Decleva, P. (1987a). J. Phys. B: At. Mol. Phys., 20, 457-464.
A CI study of the correlation effects in the $(np)^{-1}$ and $(ns)^{-1}$ photoelectron spectra of the alkali atoms

5074 Lisini, A. and Decleva, P. (1987b). J. Phys. B: At. Mol. Phys., 20, L117-L119.
On the satellite structure in the $(5s)^{-1}$ photoelectron spectrum of caesium

5075 Lister, G.M.S. and Judd, B.R. (1992).
Unexpected relations between matrix elements of the spin-other-orbit operators in the atomic f shell
In "Group Theory and Special Symmetries in Nuclear Physics", Ed. J.P. Draayer, J. Janecke, World Scientific, Singapore, pp.258-262.

5076 Littlejohn, R.G. and Flynn, W.G. (1992). Phys. Rev. A, 45, 7697-7717.
Semiclassical theory of spin-orbit coupling

5077 Litzén, U. and Zeng, X.-T. (1991). Phys. Scr., 43, 262-265.
The $4s^2 4p$-$4s4p^2$ transition array and energy levels of the Ga-like ions Ru XIV-In XIX

5078 Liu, A.Y., García, A., Cohen, M.L., Godwal, B.K. and

Jeanloz, R. (1991). Phys. Rev. B, 43, 1795-1798.
Theory of high-pressure phases of Pb

5079 Liu, A.Y., Wentzcovitch, R.M. and Cohen, M.L. (1988).
Phys. Rev. B, 38, 9483-9489.
Structural and electronic properties of WC

5080 Liu, L. and Li, J.-M. (1988). Acta Phys. Sin., 37,
2053-2056.
Structure of excited francium atom (in Chinese)

5081 Liu, L. and Li. J.-M. (1991). Acta Phys. Sin., 40,
1922-1928.
Amplitudes of electronic wavefunctions at nuclei for atoms:
Pt. 1, High energy photoionization processes of atomic ions

5082 Liu, L., Tong, X.-M. and Li, J.-M. (1988). Acta Phys. Sin.,
37, 1800-1806.
Photoionization of ionized iron atoms (in Chinese)

5083 Liu, Z.W. and Kelly, H.P. (1991). Phys. Rev. A, 43,
3305-3316.
Atomic many-body perturbation method based on
multiconfiguration Dirac-Fock wave functions

5084 Liu, Z.W. and Kelly, H.P. (1992). Phys. Rev. A, 45,
R4210-R4213.
Analysis of atomic electric dipole moment in thallium
by all-order calculations in many-body perturbation
theory

5085 Loder Jr., D.R. (1986). Diss. Abstr. Int. B, 47, 1052.
High-temperature photoelectron spectroscopy of
thallium(I) salts

5086 Loder Jr., D.R., Sherrod, R.E., Yih, P.C., Bloor, J.E.
and Schweitzer, G.K. (1988). Inorg. Chim. Acta, 142,
247-253.
Relativistic MS-Xα calculations for some thallium(I)
compounds

5087 Loewe, M. and Sanhueza, M. (1990). J. Phys. A, 23,
553-561.
Supercritical effects and the δ potential

5088 Loginov, A.V. (1992). Opt. Spectrosc., 72, 161-164;
Opt. Spektrosk., 72, 296-301.
Magnetic effective interaction and radiation constants
in spectra of Tl IV - Bi VI

5089 Lohmann, B. (1992). J. Phys. B: At. Mol. Opt. Phys.,
25, 4163-4167.
Angular distribution, spin polarization and relative
intensities of Auger electrons from mercury atoms

5090 Lohr, L.L. (1987). Inorg. Chem., 26, 2005-2009.
Relativistically parameterized extended Hückel
calculations. 11. Energy bands for elemental tellurium
and polonium

5091 Lohr, L.L. (1991). Int. J. Quantum Chem., S25, 121-130. Relativistically parameterized extended Hückel calculations. 13. Energy bands for uranium compounds UB_2, UB_4, UC, UBC, and UPt_3

5092 Lohr, L.L. and Jia, Y.Q. (1986). Inorg. Chim. Acta, 119, 99-105. Relativistically parameterized extended Hückel calculations. 10. Lanthanide trihalides

5093*Lohr Jr., L.L. (1966). J. Chem. Phys., 45, 1362-1363. Spin-forbidden electric-dipole transition moments

5094*Lohr Jr., L.L. (1972). Coord. Chem. Rev., 8, 241-259. Spin-forbidden electronic excitations in transition metal complexes

5095 Lombardi, J.C. (1987). Phys. Rev. A, 36, 1445-1448. Determination of core polarizabilities, quantum defects, and fine structure in sodium

5096*Loucks, T.L. (1966). Phys. Rev., 143, 506-512. Relativistic electronic structure in crystals. II. Fermi surface of tungsten

5097 Louwen, J.N., Hengelmolen, R., Grove, D.M., Stufkens, D.J. and Oskam, A. (1986). J. Chem. Soc. Dalton Trans., 141-149. He I and He II photoelectron spectra of trans-[MCl_2L_2] (M = Pd, L = PEt_3 or $AsEt_3$; M = Pt, L = NMe_3, PMe_3, PEt_3, or $AsEt_3$) and trans-[PtH(Cl)$(PEt_3)_2$], and Hartree-Fock-Slater calculations on model compounds

5098 Lovas, I. and Molnár, L. (1989). Acta Phys. Hung., 65, 331-334. Dirac particle in a periodic field

5099*Low, J.J. and Goddard III, W.A. (1984b). J. Am. Chem. Soc., 106, 8321-8322. Reductive coupling of H-H, H-C and C-C bonds from Pd complexes

5100 Low, J.J. and Goddard III, W.A. (1986a). Organometallics, 5, 609-622. Theoretical studies of oxidative addition and reductive elimination. 2. Reductive coupling of H-H, H-C, and C-C bonds from palladium and platinum complexes

5101 Low, J.J. and Goddard III, W.A. (1986b). J. Am. Chem. Soc., 108, 6115-6128. Theoretical studies of oxidative addition and reductive elimination. 3. C-H and C-C reductive coupling from palladium and platinum bis(phosphine) complexes

5102 Lozada, A. (1992). J. Phys. A, 25, 4043-4055. Multipole in an external electromagnetic field

5103 Lu, J.-F., Yan, Z.-Q. and Shi, Z.-J. (1991). Chi. J. Lasers, 18, 275-281. Scattering effect of relativistic electrons and high power laser light (in Chinese)

5104 Lu, Z.W., Singh, D.J. and Krakauer, H. (1989a). Phys. Rev. B, 39, 4921-4925.
Ground-state properties of fcc and bcc lanthanum

5105 Lu, Z.W., Singh, D.S. and Krakauer, H. (1989b). Phys. Rev. B, 39, 10154-10161.
Total-energy study of the equation of state of HgTe and HgSe

5106 Lu, Z.W., Wei, S.-H. and Zunger, A. (1991). Phys. Rev. B, 44, 10470-10484.
Electronic structure of random $Ag_{0.5}Pd_{0.5}$ and $Ag_{0.5}Au_{0.5}$ alloys

5107 Lucha, W., Rupprecht, H. and Schöberl, F.F. (1991). Phys. Rev. D, 44, 242-249.
Relativistic treatment of fermion-antifermion bound states

5108 Lucha, W., Rupprecht, H. and Schöberl, F.F. (1992). Phys. Rev. D, 46, 1088-1095.
Significance of relativistic wave equations for bound states

5109 Lucha, W. and Schöberl, F.F. (1990a). Phys. Rev. Lett., 64, 2733-2735.
Relativistic virial theorem

5110 Lucha, W. and Schöberl, F.F. (1990b). Mod. Phys. Lett. A, 5, 2473-2483.
Relativistic virial theorems

5111 Lucha, W. and Schöberl, F.F. (1992). Int. J. Mod. Phys. A, 7, 6431-6456.
Quark-antiquark bound states: relativistic versus nonrelativistic point of view

5112 Lundell, J. and Kunttu, H. (1992). J. Phys. Chem., 96, 9774-9781.
Structure, spectra, and stability of Ar_2H^+, Kr_2H^+, and Xe_2H^+: An effective core potential approach

5113 Luo, D., Pradhan, A.K. and Shull, J.M. (1988). Astrophys. J. 335, 498-502.
Oscillator strengths for Si II

5114 Luo, Z.-F. and Qiu, X.-J. (1992). J. Phys. G, 18, 221-224.
General virial theorem

5115 Lyudchik, A.M., Borkovskii, N.B. and Kovrikov, A.B. (1986). Koord. Khim., 12, 1038-1043.
Semiempirical calculation method for relativistic effects in coordination compounds

5116 Ma, Z.-Q. (1986). Phys. Rev. D, 33, 1745-1754.
Levinson's theorem for potentials with Coulomb tails

5117 Ma, Z.-Q. and Dai, A.-Y. (1988). J. Math. Phys., 29, 1321-1324.
Integral equations with symmetrical kernel applied to a system with a Dirac-type spectrum

5118 Ma, Z.-X. and Balasubramanian, K. (1991). Chem. Phys. Lett., 181, 467-473.
Electronic states of WH

5119 Ma, Z.-X. and Dai, S.-S. (1986). Huaxue Tongbao, 11, 35-36.
Relativistic quantum chemistry study on hydrogen halides and halogen molecules (in Chinese)

5120 Ma, Z.-X. and Dai, S.-S. (1988a). Acta Phys.-Chim. Sin., 4, 347-348.
Ab initio calculation of Mo_2

5121 Ma, Z.-X. and Dai, S.-S. (1988b). Chi. Sci. Bull., 33, 1316-1318; Chi. Sci. Bull.(Engl.), 34, 468-470.
Ab initio study in relativistic effective potential -The electronic structure of KX^- (X=Cl, Br, I)

5122 Ma, Z.-X. and Dai, S.-S. (1989). Acta Phys.-Chim. Sin., 5, 551-557.
Ab initio study of compounds containing mercury using relativistic pseudopotentials

5123 Ma, Z.-X. and Dai, S.-S. (1990a). Chem. Phys. Lett., 166, 404-407.
Electronic structure of BaCl

5124 Ma, Z.-X. and Dai, S.-S. (1990b). Chi. Sci. Bull., 35, 1186-1188.
The relativistic pseudopotential MCSCF calculations for some low-lying states of $TiCl_2$

5125 Ma, Z.-X., Li, X.-P. and Dai, S.-S. (1990). Chem. J. Chi. Univ., 11, 49-53.
Relativistic pseudopotential ab initio study of boron halides

5126 Macêdo, A.M.S. and Chalker, J.T. (1992). Phys. Rev. B, 46, 14985-14994.
Effects of spin-orbit interactions in disordered conductors: a random-matrix approach

5127 Maciá, E. and Domínguez-Adame, F. (1991). J. Phys. A, 24, 59-69.
Scattering states of relativistic point interaction potentials

5128 Mackay, K.M. and Mackay, R.A. (1989).
Introduction to Modern Inorganic Chemistry (4th edition) Blackie, Glasgow and London.

5129 MacLaren, J.M., Clougherty, D.P. and Albers, R.C. (1990). Phys. Rev. B, 42, 3205-3208.
Local-spin-density calculations for iron: Effect of spin interpolation on ground-state properties

5130 Maier, T.J. and Dreizler, R.M. (1992). Phys. Rev. A, 45, 2974-2977.
Poles of the S matrix for a Dirac square-well potential

5131 Mains, G.J. and White, J.M. (1991). J. Phys. Chem., 95, 112-118.
All-electron and relativistic effective core potential study of rhodium compounds

5132 Makhyoun, M.A. (1987). Inorg. Chem., 26, 3592-3595.
Electronic structure and ligand field states of $NpO_2Cl_4^{2-}$ and $NpO_2(NO_3)_3^-$: A relativistic MS-Xα study

5133 Makhyoun, M.A. (1988a). Polyhedron, 7, 2675-2678.
On the electronic structure of $[W(WS_4)_2]^{2-}$

5134 Makhyoun, M.A. (1988b). J. Chim. Phys., 85, 917-924.
On the electronic structure of $5g^1$ complexes of element 125: A quasi-relativistic MS-Xα study

5135 Makhyoun, M.A. and El-Ezaby, M.S. (1988). Inorg. Chim. Acta, 141, 295-300.
The MS-Xα treatment of the energy transitions in uranium(V) complexes

5136 Makhyoun, M.A., El-Issa, B.D. and Salsa, B.A. (1987). J. Mol. Struct.(Theochem), 153, 241-248.
A molecular orbital study of the bonding between Th and Ni in a novel compound

5137 Makhyoun, M.A. and Hasanein, A.A. (1988). Gazz. Chim. Ital., 118, 145-148.
Relativistic SCF-MS-Xα calculations of the electronic structure of $UO_2Cl_4^{2-}$

5138 Makhyoun, M.A. and Sawan, Y. (1989). J. Mol. Struct. (Theochem), 183, 191-196.
Molecular orbital calculations of the cluster $Cu_3AsSe_6^{6-}$

5139 Malenfant, J. (1988). Phys. Rev. D, 38, 3295-3308.
Reduction of the Breit Coulomb equation to an equivalent Schrödinger equation, and investigation of the behavior of the wave function near the origin

5140 Malenfant, J. (1989). Phys. Rev. D, 39, 1787-1790.
Normalization of bound-state solutions to the Breit equation

5141 Malenfant, J. (1991). Phys. Rev. A, 43, 1233-1247.
Some approximate positronium solutions to the Breit-Coulomb equation

5142 Malfliet, R. (1989).
The relativistic nuclear many-body problem
In "Heavy Ions in Nuclear and Atomic Physics. Proceedings of the 20th Mikolajki Summer School on Nuclear Physics Mikolajki, Poland 1-10 Sept. 1988", Ed. Z. Wilhelmi and G. Szeflinska, Adam Hilger, Bristol, 377-390.

5143 Malik, G.P., Singh, S. and Varma, V.S. (1992). Pramana, 38, 319-327.
Relativistic remnants in the reduction of the Bethe-Salpeter equation to the Schrödinger equation

5144 Malinovskaya, S.V. (1988). Opt. Spectrosc., 64, 263-265.
Calculation and extrapolation of energy levels of Cl, Ar, and K-like ions by the model-potential method

5145 Malinovskaya, S.V. and Glushkov, A.V. (1992). Izv. VUZov, Fiz., 35(11), 3-9.
Theoretical determination of spectra of K-like multicharged ions

5146*Malli, G. and Saxena, K.M.S. (1971).
Intra-atomic magnetic interactions in many-electron systems: Spin-spin interaction in f^4 electron configurations
In "Theory of Electronic Structure of Atoms and Molecules", Ed. J. Batarunas et al, Mintis, Vilnius, pp. 65-72.

5147 Malli, G.L. (1988).
Ab-initio relativistic quantum chemistry
In "Molecules in Physics, Chemistry and Biology", Ed. J. Maruani, Vol.2, Kluwer, Dordrecht, pp. 85-144.

5148 Malli, G.L. (1989).
Ab initio relativistic quantum chemistry of third-row transition elements and actinides
In "The Challenge of d and f Electrons", Ed. D.R. Salahub and M.C. Zerner, ACS Symposium Series, 394, 291-308.

5149 Malli, G.L. (1992). Can. J. Chem., 70, 421-426.
Ab initio relativistic molecular orbital self-consistent field calculations for the thorium-platinum diatomic

5150 Malli, G.L., Da Silva, A.B.F. and Ishikawa, Y. (1992). Can. J. Chem., 70, 1822-1826.
Universal Gaussian basis functions in relativistic quantum chemistry: atomic Dirac-Fock-Coulomb and Dirac-Fock-Breit calculations

5151 Malli, G.L. and Pyper, N.C. (1986). Proc. R. Soc. Lond. A, 407, 377-404.
Ab initio fully relativistic molecular calculations: Bonding in gold hydride

5152 Malli, G.L., Pyper, N.C. and Ramos, A.F. (1991). Phys. Rev. A, 43, 2583-2584.
Reply to "Comment on 'relativistic effects in bonding and dipole moments for the diatomic hydrides of sixth-row elements'"

5153 Malvetti, M. (1987). Phys. Rev. A, 35, 1490-1492.
Recoil corrections to the hyperfine structure of the Dirac equation

5154 Malvetti, M. and Pilkuhn, H. (1987). Z. Phys. D., 6, 119-124.
Spinless particles in atoms

5155 Malvetti, M. and Pilkuhn, H. (1989). Phys. Lett. A, 136, 418-422.
From Dirac-Coulomb to Kramers-Breit

5156 Malvetti, M. and Pilkuhn, H. (1990). J. Phys. B: At. Mol. Opt. Phys., 23, 3469-3474.
The Breit-Brown operator of the atomic Kramers equation

5157 Malzacher, P. and Dreizler, R.M. (1986). Z. Phys. D, 2, 37-42.
Remarks on the nonintegrable density of the relativistic Thomas-Fermi and Thomas-Fermi-Weizsäcker models

5158 Mamaev, V.M., Topaler, M.S. and Ustynyuk, Yu.A. (1992). Metalloorg. Khim., 5, 35-52.
The dynamical quantum chemical models for the activation of the C-H bonds in metallocomplex catalysis (in Russian)

5159 Manakov, N.L., Nekipelov, A.A. and Fainshtein, A.G. (1987). Yad. Fiz., 45, 1091-1098.
Sturm expansion of the Coulomb Green function in QED: photon scattering by a bound electron

5160 Manakov, N.L., Nekipelov, A.A. and Fainshtein, A.G. (1989). Zh. Eksp. Teor. Fiz., 95, 1167-1177; Sov. Phys. JETP, 68, 673-679.
Vacuum polarization by a strong Coulomb field and its contribution to the spectra of multiply-charged ions

5161 Mandelzweig, V.B. and Wallace, S.J. (1987). Phys. Lett. B, 197, 469-473.
QED based two-body Dirac equation

5162*Mann, D.B. (1976).
Electronic structure of superheavy elements (in Russian) In "Predictions in the Study of Periodicity", Ed. B.M. Kedrov and D.N. Trifonov, Nauka, Moscow, pp. 161-201.

5163 Mannervik, S. and Cederquist, H. (1986). J. Phys. B, 19, L845-L847.
Comment on 'Mechanisms of configuration interaction in relativistic autoionisation: threshold phenomena and the Li ^4P_J levels'

5164 Mansfield, M. and Needs, R.J. (1991). Phys. Rev. B, 43, 8829-8833.
Surface energy and stress of lead (111) and (110) surfaces

5165 Mansour, N.B., Dinneen, T., Young, L. and Cheng, K.T. (1989). Phys. Rev. A, 39, 5762-5767.
Laser-rf double-resonance measurements of the hyperfine structure in Sc II

5166 Mäntykenttä, A., Aksela, H., Aksela, S., Yagishita, A. and Shigemasa, E. (1992). J. Phys. B: At. Mol. Opt. Phys., 25, 5315-5326.
Configuration interaction and shakeup satellites in 4d, 5s and 5p photoelectron spectra of Cs and Ba

5167 March, N.H. (1986). Phys. Rev. A, 34, 5104-5105.
Relativistic correction to equilibrium bond lengths for heavy, almost spherical molecules

5168 March, N.H. (1988). Phys. Rev. A, 37, 1869-1871.
Direct generation of approximate relativistic electron densities from the nonrelativistic limit

5169 March, N.H. (1989). J. Mol. Struct.(Theochem), 199, 75-83.
Relativistic density functional theory

5170 Marcos, S., Van Giai, N. and Savushkin, L.N. (1992).
Nucl. Phys. A, 549, 143-154.
Coulomb displacement energies in relativistic and non-relativistic self-consistent models

5171 Marian, C.M. (1990a). J. Chem. Phys., 93, 1176-1186.
Quasirelativistic calculation of the vibronic spectra of NiH and NiD

5172 Marian, C.M. (1990b). Chem. Phys. Lett., 173, 175-180.
On the dependence of correlation and relativity: The electron affinity of the copper atom

5173 Marian, C.M. (1991a).
Theoretische Spektroskopie zweiatomiger Übergangsmetallverbindungen
Habilitationsschrift, Bonn, 163p.

5174 Marian, C.M. (1991b). J. Chem. Phys., 94, 5574-5585.
Theoretical study of the spectra of CuH and CuD

5175 Marian, C.M., Blomberg, M.R.A. and Siegbahn, P.E.M. (1989).
J. Chem. Phys., 91, 3589-3595.
Multireference and relativistic effects in NiH

5176 Marian, C.M., Hippe, D., Hess, B.A. and Peyerimhoff, S.D. (1992). Theor. Chim. Acta, 81, 375-390.
Relativistic treatment of excited electronic states of atomic copper

5177 Marian, C.M., Wahlgren, U., Gropen, O. and Pyykkö, P. (1988). J. Mol. Struct.(Theochem), 169, 339-354.
Bonding and electronic structure in diatomic ThO:
Quasirelativistic effective core potential calculations

5178 Marino, M.M. (1991). Diss. Abstr. Int. B, 51, 4363-4364.
Ab initio studies of heterogeneous atomic clusters

5179 Marino, M.M. and Ermler, W.C. (1991). Chem. Phys. Lett., 176, 36-40.
Ab initio study of the effects of cesium, hydrogen and oxygen adsorption on the work function of beryllium

5180 Marino, M.M., Ermler, W.C., Kern, C.W. and Bondybey, V.E. (1992). J. Chem. Phys., 96, 3756-3766.
Spin-orbit configuration-interaction study of valence and Rydberg states of LiBe

5181 Marino, M.M., Ermler, W.C., Tompa, G.S. and Seidl, M.

(1989a). Surface Sci., 208, 189-204.
Effects of hydrogen and of cesium adsorption on a beryllium surface: A theoretical and experimental study

5182 Marino, M.M., Sawamura, M., Ermler, W.C. and Sandroff, C.J. (1989b). Chem. Phys. Lett., 163, 202-206.
Ab initio study of the layered semiconductor cluster Bi_6I_{18}

5183 Marino, M.M., Sawamura, M., Ermler, W.C. and Sandroff, C.J. (1990). Phys. Rev. B, 41, 1270-1273.
Ab initio study of the geometry and electronic structure of lead iodide semiconductor clusters

5184 Mark, F. (1986). Theor. Chim. Acta, 70, 165-188.
Relativistic Gaussian basis sets for the atoms hydrogen to neon

5185 Mark, F. and Becker, U. (1987). Phys. Scr., 36, 393-396.
The relativistic energy correction of the H_2^+ molecule

5186 Marketos, P. (1989). Z. Phys. D, 13, 83-88.
Atomic basis set calculations with ab-initio relativistic pseudopotentials

5187 Marketos, P. and Lambropoulos, P. (1990). Z. Phys. D, 15, 185-198.
A relativistic C.I. study on the J=1/2 even spectrum of Sr II

5188 Markowski, R. and Podgorny, M. (1992). J. Phys.: Cond. Matt., 4, 2505-2515.
Calculated optical properties of zincblende semiconductors ZnTe, CdTe and HgTe

5189 Márquez, A. and Fernández Sanz, J. (1992a). J. Am. Chem. Soc., 114, 2903-2909.
Electronic structure of the transition-metal-carbene-like complexes $(CO)_5Mo-M'H_2$ (M'=C, Si, Ge, and Sn). A theoretical study based on ab initio CASSCF calculations

5190 Márquez, A. and Fernández Sanz, J. (1992b). J. Am. Chem. Soc., 114, 10019-10024.
Ab initio CASSCF study of the electronic structure of the transition-metal alkylidene-like complexes $Mo-M'H_2$ (M' = C, Si, Ge, and Sn)

5191 Marshakov, A.V. and Fainberg, V.Ya. (1987). Pis'ma Zh. Eksp. Teor. Fiz., 46, 253-255; JEPT Lett., 46, 319-322.
Representation of a Dirac particle as a sum over paths

5192 Mårtensson-Pendrill, A.-M. (1987). Phys. Scr., 36, 122-128.
Comparisons between different calculation of PNC in Cs and Tl

5193 Mårtensson-Pendrill, A.-M. (1989a).
Numerical determination of non-relativistic and relativistic pair correlation

In "Numerical Determination of the Electronic Structure of Atoms, Diatomic and Polyatomic Molecules", Ed. M. Defranceschi and J. Delhalle, Kluwer, Dordrecht, 131-160.

5194 Mårtensson-Pendrill, A.-M. (1989b).
The no-pair equation - Fundamental problems, numerical solutions and applications
In "Relativistic, Quantum Electrodynamic and Weak Interaction Effects in Atoms", Ed. W. Johnson, P. Mohr and J. Sucher, AIP Conf. Proc., 189, 305-333.

5195 Mårtensson-Pendrill, A.-M. (1992a).
Atomic spectroscopy as a probe of parity non-conserving effects in the nucleus
In "Nuclear Shapes and Nuclear Structure at Low Excitation Energies", Ed. M. Vergnes et al., Plenum, New York, pp. 87-92.

5196 Mårtensson-Pendrill, A.-M. (1992b). Meth. Comp. Chem., 5, 99-156.
Calculation of P and T violating properties in atoms and molecules

5197 Mårtensson-Pendrill, A.-M., Alexander, S.A., Adamowicz, L., Oliphant, N., Olsen, J., Öster, P., Quiney, H.M., Salomonson, S. and Sundholm, D. (1991). Phys. Rev. A, 43, 3355-3364.
Beryllium atom reinvestigated: A comparison between theory and experiment

5198 Mårtensson-Pendrill, A.-M. and Lindroth, E. (1991). Europhys. Lett., 15, 155-160.
Limit on a P- and T-violating electron-nucleon interaction

5199 Mårtensson-Pendrill, A.-M. and Öster, P. (1987). Phys. Scr., 36, 444-452.
Calculations of atomic electric dipole moments

5200*Mårtensson-Pendrill, A.-M. and Salomonsson, S. (1984). Phys. Rev. A, 30, 712-721.
Hyperfine structure of the 4s, 4p, and 3d states in Ca^+ evaluated by many-body perturbation theory

5201 Mårtensson-Pendrill, A.-M. and Ynnerman, A. (1992). J. Phys. B: At. Mol. Opt. Phys., 25, L551-L559.
Isotope shift and nuclear charge radii of barium isotopes

5202 Mårtensson-Pendrill, A.-M., Ynnerman, A., Warston, H., Vermeeren, L., Silverans, R.E., Klein, A., Neugart, R., Schulz, C., Lievens, P. and The ISOLDE Collaboration. (1992). Phys. Rev. A, 45, 4675-4681.
Isotope shifts and nuclear-charge radii in singly ionized $^{40-48}Ca$

5203 Martin, I. (1990). Ana. Fis. A, 86, 108-110.
Quantum defect orbitals and their relativistic formulation (in Spanish)

5204 Martin, I. and Karwowski, J. (1991). J. Phys. B: At. Mol. Opt. Phys., 24, 1539-1542.

Quantum defect orbitals and the Dirac second-order equation

5205 Martin, I., Karwowski, J., Lavin, C. and Diercksen, G.H.F. (1991). Phys. Scr., 44, 567-573.
Quantum defect orbital study of the sodium isoelectronic sequence

5206 Martin, I., Lavin, C. and Barrientos, C. (1992). Int. J. Quantum Chem., S26, 465-474.
Fine-structure oscillator strengths for excited-state transitions in Cu-like ions

5207 Martin, R.L. (1987). J. Chem. Phys., 86, 5027-5031.
AgH, Ag_2, and AgO revisited: Basis set extensions

5208 Martin, W.C. (1987). Phys. Rev. A, 36, 3575-3589.
Improved ^{4}He I 1snl ionization energy, energy levels, and Lamb shifts for 1sns and 1snp terms

5209 Martinez-Magadan, J.M., Ramirez-Solis, A. and Novaro, O. (1991). Chem. Phys. Lett., 186, 107-112.
MCSCF + MRCI study of the interaction of Zn, Zn^+ and Zn^{2+} with the hydrogen molecule

5210 Martínez y Romero, R.P., Moreno, M. and Zentella, A. (1991). Phys. Rev. D, 43, 2036-2040.
Supersymmetric properties and stability of the Dirac sea

5211 Martínez-y-Romero, R.P. and Salas-Brito, A.L. (1992). J. Math. Phys., 33, 1831-1836.
Conformal invariance in a Dirac oscillator

5212 Martynenko, A.P. and Faustov, R.N. (1987). Yad. Fiz., 45, 770-776; Sov. J. Nucl. Phys., 45, 479-482.
Calculation of the relativistic corrections to the hyperfine structure of muonium

5213 Maseras, F., Duran, M., Lledós, A. and Bertrán, J. (1989). Inorg. Chem., 28, 2984-2988.
Molecular hydrogen complex vs dihydride in ML_4+H_2 systems. Influence of the ML_4 fragment geometry

5214 Maseras, F., Lledós, A., Duran, M. and Bertrán, J. (1992). J. Chem. Soc. Faraday Trans., 88, 1111-1117.
Ab initio calculations on the [Rh(PH_3)$_3$Cl] system. Influence of the basis set on the structural and reactivity trends of transition-metal complexes

5215 Maseras, F. and Morokuma, K. (1992). Chem. Phys. Lett., 195, 500-504.
Application of the natural population analysis to transition-metal complexes. Should the empty metal p orbitals be included in the valence space?

5216*Masmanidis, C.A., Jaffé, H.H. and Ellis, R.L. (1975). J. Phys. Chem., 79, 2052-2061.
Spin-orbit coupling in organic molecules

5217 Masovic, D.R. (1989). J. Phys.: Condens. Matter, 1,

7315-7327.
Application of the chemical pseudopotential method to the calculation of the energy band structure of CuI V

5218 Masovic, D.R. (1990). Physica B, 164, 335-338.
The chemical pseudopotential of the spin-orbit interaction

5219 Massey, A.G. (1990).
Main Group Chemistry
Ellis Horwood, New York, 537 p.

5220 Mathur, H. (1991). Phys. Rev. Lett., 67, 3325-3327.
Thomas precession, spin-orbit interaction, and Berry's phase

5221 Matsumoto, M., Staunton, J.B. and Strange, P. (1990). J. Phys.: Cond. Matter, 2, 8365-8379.
A new formalism for the paramagnetic spin susceptibility of metals using relativistic spin-polarized multiple-scattering theory: a temperature-dependent anisotropy effect

5222 Matsumoto, M., Staunton, J.B. and Strange, P. (1991). J. Phys.: Cond. Matter, 3, 1453-1460.
Relativistic spin polarized ab initio calculations of the anisotropy and temperature dependence of the paramagnetic spin susceptibility of scandium and yttrium

5223 Matsunaga, N., Cundari, T.R., Schmidt, M.W. and Gordon, M.S. (1992). Theor. Chim. Acta, 83, 57-68.
A comparative study of the bonding in heteroatom analogues of benzene

5224 Matsuoka, O. (1987). Chem. Phys. Lett., 140, 362-366.
Nuclear attraction integrals in the homogeneously charged sphere model of the atomic nucleus

5225 Matsuoka, O. (1989). Chem. Phys. Lett., 155, 544-546.
Comments on the basis spinors in Dirac-Fock-Roothaan calculations

5226 Matsuoka, O. (1990a). Chem. Phys. Lett., 172, 175-179.
Basis-set expansion method for relativistic atoms in the atomic-nucleus model of a finite sphere of constant electric field

5227 Matsuoka, O. (1990b). J. Chem. Phys., 92, 4364-4371.
Molecular integrals over spherical Laguerre Gaussian-type functions

5228 Matsuoka, O. (1992a). Chem. Phys. Lett., 195, 184-188.
Segmented contractions of relativistic Gaussian basis sets

5229 Matsuoka, O. (1992b). J. Chem. Phys., 97, 2271-2275.
Relativistic self-consistent-field methods for molecules. III. All-electron calculations on diatomics HI, HI^+, AtH and AtH^+

5230 Matsuoka, O., Fujimura, N. and Hasegawa, T. (1991). Int. J. Quantum Chem., 39, 813-821.

Hydrogenlike atoms in uniformly charged sphere model of atomic nucleus. II. Application of basis-set expansion method

5231 Matsuoka, O. and Huzinaga, S. (1987). Chem. Phys. Lett., 140, 567-571.
Relativistic well-tempered Gaussian basis sets

5232 Matsuoka, O. and Okada, S. (1989). Chem. Phys. Lett., 155, 547-549.
Dirac-Fock-Roothaan calculations on the sixth-row elements Tl-Rn

5233 Matsuoka, O. and Okamura, K. (1989). Bull. Univ. Electro-Commun., 2, 313-319.
Accurately energy-optimized Gaussian basis sets for hydrogen atom

5234 Matsushita, T., Klotz, R., Marian, C.M. and Peyerimhoff, S.D. (1987). Mol. Phys., 62, 1385-1420.
Ab initio calculations of the radiative lifetime of the a $^1\Delta$ and b $^1\Sigma^+$ states in the SeO molecule

5235 Matsutani, S. (1992). J. Phys. Soc. Japan, 61, 3825-3826.
Berry phase of Dirac particle in thin rod

5236 Mattar, S.M. and Ozin, G.A. (1986). J. Phys. Chem., 90, 1355-1360.
Magnetic inequivalency in the EPR spectra of small metal clusters

5237 Mattheiss, L.F. (1992). Phys. Rev. B, 45, 3252-3259.
Calculated structural properties of $CrSi_2$, $MoSi_2$, and WSi_2

5238 Maurer, R.J. and Watson, R.L. (1986). At. Data Nucl. Data Tables, 34, 185-199.
Dirac-Fock energies for multiple K-plus L-shell vacancy state of ions with Z=9 to 18

5239 Mayol, R. and Salvat, F. (1990). J. Phys. B: At. Mol. Opt. Phys., 23, 2117-2130.
Cross sections for K-shell ionisation by electron impact

5240 Mazin, I.I. and Savrasov, S.Yu. (1988). Dokl. Akad. Nauk SSSR, 301, 92-94; Sov. Phys. Dokl., 33, 514-515.
Pressure dependence of the dielectric gap in Xe

5241 Mazzoni, M. and Joshi, Y.N. (1990). J. Opt. Soc. Am. B, 7, 914-917.
Photoexcitation of the inner 3d electron in As I

5242 McAdon, M.H. and Goddard III, W.A. (1988a). J. Chem. Phys., 88, 277-302.
Charge density waves, spin density waves, and Peierls distortions in one-dimensional metals. 1. Hartree-Fock studies of Cu, Ag, Au, Li, and Na

5243 McAdon, M.H. and Goddard III, W.A. (1988b). J. Phys. Chem., 92, 1352-1365.

Charge density waves, spin density waves, and Peierls distortions in one-dimensional metals. 2. Generalized valence bond studies of Cu, Ag, Au, Li, and Na

5244 McCarroll, R. (1987).
Relativistic effects in electron capture
In "Recent Studies in Atomic and Molecular Processes", Ed. A.E. Kingston, Plenum, New York.

5245 McEachran, R.P. and Stauffer, A.D. (1986). J. Phys. B, 19, 3523-3538.
Spin polarisation of electrons elastically scattered from xenon

5246 McEachran, R.P. and Stauffer, A.D. (1987a). J. Phys. B: At. Mol. Phys., 20, 3483-3486.
Relativistic low-energy elastic and momentum transfer cross sections for electron scattering from xenon

5247 McEachran, R.P. and Stauffer, A.D. (1987b). J. Phys. B: At. Mol. Phys., 20, 5517-5528.
Spin polarisation of electrons elastically scattered from mercury

5248 McEachran, R.P. and Stauffer, A.D. (1992). J. Phys. B: At. Mol. Opt. Phys., 25, 1527-1532.
Spin polarization of electrons elastically scattered from cadmium and zinc

5249*McGlynn, S.P., Azumi, T. and Kinoshita, M. (1969).
Molecular Spectroscopy of the Triplet State
Prentice-Hall, Englewood Cliffs, NJ, 434 p.

5250 McKee, M.L. (1987). J. Chem. Phys., 87, 3143-3152.
Ab initio study of O, O_2, and C_2H_4 interactions with Ag_2 and Ag_2^+. A model for surface-adsorbate interactions

5251 McKee, M.L. and Worley, S.D. (1988). J. Phys. Chem., 92, 3699-3700.
A theoretical study of Rh/CO species

5252 McKellar, B.H.J. and Stephenson Jr., G.J. (1987). Phys. Rev. A, 36, 2566-2569.
Klein paradox and the Dirac-Kronig-Penney model

5253 McKenzie, D.K. and Drake, G.W.F. (1991). Phys. Rev. A, 44, R6973-R6976.
Variational calculation for the ground state of lithium and QED corrections for Li-like ions

5254 McKeon, D.G.C. and Ord, G.N. (1992). Phys. Rev. Lett., 69, 3-4.
Time reversal in stochastic processes and the Dirac equation

5255 McMichael Rohlfing, C., Hay, P.J. and Martin, R.L. (1986). J. Chem. Phys., 85, 1447-1455.
An effective core potential investigation of Ni, Pd, and Pt and their monohydrides

5256 McNeil, J.A., Furnstahl, R.J., Rost, E. and Shepard, J.R. (1989). Phys. Rev. C, 40, 399-406.
Finite nucleus Dirac mean field theory and random phase approximation using finite B splines

5257*McVoy, K.W. and van Hove, L. (1963). Phys. Rev., 125, 1034-1043.
Inelastic electron-nucleus scattering and nucleon-nucleon correlations

5258*Meath, W.J. (1966). J. Chem. Phys., 45, 4519-4527.
Relativistic interaction energies between atoms in degenerate states

5259*Meath, W.J. and Hirschfelder, J.O. (1966a). J. Chem. Phys., 44, 3197-3209.
Relativistic intermolecular forces, moderately long range

5260*Meath, W.J. and Hirschfelder, J.O. (1966b). J. Chem. Phys., 44, 3210-3215.
Long-range (retarded) intermolecular forces

5261 Mehler, G. (1988).
Inner-Shell Ionization in Ion-Atom Collisions
GSI-Rep. GSI-88-04, 198 pp.

5262 Mehler, G., Soff, G., Rumrich, K. and Greiner, W. (1989). Z. Phys. D, 13, 193-202.
K-shell ionization in relativistic ion-atom collisions

5263 Mei, J. and Davenport, J.W. (1992). Phys. Rev. A, 45, 1497-1501.
First-order perturbation calculations for transition-metal atoms and ions

5264 Meier, U., Peyerimhoff, S.D. and Grein, F. (1991). Chem. Phys., 150, 331-351.
Ab initio MRD-CI study of $GaAs^-$, $GaAs_2^{(\pm)}$, $Ga_2As_2^{(\pm)}$, $As_2^{(\pm)}$ and As_4 clusters

5265*Meister, H.J. and Weiss, H.F. (1968). Z. Phys., 216, 165-171.
Numerical comparison between relativistic and nonrelativistic theory for scattering of slow electrons from atoms

5266 Membrado, M., Pacheco, A.F. and Segui, A. (1988). Phys. Rev. A, 38, 1069-1071.
Electron surface of the planar uniform-background model: Relativistic Thomas-Fermi solution

5267 Mendel, R.R. and Trottier, H.D. (1992). Phys. Rev. D, 46, 2068-2079.
Matrix elements of pseudoscalar $Q\bar{q}$ mesons including relativistic and asymptotic-freedom effects

5268 Méndez, B.and Dominguez-Adame, F. (1991). J. Phys. A, 24, L331-L336.
A simple numerical method for the determination of

relativistic one-dimensional band structures

5269 Méndez, B.and Dominguez-Adame, F. (1992). J. Phys. A, 25, 2065-2070.
Level shift under the influence of relativistic point interaction potentials

5270 Menendez, M.I., Lopez, R., Sordo, T.L. and Sordo, J.A. (1988). J. Mol. Struct.(Theochem), 166, 75-80.
An alternative method to assess the quality of Hartree-Fock-Roothaan wave functions

5271 Menon, V.J. (1992). Am. J. Phys., 60, 928-929.
Spin conservation in one-particle Dirac theory

5272 Merchán, M., Daudey, J.-P., González-Luque, R. and Nebot-Gil, I. (1990). Chem. Phys., 148, 69-76.
On the dissociation energy of Cu_2 and CuH using a differential correlation approach

5273 Merchán, M., Nebot-Gil, I., González-Luque, R. and Ortí, E. (1987). J. Chem. Phys., 87, 1690-1700.
A CI study of the CuCO and $CuCO^+$ complexes

5274*Merzbacher, E. (1985). NATO ASI, B134, "Fundam. processes At. Collision Phys.", 31-49.
Introduction to fundamental interactions in atomic collisions

5275 Methfessel, M., Hennig, D. and Scheffler, M. (1992). Phys. Rev. B, 46, 4816-4829.
Trends of the surface relaxations, surface energies, and work functions of the 4d transition metals

5276 Meyer, J. (1988). Int. J. Quantum Chem., 33, 445-465.
Construction of linearly independent relativistic symmetry orbitals for finite double-point groups including time reversal symmetry

5277 Meyer, J., Sepp, W.-D., Fricke, B. and Rosén, A. (1989). Comp. Phys. Comm., 54, 55-73.
Computation of relativistic symmetry orbitals for finite double point groups

5278 Meyer, M., von Raven, E., Richter, M., Sonntag, B., Cowan, R.D. and Hansen, J.E. (1989). Phys. Rev. A, 39, 4319-4322.
Autoionization of the Ca $2p^5 3d$ core resonances: Breakdown of the spectator model

5279 Mian, S.N. (1991). Nuovo Cim. D, 13D, 187-194.
Phase shift calculation for low-energy electron scattering from mercury

5280*Michels, H.H., Hobbs, R.H. and Wright, L.A. (1978). J. Chem. Phys., 69, 5151-5162.
Electronic structure of the noble gas dimer ions. I. Potential energy curves and spectroscopic constants

5281 Miecznik, G., Brage, T. and Froese Fischer, C. (1992). Phys. Scr., 45, 436-444.

Transition energies and probabilities in high-spin states of the boron sequence and an electron affinity for negative beryllium

5282 Migdalek, J. (1990).
Model-Potential Methods in Atomic Structure Calculations
Acta Sci Litt. Univ. Iagellonica, Physica, 32, 106 p.

5283 Migdalek, J. and Baylis, W.E. (1986a). J. Phys. B: At. Mol. Phys., 19, 1-16.
Relativistic oscillator strengths and excitation energies for the $ns^2\ ^1S_0$-$nsnp\ ^3P_1$, $^1P_1$ transitions: II. Cadmium isoelectronic sequence

5284 Migdalek, J. and Baylis, W.E. (1986b). Phys. Rev. A, 33, 1417-1420.
Correlation effects in a relativistic calculation of the $6s^2\ ^1S_0$-$6s6p\ ^1P_1$ transition in ytterbium

5285 Migdalek, J. and Baylis, W.E. (1987a). J. Quant. Spectrosc. Radiat. Transfer, 37, 183-191.
Relativistic SCF calculations of transitions in the principal and intercombination series of mercury

5286 Migdalek, J. and Baylis, W.E. (1987b). J. Quant. Spectrosc. Radiat. Transfer, 37, 521-526.
A multiconfiguration Dirac-Fock study of the $6s^2\ ^1S_0$-$6s6p\ ^3P_1$, $^1P_1$ transitions in the Yb isoelectronic sequence

5287 Migdalek, J. and Baylis, W.E. (1987c). Phys. Rev. A, 35, 3227-3232.
Correlation effects in a relativistic calculation of the $6s^2\ ^1S_0$-$6s6p\ ^3P_1$, $^1P_1$ transitions in barium

5288 Migdalek, J. and Baylis, W.E. (1987d). Can. J. Phys., 65, 1612-1619.
Core-polarization-corrected multiconfiguration Dirac-Fock oscillator strengths and excitation energies for the $5s^2\ ^1S_0$-$5s5p\ ^3P_1$, $^1P_1$ transitions in strontium (Sr I) and yttrium (Y II) spectra

5289 Migdalek, J. and Baylis, W.E. (1990). Phys. Rev. A, 42, 6897-6899.
Multiconfiguration Dirac-Fock calculations of two electric quadrupole transitions in neutral barium

5290 Migdalek, J. and Baylis, W.E. (1991a). J. Phys. B: At. Mol. Opt. Phys., 24, L99-L102.
Relativistic transition probabilities and lifetimes of low-lying levels in ytterbium

5291 Migdalek, J. and Baylis, W.E. (1991b). Phys. Rev. A, 43, 4625-4628.
Multiconfiguration Dirac-Fock study of the 5d6p $^3F^\circ_4$ lifetime in singly ionized lanthanum

5292 Migdalek, J. and Bojara, A. (1987). J. Phys. B: At. Mol. Phys., 20, L1-L6.
Comparison of relativistic CI and MCRHF calculations of

oscillator strengths and excitation energies for the $ns^2\ ^1S_0$-nsnp $^3P_1$, $^1P_1$ transitions in neutral cadmium and mercury

5293 Migdalek, J. and Bojara, A. (1988). J. Phys. B: At. Mol. Opt. Phys., 21, 2221-2236.
Relativistic CI calculations for the $ns^2\ ^1S_0$-nsnp $^3P_1$, $^1P_1$ transitions in the cadmium and mercury isoelectronic sequences

5294 Migdalek, J. and Stanek, M. (1989). J. Quant. Spectrosc. Radiat. Tranfer, 42, 585-592.
Relativistic MCDF oscillator strengths for the $4s^2\ ^1S_0$-4s4p $^3P_1$, $^1P_1$ transitions in zinc isoelectronic sequence

5295 Migdalek, J. and Stanek, M. (1990). Phys. Rev. A, 41, 2869-2872.
Comparison of "optimal-level" and "average-level" multiconfigurational Dirac-Fock as well as of relativistic configuration-interaction calculations for the $ns^2\ ^1S_0$-nsnp $^3P_1$, $^1P_1$ transitions

5296 Migdalek, J. and Wyrozumska, M. (1987). J. Quant. Spectrosc. Radiat. Tranfer, 37, 581-589.
Relativistic oscillator strengths for the Cs isoelectronic sequence and collapse of f and d orbitals

5297*Mikhailov, A.I. (1969). Zh. Eksp. Teor. Fiz., 55, 625-634; Sov. Phys. JETP, 28, 326-330.
Relativistic atomic photoeffect from the K-shell near threshold

5298 Mikhailov, A.I. (1990). Zh. Eksp. Teor. Fiz., 98, 838-846; Sov. Phys. JETP, 71, 465-469.
Relativistic atomic photoelectric effect from the K shell near the threshold in the screened nuclear field: Analytic calculation

5299 Mikhailov, A.I. and Porsev, S.G. (1992). J. Phys. B: At. Mol. Opt. Phys., 25, 1097-1101.
Double ionization of the atomic K shell in the annihilation of a positron with a bound K electron

5300 Millack, T. (1988). Z. Phys. D, 8, 119-127.
g-Hartree ab-initio calculation of hyperfine-constants for atoms

5301 Millack, T. (1989). Z. Phys. D, 11, 181-185.
Second order g-Hartree calculations of the dipole-constants of Li and B

5302 Miller, D., Milton, K.A. and Siegemund-Broka, S. (1992). Phys. Rev. D, 46, 806-813.
Finite-element quantum electrodynamics: Canonical formulation, unitarity, and the magnetic moment of the electron

5303 Miller, J.H. and Manson, S.T. (1991). Phys. Rev. A, 44, 4321-4327.
Relativistic model of secondary-electron energy spectra

in electron-impact ionization

5304*Miller, L.D. (1974). Phys. Rev. C, 9, 537-554.
Exchange potentials in relativistic Hartree-Fock theory of closed-shell nuclei

5305*Miller, L.D. (1975a). Phys. Rev. C, 12, 710-712.
State-dependent equivalent local potentials for the Dirac equation

5306*Miller, L.D. (1975b). Ann. Phys. (NY), 91, 40-57.
Relativistic single-particle potentials for nuclei

5307*Miller, L.D. (1976). Phys. Rev. C, 14, 706-717.
Elastic electron scattering formalism for relativistic nuclear models

5308*Miller, L.D. (1983a). Phys. Rev. Lett., 51, 1733-1736.
Impulse-approximation nucleon-nucleus optical-model potential for the Dirac equation

5309*Miller, L.D. (1983b). Phys. Rev. Lett., 51, 1807.
Comment on nucleon-nucleus impulse-approximation optical-model potentials for the Dirac equation

5310 Miller, L.D. (1987). J. Phys. B: At. Mol. Phys., 20, 4309-4316.
A Foldy-Wouthuysen reduction of the atomic relativistic Hartree-Fock equations

5311 Miller, P.K., Abney, K.D., Rappé, A.K., Anderson, O.P. and Strauss, S.H. (1988). Inorg. Chem., 27, 2255-2261.
Electronic and molecular structure of $OTeF_5^-$

5312 Miller, R.M., Tinti, D.S. and Case, D.A. (1989). Inorg. Chem., 28, 2738-2743.
Comparison of the electronic structures of chromate, halochromates, and chromyl halides by the Xα method

5313 Milman, V.Yu., Antonov, V.N. and Nemoshkalenko, V.V. (1990). J. Phys.: Cond. Matter, 2, 7101-7114.
Relativistic pseudopotentials for transition metals: I. Construction and application to atoms

5314 Milonni, P.W., Cook, R.J. and Goggin, M.E. (1988). Phys. Rev. A, 38, 1621-1623.
Radiation pressure from the vacuum: Physical interpretation of the Casimir force

5315 Milonni, P.W. and Shih, M.-L. (1992). Phys. Rev. A, 45, 4241-4253.
Source theory of the Casimir force

5316*Mil'shtein, A.I. and Strakhovenko, V.M. (1982). Phys. Lett., 90A, 447-450.
The O(2,1) algebra and electron Green function in a Coulomb field

5317 Milstein, A.I. and Yelkhovsky, A.S. (1990). Phys. Lett. B, 241, 184-188.

Vacuum polarization and the quadrupole moment of a heavy nucleus

5318 Min, B.I. and Jang, Y.-R. (1991). J. Phys.: Condens. Matter, 3, 5131-5141.
The effect of the spin-orbit interaction on the electronic structure of magnetic materials

5319*Minaev, B.F. (1980). Int. J. Quantum Chem., 17, 367-374.
Intensities of spin-forbidden transitions in molecular oxygen and selective heavy-atom effects

5320 Minaev, B.F., Buketova, A.E. and Muldahmetov, Z.M. (1989). Spectrosc. Lett., 22, 211-236.
Quantum-chemical study of the diatomic hydrides electronic structure

5321*Minaev, B.F. and Muldakhmetov, Z.M. (1984). Opt. Spectrosc., 56, 29-31; Opt. Spektrosk., 56, 48-52.
Influence of spin-orbit interaction on the intensity of optical doublet-doublet and triplet-triplet transitions in molecules

5322*Minaev, B.F. and Terpugova, A.F. (1969).
Spin-orbit interaction in the complexes with charge-transfer
In "Theory of Electronic Structure of Atoms and Molecules", Ed. J. Batarunas et al., Mintis, Vilnius, pp. 321-324.

5323*Mingos, D.M.P. (1976). J. C. S. Dalton, 1163-1169.
Molecular-orbital calculations on cluster compounds of gold

5324 Mingos, D.M.P. and Kanters, R.P.F. (1990). J. Organomet. Chem., 384, 405-415.
Molecular orbital analysis of mono- and di-carbido-gold cluster compounds

5325 Minot, C., Bigot, B. and Hariti, A. (1986a). Nouv. J. Chim., 10, 461-472.
Thermodynamic versus kinetic control in the selective hydrogenation of the small platinum cluster; A theoretical investigation

5326 Minot, C., Bigot, B. and Hariti, A. (1986b). J. Am. Chem. Soc., 108, 196-206.
A theoretical study of successive hydrogenations of small platinum cluster: Structure and energetics

5327 Miralles, J., Daudey, J.-P. and Caballol, R. (1992). Chem. Phys. Lett., 198, 555-562.
Variational calculation of small energy differences. The singlet-triplet gap in $[Cu_2Cl_6]^{2-}$

5328 Miralles-Sabater, J., Garcia-Cuesta, I., Merchán, M. and Nebot-Gil, I. (1989). J. Mol. Struct.(Theochem), 188, 1-13.
A MO-CI study of the Cu/H_2O and Cu^+/H_2O systems

5329 Miralles-Sabater, J., Merchán, M., Nebot-Gil, I. and

Viruela-Martin, P.M. (1988). J. Phys. Chem., 92, 4853-4859.
A theoretical study of the interaction of acetylene with copper and silver monoions

5330*Miramontes, J.L. and Pajares, C. (1984). Nuovo Cim., 84B, 10-18.
On the large-N limits of relativistic equations

5331 Mishchenko, A.V. and Sitenko, Yu.A. (1991). Ukr. Fiz. Zh., 36, 1865-1876.
Non-local boundary conditions and zero modes of the Dirac operator in a two-dimensional surface in a magnetic field (in Russian)

5332 Mishra, B., Das, L.K., Sahu, T., Tripathi, G.S. and Misra, P.K. (1990). J. Phys.: Cond. Matter, 2, 9891-9905.
A pseudopotential calculation of the Knight shift in simple metals

5333 Mishustin, I.N. (1990). Yad. Fiz., 52, 1135-1139.
Instability of Dirac sea of nucleons in relativistic nuclear collisions

5334 Misra, P.K. and Callaway, J. (1987). Phys. Rev. B, 36, 3731-3738.
Relativistic theory of neutron form factors

5335 Misra, S.P., Naik, S. and Panda, A.R. (1987). Pramana, 28, 131-158.
Relativistic effects on meson spectra

5336 Misra, U.D., Sah, M. and Gokhale, B.G. (1989). J. Phys. B, 22, 1505-1513.
The Sommerfeld screening parameter σ_2 and theoretically calculated energies of atomic inner shells

5337 Misra, U.D., Sah, M. and Gokhale, B.G. (1992). J. Phys. B, 25, 4107-4117.
Precise energy values of x-ray spin doublet levels

5338 Missana, M. (1989).
Exact solution of the Dirac equation
In "Large Scale Structure and Motions in the Universe", Ed. M. Mezzetti, G. Giuricin, F. Mardirossian and M. Ramella, Kluwer, Dordrecht, pp. 411-412.

5339 Mitás, L., Shirley, E.L. and Ceperley, D.M. (1991). J. Chem. Phys., 95, 3467-3475.
Nonlocal pseudopotentials and diffusion Monte Carlo

5340 Mitchell, D.B., Nogami, Y. and Whelan, N.D. (1989). Can. J. Phys., 67, 583-586.
Relativistic versus nonrelativistic Hartree-Fock calculations: solvable examples in one dimension

5341 Mittleman, M.H. (1989). Phys. Rev. A, 39, 1-8.
Three-electron atoms from a Bethe-Salpeter equation approach

5342 Mittleman, M.H. (1991). Phys. Rev. A, 44, 6150.
Comment on "Approximate solution of the hydrogenlike atoms in intense laser radiation"

5343 Miyazawa, H. and Tanaka, K. (1992). Prog. Theor. Phys., 87, 1457-1466.
Relativistic bound-state problem

5344 Miyoshi, E. and Sakai, Y. (1988a). J. Comp. Chem., 9, 719-727.
Applications of the model potential method to transition metal compounds

5345 Miyoshi, E. and Sakai, Y. (1988b). J. Chem. Phys., 89, 7363-7366.
Theoretical study on electronic structures of AuF_6 and its anions

5346 Miyoshi, E., Sakai, Y., Murakami, A., Iwaki, H., Terashima, H., Shoda, T. and Kawaguchi, T. (1988). J. Chem. Phys., 89, 4193-4198.
On the electron affinities of hexafluorides CrF_6, MoF_6, and WF_6

5347 Mladenov, I.M. (1988). Phys. Stat. Sol. B, 145, K115-K119.
Generalization of the Saxon-Hutner theorem in the relativistic domain

5348 Mladenov, I.M. (1989). Phys. Lett. A, 137, 313-318.
An extension of the Saxon-Hutner theorem in the relativistic domain

5349 Mohamed, A. and Parisse, B. (1992). Annales de l'I.H.P. Phys. Theor., 56, 235-277.
Approximation of the eigenvalues of some singular perturbations and application to the Dirac operator

5350 Mohan, M. and Hibbert, A. (1987). J. Phys. B: At. Mol. Phys., 20, 907-918.
Model-potential calculations for low-lying states of mercury

5351 Mohan, M. and Le Dourneuf, M. (1990). Phys. Rev. A, 41, 2862-2864.
Radiative and collisional excitation rates for transitions between the fine-structure levels within the ground-state term of Si^{5+}

5352 Mohanty, A.K. (1992). Int. J. Quantum Chem., 42, 627-662.
A Dirac-Fock self-consistent field method for closed-shell molecules including Breit interaction

5353 Mohanty, A.K. and Clementi, E. (1989a). Chem. Phys. Lett., 157, 348-352.
Kinetically balanced geometric Gaussian basis set calculations for relativistic many-electron atoms with finite nuclear size

5354 Mohanty, A.K. and Clementi, E. (1989b).

Kinetically balanced geometric Gaussian basis set calculations for relativistic many-electron atoms
In "Modern Techniques in Computational Chemistry", Ed. E. Clementi, ESCOM, Leiden, pp. 169-206.

5355 Mohanty, A.K. and Clementi, E. (1990a). J. Chem. Phys., 93, 1829-1833.
Kinetically balanced geometric Gaussian basis sets for relativistic atoms

5356 Mohanty, A.K. and Clementi, E. (1990b).
Dirac-Fock self-consistent field calculations for closed-shell molecules with kinetic balance and finite nuclear size
In "MOTECC-90", Ed. E. Clementi, ESCOM, Leiden, pp. 693-730.

5357 Mohanty, A.K. and Clementi, E. (1991a). Int. J. Quantum Chem., 39, 487-517.
Dirac-Fock self-consistent field method for closed-shell molecules with kinetic balance and finite nuclear size; Erratum: ibid. 40, 429-432.

5358 Mohanty, A.K. and Clementi, E. (1991b). AIP Conf. Proc., 233, 537-539.
Dirac-Fock basis set calculations for atoms and molecules

5359 Mohanty, A., Panigrahy, S. and Clementi, E. (1991a).
Dirac-Fock self-consistent field calculation for closed-shell molecules with kinetic balance and finite nuclear size
In "MOTECC-91", Ed. E. Clementi, ESCOM, Leiden, pp. 647-694.

5360 Mohanty, A.K., Parpia, F.A. and Clementi, E. (1991b).
Kinetically balanced geometric Gaussian basis set calculations for relativistic many-electron atoms
In "MOTECC-91", Ed. E. Clementi, ESCOM, Leiden, pp. 167-209.

5361 Mohanty, A.K. and Sampson, D.H. (1987). Phys. Scr., 35, 645-649.
A comparison of relativistic and quasirelativistic line strengths

5362 Mohanty, A.K. and Sampson, D.H. (1988). Phys. Scr., 38, 557-559.
Reply to "Comment on 'A comparison of relativistic and quasirelativistic line strengths'"

5363*Mohr, P.J. (1983b). At. Data Nucl. Data Tables, 29, 453-466.
Energy levels of hydrogen-like atoms predicted by quantum electrodynamics, $10 \leq Z \leq 40$

5364 Mohr, P.J. (1986).
Relativistic and QED calculations for many-electron and few-electron atoms
In "Atoms in Unusual Situations", Ed. J. P. Briand, Plenum, New York, pp. 301-307.

5365 Mohr, P.J. (1988a).
Quantum electrodynamics calculations
In "The Spectrum of Atomic Hydrogen: Advances", Ed. G. Series, World Scientific, Singapore, pp. 113-136.

5366 Mohr, P.J. (1988b). Nucl. Intrum. Meth. B, 31, 1-6.
Structure of high-Z one- and two-electron atoms

5367 Mohr, P.J. (1989a).
Quantum electrodynamics perturbation theory
In "Relativistic, Quantum Electrodynamic and Weak Interaction Effects in Atoms", Ed. W. Johnson, P. Mohr and J. Sucher, AIP Conf. Proc., 189, 47-62.

5368 Mohr, P.J. (1989b).
Quantum electrodynamics of high-Z few-electron atoms
In "Physics of Highly-Ionised Atoms. Proceedings of a NATO Advanced Study Institute", Ed. R. Marrus, Plenum, New York, pp. 111-142.

5369 Mohr, P.J. (1991). Phys. Rev. A, 44, R4089-R4090.
Lamb shift in the n=4 state of He^+

5370 Mohr, P.J. (1992). Phys. Rev. A, 46, 4421-4424.
Self-energy correction to one-electron energy levels in a strong Coulomb field

5371 Mohr, P.J. and Kim, Y.-K. (1992). Phys. Rev. A, 45, 2727-2735.
Self-energy of excited states in a strong Coulomb field

5372 Mohr, P.J. and Safronova, U.I. (1986). Protsessy vo Vnutr. At. Obolochkakh, Moscow, 140-161.
Perturbation theory in a relativistic basis for energies of two-electron systems

5373 Moine, B., Chermette, H. and Pedrini, C. (1986). J. Chem. Phys., 85, 2784-2790.
MSXα calculation of the electronic structure of Cu^+ and Ag^+ ions in lithium chloride single crystal

5374 Moiseiwitsch, B.L. (1986). J. Phys. B: At. Mol. Phys., 19, 3733-3737.
Relativistic symmetric eikonal approximation for electron capture

5375 Moiseiwitsch, B.L. (1987). J. Phys. B: At. Mol. Phys., 20, 4111-4114.
Relativistic eikonal phase factors for electron capture

5376 Moiseiwitsch, B.L. (1988). J. Phys. B: At. Mol. Opt. Phys., 21, 603-610.
Relativistic second-order OBK approximation for electron capture

5377 Moiseiwitsch, B.L. (1990). Adv. At. Mol. Opt. Phys., 26, 51-77.
Electron capture at relativistic energies

5378 Moiseiwitsch, B.L. (1992a). J. Phys. B: At. Mol. Opt. Phys., 25, 3015-3020.
Fine structure constant expansions for electron capture

5379 Moiseiwitsch, B.L. (1992b). J. Phys. B: At. Mol. Opt. Phys., 25, L487-L489.
Ultra-high relativistic energy limit for electron capture

5380 Molnár, L. and Lovas, I. (1991). Acta Phys. Hung., 69, 225-232.
Dirac particle in a periodic field II

5381 Momberger, K., Grün, N. and Scheid, W. (1991). Z. Phys. D, 21, S183-184.
Coupled channel analysis of electron-positron pair production in relativistic heavy ion collisions

5382 Momberger, K., Grün, N., Scheid, W. and Becker, U. (1989). J. Phys. B: At. Mol. Opt. Phys., 22, 3269-3273.
Angular distribution of electrons emitted from $1s_{1/2}$ and $2s_{1/2}$ states in relativistic heavy-ion collisions

5383 Mooney, T., Lindroth, E., Indelicato, P., Kessler Jr., E.G. and Deslattes, R.D. (1992). Phys. Rev. A, 45, 1531-1543.
Precision measurements of K and L transitions in xenon: Experiment and theory for the K, L, and M levels

5384 Moore, R.A. and Scott, T.C. (1986). Can. J. Phys., 64, 297-302.
Approximate solutions to the one-particle Dirac equation: Numerical results

5385 Moore, R.A. and Scott, T.C. (1988). Can. J. Phys., 66, 365-368.
A class of physically acceptable, classical, relativistic, many-particle Lagrangians

5386 Moore, R.A. and Scott, T.C. (1989). Nucl. Phys. B, 6, 458-460.
A class of physically acceptable, relativistic many-particle Lagrangians

5387 Moore, R.A. and Scott, T.C. (1992). Phys. Rev. A, 46, 3637-3645.
Quantization of second-order Lagrangians: The Fokker-Wheeler-Feynman model of electrodynamics

5388 Moore, R.A., Scott, T.C. and Monagan, M.B. (1987). Phys. Rev. Lett., 59, 525-527.
Relativistic, many-particle Lagrangean for electromagnetic interactions

5389 Moore, R.A., Scott, T.C. and Monagan, M.B. (1988). Can. J. Phys., 66, 206-211.
A model for a relativistic, many-particle Lagrangian with electromagnetic interactions

5390 Moores, D.L. (1991).
Relativistic calculations of electron impact ionisation

cross-sections of highly charged ions
In "The Effects of Relativity in Atoms, Molecules, and the Solid State", Ed. S. Wilson, I.P. Grant and B.L. Gyorffy, Plenum, New York and London, pp. 125-131.

5391 Moores, D.L. and Pindzola, M.S. (1990a). Phys. Rev. A, 41, 3603-3606.
Electron-impact ionization of highly charged hydrogenic atomic ions

5392 Moores, D.L. and Pindzola, M.S. (1990b). Phys. Rev. A, 42, 5384-5388.
Electron-impact ionization of highly charged ions in a level-to-level multiconfiguration Dirac-Fock approximation

5393 Moores, D.L. and Pindzola, M.S. (1991). Z. Phys. D, 21, S187-188.
Relativistic calculations of inelastic scattering of electrons by highly-charged ions

5394 Moores, D.L. and Pindzola, M.S. (1992). J. Phys. B, 25, 4581-4591.
Electron impact excitation of highly charged hydrogenic ions

5395 Moreno, M., Martínez, R. and Zentella, A. (1990). Mod. Phys. Lett. A, 5, 949-954.
Supersymmetry, Foldy-Wouthuysen transformation and stability of the Dirac sea

5396 Moreno, M. and Zentella, A. (1989). J. Phys. A, 22, L821-L825.
Covariance, CPT and the Foldy-Wouthuysen transformation for the Dirac oscillator

5397 Moriarty, J.A. (1986). Phys. Rev. B, 34, 6738-6745.
First-principles phonon spectrum in bcc Ba: Three-ion forces and transition-metal behavior

5398 Moriarty, J.A. (1988). Phys. Lett. A, 131, 41-46.
High-pressure structural phase stability in Hg to 1 TPa (10 Mbar)

5399 Morin, M., Salahub, D.R., Nour, S., Mehadji, C. and Chermette, H. (1989). Chem. Phys. Lett., 159, 472-478.
Spin-orbit coupling in the local spin density-scattered wave method

5400 Morita, K. (1986). Prog. Theor. Phys., 75, 220-223.
A role of quaternions in the Dirac theory

5401 Morokuma, K. and Borden, W.T. (1991). J. Am. Chem. Soc., 113, 1912-1914.
Ab initio calculations on $(H_3P)_2Pt(C_2H_4)$. The effect of alkene pyramidalization on internal rotation and alkene binding energies

5402*Moses, H.E. (1959). Phys. Rev., 113, 1670-1679.
Solution of Maxwell's equations in terms of a spinor notation: the Dirac and inverse problem

5403 Mösges, G., Hampel, F., Kaupp, M. and Schleyer, P.v.R. (1992). J. Am. Chem. Soc., 114, 10880-10889.
Experimental and theoretical studies of alkaline-earth metal coordination: X-ray crystal structures of calcium, strontium, and barium carbazoles and ab initio model calculations

5404 Moshinsky, M., Benet, L., Loyola, G. and Salinas, A. (1992). Rev. Mex. Fís., 38, 778-790.
Comparison of perturbative and variational procedures in a relativistic problem

5405 Moshinsky, M. and López Laurrabaquio, G. (1991). J. Math. Phys., 32, 3519-3528.
Relativistic interactions by means of boundary conditions: The Breit-Wigner formula

5406 Moshinsky, M., Loyola, G. and Villegas, C. (1991). J. Math. Phys., 32, 373-381.
Anomalous basis for representations of the Poincaré group

5407 Moshinsky, M. and Szczepaniak, A. (1989). J. Phys. A, 22, L817-L819.
The Dirac oscillator

5408 Moss, R.E. (1987). Am. J. Phys., 55, 397-401.
The hydrogen atom in one dimension

5409 Moss, R.E. (1990a). Chem. Phys. Lett., 171, 513-516.
Calculated transition frequencies for microwave and far-infrared electronic transitions of the hydrogen molecular ion

5410 Moss, R.E. (1990b). Chem. Phys. Lett., 172, 458-460.
Calculations on the 0,1 and 0,3 vibration-rotation levels of H_2^+

5411*Moss, R.E. and Okninski, A. (1976). Phys. Rev. D, 14, 3358-3361.
Diagonal forms of the Dirac Hamiltonian

5412 Moss, R.E. and Sadler, I.A. (1989). Mol. Phys., 68, 1015-1031.
Variational calculations for high J, low v levels of HD^+

5413 Mostepanenko, V.M. and Trunov, N.N. (1988). Usp. Fiz. Nauk, 156, 385-426; Sov. Phys. Usp., 31, 965-987.
The Casimir effect and its applications

5414*Mott, N.F. (1949). Proc. Phys. Soc. London, 62A, 416-422.
The basis of the electron theory of metals, with special reference to the transition metals

5415 Moullet, I., Andreoni, W. and Giannozzi, P. (1989). J. Chem. Phys., 90, 7306-7312.
Ground state properties of cesium dimers from ab initio pseudopotential approaches

5416 Mourad, H.A. and Vashakidze, I. Sh. (1992). Acta Phys.

Hung., 71, 61-65.
Relativistic Schrödinger equation including spin-orbit interaction

5417 Movaghar, B. and Cochrane, R.W. (1991). Z. Phys. B, 85, 217-225.
The connection between Hall effect and magnetism

5418 Mrówczynski, S. (1987). Phys. Rev. D, 36, 1520-1528.
Interaction of relativistic elementary atoms with matter. I. General formulas

5419 Mrówczynski, S. (1989). Postepy Fizyki, 40, 27-38.
Relativistic elementary atoms (in Polish)

5420 Mu, X.-D. and Crasemann, B. (1986). Phys. Rev. Lett., 57, 3039-3041.
Two-photon transitions in atomic inner shells: relativistic self-consistent-field calculation for xenon

5421 Mu, X.-D. and Crasemann, B. (1988). Phys. Rev. A, 38, 4585-4596.
Two-photon transitions in atomic inner shells: A relativistic self-consistent-field calculation with applications to Mo, Ag, and Xe

5422 Mueller, W. (1989). Physica C, 162-164, 1357-1358.
Relativistic cluster results for $YBa_2Cu_3O_7$

5423 Mukherjee, S., Pastor, G.M. and Bennemann, K.H. (1990). Phys. Rev. B, 42, 5327-5330.
Theoretical study of the metastability of Au_2^{2+} clusters

5424 Mukoyama, T. (1986). Bull. Inst. Chem. Res., Kyoto Univ., 64, 12-19.
Relativistic calculations of the excitation cross sections of hydrogen-like ions by heavy charged-particle impact

5425 Mukoyama, T. (1989a). Bull. Inst. Chem. Res., Kyoto Univ., 66, 624-630.
M-shell X-ray emission rates for Dy in Dirac-Fock approximation

5426 Mukoyama, T. (1989b). Acta Phys. Hung., 65, 165-169.
Wave function effects in K-shell ionization

5427 Mukoyama, T. and Adachi, H. (1986). Bull. Inst. Chem. Res., Kyoto Univ., 64, 133-139.
Relativistic molecular orbital calculations for diatomic molecules

5428 Mukoyama, T. and Adachi, H. (1989). Bull. Inst. Chem. Res., Kyoto Univ., 67, 15-22.
K X-ray emission rates in superheavy elements

5429 Mukoyama, T., Ito, S., Sulik, B. and Hock, G. (1991). Bull. Inst. Chem. Res., Kyoto, 69, 15-28.
Relativistic effect on the ionization probability in the geometrical model

5430 Mukoyama, T., Kaji, H. and Yoshihara, K. (1990). Bull. Inst. Chem. Res., Kyoto Univ., 68, 177-184.
K-shell ionization of atoms by mesons

5431 Mukoyama, T. and Lin, C.-D. (1987a). J. Phys.(Paris), Coll., 1, C9-523 - C9-526.
Basis-set expansion of the Dirac equation for atoms

5432 Mukoyama, T. and Lin, C.-D. (1987b). Phys. Rev. A, 35, 4942-4945.
Finite-basis-set approach to the Dirac equation for atoms in a Dirac-Fock-Slater potential

5433 Mukoyama, T. and Lin, C.-D. (1987c). Nucl. Instrum. Methods Phys. Res. A, 262, 15-18.
Finite-base-set expansion method for inner-shell ionization in ion-atom collisions

5434 Mukoyama, T. and Lin, C.-D. (1991). Z. Phys. D, 21, S237-238.
Distortion approximation for inner-shell ionization cross sections in ion-atom collisions using relativistic pseudostates

5435 Mukoyama, T. and Sarkadi, L. (1988). Bull. Inst. Chem. Res., Kyoto Univ., 66, 11-18.
Relativistic Coulomb continuum wave function at zero kinetic energy in ion-atom collisions

5436 Müller, B. (1987).
The many facets of the Dirac vacuum
In "Physics of Strong Fields", NATO ASI Ser. B, 153, 43-57.

5437 Müller, B. and Reinhold, J. (1992). Chem. Phys. Lett., 196, 363-367.
The nature of gas-phase complexes R_3P-H-X (R=H, CH_3; X=Cl, Br, I)

5438 Müller, B., Reinhold, J., Graf, O., Graf, S., Ionescu, C., Schramm, S., Grabiak, M., Greiner, W., Schäfer, A. and Soff, G. (1989).
Theoretical attempts to explain the GSI positron lines
In "Relativistic, Quantum Electrodynamic and Weak Interaction Effects in Atoms", Ed. W. Johnson, P. Mohr and J. Sucher, AIP Conf. Proc., 189, pp. 479-513.

5439 Müller, H., Engel, E. and Dreizler, R.M. (1989). Phys. Rev. A, 40, 5542-5547.
Extension of the relativistic Thomas-Fermi-Dirac-Weizsäcker model to arbitrary external fields

5440 Muller, R.A. (1992). Am. J. Phys., 60, 313-317.
Thomas precession: where is the torque?

5441 Müller, W. (1989). IEEE Trans. El. Devices, 36, 180-187.
Electronic structure of BaO/W cathode surface

5442 Musolino, V., Toscano, M. and Russo, N. (1990). J. Comp. Chem., 11, 924-929.

Electronic structure and photoelectron spectra of Sb_2 and Sb_4 from local-spin-density calculations. Model potential for Sb

5443 Mustafa, O. and Sever, R. (1991). Phys. Rev. A, 44, 4142-4144.
Shifted 1/N expansion for the Klein-Gordon equation with vector and scalar potentials

5444 Mustre de Leon, J., Rehr, J.J., Zabinsky, S.I. and Albers, R.C. (1991). Phys. Rev. B, 44, 4146-4156.
Ab initio curved-wave x-ray-absorption fine structure

5445 Nagase, S. (1989). Angew. Chem., 101, 340-341.
Spannungsarme Cuban-Analoga mit Si-, Ge-, Sn- and Pb-Gerüsten

5446 Nagase, S. and Kudo, T. (1990). J. Chem. Soc., Chem. Commun., 630-632.
Theoretical study of bicyclo[2.2.0]hexaplumbane. A more flexible and less strained Pb skeleton compared with C, Si, Ge, and Sn skeletons

5447 Nagase, S., Suzuki, S. and Kurakake, T. (1990). J. Chem. Soc., Chem. Commun., 1724-1726.
Do distibene(HSb=SbH) and dibismuthene(HBi=BiH) feature double bonding? A Theoretical comparison with diphosphene (HP=PH) and diarsene(HAs=AsH)

5448 Nagiyev, Sh.M. (1988). J. Phys. A, 21, 2559-2564.
On a hidden symmetry of a relativistic Coulomb problem in the quasipotential approach

5449 Nahar, S.N. (1991). Phys. Rev. A, 43, 2223-2235.
Cross sections and spin polarizations for $e^{\pm}$ scattering from cadmium

5450 Nahar, S.N. and Wadehra, J.M. (1991). Phys. Rev. A, 43, 1275-1289.
Relativistic approach for $e^{\pm}$ scattering from argon

5451 Nakamatsu, H., Adachi, H. and Mukoyama, T. (1992). Bull. Inst. Chem. Res., Kyoto Univ., 70 (1), 16-27.
Relativistic molecular orbital calculations for polyatomic molecules

5452 Nakamura, E., Miyachi, Y., Koga, N. and Morokuma, K. (1992). J. Am. Chem. Soc., 114, 6686-6692.
Theoretical studies of heteroatom-directed carbometalation. Addition of MeCu, Me_2Cu^-, and MeLi to substituted acetylenes

5453 Nakamura, S. and Morokuma, K. (1988). Organometallics, 7, 1904-1909.
Theoretical investigation of the platinum ketene complex $Pt(PR_3)_2(CH_2{=}C{=}O)$ and carbene complex $Pt(PR_3)_2(CH_2)(CO)$

5454 Nakanishi, N. (ed.) (1988). Prog. Theor. Phys., Suppl. 95, 1-111.
Behaviour of the solutions to the Bethe-Salpeter equation

5455 Nakatsuji, H. and Ehara, M. (1990). Chem. Phys. Lett., 172, 261-264.
Collision-induced absorption spectra of CsXe system studied by SAC-CI theory

5456 Nakatsuji, H., Hada, M. and Kawashima, A. (1992a). Inorg. Chem., 31, 1740-1744.
Electronic structures of dative metal-metal bonds: ab initio molecular orbital calculations of $(OC)_5Os\text{-}M(CO)_5$ (M=W, Cr) in comparison with $(OC)_5M\text{-}M(CO)_5$ (M=Re, Mn)

5457 Nakatsuji, H., Hada, M. and Yonezawa, T. (1987). J. Am. Chem. Soc., 109, 1902-1912.
Theoretical study on the chemisorption of a hydrogen molecule on palladium

5458 Nakatsuji, H., Matsuzaki, Y. and Yonezawa, T. (1988). J. Chem. Phys., 88, 5759-5769.
Ab initio theoretical study on the reactions of a hydrogen molecule with small platinum clusters: A model for chemisorption on a Pt surface

5459 Nakatsuji, H. and Nakai, H. (1990). Chem. Phys. Lett., 174, 283-286.
Theoretical study on molecular and dissociative chemisorptions of an O_2 molecule on an Ag surface: dipped adcluster model combined with symmetry-adapted cluster-configuration interaction method

5460 Nakatsuji, H. and Nakai, H. (1992). Can. J. Chem., 70, 404-408.
Dipped adcluster model study for the end-on chemisorption of O_2 on an Ag surface

5461 Nakatsuji, H. and Saito, S. (1990). J. Chem. Phys., 93, 1865-1875.
Theoretical study for the excited states of $MoO_{4-n}S_n^{2-}$ (n=0-4) and $MoSe_4^{2-}$

5462 Nakatsuji, H. and Saito, S. (1991). Int. J. Quantum Chem., 39, 93-113.
Excited and ionized states of RuO_4 and OsO_4 studied by SAC and SAC-CI theories

5463 Nakatsuji, H., Sugimoto, M. and Saito, S. (1990). Inorg. Chem., 29, 3095-3097.
Electronic origin of ^{95}Mo NMR chemical shifts in molybdenum complexes. Relationship between excitation energy and chemical shift

5464 Namikawa, K., Ando, M., Nakajima, T. and Kawata, H. (1990). AIP Conf. Proc., 215, 760-769.
X-ray magnetic scattering from atomic inner shell electron

5465 Nandi, S. and Shastry, C.S. (1992). Pramana, 36, 271-288.
Classical limit of relativistic quantal system with attractive Coulomb interaction

5466*Narnhofer, H. (1974). Acta Phys. Austr., 40, 306-322.

Quantum theory for $1/r^2$-potentials

5467 Nasreen, G., Manson, S.T. and Deshmukh, P.C. (1989). Phys. Rev. A, 40, 6091-6094.
Photoionization of inner shells of the Mg and Ar isonuclear sequences

5468 Ndimubandi, J. and Ntibashirakandi, L. (1991). Bull. Soc. Roy. Sci. Liège, 60, 429-436.
On relativistic oscillators through supersymmetry and Foldy-Wouthuysen transformations

5469 Nefiodov, A.V. (1990). Vestn. Leningr. Univ., Fiz. Khim., (1), 85-88.
Form of the spectral line shape of the relativistic atom in the non-resonant laser field (in Russian)

5470 Neisler, R.P. (1987). Diss. Abstr. Int. B, 48, 1367.
The mercurous di-ion (Hg_2^{+2}), a theoretical study

5471 Neisler, R.P. and Pitzer, K.S. (1987). J. Phys. Chem., 91, 1084-1087.
The dipositive dimeric ion Hg_2^{2+}: A theoretical study

5472 Németh, Z. (1992). Nucl. Instr. Meth. Phys. Res. A, 312, 296-302.
Comparison of experimental and Dirac-Fock calculated high-multipole internal conversion coefficients

5473 Nemoshkalenko, V.V., Antonov, V.N., Antonov, Vl.N., Baglyuk, A.I. and Perlov, A.Ya. (1990). Metallofizika, 12(2), 3-6.
The band structure and optical properties of thorium

5474 Nemoshkalenko, V.V., Antonov, V.N., Antonov, Vl.N., Knyazev, Yu.V., Kirillova, M.M. and Krasovskii, A.E. (1989). Fiz. Met. Metalloved., 67, 501-507.
Energy spectrum and optical properties of a lutetium single crystal (in Russian)

5475 Nemoshkalenko, V.V., Mil'man, V. Yu. and Antonov, V.N. (1987). Ukr. Fiz. Zh., 32, 1238-1245.
Relativistic pseudopotential for transition metals (in Russian)

5476 Neugart, R., Otten, E.W., Wendt, K., Ahmad, S., Panigrahy, S.N., Dougherty, R.W., Mishra, K.C., Das, T.P. and Andriessen, J. (1989). Hyperfine Interactions, 59, 145-148.
Evaluation of the magnetic moments of radium isotopes

5477 Neumann, T., Borstel, G., Scharfschwerdt, C. and Neumann, M. (1992). Phys. Rev. B, 46, 10623-10628.
Electronic structure of $KNbO_3$ and $KTaO_3$

5478 Nguyen, K.A., Carroll, M.T. and Gordon, M.S. (1991). J. Am. Chem. Soc., 113, 7924-7929.
Structures and bonding of group IV sulfur and oxygen propellane derivatives

5479*Ni, G.-J. (1979). Phys. Energ. Fort. Phys. Nucl., 3, 449.
The Levinson theorem and its generalization in relativistic quantum mechanics

5480 Ni, G.-J. (1986). Fudan Xuebao, Ziran Kexuebao, 25, 455-460.
Exact solution of the Dirac equation under a constant uniform electric field and the electron-positron pair creation rate in vacuum (in Chinese)

5481*Nickisch, L.J., Durand, L. and Durand, B. (1984). Phys. Rev. D, 30, 660-670.
Salpeter equation in position space: Numerical solution for arbitrary confining potentials

5482*Nicolaides, C.A. (1972). Chem. Phys. Lett., 17, 436-438.
Theoretical lifetimes of the metastable $^5S_2{}^0$ states in atomic carbon and oxygen

5483*Nicolaides, C.A. (1983). IEEE J. Quantum Electronics, QE-19, 1781-1785.
Radiative autoionization - A proposal for the construction of short-wavelength lasers

5484 Nicolaides, C.A. (1988). Studies in Phys. Theor. Chem., 62, 343-360.
State-specific theory of quantum chemistry

5485 Nicolaides, C.A. and Aspromallis, G. (1986). J. Phys. B: At. Mol. Phys., 19, L841-L844.
Mechanisms of configuration interaction in relativistic autoionisation: threshold phenomena and the Li ^4P_J levels

5486 Nicolaides, C.A. and Aspromallis, G. (1988). Phys. Scr., 38, 55-58.
The threshold effect on the lifetime of relativistically autoionizing metastable states: The doubly excited spectrum of the fictitious Z= 2.05 atom

5487 Nicolaides, C.A. and Aspromallis, G. (1991). Phys. Rev. A, 44, 2217-2219.
Binding of an electron by He and Xe

5488 Nicolaides, C.A., Aspromallis, G. and Beck, D.R. (1989). J. Mol. Struct.(Theochem), 199, 283-302.
Bound excited states of atomic negative ions

5489*Nicolaides, C.A. and Beck, D.R. (1978).
Many-body theory of photoabsorption in atoms and molecules In "Excited States in Quantum Chemistry", Ed. C.A. Nicolaides and D.R. Beck, Dordrecht, Holland, pp. 143-182.

5490 Nicolas, G. and Barthelat, J.C. (1986). J. Phys. Chem., 90, 2870-2877.
Electronic structure and optical spectrum of the copper-ethylene complex

5491 Nicolas, G. and Spiegelmann, F. (1990). J. Am. Chem. Soc., 112, 5410-5419.
Theoretical study of ethylene-noble metal complexes

5492 Nielsen, U., Aaen Andersen, L.U., Poulsen, O. and Riis, E. (1987). J. Phys. B: At. Mol. Phys., 20, 1697-1708.
Hyperfine structure in erbium (^{167}Er I + II + III): Multiconfiguration Dirac-Fock interpretation of measurements

5493 Nielsen, U. and Pfeufer, V. (1986). J. Phys. B: At. Mol. Opt. Phys., 19, L645-L650.
Ab initio state-dependent multiconfiguration Dirac-Fock calculations of the electron density at the nucleus for low-lying states in the $4f^{11}5d6s^2$ Er I configuration

5494*Nieto, M.M. (1979). Am. J. Phys., 47, 1067-1072.
Hydrogen atom and relativistic pi-mesic atom in N-space dimensions

5495 Nikolaev, A.V. and Ionova, G.V. (1991). Phys. Stat. Sol. B, 167, 613-623.
Trend to localization of f-states in actinide metals

5496 Nikolaus, B.A., Hoch, T. and Madland, D.G. (1992). Phys. Rev. C, 46, 1757-1781.
Nuclear ground state properties in a relativistic point coupling model

5497 Nilsen, J. (1989). At. Data Nucl. Data Tables, 41, 131-177.
Dielectronic satellite spectra for neon-like ions

5498 Nilsen, J., Safronova, U.I. and Safronova, M.S. (1990). J. Quant. Spectrosc. Radiat. Transfer, 43, 445-450.
Z-dependences of the atomic characteristics for selected 2l4l′ states

5499 Ning, Y.-M., Xie, J.-K., Zhao, S.-B. and Wu, P.-F. (1991). Phys. Stat. Sol. B, 165, K69-K71.
Theoretical calculation for the spectra and the cubic zero-field splitting of Mn^{2+} ions in the CaF_2: Mn^{2+} system

5500 Nizam, M., Allavena, M., Bouteiller, Y., Suits, B.H. and White, D. (1989). J. Magn. Res., 82, 441-453.
NMR chemical shifts and the electronic structure of lead in lead halides

5501 Nizam, M. and Bouteiller, Y. (1990). J. Mol. Struct. (Theochem), 206, 99-107.
Ab initio calculations of potential barriers for ionic conductivity of the β-PbF_2 compound

5502 Nizam, M., Bouteiller, Y. and Allavena, M. (1987). J. Mol. Struct., 159, 365-370.
Determination of structural and vibrational properties of some lead compounds by pseudo-potential methods

5503 Nizam, M., Bouteiller, Y., Silvi, B., Pisani, C., Causà, M., and Dovesi, R. (1988). J. Phys. C, 21, 5351-5359.
A theoretical investigation of the electronic structure and some thermodynamic properties of β-PbF_2

5504 Nogami, Y. and Toyama, F.M. (1992). Phys. Rev. A, 45, 5258-5261.
Transparent potential for the one-dimensional Dirac equation

5505 Nogami, Y. and van Dijk, W. (1986). Phys. Rev. C, 34, 1855-1859.
Dirac equation with a separable potential

5506 Nogueira, S.R. and Guenzburger, D. (1991). Phys. Rev. A, 44, 5558-5568.
Valence-orbital ionization potentials for third-row transition elements

5507 Nogueira, S.R., Vugman, N.V. and Guenzburger, D. (1992). Chem. Phys., 164, 229-245.
Molecular orbital calculations of the electronic structure and g-tensor for divalent paramagnetic iridium complexes

5508 Norman, M.R. (1987).
Density functional theory and f electron systems
In "Condensed Matter Theories", Vol. 2, Ed. P. Vashishta, R.K. Kalia and R.F. Bishop, Plenum, New York, pp. 113-119.

5509 Norrby, L.J. (1991). J. Chem. Ed., 68, 110-113.
Why is mercury liquid?

5510 Norrington, P.H. and Grant, I.P. (1987). J. Phys. B: At. Mol. Phys., 20, 4869-4881.
Low-energy electron scattering by Fe XXIII and Fe VII using the Dirac R-matrix method

5511 Noskov, M.D. and Shvartsman, Sh.M. (1989). Izv. Vyssh. Uche. Zave., Fiz., 32(6), 70-74; Sov. Phys. J., 32, 468-471.
Investigation of Green's functions in an external electromagnetic field by the eigenfunction method

5512 Nour, S., Mehadji, C., Chermette, H. and Krupa, J.-C. (1992). Int. J. Quantum Chem., 41, 421-435.
Quasi-relativistic MS-Xα approach to the electronic structures of protactinium halide ions of $[PaX_8]^{4-}$ type

5513 Novak, I. (1990). Bull. Chem. Soc. Jap., 63, 1278-1280.
Electronic structure of chiral halomethanes

5514 Novaro, O., Garcia-Prieto, J., Poulain, E. and Ruiz, M.E. (1986). J. Mol. Struct.(Theochem), 135, 79-91.
Molecular orbital study of metal dihydride complexes. SCF and CI studies of Li-, Cu-, Pd-, Ag- and Pt-H_2 systems

5515 Novaro, O. and Jarque, C. (1991). Theor. Chim. Acta, 80, 19-27.
Potential energy surface for the Pd-H_2 system: Comparison with matrix isolation experiments

5516 Novaro, O.A. (1989).
Activation of small molecules by transition metal atoms

Theoretical interpretation of low-temperature experiments with Cu, Pd, and Pt Atoms
In "The Challenge of d and f Electrons", Ed. D.R. Salahub and M.C. Zerner, ACS Symposium Series, 394, 106-124.

5517 Novo-Gradac, K.J. (1989). Diss. Abstr. Int. B, 50, 5615.
Molecular orbital studies of metal-metal bonding in bimetallic organoactinide compounds

5518 Ohanessian, G., Brusich, M.J. and Goddard III, W.A. (1990). J. Am. Chem. Soc., 112, 7179-7189.
Theoretical study of transition-metal hydrides. 5. HfH^+ through HgH^+, BaH^+, and LaH^+

5519 Ohanessian, G. and Goddard III, W.A. (1990). Acc. Chem. Res., 23, 386-392.
Valence-bond concepts in transition metals: Metal hydride diatomic cations

5520 Ohlsén, H. and Calais, J.L. (1987). Phys. Rev. B, 35, 7914-7921.
Effect of spin-orbit coupling on the conduction-electron Zeeman splitting in platinum-group metals

5521 Ohno, M. (1987a). Phys. Rev. A, 35, 3350-3353.
g-Hartree ab initio calculation of ionization energies of Ca (Z=20)

5522 Ohno, M. (1987b). Z. Phys. D, 6, 13-15.
g-Hartree ab-initio calculation of atomic ionization energy of small atoms

5523 Ohno, M. (1988). Phys. Rev. A, 37, 2953-2957.
g-Hartree ab initio calculations of the ionization energies of the four-electron isoelectronic series

5524 Ohtsuki, K. and Hijikata, K. (1988). J. Phys. Soc. Jap., 57, 4150-4159.
The calculation of the fine and the hyperfine structures of $2\,^3P$ and $2\,^1P$ states of the helium isoelectronic sequence up to Z=10

5525 Oja, A.S., Wang, X.W. and Harmon, B.N. (1989). Phys. Rev. B, 39, 4009-4021.
First-principles study of the conduction-electron-mediated interactions between nuclear spins in copper metal

5526 Okada, S. and Matsuoka, O. (1989). J. Chem. Phys., 91, 4193-4196.
Relativistic well-tempered Gaussian basis sets for helium through mercury

5527 Okada, S., Shinada, M. and Matsuoka, O. (1990). J. Chem. Phys., 93, 5013-5019.
Relativistic well-tempered Gaussian basis sets for helium through mercury. Breit interaction included

5528 Okun, L.B. (1989a). Phys. Today, 42, 31-36.
The concept of mass

5529 Okun, L.B. (1989b). Sov. Phys. Usp., 32, 629-638.
The concept of mass (mass, energy, relativity)

5530 Olsson, T. and Lindgren, I. (1987). Z. Phys. D, 7, 95-106.
Many-body aspects of the parameterization of the hyperfine and other atomic interactions

5531 Omote, M., Kamefuchi, S., Takahashi, Y. and Ohnuki, Y. (1989). Fortschr. Phys., 37, 933-950.
Galilean covariance and the Schrödinger equation

5532 Onishchuk, V.A. (1990). Khim. Fiz., 9, 1723-1727.
Chirality of valence electrons on atomic nuclei of asymmetrical molecules. On the influence of molecular asymmetry on radioactivity (in Russian)

5533 Onoe, J., Takeuchi, K., Nakamatsu, H., Mukoyama, T., Sekine, R. and Adachi, H. (1992a). Chem. Phys. Lett., 196, 636-640.
Theoretical study on the UV excited states of UF_6 using the discrete-variational Dirac-Slater calculation method

5534 Onoe, J., Takeuchi, K., Nakamatsu, H., Mukoyama, T., Sekine, R. and Adachi, H. (1992b). J. El. Spectr. Rel. Phen., 60, 29-36.
Discrete-variational Dirac-Slater calculation on XPS for the valence levels of UF_6

5535 Oreg, J., Goldstein, W.H., Bar-Shalom, A. and Klapisch, M. (1989). Phys. Rev. A, 39, 4599-4609.
Configuration-average energy shift owing to configuration interaction

5536 Oreg, Y. and Entin-Wohlman, O. (1992). Phys. Rev. B, 46, 2393-2399.
Transmissions through low-dimensional mesoscopic systems subject to spin-orbit scattering

5537 Orita, N., Niizeki, K., Shindo, K. and Tanaka, H. (1992). J. Phys. Soc. Japan, 61, 4502-4510.
The band structure of solid iodine under pressure and the mechanism of the pressure-induced insulator-to-metal transition

5538 Orlando, R., Pisani, C., Roetti, C. and Stefanovich, E. (1992). Phys. Rev. B, 45, 592-601.
Ab initio Hartree-Fock study of tetragonal and cubic phases of zirconium dioxide

5539 Ortiz, G. and Ballone, P. (1991a). Phys. Rev. B, 43, 6376-6387.
Pseudopotentials for non-local density functionals

5540 Ortiz, G. and Ballone, P. (1991b). Phys. Rev. B, 44, 5881-5884.
Metastability of doubly charged transition-metal dimers in density-functional theory

5541 Ortiz, G. and Ballone, P. (1991c). Z. Phys. D, 19, 169-171.

Bonding properties of IIA and IIB dimers in density functional theory with gradient corrections to the local density approximation

5542 Ortiz, J.V. (1988). J. Chem. Phys., 89, 6348-6352.
Electron binding energies of anionic alkali metal atoms from partial fourth order electron propagator theory calculations

5543 Ortiz, J.V., Hay, P.J. and Martin, R.L. (1992). J. Am. Chem. Soc., 114, 2736-2737.
Role of d and f orbitals in the geometries of low-valent actinide compounds. Ab initio studies of $U(CH_3)_3$, $Np(CH_3)_3$, and $Pu(CH_3)_3$

5544 Ostanin, S.A. and Shirokovskii, V.P. (1989a). Fiz. Met. Metalloved., 67, 244-248.
Optical absorption in molybdenum. Relativistic calculation (in Russian)

5545 Ostanin, S.A. and Shirokovskii, V.P. (1989b). Fiz. Met. Metalloved., 68, 1223-1226.
Optical conductivity of niobium and tantalum in relativistic model (in Russian)

5546 Ostanin, S.A. and Shirokovskii, V.P. (1990). J. Phys.: Cond. Matt., 2, 7585-7595.
A simplified method of ab initio calculation of electron states in relativistic magnetics. I. Ferromagnets

5547 Ostanin, S.A. and Shirokovskii, V.P. (1991). J. Phys.: Condens. Matter, 3, 5287-5295.
A simplified method of ab initio calculation of electron states in relativistic magnetics: II. Ferrimagnets and antiferromagnets

5548 Ostanin, S.A., Trubitsyn, V. Yu. and Shilkova, N.A. (1987). Fiz. Met. Metalloved., 64, 421-425.
Calculation of normalized wave functions in molybdenum. Role of relativistic interactions

5549 Ostrovskii, V.N. and Sheinerman, S.A. (1989). Opt. Spektrosk., 67, 16-19; Opt. Spectrosc., 67, 8-10.
Radiative transitions in the outer shells of the Hg^+ ion

5550 O'Sullivan, G. (1989). J. Phys. B: At. Mol. Opt. Phys., 22, 987-996.
The emission spectrum of a laser-produced plasma of rubidium in the 320-450 Å region

5551 O'Sullivan, G. and Kane, M. (1989). Phys. Scr., 39, 317-320.
$4s^2p^3$ and $4s^24p^25s$ configurations in Ru XII and Rh XIII

5552 Otto, M., Lotz, S.D. and Frenking, G. (1992). Inorg. Chem., 31, 3647-3655.
Quantum mechanical ab initio studies of the structures and stabilities of halogen azides XN_3 (X=F, Cl, Br, I)

5553 Otto, P., Wagner, F. and Beck, H.P. (1992). Phys. Rev.

B, 45, 2654-2661.
Ab initio molecular-cluster study of the $(TlI)_n$ crystal

5554*Overhof, H., Knecht, J., Fischer, R. and Hensel, F. (1978). J. Phys. F, 8, 1607-1616.
Photoelectron spectroscopy studies and calculation of the energy bands of CsAu and RbAu

5555*Overhof, H., Uchtmann, H. and Hensel, F. (1976). J. Phys. F, 6, 523-529.
Band theoretical study of the optical gap in expanded fluid mercury

5556 Pacchioni, G. and Bagus, P.S. (1990). J. Chem. Phys., 93, 1209-1214.
Cluster calculations of CO chemisorbed on the bridge site of Pd(100)

5557 Pacchioni, G. and Bagus, P.S. (1992). Inorg. Chem., 31, 4391-4398.
Metal-phosphine bonding revisited. -basicity, -acidity, and the role of phosphorus d orbitals in zerovalent metal-phosphine complexes

5558 Pacheco, A.F. and Sañudo, J. (1989). Phys. Rev. A, 39, 380-382.
Energy shifts in the x rays of strange atoms

5559 Pacher, M.C., González, A.D. and Miraglia, J.E. (1987). Phys. Rev. A, 35, 4108-4113.
Retardation effects in radiative electron capture

5560 Pachucki, K. (1991). Phys. Rev. A, 44, 5407-5413.
Time evolution in quantum electrodynamics

5561 Pachucki, K. (1992). Phys. Rev. A, 46, 648-651.
Analytical evaluation of higher-order binding corrections to the Lamb shift

5562 Padma, R. and Deshmukh, P.C. (1992). Phys. Rev. A, 46, 2513-2518.
Calculations of generalized oscillator strength for electron-impact excitations of krypton and xenon using a relativistic local-density potential

5563 Pak, K., Ermler, W.C., Kern, C.W. and Bondybey, V.E. (1991). J. Cluster Sci., 2, 19-28.
Full spin-orbit configuration interaction calculations on electronic states of LiBe

5564 Pal'chikov, V.G. (1990). Zh. Prik. Spektrosk., 52, 742-748; J. Appl. Spectrosc., 5, 488-492.
Effects of electron interactions in calculations of probabilities of two-photon transitions in ionized atoms

5565 Pal'chikov, V.G. and Safronova, U.I. (1990). Opt. Spektrosk., 68, 481-486; Opt. Spectrosc., 68, 281-284.
Analysis of Dirac-Fock calculations based on perturbation theory in reference to 1/Z

5566 Pal'chikov, V.G. and Tkachev, A.N. (1989). Opt. Spektrosk., 67, 995-997; Opt. Spectrosc., 67, 586-588.
Calculation of dipole susceptibility of a heliumlike ion taking into account electron interaction

5567 Palladino, B.E. and Leal Ferreira, P. (1987). Phys. Lett. B, 185, 118-120.
Level order in a relativistic potential model

5568 Pan, H.-Y. and Zhao, Z.-S. (1989). Int. J. Quantum Chem., 36, 15-18.
On the validity of the mass-velocity operator in quantum chemistry

5569 Paniagua, J.C. and Illas, F. (1990). Chem. Phys. Lett., 170, 561-564.
Chemisorption of atomic hydrogen on rhodium: An ab initio cluster-model approach

5570 Panigrahy, S.N., Dougherty, R.W., Ahmad, S., Mishra, K.C., Das, T.P., Andriessen, J., Neugart, R., Otten, E.W. and Wendt, K. (1991a). Phys. Rev. A, 43, 2215-2222.
Evaluation of the magnetic moment of ^{213}Ra

5571 Panigrahy, S.N., Dougherty, R.W., Das, T.P. and Andriessen, J. (1991b). Phys. Rev. A, 44, 121-125.
Theory of hyperfine interactions in positive ions isoelectronic with the noble-metal atoms

5572 Panigrahy, S.N., Dougherty, R.W., Das, T.P. and Andriessen, J. (1989). Phys. Rev. A, 40, 1765-1774.
Theory of hyperfine interactions in lithiumlike systems

5573 Panja, M.M., Bag, M., Dutt, R. and Varshni, Y.P. (1992). Phys. Rev. A, 45, 1523-1530.
Large-N expansion method for a spin-1/2 particle in the presence of vector and scalar potentials

5574 Panja, M.M. and Dutt, R. (1988). Phys. Rev. A, 38, 3937-3943.
Shifted large-N expansion for the energy levels of relativistic particles

5575 Panja, M.M, Dutt, R. and Varshni, Y.P. (1990). Phys. Rev. A, 42, 106-115.
Shifted large-N expansion for a relativistic spin-½ particle in screened Coulomb potentials

5576 Pankratov, O.A. (1987). Pis'ma Zh. Eksp. Teor. Fiz., 45, 82-85.
Types of heterogeneous structures, boundary electronic states and chiral anomaly in semiconductors with Dirac bands

5577 Pápai, I., Goursot, A., St-Amant, A. and Salahub, D.R. (1992). Theor. Chim. Acta, 84, 217-235.
Molecular structure of mono- and dicarbonyls of rhodium and palladium

5578 Papanikolaou, N., Stefanou, N., Dederichs, P.H., Geier, S.

and Bergmann, G. (1992). Phys. Rev. Lett., 69, 2110-2113.
First-principles calculations of the spin-orbit scattering cross section of sp impurities in Mg

5579 Papp, E. (1986a). Phys. Rev. A, 33, 719-722.
General properties of ground-state energies of relativistic two-body Hamiltonians with Coulomb and Yukawa potentials

5580 Papp, E. (1986b). Phys. Rev. A, 34, 47-55.
Quasiclassical ground-state energies for quarkonia potentials

5581 Papp, E. (1986c). Phys. Reports, 136, 103-151.
Quasiclassical approach to the virial theorem and to the evaluation of the ground-state energy

5582 Papp, E. (1991a). Phys. Lett. B, 259, 19-23.
Closed 1/N mass formulae for relativistic two-body Q_1Q_2 systems

5583 Papp, E. (1991b). Ann. Phys.(Leipzig), 48, 319-326.
Quasiclassical 1/N energy-formulas for relativistic two-body systems

5584 Papp, E. (1992). Phys. Lett. B, 288, 148-152.
On the fixed point structure of the quantum mechanical Dirac-Coulomb system

5585 Papp, E. and Micu, C. (1991). Z. Naturforsch., 46a, 729-732.
1/N-energy-solutions to relativistic two-boson systems with an exponential potential

5586 Parera, J.J. and Grave de Peralta, L. (1987). Rev. Cubana Quim., 3(2), 17-25.
Relativistic electron pair in a multi-electron atom: Configuration-space formulation

5587 Parera, J.J. and Lopez Boada, R. (1986). Rev. Cubana Fis., 6(2), 53-58.
Relativistic study of valence electrons in alkaline earth metal atoms. I. Scheme for calculating their energy

5588 Parera López, J.J. (1990). Phys. Lett. A, 143, 70-73.
Analytical variational calculation for the ground state of the relativistic high-Z helium-like ion with magnetic interaction

5589*Parera López, J.J., Roca Oria, E. and De Peralta Morell, L.G. (1985). Rev. Cubana Fis., 5(2-3), 51-62.
Relativistic electron pair in a multielectron atom. I. Generalities

5590 Parisse, B. (1992a). Helv. Phys. Acta, 64, 557-591.
Résonances pour l'opérateur de Dirac

5591 Parisse, B. (1992b). Helv. Phys. Acta, 65, 1077-1118.
Resonances for the Dirac operator. II

5592 Parker, G.W. (1986). Phys. Rev. A, 33, 799-803.
Energy levels of inner electrons in atoms

5593 Parlant, G., Senekowitsch, J., ONeil, S.V. and Yarkony, D.R. (1991). J. Chem. Phys., 94, 7208-7211.
Spin-forbidden decay of the dication HS^{2+}

5594 Parpia, F.A. and Grant, I.P. (1990). J. Phys. B: At. Mol. Opt. Phys., 23, 211-217.
Accurate Dirac-Coulomb energies for the ground states of helium-like atoms

5595 Parpia, F.A. and Grant, I.P. (1991a). J. Phys.(Paris), Coll., IV, 1, C1-33 - C1-46.
Software for relativistic atomic theory: The GRASP project at Oxford

5596 Parpia, F.A. and Grant, I.P. (1991b). AIP. Conf. Proc., 233, 528-530.
Software for relativistic atomic structure theory: the GRASP project at Oxford

5597 Parpia, F.A. and Mohanty, A.K. (1991).
A relativistic multiconfiguration self-consistent-field method for atoms
In "MOTECC-91", Ed. E. Clementi, ESCOM, Leiden, pp. 211-231.

5598 Parpia, F.A. and Mohanty, A.K. (1992). Phys. Rev. A, 46, 3735-3745.
Relativistic basis-set calculations for atoms with Fermi nuclei

5599 Parpia, F.A., Mohanty, A.K. and Clementi, E. (1992a). J. Phys. B: At. Mol. Opt. Phys., 25, 1-16.
Relativistic calculations for atoms: self-consistent treatment of Breit interaction and nuclear volume effect

5600 Parpia, F.A., Perger, W.F. and Das, B.P. (1988). Phys. Rev. A, 37, 4034-4036.
Parity-nonconserving transitions in Cs and Tl: A semiempirical treatment of core-polarization effects

5601 Parpia, F.A., Tong, M. and Fischer, C.F. (1992b). Phys. Rev. A, 46, 3717-3724.
Relativistic calculations of nuclear motional effects in many-electron atoms

5602 Partridge, H., Bauschlicher Jr., C.W. and Langhoff, S.R. (1990). Chem. Phys. Lett., 175, 531-535.
Theoretical study of the positive ions of the dimers and trimers of the group IB metals (Cu, Ag, and Au)

5603 Partridge, H., Bauschlicher Jr., C.W. and Langhoff, S.R. (1992). J. Phys. Chem., 96, 5350-5355.
Theoretical study of metal ions bound to He, Ne, and Ar

5604 Passante, R. and Power, E.A. (1987). Phys. Lett. A, 122, 14-16.
The Lamb shift in non-relativistic quantum electrodynamics

5605 Pastor, G.M. and Reindl, S.L. (1992). Ber. Bunsenges.,

96, 1299-1301.
Study of electron correlations in Pb_n^{2+}-clusters

5606 Pastor, G.M., Reindl, S.L. and Bennemann, K.H. (1991). Phys. Rev. Lett., 67, 1250-1253.
Metastability of Pb_n^{2+} clusters: Role of correlation effects

5607 Pastusiak, W. and Kuriata, J. (1986). Phys. Stat. Sol. B, 135, K69-K73.
On the microscopic interpretation of tetragonal second-order spin-lattice coefficients of $4f^7$ ions in the dynamic relativistic crystal field approach

5608 Pavolini, D., Gustavsson, T., Spiegelmann, F. and Daudey, J.-P. (1989). J. Phys. B: At. Mol. Opt. Phys., 22, 1721-1731.
Theoretical study of the excited states of the heavier alkali dimers: I. The RbCs molecule

5609 Pavsic, M. (1991). Found. Phys., 21, 1005-1019.
On the interpretation of the relativistic quantum mechanics with invariant evolution parameter

5610 Pedash, Yu.F., Babich, E.N., Ponomarev, O.A. and Luzanov, A.V. (1990). Opt. Spektrosk., 69, 550-553; Opt. Spectrosc., 69, 329-330.
Spin-orbit interaction in the π-electron approximation

5611 Pekeris, C.L. and Frankowski, K. (1989). Phys. Rev. A, 39, 518-529.
Hyperfine splitting in muonium, positronium, and hydrogen, deduced from a solution of Dirac's equation in Kerr-Newman geometry

5612 Pellow, R. and Vala M. (1989). J. Chem. Phys., 90, 5612-5621.
The external heavy atom effect: Theory of spin-orbit coupling of alkali and noble metals in rare gas matrices

5613 Pepper, M. and Bursten, B.E. (1990). J. Am. Chem. Soc., 112, 7803-7804.
Ab initio studies of the electronic structure of the diuranium molecule

5614 Pepper, M. and Bursten, B.E. (1991). Chem. Rev., 91, 719-741.
The electronic structure of actinide-containing molecules: A challenge to applied quantum chemistry

5615 Percoco, U. and Villalba, V.M. (1989a). Phys. Lett. A, 140, 105-107.
Aharonov-Bohm effect for a relativistic Dirac electron

5616 Percoco, U. and Villalba, V.M. (1989b). Phys. Lett. A, 141, 221-222.
On the Dirac equation with a one-dimensional scalar potential

5617 Perger, W.F., Cai, Z.-Y. and Beck, D.R. (1991). J. Phys.:

At. Mol. Opt. Phys., 24, 4863-4875.
Numerical calculation of relativistic atomic continuum wavefunctions in a frozen-core Dirac-Fock potentiual: application to the $6p^2$ resonances in Hg

5618 Perger, W.F. and Das, B.P. (1987a). Phys. Rev. A, 35, 3942-3944.
Influence of the Dirac-Hartree-Fock starting potential on the parity-nonconserving electric-dipole-transition amplitudes in cesium and thallium

5619 Perger, W.F. and Das, B.P. (1987b). J. Phys. B: At. Mol. Phys., 20, 665-670.
Multiconfigurational Dirac-Fock calculations of fine-structure intervals in the beryllium isoelectronic sequence

5620 Perger, W.F. and Karighattam, V. (1991). Comp. Phys. Comm., 66, 392-402.
Relativistic continuum wavefunction solver

5621*Perkins, G.A. (1976). Found. Phys., 6, 237-248.
The direction of the Zitterbewegung: A hidden variable

5622 Perry, J.K., Goddard III, W.A. and Ohanessian, G. (1992). J. Chem. Phys., 97, 7560-7572.
Inequivalence of equivalent bonds: symmetry breaking in $Co(CH_3)_2^+$

5623 Pershina, V., Sepp, W.-D., Bastug, T., Fricke, B. and Ionova, G.V. (1992c). J. Chem. Phys., 97, 1123-1131.
Relativistic effects in physics and chemistry of element 105. III. Electronic structure of hahnium oxyhalides as analogs of group 5 elements oxyhalides

5624 Pershina, V., Sepp, W.-D., Fricke, B., Kolb, D., Schädel, M. and Ionova, G.V. (1992b). J. Chem. Phys., 97, 1116-1122.
Relativistic effects in physics and chemistry of element 105. II. Electronic structure and properties of group 5 elements bromides

5625 Pershina, V., Sepp, W.-D., Fricke, B. and Rosén, A. (1992a). J. Chem. Phys., 96, 8367-8378.
Relativistic effects in physics and chemistry of element 105. I. Periodicities in properties of group 5 elements. Electronic structure of the pentachlorides

5626 Pershina, V.G., Ionova, G.V. and Suraeva, N.I. (1989). Croatica Chim. Acta, 62, 763-773.
Electronic structure of superheavy elements and their compounds

5627 Petelenz, P. and Smith Jr., V.H. (1989). Phys. Rev. A, 39, 1016-1019.
Vacuum-polarization correction in muonic molecules

5628 Pettersson, L.G.M., Bauschlicher Jr., C.W., Langhoff, S.R. and Partridge, H. (1987). J. Chem. Phys., 87, 481-492.
Positive ions of the first- and second-row transition metal hydrides

5629 Pettersson, L.G.M. and Langhoff, S.R. (1986). Chem. Phys. Lett., 125, 429-432.
Theoretical electric dipole moments of SiH, GeH and SnH

5630 Pierloot, K., Renders, A., Goodman, G.L., Devoghel, D. and Görller-Walrand, C. (1991). J. Chem. Phys., 94, 2928-2938.
Interpretation of photoelectron spectra for ThF_4, UF_4, $ThCl_4$, and UCl_4 in terms of relativistic local-density molecular orbital calculations

5631*Pilkuhn, H. (1979).
Relativistic Particle Physics
Springer, Berlin.

5632 Pilkuhn, H. (1987). J. Phys. B: At. Mol. Phys., 20, L11.
Reply to 'On the quadrupole moment of the hydrogen atom'

5633 Pilkuhn, H. (1988). Z. Phys. D, 9, 105-115.
Kramers equations for atoms

5634 Pilkuhn, H. (1992). J. Phys. B: At. Mol. Opt. Phys., 25, 299-315.
Equal-time relativistic two-body equations

5635 Pilkuhn, H. and Soerensen, A.H. (1987). NATO ASI B, 165, 185-189.
Dirac equation for electron channeling

5636 Pimentel, L.O. (1989). Phys. Lett. A, 137, 93-95.
More solutions with local rotational symmetry in which the Dirac equation separates

5637 Pindzola, M.S. (1987). Phys. Rev. A, 35, 4548-4550.
Relativistic effects on giant resonances in electron-impact double ionization

5638 Pindzola, M.S. (1992). Phys. Rev. A, 45, 4530-4534.
Postitron-impact ionization of highly charged ions in lowest-order QED theory

5639 Pindzola, M.S. and Badnell, N.R. (1990). Phys. Rev. A, 42, 6526-6530.
Dielectronic capture processes in highly charged uranium ions

5640 Pindzola, M.S., Badnell, N.R. and Griffin, D.C. (1992). Phys. Rev. A, 45, 7659-7662.
Dielectronic recombination between hyperfine levels of the ground state of Bi^{82+}

5641 Pindzola, M.S., Badnell, N.R., Moores, D.L. and Griffin, D.C. (1991). Z. Phys. D, 21, S23-27.
Electron scattering theory for highly charged ions

5642 Pindzola, M.S. and Buie, M.J. (1988). Phys. Rev. A, 37, 3232-3236.
Fully relativistic electron-impact ionization of atomic ions

5643 Pindzola, M.S. and Buie, M.J. (1989). Phys. Rev. A, 39, 1029-1032.
Electron-impact ionization of uranium atomic ions

5644 Pindzola, M.S., Griffin, D.C. and Bottcher, C. (1988). Electron-impact ionization of heavy atomic ions In "Atomic Processes in Plasmas", Ed. A. Hauer and A.L. Merts, AIP Conf. Proc., 168.

5645 Pindzola, M.S., Moores, D.L. and Griffin, D.C. (1989). Phys. Rev. A, 40, 4941-4946.
Electron-impact ionization of highly charged ions in lowest-order QED theory

5646 Pineda de Castro, L.F. and Müller, H. (1990). J. Less-Common Met., 162, 87-97.
Zum bindungsmechanismus des interstitiellen Fe-atoms in $FeTh_6Br_{15}$ (in German)

5647 Pinnington, E.H. and Baylis, W.E. (1992). Phys. Rev. A, 46, 7325-7327.
Comment on "Relativistic excitation energies and oscillator strengths for the $6s^2\ ^1S_0$ --> $6s6p\ ^1P_1, ^3P_1$ transitions in Hg-like ions"

5648 Pitzer, R.M. and Winter, N.W. (1988). J. Phys. Chem., 92, 3061-3063.
Electronic-structure methods for heavy-atom molecules

5649 Pitzer, R.M. and Winter, N.W. (1991). Int. J. Quantum Chem., 40, 773-780.
Spin-orbit (core) and core potential integrals

5650 Planelles, J., Merchán, M., Nebot-Gil, I. and Tomás, F. (1989). J. Phys. Chem., 93, 6596-6601.
A MO-CI study on the interaction of copper(I) with methylene: The $(CuCH_2)^+$ carbene-like ground state

5651 Planelles, J., Merchán, M. and Tomás, F. (1988). Chem. Phys. Lett., 149, 222-227.
On the low-lying states of the copper methylene $(CuCH_2)$ complex. A MO CI study

5652 Plass, W., Savin, A., Stoll, H., Preuss, H., Nesper, R. and von Schnering, H.-G. (1990). Inorg. Chem., 29, 860-868.
Pseudopotential investigations on the molecules Cu_2Si_2, Cu_2Sn_2, Cu_4Si_4, and Cu_4Sn_4

5653 Plunien, G., Müller, B. and Greiner, W. (1986). Phys. Reports, 134, 87-193.
The Casimir effect

5654 Plunien, G., Müller, B., Greiner, W. and Soff, G. (1989). Phys. Rev. A, 39, 5428-5431.
Nuclear polarization contribution to the Lamb shift in heavy atoms

5655 Plunien, G., Müller, B., Greiner, W. and Soff, G. (1991). Phys. Rev. A, 43, 5853-5866.
Nuclear polarization in heavy atoms and superheavy quasiatoms

5656 Plyushchay, M.S. (1992). Int. J. Mod. Phys. A, 7, 7045-7064.
The model of a free relativistic particle with fractional spin

5657 Pohlner, W.F. and Dreizler, R.M. (1991). Phys. Rev. A, 44, 7165-7169.
Fourth-order gradient corrections to the relativistic Thomas-Fermi-Weizsäcker model

5658 Polasik, M. (1989a). Phys. Rev. A, 39, 616-627.
Theoretical multiconfiguration Dirac-Fock method study on the x-ray spectra of multiply ionized heavy atoms: The structure of the $K\alpha L^n$ lines

5659 Polasik, M. (1989b). Phys. Rev. A, 39, 5092-5097.
Theoretical simulation of the x-ray spectra of multiply ionized heavy atoms: The $K\alpha L^n$ spectra of molybdenum

5660 Polasik, M. (1989c). Phys. Rev. A, 40, 4361-4368.
Theoretical multiconfiguration Dirac-Fock method study on the x-ray spectra of multiply ionized heavy atoms: The structure of the $K\alpha L^0 M^r$ lines

5661 Polasik, M. (1990). Phys. Rev. A, 41, 3689-3697.
Theoretical multiconfiguration Dirac-Fock method study on the x-ray spectra of multiply ionized heavy atoms: The structure of the $K\alpha L^1 M^r$ satellite lines

5662 Polyzou, W.N. (1987). J. Comp. Phys., 70, 117-137.
Nonperturbative method for treating square root operators in relativistic quantum theories

5663 Polyzou, W.N. (1989). Ann. Phys.(NY), 193, 367-418.
Relativistic two-body models

5664*Pomeranchuk, I. and Smorodinsky, J. (1945). J. Phys.(USSR), 9, 97-99.
On the energy levels of systems with Z > 137

5665 Pospelov, Yu.A. and Novikov, S.G. (1988). Fiz. Nizk. Temp., 14, 1247-1252; Sov. J. Low Temp. Phys., 14, 689-692.
The influence of spin-orbit interaction on electron structures of Sb and As

5666 Pouchan, C., Lespes, G. and Dargelos, A. (1988). J. Phys. Chem., 92, 28-33.
Vibrational spectra of methylstannane: A molecular force field and dipole moment derivatives from ab initio second-order Møller-Plesset calculations

5667 Poulain, E., Colmenares, F., Castillo, S. and Novaro, O. (1990). J. Mol. Struct.(Theochem), 210, 337-351.
Activation of the hydrogen bond by transition metals

5668 Poulain, E., Garcia-Prieto, J., Ruiz, M.E. and Novaro, O. (1986). Int. J. Quantum Chem., 29, 1181-1190.
Some theoretical studies on hydrogen molecule capture by platinum atoms

5669 Pratap, R. (1990). Phys. Lett. A, 143, 99-102.
Dirac electron in axial-wiggler fields

5670*Pratt, R.H., Levee, R.D., Pexton, R.L. and Aron, W. (1964a). Phys. Rev., 134, A898-A915.
K-shell photoelectric cross sections from 200 keV to 2 MeV

5671*Pratt, R.H., Levee, R.D., Pexton, R.L. and Aron, W. (1964b). Phys. Rev., 134, A916-A922.
Polarization correlations in atomic photoeffect

5672 Price, C.E., Rost, E., Shepard, J.R. and McNeil, J.A. (1992). Phys. Rev. C, 45, 1089-1097.
Toward a consistent random phase approximation based on the relativistic Hartree approximation

5673 Probst, M.M. (1992). J. Mol. Struct.(Theochem), 253, 275-285.
A quantum chemical study of the binding energies in the hydrates of Zn^{2+}, Cd^{2+} and Hg^{2+}

5674 Pyper, N.C. (1986). Phil. Trans. R. Soc. Lond. A, 320, 107-158.
Relativistic ab initio calculations of the properties of ionic solids

5675 Pyper, N.C. (1988). Mol. Phys., 64, 933-961.
Exact relativistic analogues of the non-relativistic hyperfine structure operators I. Theory

5676 Pyper, N.C. (1990). Mol. Simul., 5, 23-81.
Four important factors in the ab initio determination of accurate inter-ionic potentials

5677 Pyper, N.C., Marketos, P. and Malli, G.L. (1987). J. Phys. C, 20, 4711-4724.
Relativistic modifications of inter-ionic potentials in lead fluoride

5678 Pyykkö, P. (1986a).
Relativistic Theory of Atoms and Molecules. A Bibliography 1916-1985
Lecture Notes in Chem., 41, 389 pp.

5679 Pyykkö, P. (1986b). Report HUKI 1-86.
Relativistic extended Hueckel program ITEREX - 85

5680 Pyykkö, P. (1988a). Chem. Rev., 88, 563-594.
Relativistic effects in structural chemistry

5681 Pyykkö, P. (1988b).
Semiempirical relativistic molecular structure calculations In "Relativistic Effects in Atoms and Molecules", Ed. S. Wilson, Meth. Comp. Chem., 2, pp. 137-226 + the ITEREX diskette, Plenum, New York and London

5682 Pyykkö, P. (1991a).
Relativistic effects on periodic trends
In "The Effects of Relativity in Atoms, Molecules, and the Solid State", Ed. S. Wilson, I.P. Grant, and B.L. Gyorffy, Plenum, New York, and London, pp. 1-13.

5683 Pyykkö, P. (1991b).
Comment on "Basis set expansion Dirac-Fock SCF calculations and MBPT refinement" by Y. Ishikawa
In "The Effects of Relativity in Atoms, Molecules, and the Solid State", Ed. S. Wilson, I.P. Grant, and B.L. Gyorffy, Plenum, New York, and London, pp. 163-164.

5684 Pyykkö, P., Görling, A. and Rösch, N. (1987). Mol. Phys., 61, 195-205.
A transparent interpretation of the relativistic contribution to the N.M.R. 'heavy atom chemical shift'

5685 Pyykkö, P. and Jové, J. (1991). New J. Chem., 15, 717-720.
Relativistic extended Hückel (REX) interpretation of ^{237}Np nuclear quadrupole coupling and isomer shifts in neptunyl compounds

5686 Pyykkö, P., Laakkonen, L.J. and Tatsumi, K. (1989). Inorg. Chem., 28, 1801-1805.
REX calculations. 12. Iteration parameters for the 5f-element organometallics of thorium-neptunium. Geometries of ThO_2 and UO_2^{2+} revisited

5687 Pyykkö, P. and Zhao, Y.-F. (1991a). Chem. Phys. Lett., 177, 103-106.
Relativistic pseudopotential calculation of bonding trends in XAu_n^{m+} clusters (X=B-N, Al-S; n=4-6)

5688 Pyykkö, P. and Zhao, Y.-F. (1991b). Angew. Chem., 103, 622-623; Angew. Chem. Int. Ed. Engl., 30, 604-605.
Ab initio calculations on the $(ClAuPH_3)_2$ dimer with relativistic pseudopotential: Is the "Aurophilic attraction" a correlation effect?

5689 Pyykkö, P. and Zhao, Y.-F. (1991c). Inorg. Chem., 30, 3787-3788.
The large range of uranyl bond lengths: Ab initio calculations on simple uranium-oxygen clusters

5690 Pyykkö, P. and Zhao, Y.-F. (1991d). Int. J. Quantum Chem., 40, 527-544.
The elements of Flatland: Hartree-Fock atomic ground states in two dimensions for Z = 1-24

5691 Qi, B.-Y., Perez, I., Ansari, P.H., Lu, F. and Croft, M. (1987). Phys. Rev. B, 36, 2972-2975.
L_2 and L_3 measurements of transition-metal 5d orbital occupancy, spin-orbit effects, and chemical bonding

5692 Qian, S.-W., Huang, X.-Y. and Wang, J.-S. (1986). J. Phys. A, 19, L849-L851.
Double wave description of a single relativistic free electron

5693 Qian, W.-J., Kim, Y.-K. and Desclaux, J.-P. (1989). Phys. Rev. A, 39, 4509-4517.
Relativistic distorted-wave cross sections for electron-impact excitations of berylliumlike ions

5694 Qiu, Y.-H. and Li, S.-C. (1990). Chin. J. Comp. Phys., 7, 61-68.
On the energy levels of the compressed atom (in Chinese)

5695 Quesne, C. (1991). Int. J. Mod. Phys. A, 6, 1567-1589.
Supersymmetry and the Dirac oscillator

5696 Quesne, C. and Moshinsky, M. (1990). J. Phys. A, 23, 2263-2272.
Symmetry Lie algebra of the Dirac oscillator

5697 Quinet, P. and Biemont, E. (1990). Bull. Soc. Roy. Sci. Liège, 59, 307-311.
A theoretical survey of resonance transitions in nickel-like ions (Ge V-La XXX)

5698 Quinet, P. and Biemont, E. (1991). Phys. Scr., 43, 150-157.
Transition rates and wavelengths in the X-ray spectra of nickel-like ions (Ag XX-Pb LV)

5699 Quinet, P., Gorlia, T. and Biemont, E. (1991). Phys. Scr., 44, 164-183.
Allowed and forbidden transitions in highly ionized neon-like atoms (Ni XIX-U LXXXIII)

5700 Quiney, H.M. (1987). Thesis, Oxford, England.
Finite-basis-set studies of the Dirac equation

5701 Quiney, H.M. (1988).
Relativistic many-body perturbation theory
In "Relativistic Effects in Atoms and Molecules", Ed. S. Wilson, Meth. Comp. Chem., 2, pp. 227-278, Plenum, New York and London

5702 Quiney, H.M. (1990a).
Relativistic atomic structure calculations I: Basic theory and the finite basis set approximation
In "Supercomputational Science", Ed. R.G. Evans and S. Wilson, Plenum Press, New York, pp. 159-183.

5703 Quiney, H.M. (1990b).
Relativistic atomic structure calculations II: Computational aspects of the finite basis set method
In "Supercomputational Science", Ed. R.G. Evans and S. Wilson, Plenum Press, New York, pp. 185-199.

5704 Quiney, H.M. (1991).
High precision relativistic atomic structure calculations using the finite basis set approximation
In "The Effects of Relativity in Atoms, Molecules, and the Solid State", Ed. S. Wilson, I.P. Grant, and B.L. Gyorffy, Plenum, New York and London, pp. 83-123.

5705 Quiney, H.M., Grant, I.P. and Wilson, S. (1987a). Phys. Scr., 36, 460-463.
The Dirac equation in the algebraic approximation

5706 Quiney, H.M., Grant, I.P. and Wilson, S. (1987b). J. Phys. B: At. Mol. Phys., 20, 1413-1422.
The Dirac equation in the algebraic approximation: V. Self consistent field studies including the Breit interaction

5707 Quiney, H.M., Grant, I.P. and Wilson, S. (1989a). J. Phys. B: At. Mol. Opt. Phys., 22, L15-L19.
On the accuracy of Dirac-Hartree-Fock calculations using analytic basis set

5708 Quiney, H.M., Grant, I.P. and Wilson, S. (1989b)
On the accuracy of the algebraic approximation in relativistic electronic structure calculations
In "Numerical Determination of the Electronic Structure of Atoms, Diatomic and Polyatomic Molecules", Ed. M. Defranceschi and J. Delhalle, Kluwer, Dordrecht, pp. 321-327.

5709 Quiney, H.M., Grant, I.P. and Wilson, S. (1989c). Lecture Notes Chem., 52, 307-344.
On the relativistic many-body perturbation theory of atomic and molecular electronic structure

5710 Quiney, H.M., Grant, I.P. and Wilson, S. (1990). J. Phys. B: At. Mol. Opt. Phys., 23, L271-L278.
Relativistic many-body perturbation theory using analytic basis functions

5711*Rabe, K.M. and Joannopoulos, J.D. (1985). Phys. Rev. B, 32, 2302-2314.
Ab initio relativistic pseudopotential study of the zero-temperature structural properties of SnTe and PbTe

5712*Rabin, J.M. (1982). Nucl. Phys. B, 201, 315-332.
Homology theory of lattice fermion doubling

5713 Radojevic, V. (1987). Phys. Rev. A, 36, 425-427.
Low-energy autoionizing states in Pb

5714 Radojevic, V. and Kelly, H.P. (1992). Phys. Rev. A, 46, 662-665.
Photodetachment of the negative iodine ion including relaxation effects

5715 Radojevic, V., Kelly, H.P. and Johnson, W.R. (1987). Phys. Rev. A, 35, 2117-2121.
Photodetachment of negative halogen ions

5716 Radojevic, V., Kutzner, M. and Kelly, H.P. (1989). Phys. Rev. A, 40, 727-734.
Photoionization of the barium 4d subshell including relativistic and relaxation effects

5717 Radojevic, V. and Talman, J.D. (1990). J. Phys. B: At. Mol. Opt. Phys., 23, 2241-2246.
Calculation of Beutler-Fano autoionizing resonances in

neon: $1s^2 2s^2 2p^5 (^2P_{1/2})$ns and $nd_{3/2}$ series

5718 Radozycki, T. (1990a). J. Phys. A, 23, 4911-4923.
The electromagnetic virtual cloud of the ground-state hydrogen atom - a quantum field theory approach

5719 Radozycki, T. (1990b). J. Phys. A, 23, 4925-4952.
The properties of the electromagnetic field around the excited hydrogen atom - a quantum field theory approach

5720 Radzio, E. and Andzelm, J. (1987). Chem. Phys., 117, 105-111.
Non-additive interactions between noble gas atoms by the model potential method

5721 Raghavachari, K. and Trucks, G.W. (1989a). J. Chem. Phys., 91, 1062-1065.
Highly correlated systems. Excitation energies of first row transition metals Sc-Cu

5722 Raghavachari, K. and Trucks, G.W. (1989b). J. Chem. Phys., 91, 2457-2460.
Highly correlated systems. Ionization energies of first row transition metals Sc-Zn

5723*Ramana, M.V. and Rajagopal, A.K. (1981c). J. Phys. C, 14, 4291-4302.
Theory of spin polarised inhomogeneous relativistic electron gas

5724 Ramírez-Solís, A. and Daudey, J.P. (1989). Chem. Phys., 134, 111-118.
Theoretical study of the spectroscopy of low-lying electronic states of the CuF molecule

5725 Ramírez-Solís, A. and Daudey, J.P. (1990). Phys. Rev. A, 42, 5168-5174.
Ab initio calculation of the radiative lifetimes of the $^1\Sigma^+$ and $^1\pi$ excited states of CuCl

5726 Ramírez-Solís, A., Daudey, J.P. and Teichteil, C. (1990). J. Chem. Phys., 93, 7277-7283.
Multiconfigurational self-consistent field-multireference configuration interaction study of the spectroscopy of CuI including spin-orbit couplings

5727 Ramírez-Solís, A., Schamps, J. and Delaval, J.M. (1992). Chem. Phys. Lett., 188, 599-603.
MCSCF+MRCI calculation of diagonal and transition dipole moments in CuCl

5728 Ramondo, F., Rossi, V. and Bencivenni, L. (1989). J. Mol. Struct.(Theochem), 188, 105-110.
Molecular structure and vibrational frequencies of lead and tin dihalides: A SCF study

5729 Ramos, A.F., Arratia-Perez, R. and Malli, G.L. (1987). Phys. Rev. B, 35, 3790-3798.
Dirac scattered-wave calculations on an icosahedral Au_{13} cluster

5730 Ramos, A.F., Pyper, N.C. and Malli, G.L. (1988). Phys. Rev. A, 38, 2729-2739.
Relativistic effects in bonding and dipole moments for the diatomic hydrides of the sixth-row heavy elements

5731 Rao, R.S., Godwal, B.K. and Sikka, S.K. (1992). Phys. Rev. B, 46, 5780-5782.
Thorium: a 5f-band metal at ultrahigh pressures

5732 Rappé, A.K. (1986). J. Chem. Phys., 85, 6576-6578.
Calculation of the electron affinity of TeF_6

5733 Rasch, G. and Tobisch, S. (1988). Z. Phys. Chem. Leipzig, 269, 1201-1212.
Ab Initio Molecular-Orbital-Berechnungen an Pt, PtH und PtH_2 unter Verwendung eines relativistischen Pseudopotentials für Pt

5734 Raseev, G., Keller, F. and Lefebvre-Brion, H. (1987). Phys. Rev. A, 36, 4759-4774.
Theoretical spin-polarization parameters of molecular photoelectrons: Application to hydrogen halides

5735 Rashba, E.I. and Sherman, E.Ya. (1988). Phys. Lett. A, 129, 175-179.
Spin-orbital band splitting in symmetric quantum wells

5736 Rashid, K. and Ahmad, N. (1992). Opt. Commun., 93, 300-305.
Wavelengths for 4s-4p, 5p; 4p-5d; 4p-5s; transition in potassium-like ions with additional charges ±(1/3)e, ±(2/3)e

5737 Rashid, K., Fricke, B., Blanke, J.H. and Desclaux, J.-P. (1989). Z. Phys. D, 11, 99-103.
Ground state correlation energy of the Be-sequence for Z=4-20 in MCDF approximation

5738 Rashid, K., Saadi, M.Z. and Yasin, M. (1988). At. Data Nucl. Data Tables, 40, 365-378.
Dirac-Fock total energies, ionization energies, and orbital energies for uranium ions U I to U XCII

5739 Rashid, S. (1989). Phys. Rev. A, 40, 4242-4244.
Approximate solution of the hydrogenlike atoms in intense laser radiation

5740 Raspini, A. (1989). Int. J. Theor. Phys., 28, 1359-1369.
Relativistic Zeeman effect in positronium, n=2

5741 Rastall, P. (1988). Nuovo Cim. B, 101, 479-494.
New forms of the Dirac equation

5742 Rawnsley, J. (1992). Lett. Math. Phys., 24, 331-336.
Exact symplectic structures and a classical model for the Dirac electron

5743 Reed, K.J. (1988). Phys. Rev. A, 37, 1791-1794.
Relativistic distorted-wave collision strengths for Ba^{46+}

5744 Reed, K.J. and Chen, M.H. (1992). Phys. Rev. A, 45, 4519-4524.
Distorted-wave calculations of cross sections for electron-impact ionization of lithiumlike ions

5745 Reed, K.J., Chen, M.H. and Hazi, A.U. (1988). Phys. Rev. A, 38, 3319-3326.
Autoionizing resonances in electron-impact excitation of oxygenlike selenium

5746 Reed, K.J., Chen, M.H. and Moores, D.L. (1990a). Phys. Rev. A, 41, 550-553.
Excitation-autoionization processes in electron-impact ionization of Au^{68+}

5747 Reed, K.J., Chen, M.H. and Moores, D.L. (1990b). Phys. Rev. A, 42, 5315-5318.
Indirect processes in electron-impact ionization of Xe^{43+}

5748 Reed, K.J., Chen, M.H. and Moores, D.L. (1991). Phys. Rev. A, 44, 4336-4344.
Excitation-autoionization in highly charged sodiumlike ions

5749 Rehr, J.J., Mustre de Leon, J., Zabinsky, S.I. and Albers, R.C. (1991). J. Am. Chem. Soc., 113, 5135-5140.
Theoretical X-ray absorption fine structure standards

5750 Reiche, I. and Zschornack, G. (1991). Z. Phys. D, 21, S153-154.
X-ray emission rates from MCDF wave functions including electron exchange corrections

5751 Reinhard, P.-G. (1989). Rep. Prog. Phys., 52, 439-514.
The relativistic mean-field description of nuclei and nuclear dynamics

5752 Reinhold, J., Steinfeldt, N., Schüler, M. and Steinborn, D. (1992). J. Organomet. Chem., 425, 1-8.
Investigations on the electronic influence of organyl ligands. IX. Trans-influence in mercury organyl compounds. A pseudopotential ab initio study

5753 Reiss, H.R. (1990). J. Opt. Soc. Am. B, 7, 574-586.
Relativistic strong-field photoionization

5754 Reiss, H.R. (1992). Prog. Quant. Electr., 16, 1-71.
Theoretical methods in quantum optics: S-matrix and Keldysh techniques for strong-field processes

5755 Reizer, M.Yu. (1989a). Phys. Rev. B, 40, 9283-9285.
Relativistic electron-electron interaction in impure superconductors

5756 Reizer, M.Yu. (1989b). Phys. Rev. B, 40, 11571-11575.
Relativistic effects in the electron density of states, specific heat, and the electron spectrum of normal metals

5757 Ren, T. (1991). Diss. Abstr. Int. B, 52, 828.
Molecular, electronic, and magnetic structures of some diruthenium and diosmium compounds

5758 Reuter, W. and Peyerimhoff, S.D. (1992). Chem. Phys., 160, 11-24.
Ab initio study of the vibrational structure of the X $^2A_1$ and A $^2B_1$ electronic states of CH_2^+

5759 Reynolds, C.H. (1992). J. Am. Chem. Soc., 114, 8676-8682.
Structure and relative stability of halogenated carbocations: The $C_2H_4X^+$ and $C_4H_8X^+$ (X=F, Cl, Br) cations

5760 Rhoades-Brown, M.J., Bottcher, C. and Strayer, M.R. (1989). Phys. Rev. A, 40, 2831-2834.
Feynman-Monte Carlo calculations of electron capture at relativistic collider energies

5761 Ricart, J.M., Rubio, J. and Illas, F. (1986). Chem. Phys. Lett., 123, 528-532.
Molecular structure, vibrational frequencies and ionization potential of tin dihalides. An ab initio SCF and CI study

5762 Richter, M. and Eschrig, H. (1989). Solid State Comm., 72, 263-265.
Spin polarized relativistic LCAO formalism: fundamentals and first results

5763 Riehl, J.-F., Jean, Y., Eisenstein, O. and Pélissier, M. (1992). Organometallics, 11, 729-737.
Theoretical study of the structures of electron-deficient d^6 ML_5 complexes. Importance of a π-donating ligand

5764 Riisager, K. and Rosén, A. (1986). Phys. Rev. A, 34, 3454-3456.
Energy shifts of M x rays

5765*Rinker, G.A. (1979). Comp. Phys. Comm., 16, 221-242.
Static and dynamic muonic-atom codes-MUON and RURP

5766 Ritchie, B. (1989). Phys. Rev. A, 40, 1310-1314.
Relativistic theory for electron-ion scattering

5767 Rochefort, A., Andzelm, J., Russo, N. and Salahub, D.R. (1990). J. Am. Chem. Soc., 112, 8239-8247.
Chemisorption and diffusion of atomic hydrogen in and on cluster models of Pd, Rh, and bimetallic PdSn, RhSn, and RhZn catalysts

5768 Rogemond, F., Pedrini, C. and Chermette, H. (1990). Chem. Phys., 146, 129-137.
Electronic structure calculations of $Bi_4Ge_3O_{12}$ with the MSXα method. Comparison with experimental data

5769 Rogers, F.J., Wilson, B.G. and Iglesias, C.A. (1988). Phys. Rev. A, 38, 5007-5020.
Parametric potential method for generating atomic data

5770 Romanov, D.A. (1992). JETP Lett., 55, 738-741; Pis'ma Zh.

Eksp. Teor. Fiz., 55, 706-708.
Spin-orbit splitting in cylindrical quantum wires

5771 Romanovskaya, T.B. (1986).
History of the Quantum-Mechanical Interpretation of Periodicity (in Russian)
Nauka, Moscow.

5772*Roos, B.W. and Sangren, W.C. (1962). J. Math. Phys., 3, 882-890.
Spectral theory of Dirac's radial relativistic wave equation

5773 Roothaan, C.C.J. (1989). Studies in Phys. Theor. Chem., 62, 291-295.
Relativistic effects in N-electron systems

5774 Rösch, N., Görling, A., Ellis, D.E. and Schmidbaur, H. (1989). Angew. Chem., 101, 1410-1412.
Aurophilie als konzertierter Effekt: Relativistische MO-Berechnungen für Kohlenstoff-zentrierte Goldcluster

5775 Rösch, N. and Häberlen, O.D. (1992). J. Chem. Phys., 96, 6322-6323.
Reply to the comment on: Relativistic linear combination of Gaussian-type orbitals density functional method based on a two-component formalism with external field projectors

5776 Rösch, N. and Pyykkö, P. (1986). Mol. Phys., 57, 193-200.
On the relativistic theory of nuclear spin-spin coupling constants
Time-reversal symmetry aspects and applications to some heavy-element fluorides

5777*Rose, M.E. and Holmes, D.K. (1951). Phys. Rev., 83, 190-191.
The effect of finite nuclear size in beta-decay

5778*Rosen, B. (1963). J. Math. Phys., 4, 392-404.
Relativistic Coulomb scattering

5779 Rosenberg, L. (1986). Phys. Rev. A, 33, 164-169.
Relativistic scattering in an external quasistatic electric field

5780 Rosenberg, L. (1989). Phys. Rev. A, 39, 4377-4386.
Virtual-pair effects in atomic structure theory

5781 Rosenberg, L. (1990). Phys. Rev. A, 41, 1383-1391.
Dynamics of neutral vacuum decay

5782 Rosenberg, L. (1991). Phys. Rev. A, 44, 2949-2954.
Relativistic multiphoton bremsstrahlung

5783 Rosenberg, L. and Spruch, L. (1986). Phys. Rev. A, 34, 1720-1726.
Extremum principles for the determination of relativistic bound-state energies

5784 Rosenberg, L. and Zhou, F. (1991). Phys. Rev. A, 44,

7283-7289.
Soft-photon approximation for bound state Compton scattering

5785 Rosenhouse, A. and Katriel, J. (1987). Lett. Math. Phys., 13, 141-146.
Isospectral transformations in relativistic and nonrelativistic mechanics

5786 Rosi, M. and Bauschlicher Jr., C.W. (1990). Chem. Phys. Lett., 166, 189-194.
On the bonding of La^+ and La^{2+} to C_2H_2, C_2H_4, and C_3H_6

5787 Rosi, M., Bauschlicher Jr., C.W., Langhoff, S.R. and Partridge, H. (1990). J. Phys. Chem., 94, 8656-8663.
Theoretical studies of the first- and second-row transition-metal dimethyls and their positive ions

5788 Ross, A., Bussery, B., Jeung, G.-H., Bacchus-Montabonel, M.-C. and Aubert-Frécon, M. (1987). J. Chim. Phys., 84, 745-750.
Courbes d'énergie potentielle calculées pour les états électroniques les plus bas de KH et RbH

5789 Ross, A.W. and Fink, M. (1988). Phys. Rev. A, 38, 6055-6058.
Atomic scattering factor and spin-polarization calculations

5790 Ross, R.B. (1989). Diss. Abstr. Int. B, 49, 2672.
Ab initio studies including spin-orbit coupling and other relativistic effects in atoms, molecules and clusters

5791 Ross, R.B., Ermler, W.C. and Christiansen, P.A. (1986). J. Chem. Phys., 84, 3297-3300.
Ab initio calculations of spin-orbit coupling in the Group 13 and 17 atoms

5792 Ross, R.B., Powers, J.M., Atashroo, T., Ermler, W.C., LaJohn, L.A. and Christiansen, P.A. (1990). J. Chem. Phys., 93, 6654-6670.
Ab initio relativistic effective potentials with spin-orbit operators. IV. Cs though Rn

5793 Roszak, S. and Chojnacki, H. (1988). Acta Phys. Pol. A, 73, 627-630.
Molecular relativistic Hartree-Fock-Roothaan calculations within the modified hydrogen-like basis set

5794 Roszak, S. and Lipinski, J. (1992). Int. J. Quantum Chem., 44, 831-837.
The modified all-valence INDO method with the inclusion of spin-orbit coupling

5795*Rotenberg, M. (1966). Phys. Rev. Lett., 16, 969-971.
Importance of relativistic effects in the scattering of slow electrons

5796 Roth, L.M., Freiser, B.S., Bauschlicher Jr., C.W., Partridge, H. and Langhoff, S.R. (1991). J. Am. Chem. Soc., 113, 3274-3280.

Heteronuclear diatomic metal cluster ions in the gas phase: Theoretical treatment of $MgFe^+$ and study of its reactions with hydrocarbons

5797 Roy, B. and Roychoudhury, R. (1990a). J. Phys. A, 23, 3555-3561.
Shifted 1/N expansion for the Dirac equation for vector and scalar potentials

5798 Roy, B. and Roychoudhury, R. (1990b). J. Phys. A, 23, 5095-5102.
Dirac equation with Hulthen potential: an algebraic approach

5799 Roy, C.L. (1986). Phys. Lett. A, 118, 32-36.
A note on relativistic electronic states in disordered systems

5800 Roy, C.L. (1988). Phys. Lett. A, 130, 203-205.
A note on spatial behaviour of solutions of the one-dimensional Dirac equation

5801 Roy, C.L. (1989). J. Phys. Chem. Solids, 50, 111-115.
Relativistic electronic states of disordered systems

5802 Roy, C.L. (1991). Surface Sci., 247, 168-174.
Some aspects of relativistic surface states

5803 Roy, C.L. and Basu, C. (1990). J. Phys. Chem. Solids, 51, 147-155.
A study of the relativistic electronic state of disordered systems

5804 Roy, C.L. and Basu, C. (1991). J. Phys. Chem. Solids, 52, 745-753.
A relativistic treatment about the existence of band gaps in one-dimensional disordered systems

5805 Roy, C.L. and Basu, C. (1992). Phys. Rev. B, 45, 14293-14301.
Relativistic study of electrical conduction in disordered systems

5806 Roy, C.L. and Khan, A. (1992). Indian J. Pure Appl. Phys., 30, 432-438.
A relativistic study of tunnelling through double barriers

5807 Roy, C.L. and Pandey, J.S. (1986a). Physica B, 141, 289-301.
Relativistic Kronig-Penney equation and relativistic surface states of a disordered surface

5808 Roy, C.L. and Pandey, J.S. (1986b). Physica A, 137, 389-406.
Relativistic surface states of a finite crystal

5809 Roy, T. (1989). Bull. Calcutta Math. Soc., 81, 491-494.
The energy eigenvalues of Dirac hydrogen atom

5810 Roychoudhury, R. and Varshni, Y.P. (1989). Phys. Rev. A,

39, 5523-5527.
Relativistic 1/N expansion for the Dirac equation

5811*Rozhanskii, I.D. and Frenkel', Ya.I. (1938). Zh. Eksp. Teor. Fiz., 8, 127-138.
Quantum theory of 'spiral' electron orbits in a Coulomb field (in Russian)

5812 Rozing, G.J., Mijnarends, P.E. and Benedek, R. (1991). Phys. Rev. B, 43, 6996-7002.
Fully relativistic calculation of the two-photon momentum distribution for positron annihilation in tungsten

5813 Rozsnyai, B.F. (1990). Phys. Rev. A, 42, 286-290.
Inner-shell photoionization of barium, xenon, and krypton

5814 Rubio, A., Balbás, L.C. and Vega, A. (1989). Z. Phys. D, 12, 209-211.
Theoretical study of the stability of X_N^n (n=-1, 0, +1, +2; X=Ag, Cu; N≤25) clusters as a function of size using a non-local density functional formalism

5815 Rudra, P. (1986). J. Phys. A, 19, 3201-3210.
Maximal symmetry groups of quantum relativistic equations

5816 Rudzikas, Z.B. and Ryabtsev, A.N. (1989). Izv. Akad. Nauk SSSR, Ser. Fiz., 53, 1709-1714; Bull. Acad. Sci. USSR, Phys. Ser., 53(9), 56-60.
Identification of the spectra of multielectron ions

5817 Ruiz, M.E., Garcia-Prieto, J., Poulain, E., Ozin, G.A., Poirier, R.A., Matta, S.M., Czismadia, I.G., Gracie, C. and Novaro, O. (1986). J. Phys. Chem., 90, 279-282.
Theoretical study on the CuH + H --> Cu + H_2 reaction pathway

5818 Rumrich, K., Greiner, W. and Soff, G. (1987). Phys. Lett. A, 125, 394-398.
The influence of strong magnetic fields on positron production in heavy-ion collisions

5819 Rumrich, K., Greiner, W. and Soff, G. (1990). Phys. Lett. A, 149, 17-20.
Coupled-channel equations and gauge transformations

5820 Rumrich, K., Greiner, W., Soff, G., Wietschorke, K.-H. and Schlüter, P. (1989). J. Phys. B: At. Mol. Opt. Phys., 22, 165-187.
Continuum states of the two-centre Dirac equation

5821 Ruoff, H. and Nakel, W. (1987). J. Phys. B: At. Mol. Phys., 20, 2299-2306.
Absolute triply differential cross section for K shell ionization by relativistic electron impact for high atomic number

5822 Ruscic, B., Goodman, G.L. and Berkowitz, J. (1986). J. Electron Spectr. Rel. Phen., 41, 357-384.
Photoelectron spectra of Ga_2O, In_2O and Tl_2O

5823 Russo, N., Andzelm, J. and Salahub, D.R. (1987). Chem. Phys., 114, 331-338.
LCGTO MP LSD calculations for the diatomics PdX (X=C, Si, Ge, Sn)

5824 Rustgi, M.L., Leung, P.T. and Long, S.A.T. (1988a). Phys. Rev. A, 37, 1775-1778.
Limitation of the semirelativistic approach in sum rules and related calculations in atomic physics

5825 Rustgi, M.L., Leung, P.T. and Long, S.A.T. (1988b). Phys. Rev. A, 37, 3169-3171.
Ionization of K-shell electrons by highly relativistic protons

5826 Rutkowski, A. (1986a). J. Phys. B: At. Mol. Phys., 19, 149-158.
Relativistic perturbation theory: I. A new perturbation approach to the Dirac equation

5827 Rutkowski, A. (1986b). J. Phys. B: At. Mol. Phys., 19, 3431-3441.
Relativistic perturbation theory: II. One-electron variational perturbation calculations

5828 Rutkowski, A. (1986c). J. Phys. B: At. Mol. Phys., 19, 3443-3455.
Relativistic perturbation theory: III. A new perturbation approach to the two-electron Dirac-Coulomb equation

5829 Rutkowski, A. (1988). Phys. Scr., 38, 816-820.
On the variational perturbation approach to the Dirac equation

5830 Rutkowski, A., Jankowski, K. and Mikielewicz, B. (1988). J. Phys. B: At. Mol. Opt. Phys., 21, L147-L150.
An alternative to the quasirelativistic approach

5831 Rutkowski, A. and Rutkowska, D. (1987). Phys. Scr., 36, 397-399.
Relativistic perturbation theory. Third order variational perturbation calculations for H_2^+

5832 Rutkowski, A., Rutkowska, D. and Schwarz, W.H.E. (1992). Theor. Chim. Acta, 84, 105-114.
Relativistic perturbation theory of molecular structure

5833 Rutkowski, A. and Schwarz, W.H.E. (1990). Theor. Chim. Acta, 76, 391-410.
Relativistic perturbation theory of chemical properties

5834*Ruvalds, J. and Falicov, L.M. (1968). Phys. Rev., 172, 508-513.
de Haas-van Alphen effect, exchange splitting, spin-orbit interaction, and magnetic breakdown in ferromagnetic nickel

5835 Ryzhkov, M.V. and Gubanov, V.A. (1987). Zh. Strukt. Khim. 28(1), 30-38.
Discrete variational method studies of cluster models of

rare-earth chalcogenides (in Russian)

5829 Ryzhkov, M.V. and Gubanov, V.A. (1990). J. Radioanal. Nucl. Chem., 143, 85-92.
Uranyl compounds and complexes: electronic structure, chemical bonding and spectral properties

5836 Ryzhkov, M.V., Gubanov, V.A. and Zvara, I. (1992). Radiochim. Acta, 57, 11-14.
Is element 104 a continuation of IVa subgroup

5837 Saalfrank, P., Otto, P. and Ladik, J. (1988). Chem. Phys. Lett., 153, 451-457.
Studies on the electronic structure of trigonal selenium and tellurium using the effective core potential approximation

5838 Saalfrank, P., Otto, P. and Ladik, J. (1989). Solid State Comm., 69, 99-102.
Ab initio Hartree-Fock band structures of different lead surfaces using effective potentials

5839 Sabirov, Z.M. and Bagatur'yants, A.A. (1991). Russian Chem. Rev., 60, 1059-1070; Usp. Khim., 60, 2065-2088.
Compounds of the f-elements. Chemical bonding and electronic structure

5840 Sacchetti, F. (1987). Phys. Rev. B, 36, 3147-3152.
X-ray scattering and condensed-matter experiments: Beyond the nonrelativistic approach

5841 Sadlej, A.J. (1991). J. Chem. Phys., 95, 2614-2622.
The dipole moment of AuH

5842 Sadlej, A.J. (1992a). Theor. Chim. Acta, 81, 339-354.
Medium-size polarized basis sets for high-level-correlated calculations of molecular electric properties. V. Fourth-row atoms: Sn through I

5843 Sadlej, A.J. (1992b). J. Chem. Phys., 96, 2048-2053.
Electric properties of diatomic interhalogens. A study of the electron correlation and relativistic contributions

5844 Sadlej, A.J. and Urban, M. (1991). Chem. Phys. Lett., 176, 293-302.
Mutual dependence of relativistic and electron correlation contributions to molecular properties: The dipole moment of AgH

5845 Sadlej, A.J., Urban, M. and Gropen, O. (1991). Phys. Rev. A, 44, 5547-5557.
Relativistic and electron-correlation contributions to the dipole polarizability of the alkaline-earth-metal atoms Ca, Sr, and Ba

5846 Saethre, L.J. and Gropen, O. (1992). Can. J. Chem., 70, 348-352.
Structure and bonding in square-planar chalcogen rings

5847 Safronova, U.I. (1986a). Opt. Spektrosk., 60, 886-889;

Opt. Spectrosc., 60, 546-548.
Oscillator strengths of the $1s^22l2l'-1s^22l3l''$ transitions: correlation and relativistic effects

5848 Safronova, U.I. (ed.) (1986b).
Correlation and Relativistic Effects in Atoms and Ions
Acad. Sci. USSR, Moscow.

5849 Safronova, U.I. (1989). Phys. Scr., T26, 59-73.
Spectroscopy of autoionization states of atomic systems

5850*Safronova, U.I., Ivanova, A.I., Rabin'kina, N.V. and Zvereva, L.I. (1971).
Theoretical spectra of S XI and S XII (in Russian)
In "Theory of Electronic Structure of Atoms and Molecules", Ed. J. Batarunas et al., Mintis, Vilnius, pp. 79-84.

5851*Safronova, U.I. and Labzovskii, L.N. (1971).
Relativistic corrections to the terms of two-electron atoms (in Russian)
In "Theory of Electronic Structure of Atoms and Molecules", Ed. J. Batarunas et al., Mintis, Vilnius, pp. 73-78.

5852 Safronova, U.I., Shlyapzeva, A.S., Vainshtein, L.A., Kato, T. and Masai, K. (1992a). Phys. Scr., 46, 409-428.
Excitation collision strengths, cross-sections and rate cofficients for O V, Si XI, Fe XXIII and Mo XXXIX by electron impact ($1s^22s^2-1s^22s2p-1s^22p^2$ transitions)

5853 Safronova, U.I., Tolstikhina, I.Yu. and Chen, M.Kh. (1990). Opt. Spektrosk., 68, 268-272; Opt. Spectrosc., 68, 156-158.
Correlational, relativistic, and radiation effects for the energy of ions of the Be isoelectronic sequence

5854 Safronova, U.I. and Wyart, J.-F. (1992). Phys. Scr., 46 134-146.
Model potential method for the calculation of the atomic characteristics of Ne-like systems

5855 Sahnoune, A., Ström-Olsen, J.O. and Fischer, H.E. (1992). Phys. Rev. B, 46, 10035-10040.
Influence of spin-orbit scattering on the magnetoresistance due to enhanced electron-electron interactions

5856 Sahu, B., Rahman Khan, M.Z., Shastry, C.S., Dey, B. and Phatak, S.C. (1989). Am. J. Phys., 57, 886-894.
Interesting features of relativistic and nonrelativistic quantal bound states in one two, and three dimensions

5857 Sai Sunder, V.V.S.S. and Sen, K.D. (1989). Chem. Phys. Lett., 162, 185-187.
Density functional calculations of the ionic radii of Be^-, Mg^-, Ca^-, Sr^-, Ba^- and some metastable mononegative ions

5858 Sakai, Y. and Miyoshi, E. (1987). J. Chem. Phys., 87, 2885-2892.
Theoretical study of the electron affinities of MF_6 and MF_6^- (M=Cr, Mo, and W) using a model potential method

5859 Sakai, Y. and Miyoshi, E. (1988). J. Chem. Phys., 89, 4452-4453.
Theoretical study of geometries and dipole moments of AsX_3 and SbX_3 (X=F, Cl, Br, and I) molecules

5860 Sakai, Y., Miyoshi, E. and Anno, T. (1992). Can. J. Chem., 70, 309-313.
A configuration interaction study of the ground and first excited $^{1}\ ^{+}$ states of the NaI molecule

5861 Sakai, Y., Miyoshi, E., Klobukowski, M. and Huzinaga, S. (1987a). J. Comp. Chem., 8, 226-255.
Model potentials for molecular calculations. I. The sd-MP set for transition metal atoms Sc through Hg

5862 Sakai, Y., Miyoshi, E., Klobukowski, M. and Huzinaga, S. (1987b). J. Comp. Chem., 8, 256-264.
Model potentials for molecular calculations. II. The spd-MP set for transition metal atoms Sc through Hg

5863 Sakaki, S. and Ieki, M. (1991). J. Am. Chem. Soc., 113, 5063-5065.
Oxidative addition of SiH_4 to $Pt(PH_3)_2$. An ab initio MO/MP4 study

5864*Sakaki, S., Maruta, K. and Ohkubo, K. (1985). Nippon Kagaku Kaishi, 3, 598.
(Title not available).

5865 Sakaki, S., Maruta, K. and Ohkubo, K. (1987). J. Chem. Soc. Dalton Trans., 361-368.
Nucleophilic attack upon ethylene co-ordinated to mercury (II): A molecular-orbital study concerning the origin of the acceleration by Hg^{II}

5866 Sakibaev, R.S. and Safronova, U.I (1988). Opt. Spektrosk., 64, 16-21; Opt. Spectrosc., 64, 8-12.
Energies of $2^3S_1-2^3P_0$ and $2^3S_1-2^3P_2$ levels as functions of Z

5867 Sakisaka, Y., Maruyama, T., Kato, H., Aiura, Y. and Yanashima, H. (1990). Phys. Rev. B, 41, 11865-11868.
Angle-resolved photoemission study of relativistic effects in the bulk electronic structure of Fe

5868*Salem, L. and Rowland, C. (1972). Angew. Chem., 84, 86-106.
Die elektronischen Eigenschaften von Diradikalen

5869 Sallhofer, H. (1986). Z. Naturforsch. A, 41A, 468-470.
Elementary derivation of the Dirac equation.

5870 Sallhofer, H. (1988). Z. Naturforsch. A, 43A, 1039-1043.
Hydrogen in electrodynamics. I. Preliminary theories

5871 Sallhofer, H. (1989a). Z. Naturforsch. A, 44A, 167-168.
Hydrogen in electrodynamics. II. Mathematical means

5872 Sallhofer, H. (1989b). Z. Naturforsch. A, 44A, 504-506.

Hydrogen in electrodynamics. III. The hydrogen spectrum

5873 Sallhofer, H. (1989c). Z. Naturforsch. A, 44A, 715-717.
Hydrogen in electrodynamics. IV. The components of the hydrogen field

5874 Sallhofer, H. (1990a). Z. Naturforsch. A, 45A, 1038-1040.
Hydrogen in electrodynamics. V. Dirac-hydrogen by means of new operators

5875 Sallhofer, H. (1990b). Z. Naturforsch. A, 45A, 1361-1366.
Hydrogen in electrodynamics. VI. The general solution

5876 Sallhofer, H. (1992). Z. Naturforsch. A, 47, 619-624.
Hydrogen in electrodynamics. VIII The half-integer spin

5877 Saloman, E.B., Hubbell, J.H. and Scofield, J.H. (1988). At. Data Nucl. Data Tables, 38, 1-197.
X-ray attenuation cross sections for energies 100 eV to 100 keV and elements Z = 1 to Z = 92

5878 Saloman, E.B. and Kim, Y.-K. (1988). Phys. Rev. A, 38, 577-583.
Effects of configuration mixing in M1 and E2 transitions between ground-state terms of sulfurlike ions

5879 Saloman, E.B. and Kim, Y.-K. (1989). At. Data Nucl. Data Tables, 41, 339-356.
Energy levels and transition probabilities in the ground-state configuration of sulfur-like ions

5880 Salomon, M. (1989). Phys. Rev. A, 39, 4935-4938.
Stochastic integration of the Bethe-Salpeter equation for two bound fermions

5881 Salomonson, S. and Öster, P. (1989). Phys. Rev. A, 40, 5548-5558.
Relativistic all-order pair functions from a discretized single-particle Dirac Hamiltonian

5882*Salpeter, E.E. (1958). Phys. Rev., 112, 1642-1648.
Some atomic effects of an electronic electric dipole moment

5883 Salvat, F. (1991). Phys. Rev. A, 43, 578-581.
Elastic scattering of fast electrons and positrons by atoms

5884 Salvat, F., Martínez, J.D., Mayol, R. and Parellada, J. (1987a). Phys. Rev. A, 36, 467-474.
Analytical Dirac-Hartree-Fock-Slater screening function for atoms (Z=1-92)

5885 Salvat, F. and Mayol, R. (1991). Comp. Phys. Comm., 62, 65-79.
Accurate numerical solution of the Schrodinger and Dirac wave equations for central fields

5886 Salvat, F., Mayol, R. and Martinez, J.D. (1987b). J. Phys. B, 20, 6597-6612.

Elastic scattering of electrons by atoms: a semiphenomenological approach

5887 Sampson, D.H. and Zhang, H.L. (1992). Phys. Rev. A, 45, 1657-1669.
Rapid-ionization approach based on the factorization method

5888 Sampson, D.H., Zhang, H.L. and Fontes, C.J. (1990). At. Data Nucl. Data Tables, 44, 209-271.
Relativistic distorted wave collision strengths and oscillator strengths for the 71 Na-like ions with $22 \leq Z \leq 92$

5889 Sampson, D.H., Zhang, H.L. and Fontes, C.J. (1991). At. Data Nucl. Data Tables, 48, 25-90.
Relativistic distorted-wave collision strengths and oscillator strengths for F-like ions with $22 \leq Z \leq 92$

5890 Sampson, D.H., Zhang, H.L. and Mohanty, A.K. (1988). Phys. Rev. A, 38, 4569-4573.
Rate coefficients for populating excited levels of Ni-like ions by inner-shell ionization of Cu-like ions

5891 Sampson, D.H., Zhang, H.L., Mohanty, A.K. and Clark, R.E.H. (1989). Phys. Rev. A, 40, 604-615.
A Dirac-Slater approach to atomic structure for highly charged ions

5892 Samzow, R. and Hess, B.A. (1991). Chem. Phys. Lett., 184, 491-496.
Spin-orbit effects in the Br atom in the framework of the no-pair theory

5893 Samzow, R., Hess, B.A. and Jansen, G. (1992a). J. Chem. Phys., 96, 1227-1231.
The two-electron terms of the no-pair Hamiltonian

5894 Samzow, R., Hess, B.A. and Jansen, G. (1992b). J. Chem. Phys., 96, 6320-6321.
Comment on: Relativistic linear combination of Gaussian-type orbitals density functional method based on a two-component formalism with external field projectors

5895 Sánchez Marcos, E., Caballol, R., Trinquier, G. and Barthelat, J.-C. (1987). J. Chem. Soc. Dalton Trans., 2373-2377.
Ab initio study of stable bis(carbon dioxide)molybdenum complexes

5896 Sandars, P.G.H. (1987). Phys. Scr., 36, 904-910.
Parity and time-reversal violation in atoms and molecules

5897 Sandin, T.R. (1991). Am. J. Phys., 59, 1032-1036.
In defense of relativistic mass

5898 Sapirstein, J. (1987). Phys. Scr., 36, 801-808.
Quantum electrodynamics of many-electron atoms

5899 Sapirstein, J. (1988). Nucl. Instrum. Meth. B, 31,

70-78.
QED of high-Z three-electron ions

5900 Sapirstein, J. (1989a).
Evaluation of two-photon exchange graphs for the helium isoelectronic sequence
In "Relativistic, Quantum Electrodynamic and Weak Interaction Effects in Atoms", Ed. W. Johnson, P. Mohr and J. Sucher, AIP Conf. Proc., 189, pp. 196-205.

5901 Sapirstein, J. (1989b). Nucl. Instrum. Meth. B, 43, 338-346.
Field theoretic effects in highly charged ions

5902 Sargent, A.L. and Hall, M.B. (1992). Inorg. Chem., 31, 317-321.
Theoretical studies of inorganic and organometallic reaction mechanisms. 4. Oxidative addition of dihydrogen to d^8 square-planar iridium complexes with trans phosphines

5903 Sargent, A.L., Hall, M.B. and Guest, M.F. (1992). J. Am. Chem. Soc., 114, 517-522.
Theoretical studies of inorganic and organometallic reaction mechanisms. 3. The origin of the difference in the barrier for the kinetic and thermodynamic products for the oxidative addition of dihydrogen to a square-planar iridium complex

5904 Saritepe, S. and Swamy, N.V.V.J. (1987). Phys. Lett. A, 126, 28-32.
Two-dimensional solutions of the Dirac equation for the motion of an electron in a helical magnetic field

5905*Saroja, D., Srinivasa Rao, K.N. and Gopala Rao, A.V. (1978). J. Phys. A, 11, 1185-1190.
New aspects of the Foldy-Wouthuysen and Cini-Touscheck transformations

5906 Sasaki, (1989). Report KEK 88-14.
Numerical tables of anomalous scattering factors calculated by the Cromer and Liberman's method

5907 Sato, T., Niwa, T., Ohtsubo, H. and Tamura, K. (1991). J. Phys. G, 17, 303-318.
Relativistic bound-state problem of a one-dimensional system

5908 Savin, A., Becke, A.D., Flad, J., Nesper, R., Preuss, H. and von Schnering, H.G. (1991). Angew. Chem., 103, 421-424.
Ein neuer Blick auf die Elektronenlokalisierung

5909 Savin, A., Preuss, H. and Stoll, H. (1987). Revue Roumaine Chim., 32, 1069-1070.
Why are caesium clusters more difficult to ionize than their monoxides?

5910 Sawamura, M. and Ermler, W.C. (1990). J. Phys. Chem., 94, 7805-7808.

Ab initio study of quantum confinement effects in $(PbI_2)_7$ semiconductor clusters

5911 Sazdjian, H. (1986). Phys. Rev. D, 33, 3401-3424.
Relativistic wave equations for the dynamics of two interacting particles

5912 Sazdjian, H. (1987). J. Math. Phys., 28, 2618-2638.
The connection of two-particle relativistic quantum mechanics with the Bethe-Salpeter equation

5913 Sazdjian, H. (1988). J. Math. Phys., 29, 1620-1633.
The scalar product in two-particle relativistic quantum mechanics

5914 Sazdjian, H. (1989). Ann. Phys. (NY), 191, 52-77.
N-body bound state relativistic wave equations

5915 Schädel, M. (1992).
Chemistry of the Heaviest Elements
GSI-92-48, Preprint.

5916 Schadler, G. (1987). Phys. Scr., 35, 365-367.
Magnetic moments in a fully relativistic spin-polarized Dirac theory for solids

5917 Schadler, G., Albers, R.C., Boring, A.M. and Weinberger, P. (1987). Phys. Rev. B, 35, 4324-4330.
Relativistic spin-polarized densities in a scattering-theory formulation: Applications to the electronic structure of plutonium

5918 Schadler, G., Weinberger, P., Boring, A.M. and Albers, R.C. (1986). Phys. Rev. B, 34, 713-722.
Relativistic spin-polarized electronic structure of Ce and Pu

5919 Schäfer, A., Soff, G., Indelicato, P., Müller, B. and Greiner, W. (1989). Phys. Rev. A, 40, 7362-7365.
Prospects for an atomic parity-violation experiment in U^{90+}

5920 Schäfers, F., Heckenkamp, Ch., Müller, M., Radojevic, V. and Heinzmann, U. (1990). Phys. Rev. A, 42, 2603-2613.
Hg 5d and 6s: Multichannel quantum-defect analysis of experimental data

5921 Schall, H., Dulick, M. and Field, R.W. (1987). J. Chem. Phys., 87, 2898-2912.
The electronic structure of LaF: A multiconfiguration ligand field calculation

5922 Scherdin, A., Schäfer, A., Greiner, W. and Soff, G. (1992). Phys. Rev. D, 45, 2982-2987.
Delbrück scattering in a strong external field

5923 Schick, A.B., Solovyev, I.V. and Gubanov, V.A. (1992). Physica B, 179, 369-376.
Electronic structure of U, Np, Pu impurities in Pd. Self-consistent spin-polarized relativistic calculations

5924 Schilling, J.B., Goddard III, W.A. and Beauchamp, J.L. (1987a). J. Am. Chem. Soc., 109, 5565-5573.
Theoretical studies of transition-metal hydrides. 3. SrH^+ through CdH^+

5925 Schilling, J.B., Goddard III, W.A. and Beauchamp, J.L. (1987b). J. Am. Chem. Soc., 109, 5573-5580.
Theoretical studies of transition-metal methyl ions, MCH_3^+ (M=Sc, Cr, Mn, Zn, Y, Mo, Tc, Pd, Cd)

5926 Schlageter, A. and Pilkuhn, H. (1991). J. Phys. B: At. Mol. Opt. Phys., 24, 2451-2454.
Relativistic polarizability in hydrogen-like systems: Spinless particles and electrons with $g \neq 2$

5927 Schleyer, P.v.R., Kaupp, M., Hampel, F., Bremer, M. and Mislow, K. (1992). J. Am. Chem. Soc., 114, 6791-6797.
Relationships in the rotational barriers of all Group 14 ethane congeners H_3X-YH_3 (X,Y=C, Si, Ge, Sn, Pb). Comparisons of ab initio pseudopotential and all-electron results

5928 Schmid, U., Christensen, N.E. and Cardona, M. (1990). Phys. Rev. B, 41, 5919-5930.
Relativistic band structure of Si, Ge, and GeSi: Inversion-asymmetry effects

5929 Schmidbaur, H., Aly, A.M., Bowmaker, G., Bissinger, P., Brachthäuser, B., Dziwok, K., Ebner v. Eschenbach, J., Ellis, D.E., Görling, A., Graf, W., Grohmann, A., Hartmann, C., Huber, B., Kolb, A., Krüger, C., Kumberger, O., Lachmann, J., Müller, G., Öller, H.-J., Paschalidis, C., Pollok, T., Riede, J., Rösch, N., Reber, G., Scherbaum, F., Schier, A., Steigelmann, O., Wagner, F.E., Weidenhiller, G. and Wilkinson, D.L. (1992).
Strukturphänomene in der Gold-Chemie und ihre Zuordnung zu Relativistischen Effekten
In "Unkonventionelle Wechselwirkungen in der Chemie metallischer Elemente", Ed. B. Krebs, VCH, Weinheim, pp. 373-386.

5930 Schmidt, H.M., Stoll, H., Preuss, H., Becker, G. and Mundt, O. (1992). J. Mol. Struct.(Theochem), 262, 171-185.
Coupled electron pair calculations for $R - C \equiv E$ molecules (E = N, P, As or Sb)

5931 Schmidt, J.M., Soff, G. and Mohr, P.J. (1989). Phys. Rev. A, 40, 2176-2179.
Higher-order vacuum-polarization corrections in muonic atoms

5932 Schmitt, K.M. and Arenhoevel, H. (1986).
Integrated photoabsorption strength for a bound Dirac particle
Proc. 5th Course Int. Sch. Intermed. Energy Nucl. Phys., Ed. R. Bergere, World Scientific, Singapore, pp. 518-520.

5933 Schmutzer, E. (1991).
Grundlagen der Theoretischen Physik, Teil III
Deutscher Verlag der Wissenschaften, Berlin.

Ch.4, Einführung in die relativistische Quantenmechanik, pp. 1430-1516.

5934 Schneider, W. and Thiel, W. (1992). Chem. Phys., 159, 49-66.
Ab initio calculation of anharmonic force fields for the methyl, silyl, germyl, and stannyl halides

5935 Schneider, W.F. (1991). Diss. Abstr. Int. B, 52, 2551.
Applications of local-density-functional methods to inorganic chemistry

5936 Schober, C., Christensen, N.E. and Ziesche, P. (1991). Solid State Comm., 77, 173-176.
Ab-initio calculations of the internal conversion coefficient for valence electrons of hcp-$^{99}T_c$ and bcc-^{235}U

5937 Schramm, S., Müller, B. and Schramm, A.J. (1992). Phys. Lett. A, 164, 28-32.
Exact Dirac propagator in a magnetic "sheet"

5938 Schröder, D., Hrusák, J. and Schwarz, H. (1992). Helv. Chim. Acta, 75, 2215-2218.
Generation of 'bare' FeF^+ by C-F bond activation in the gas phase and evaluation of thermochemical data

5939*Schüler, H. and Schmidt, Th. (1936). Z. Physik, 99, 717-728.
Das elektrische Quadrupolmoment des $^{209}_{83}Bi$-Kerns

5940 Schulman, L.S. (1992). J. Phys. A, 25, 3007-3014.
Relativistic two-time localization

5941 Schwarz, W.H.E. (1987). Phys. Scr., 36, 403-411.
Relativistic calculations of molecules. Relativity and bond lengths

5942 Schwarz, W.H.E. (1990).
Fundamentals of relativistic effects in chemistry
In "The Concept of the Chemical Bond", Vol.2, Ed. Z.B. Maksic, Springer-Verlag, Berlin, pp. 593-643.

5943 Schwarz, W.H.E., Rutkowski, A. and Collignon, G. (1991).
Nonsingular relativistic perturbation theory and relativistic changes of molecular structure
In "The Effects of Relativity in Atoms, Molecules, and the Solid State", Ed. S. Wilson, I.P. Grant and B.L. Gyorffy, Plenum, New York and London, pp. 135-147.

5944 Schwarz, W.H.E., van Wezenbeek, E.M., Baerends, E.J. and Snijders, J.G. (1989). J. Phys. B: At. Mol. Opt. Phys., 22, 1515-1530.
The origin of relativistic effects of atomic orbitals

5945 Schwerdtfeger, P. (1986).
Pseudopotentiale zur Untersuchung relativistischer Effekte
Thesis, Stuttgart

5946 Schwerftfeger, P. (1987). Phys. Scr., 36, 453-459.
Relativistic effects in molecules: Pseudopotential calcula-

tions for TlH^+, TlH and TlH_3

5947 Schwerftfeger, P. (1989). J. Am. Chem. Soc., 111, 7261-7262.
Relativistic effects in gold chemistry. II. The stability of complex halides of gold(III).

5948 Schwerftfeger, P. (1990). J. Am. Chem. Soc., 112, 2818-2820.
On the anomaly of the metal-carbon bond strength in $(CH_3)_2M$ compounds of the heavy elements M= Au^-, Hg, Tl^+, and Pb^{2+}. Relativistic effects in metal-ligand force constants

5949 Schwerdtfeger, P. (1991a). Inorg. Chem., 30, 1660-1663.
Metal-metal bonds in Tl(I)-Tl(I) compounds: Fact or fiction

5950 Schwerdtfeger, P. (1991b). Chem. Phys. Lett., 183, 457-463.
Relativistic and electron-correlation contributions in atomic and molecular properties: benchmark calculations on Au and Au_2

5951 Schwerftfeger, P. (1992a). J. Chem. Phys., 97, 3877.
Comment on: Electronic states, ionization potentials, and bond energies of TlH_n, InH_n, TlH_n^+, and InH_n^+ (n=1-3)

5952 Schwerdtfeger, P., Bowmaker, G.A., Boyd, P.D.W., Ware, D.C., Brothers, P.J. and Nielson, A.J. (1990a). Organometallics, 9, 504-508.
Scaled Hartree-Fock force field calculations for organothallium compounds: Normal-mode analysis for $TlCH_3$, $Tl(CH_3)_2^+$, $Tl(CH_3)_3$, $Tl(CH_3)_2Br$, and $Tl(CH_3)_4^-$

5953 Schwerdtfeger, P. and Boyd, P.D.W. (1992). Inorg. Chem., 31, 327-329.
Role of phosphine ligands in gold cluster chemistry. Relativistic SCF calculations on Au_2 and $Au_2(PH_3)_2$

5954 Schwerdtfeger, P., Boyd, P.D.W., Bowmaker, G.A. and Aldridge, L.P. (1990b). Struct. Chem., 1, 405-415.
Theoretical studies on CuF^+, CuF, CuF^-, CuF_2, and CuF_2^-

5955 Schwerdtfeger, P., Boyd, P.D.W., Bowmaker, G.A., Mack, H.G. and Oberhammer, H. (1989a). J. Am. Chem. Soc., 111, 15-23.
Theoretical studies on the stability of Tl-C σ-bonds in aliphatic organothallium compounds

5956 Schwerdtfeger, P., Boyd, P.D.W., Brienne, S. and Burrell, A.K. (1992b). Inorg. Chem., 31, 3411-3422.
Relativistic effects in gold chemistry. 4. Gold(III) and gold(V) compounds

5957 Schwerdtfeger, P., Boyd, P.D.W., Burrell, A.K., Robinson, W.T. and Taylor, M.J. (1990c). Inorg. Chem., 29, 3593-3607.
Relativistic effects in gold chemistry. 3. Gold(I) complexes

5958 Schwerdtfeger, P. and Dolg, M. (1991). Phys. Rev. A, 43, 1644-1647.

Anomalous high gold-metal bond stabilities: Relativistic configuration-interaction calculations for AuLa and AuLu

5959 Schwerdtfeger, P., Dolg, M., Schwarz, W.H.E., Bowmaker, G.A. and Boyd, P.D.W. (1989b). J. Chem. Phys., 91, 1762-1774.
Relativistic effects in gold chemistry. I. Diatomic gold compounds

5960 Schwerdtfeger, P., Heath, G.A., Dolg, M. and Bennett, M.A. (1992c). J. Am. Chem. Soc., 114, 7518-7527.
Low valencies and periodic trends in heavy element chemistry. A theoretical study of relativistic effects and electron correlation effects in Group 13 and Period 6 hydrides and halides

5961 Schwerdtfeger, P., Laakkonen, L.J. and Pyykkö, P. (1992a). J. Chem. Phys., 96, 6807-6819.
Trends in inversion barriers. I. Group-15 hydrides

5962 Schwerdtfeger, P., McFeaters, J.S., Moore, J.J., McPherson, D.M., Cooney, R.P., Bowmaker, G.A., Dolg, M. and Andrae, D. (1991a). Langmuir, 7, 116-125.
Theoretical studies of chemisorption and dimer model systems: Mφller-Plesset and configuration interaction calculations on PdH, PdC, PdO, PdF, Pd_2, and PdCO

5963 Schwerdtfeger, P., Silberbach, H. and Miehlich, B. (1989c). J. Chem. Phys., 90, 762-767.
Relativistic effects in molecules: Pseudopotential calculations for PbH^+, PbH, PbH_2, and PbH_4

5964 Schwerdtfeger, P., von Szentpály, L., Stoll, H. and Preuss, H. (1987). J. Chem. Phys., 87, 510-513.
Relativistic pseudopotential calculations for HBr^+, HBr, HBr^-, HI^+, HI, and HI^-

5965 Schwerdtfeger, P., von Szentpály, L., Vogel, K., Silberbach, H., Stoll, H. and Preuss, H. (1986). J. Chem. Phys., 84, 1606-1612.
Relativistic and correlation effects in pseudopotential calculations for Br, I, HBr, HI, Br_2, and I_2

5966 Scioly, A.J., Luetkens Jr., M.L., Wilson Jr., R.B., Huffman, J.C. and Sattelberger, A.P. (1987). Polyhedron, 6, 741-757.
Synthesis and characterization of binuclear tantalum hydride complexes

5967*Scofield, J. H. (1978). Phys. Rev. A, 18, 963-970.
K- and L-shell ionization of atoms by relativistic electrons

5968 Scofield, J.H. (1989). Phys. Rev. A, 40, 3054-3060.
Angular and polarization correlations in photoionization and radiative recombination

5969 Scofield, J.H. (1990). Phys. Scr., 41, 59-62.
Angular distribution of photoelectrons from polarized X-rays

5970 Scofield, J.H. and MacGowan, B.J. (1992). Phys. Scr., 46, 361-364.
Energies of nickel-like 4d to 4p laser lines

5971 Scott, T.C., Moore, R.A. and Monagan, M.B. (1989). Comp. Phys. Comm., 52, 261-281.
Resolution of many-particle electrodynamics by symbolic manipulation

5972 Scott, T.C., Moore, R.A., Monagan, M.B., Fee, G.J. and Vrscay, E.R. (1990). J. Comp. Phys., 87, 366-395.
Perturbative solutions of quantum mechanical problems by symbolic computation

5973 Scott, T.C., Shertzer, J. and Moore, R.A. (1992). Phys. Rev. A, 45, 4393-4398.
Accurate finite-element solutions of the two-body Dirac equation

5974 Seba, P. (1988). Czech. J. Phys. B, 38, 1078-1094.
On the absorption of eigenvalues by continuous spectrum in the one and three dimensional Dirac equation

5975 Seba, P. (1989). Lett. Math. Phys., 18, 77-86.
Klein's paradox and the relativistic point interaction

5976 Sebastian, K. (1986). Phys. Rev. A, 34, 3839-3846.
Relativistic invariance of the two-photon transition amplitude of a composite system

5977 Seely, J.F. (1989). Phys. Rev. A, 39, 3682-3685.
QED contributions to the 2p-2s transitions in highly charged Li-like ions

5978 Seely, J.F., Brown, C.M. and Feldman, U. (1989). At. Data Nucl. Data Tables, 43, 145-159.
Wavelengths and energy levels for the Cu I isoelectronic sequence Ru^{15+} through U^{63+}

5979 Seely, J.F. and Wagner, R.A. (1990). Phys. Rev. A, 41, 5246-5249.
QED contributions to the 3s-3p transitions in highly charged Na-like ions

5980 Seijo, L., Barandiarán, Z. and Huzinaga, S. (1991). J. Chem. Phys., 94, 3762-3773.
Ab initio model potential study of the equilibrium geometry of alkaline earth dihalides: MX_2 (M= Mg, Ca, Sr, Ba; X= F, Cl, Br, I)

5981 Seijo, L., Barandiarán, Z. and Huzinaga, S. (1992). Chem. Phys. Lett., 192, 217-220.
Atomic valence correlation energies from ab initio model potential calculations

5982*Seiler, K. (1936). Ann. Physik, 27, 329-372.
Zur atomaren Dispersion und Absorption von Röntgenstrahlen nach der relativistischen Wellenmechanik von Dirac

5983 Seitz-Beywl, J., Poxleitner, M., Probst, M.M. and Heinzinger, K. (1992). Int. J. Quantum Chem., 42, 1141-1147.
On the interaction of ions with a platinum metal surface

5984 Seke, J. (1990). J. Phys. B, 23, L659-L662.
Deviations from exponential decay for the $2P_{1/2}$ --> $1S_{1/2}$ transition in a relativistic hydrogen atom

5985 Seke, J. (1992a). Phys. Rev. A, 45, 542-545.
Finite-time deviations from exponential decay for the Lyman-α radiation in the Dirac hydrogen atom without ignoring virtual transitions

5986 Seke, J. (1992b). J. Phys. A, 25, 5415-5422.
Solution of the renormalization problem in non-relativistic quantum electrodynamics

5987 Seke, J. (1992c). J. Phys. A, 25, 691-703.
Non-Markovian treatment of the spontaneous $2P_{1/2}$ --> $1S_{1/2}$ transition in the Dirac hydrogen atom

5988 Sekino, H. and Bartlett, R.J. (1990). Int. J. Quantum Chem., S24, 241-244.
Relativistic coupled cluster calculations on neutral and highly ionized atoms

5989 Sekino, H. and Ishikawa, Y. (1989). Int. J. Quantum Chem., S23, 339-344.
Relativistic diagrammatic perturbation theory calculations on neon atom

5990 Sellers, H. (1990a). J. Phys. Chem., 94, 1338-1343.
Ab initio and relativistic effective core potential studies of niobium-nitrogen and niobium cluster systems

5991 Sellers, H. (1990b). Chem. Phys. Lett., 170, 5-12.
On modeling chemisorption processes with metal cluster systems: H, O, N and S on the Pd(111) plane

5992 Sellers, H. (1991). Chem. Phys. Lett., 178, 351-357.
On modeling chemisorption processes with metal cluster systems. II. Model atomic potentials and site specificity of N atom chemisorption on Pd(111)

5993 Sellers, H. and Ove, R. (1990). Chem. Phys. Lett., 170, 523-530.
Anisotropic effective core potentials

5994 Sellers, H. and Sudhakar, P.V. (1992). J. Chem. Phys., 97, 6644-6648.
The interaction between water and the liquid-mercury surface

5995 Selmani, A., Andzelm, J. and Salahub, D.R. (1986). Int. J. Quantum Chem., 29, 829-842.
Chemisorption of O and O_2 on Ag(110): An LCGTO-LSD cluster study

5996 Selmani, A. and Salahub, D.R. (1988a). Chem. Phys. Lett.,

146, 465-468.
Local spin density calculations for XH_3- and XH_4-type molecules (X=C, Si, Ge and Sn)

5997 Selmani, A. and Salahub, D.R. (1988b). J. Chem. Phys., 89, 1529-1532.
On the singlet-triplet splitting in SiH_2, GeH_2, and SnH_2. Local-spin-density calculations

5998 Semay, C. and Silvestre-Brac, B. (1992). Phys. Rev. D, 46, 5177-5187.
Comparison between relativistic, semirelativistic, and nonrelativistic approaches of quarkonium

5999 Semenov, V.V. (1992). J. Phys. A, 25, L619-L621.
Ground state of a spin-1/2 neutral particle with an anomalous magnetic moment in a two-dimensional electrostatic field

6000 Sen, A., Childs, W.J., Bauche, J. and Wyart, J.-F. (1990). Phys. Rev. A, 42, 2573-2577.
Measurement and analysis of isotope shifts in $^{151,153}Eu^+$ from hyperfine spectra

6001 Sen, K.D. and Politzer, P. (1989). J. Chem. Phys., 91, 5123-5124.
Approximate radii for single negative ions of 3d, 4d, and 5d metal atoms

6002 Sen, K.D. and Puri, A. (1989a). Chem. Phys. Lett., 156, 505-508.
Relativistic local spin density functional calculations of dipole oscillator strengths in alkaline earth isoelectronic series

6003 Sen, K.D. and Puri, A. (1989b). Phys. Lett. A, 137, 128-131.
Quasi-relativistic local spin density functional calculations of dipole oscilator strengths in alkali isoelectronic series

6004 Sen, K.D., Seminario, J.M. and Politzer, P. (1989a). Phys. Rev. A, 40, 2260-2264.
Electronegativities and isoelectronic energy and electronegativity differences for monatomic systems with nonintegral nuclear charges: Local-spin-density-functional calculations

6005 Sen, K.D., Seminario, J.M. and Politzer, P. (1989b). J. Chem. Phys., 90, 4373-4378.
Z-transition state calculations of energy changes and electrostatic potentials in isoelectronic atoms and molecules

6006 Sen, S. and Gautam, V.P. (1992). Indian J. Phys. A, 66, 405-408.
Non-relativistic aspects of SU(2)*U(1) parity violating Hamiltonian in the atomic systems

6007 Senekowitsch, J., Zilch, A., Carter, S., Werner, H.-J.,

Rosmus, P. and Botschwina, P.(1988). Chem. Phys., 122, 375-386.
Theoretical calculations of the vibrational transition probabilities in hydrogen selenide

6008*Sen Gupta, N.D. (1974). Phys. Stat. Sol. b, 65, 351-357.
On the relativistic electronic states of a monatomic and a diatomic one-dimensional lattice

6009 Sen Gupta, N.D. (1989). Phys. Lett. A, 135, 427-429.
The Dirac equation with a potential depending on one space variable

6010 Sen Gupta, N.D. (1991). Phys. Lett. A, 152, 453-457.
The Dirac equation in an ideal helical wiggler with guide field

6011 Sen Gupta, N.D. (1992). Acta Phys. Hung., 71, 25-27.
Dirac particle in a spatially periodic magnetic field

6012 Sepp, W. (1987).
Implicit projection operators in basis-set expansions of the molecular Dirac-Fock-Slater problem
In "Atoms in Unusual Situations", Ed. J.P. Briand, p. 310, Plenum, New York.

6013 Sepp, W.-D. and Fricke, B. (1987). Phys. Scr., 36, 268-270.
Test for basis-set errors in relativistic Dirac-Fock-Slater calculations with a numerical AO-DFS-basis

6014 Sepp, W.-D., Kolb, D., Sengler, W., Hartung, H. and Fricke, B. (1986). Phys. Rev. A, 33, 3679-3687.
Relativistic Dirac-Fock-Slater program to calculate potential-energy curves for diatomic molecules

6015 Shaarawi, A.M., Besieris, I.M. and Ziolkowski, R.W. (1990). J. Math. Phys., 31, 2511-2519.
A novel approach to the synthesis of nondispersive wave packet solutions to the Klein-Gordon and Dirac equations

6016*Shabaev, V.M. (1985b). Teor. Mat. Fiz., 63, 394-405; Theor. Math. Phys., 63, 588-596.
Mass corrections on the strong field of the nucleus

6017 Shabaev, V.M. (1988). Yad. Fiz., 47, 107-112; Sov. J. Nucl. Phys., 47, 69-72.
Nuclear recoil effect in relativistic theory of multicharged ions

6018 Shabaev, V.M. (1990a). Teor. Mat. Fiz., 82, 83-89.
Rayleigh-Schroedinger perturbation theory for relativistic atom

6019 Shabaev, V.M. (1990b). Izv. VUZov, Fiz., 33, 629-722; Sov. Phys. J., 33, 660-670.
Quantum electrodynamic theory of multiply charged ions

6020 Shabaev, V.M. (1991a). J. Phys. B: At. Mol. Opt. Phys., 24, 4479-4488.
Generalizations of the virial relations for the Dirac

equation in a central field and their applications to the Coulomb field

6021 Shabaev, V.M. (1991b). J. Phys. A, 24, 5665-5674.
Quantum electrodynamical theory for the natural shape of the spectral line

6022 Shabaev, V.M., Shabaeva, M.B. and Safina, E.B. (1990). Opt. Spektrosk., 69, 507-510; Opt. Spectrosc., 69, 303-305.
Effects of finite size of the nucleus on spectra of multiply charged ions

6023 Shabaeva, M.B. and Shabaev, V.M. (1992). Phys. Lett. A, 165, 72-78.
Hyperfine structure of the 2s ^{2}S level of Li-like multicharged ions

6024 Shachal, D. and Manassen, Y. (1991). Phys. Rev. B, 44, 11528-11531.
Spin-orbit-coupling mechanism for scanning-tunneling-microscopy observation of individual paramagnetic spins

6025 Shachal, D. and Manassen, Y. (1992). Phys. Rev. B, 46 4795-4805.
Mechanism of electron-spin resonance studied with use of scanning tunneling microscopy

6026 Shadwick, B.A., Talman, J.D. and Norman, M.R. (1989). Comp. Phys. Comm., 54, 95-102.
A program to compute variationally optimized relativistic atomic potentials

6027*Shaik, S.S. and Epiotis, N.D. (1980). J. Am. Chem. Soc., 102, 122-131.
Spin inversion in triplet Diels-Alder reactions

6028 Shanthi, N. and Deshmukh, P.C. (1989). Phys. Rev. A, 40, 2400-2403.
Xenon 4p photoionization near the 4d Cooper minimun: Interchannel coupling effects

6029 Shanthi, N. and Deshmukh, P.C. (1990). J. Phys. B: At. Mol. Opt. Phys., 23, 61-76.
Low- and intermediate-energy photoionization of palladium

6030 Shanthi, N., Deshmukh, P.C. and Manson, S.T. (1988a). Phys. Rev. A, 37, 1773-1774.
Near-threshold structure in the photoionization of krypton 3p

6031 Shanthi, N., Deshmukh, P.C. and Manson, S.T. (1988b). Phys. Rev. A, 37, 4720-4729.
Photoionization cross-section, angular-distribution, and spin-polarization calculations for the 3p and 3d subshells of krypton

6032 Shen, S.-Q., Cai, J. and Tao, R.-B. (1992a). Commun. Theor. Phys., 17, 107-110.
Exact ground state solution for anyons with spin-1/2

6033 Shen, S.-Q., Cai, J. and Tao, R.-B. (1992b). Commun. Theor. Phys., 18, 27-32.
Second virial coefficient of anyons with spins

6034 Shen, S.-Q. and Tao, R.-B. (1992). Phys. Rev. A, 45, 1357-1360.
Exact solution of the problem of a two-dimensional Dirac particle with flux 1/2

6035 Shen, S.-Q., Wu, X.-H. and Tao, R.-B. (1992c). Commun. Theor. Phys., 17, 351-364.
The atomic hydrogen in two spatial dimensions

6036*Shen, Y.-Y. (1981a). Phys. Stat. Sol. b, 108, 255-260.
Relativistic effect on a deformed semi-infinite crystalline surface

6037*Shen, Y.-Y. (1981b). Surface Sci., 108, L477-L485.
Relativistic surface states of a semi-infinite chain acted on by an electric field

6038 Sherry, G.C. (1988). Eur. J. Phys., 9, 301-302.
Zitterbewegung revisited

6039 Sherry, G.C. (1989). Found. Phys., 19, 733-741.
Pseudo-classical phase space description of the relativistic electron

6040 Shestakov, A.F. (1987). Opt. Spektrosk., 60, 1112-1115; Opt. Spectrosc., 60, 687-689.
Theoretical study of the fine structure of the $3d^94s^2\ ^2D$ term in the Cu I isoelectronic series

6041 Shi, K.-S., Gao, X.-L. and Zhao, Y.-F. (1987). Jilin Daxue Ziran Kexue Xuebao, 1, 59-64.
Relativistic calculations for the x-ray energies of the transition group elements yttrium, zirconium, niobium, molybdenum, technetium, and ruthenium (in Chinese)

6042*Shiba, H. (1976). J. Low Temp. Phys., 22, 105-119.
Effect of spin-orbit interaction on the Knight shift of normal and superconducting small particles

6043 Shilkova, N.A. and Shirokovskii, V.P. (1988a). Phys. Stat. Sol. B, 149, 195-201.
Calculation of dipole transition matrix elements in crystals by the relativistic Green's function method

6044 Shilkova, N.A. and Shirokovskii, V.P. (1988b). Phys. Stat. Sol. B, 149, 571-577.
Calculation of electron velocity in crystal ab initio Role of relativistic effects

6045 Shin, G.R., Bialynicki-Birula, I. and Rafelski, J. (1992). Phys. Rev. A, 46, 645-647.
Wigner function of relativistic spin-1/2 particles

6046 Shinada, M. (1990). J, Phys. B, 23, L347-L351.
A note on the adiabatic approximation of a relativistic

hydrogen atom in a static and uniform magnetic field

6047*Shirokov, Yu.M. (1958). Sov. Phys. JETP, 6, 919-935. A group-theoretical consideration of the basis of relativistic quantum mechanics. II. Classification of the irreducible representations of the inhomogeneous Lorentz group

6048 Shishkin, G.V. (1991). Nuovo Cim. B, 106, 1137-1142. Some generalizations of Volkov's problem

6049 Shishkin, G.V. and Cabos, W.D. (1991). J. Math. Phys., 32, 3184-3188. The Dirac equation in external fields: Variable separation in Cartesian coordinates

6050 Shishkin, G.V. and Cabos, W.D. (1992). J. Math. Phys., 33, 914-925. Dirac equation in external fields: Separation of variables in curvilinear coordinates

6051 Shishkin, G.V. and Villalba, V.M. (1989a). J. Math. Phys., 30, 2132-2142. Dirac equation in external vector fields: Separation of variables

6052 Shishkin, G.V. and Villalba, V.M. (1989b). J. Math. Phys., 30, 2373-2381. Dirac equation in external vector fields: New exact solutions

6053 Shishkin, G.V. and Yasin, M.A. (1991a). Yad. Fiz., 53, 858-868; Sov. J. Nucl. Phys., 53, 536-541. Exact description of a Dirac particle in the presence of wave fields (generalization of the Volkov problem)

6054 Shishkin, G.V. and Yasin, M.A. (1991b). Int. J. Mod. Phys. A, 6, 4831-4841. Generalized Volkov problem

6055 Shvartsman, Sh.M. (1991). Mod. Phys. Lett. A, 6, 805-810. On the lagrangian of a relativistic spinning particle in quantum electrodynamics

6056 Siegbahn, P.E.M. and Blomberg, M.R.A. (1992). J. Am. Chem. Soc., 114, 10548-10556. Theoretical study of the activation of C-C bonds by transition metal atoms

6057 Sienkiewicz, J.E. (1990). J. Phys. B: At. Mol. Opt. Phys., 23, 1869-1878. Spin polarization and differential cross sections in elastic low-energy scattering of electrons from mercury

6058 Sienkiewicz, J.E. and Baylis, W.E. (1987). J. Phys. B: At. Mol. Opt. Phys., 20, 5145-5156. A relativistic approach to the elastic scattering of electrons by argon

6059 Sienkiewicz, J.E. and Baylis, W.E. (1991a). J. Phys. B: At. Mol. Opt. Phys., 24, 265-274.
The polarization of electrons elastically scattered from xenon

6060 Sienkiewicz, J.E. and Baylis, W.E. (1991b). Phys. Rev. A, 43, 1331-1335.
Elastic scattering of positrons on mercury: A negative-energy Dirac-Fock treatment

6061 Sienkiewicz, J.E. and Baylis, W.E. (1991c). J. Phys. B: At. Mol. Opt. Phys., 24, 1739-1748.
Spin polarization of electrons elastically scattered from krypton

6062 Sigalas, M., Papaconstantopoulos, D.A. and Bacalis, N.C. (1992). Phys. Rev. B, 45, 5777-5783.
Total energy and band structure of the 3d, 4d, and 5d metals

6063*Silagadze, Z.K. and Khelashvili, A.A. (1984). Teor. Mat. Fiz., 61, 431-441.
Solvable model of the relativistic two-fermion bound state problem with infinitely rising potentials

6064 Silberbach, J. (1987). Chem. Phys. Lett., 133, 267-270.
Valence energies and polarisabilities for heavy atoms and ions by means of semi-empirical pseudopotentials

6065 Silberbach, H., Schwerdtfeger, P., Stoll, H. and Preuss, H. (1986). J. Phys. B: At. Mol. Opt. Phys., 19, 501-510.
Ground and excited states of Rb_2^+ and Cs_2^+ by means of quasi-relativistic pseudo-potentials including core polarisation

6066 Silfvast, W.T., Al-Salameh, D.Y. and Wood II, O.R. (1986). Phys. Rev. A, 34, 5164-5167.
Multielectron photoionization to the $3p^4 4p$ configuration of Ar^+: Experiment and theory

6067 Silver, J.D. (1988). Phys. Scr., 37, 720-727.
Tests of relativistic and quantum electrodynamic effects using precision spectroscopy of medium to high Z one and two electron ions

6068 Simulik, V.M. (1991). Teor. Mat. Fiz., 87, 76-85; Theor. Math. Phys., 87, 386-393.
Connection between the symmetry properties of the Dirac and Maxwell equations. Conservation laws

6069 Sinkovic, B., Friedman, D.J. and Fadley, C.S. (1991). J. Magn. Magn. Mater., 92, 301-343.
Theory of spin polarized photoelectron diffraction

6070 Sitenko, Y.A. (1989). Sov. J. Nucl. Phys., 50, 561-564.
Zero modes of the Dirac operator on a noncompact two-dimensional surface in a magnetic field

6071 Siu, G.G. (1988a). J. Phys. C, 21, 3917-3925.
The relativistic effects in the S ground-state splitting of Mn^{2+} ions

6072 Siu, G.G. (1988b). J. Phys. C, 21, 3927-3938.
Contributions to the ground-state splitting of S-state ions: II. A theoretical framework for the $3d^5\ {}^6S$ ground-state splitting of Mn^{2+}

6073 Siu, G.G. (1988c). J. Phys. C, 21, L1001-L1005.
Covalency effects in the relativistic crystal field

6074 Sklyadneva, I.Yu. and Chulkov, E.V. (1990). Metallofizika, 12, 38-43.
Self-consistent calculations of volume and surface electronic structures of hafnium by non-linear pseudopotential methods (in Russian)

6075 Smelyanskii, V.I., Krasovskii, A.E., Antonov, Vl.N. and Antonov, V.N. (1989). Metallofizika, 11, 15-20.
Energy band structure, and the photoelectron and X-ray emission spectra of uranium (in Russian)

6076 Smelyansky, V.I., Perlov, A.Ya. and Antonov, V.N. (1990). J. Phys.: Cond. Matter, 2, 9373-9379.
The calculated Fermi surface of osmium

6077 Smirnov, V.P. and Evarestov, R.A. (1988). Teor. Eksp. Khim., 24, 91-95; Theor. Exp. Chem., 24, 84-88.
Symmetry of localized orbitals in molecules with spin-orbit interaction

6078 Smit, Z. (1991). Z. Phys. D, 22, 411-417.
Semiclassical approach to K shell ionization: hydrogenic wave functions

6079 Smit, Z. (1992). Phys. Rev. A, 46, 1367-1373.
Light-ion-induced K-shell ionization in the adiabatic region

6080 Smith, D.C. and Goddard III, W.A. (1987). J. Am. Chem. Soc., 109, 5580-5583.
Bond energy and other properties of the Re-Re quadruple bond

6081 Smith, D.Y. (1987). Phys. Rev. A, 35, 3381-3387.
Anomalous x-ray scattering: Relativistic effects in x-ray dispersion analysis

6082 Smith, G.W. and Carter, E.A. (1991). J. Phys. Chem., 95, 2327-2339.
Interactions of NO and CO with Pd and Pt atoms

6083 Smith, N.V., Chen, C.T., Sette, F. and Mattheiss, L.F. (1992). Phys. Rev. B, 46, 1023-1032.
Relativistic tight-binding calculations of x-ray absorption and magnetic circular dichroism at the L_2 and L_3 edges of nickel and iron

6084 Sneddon, G.E. and Whittingham, I.B. (1990). J. Phys. B: At. Mol. Opt. Phys., 23, 2227S-2240S.
Radial matrix elements involving continuum Dirac eigenfunctions

6085 Snyderman, N.J. (1991). Ann. Phys.(NY), 211, 43-86.
Electron radiative self-energy of highly stripped heavy atoms

6086 Söderlind, P., Nordström, L., Lou, Y.-M. and Johansson, B. (1990). Phys. Rev. B, 42, 4544-4552.
Relativistic effects on the thermal expansion of the actinide elements

6087 Sodupe, M. and Bauschlicher Jr., C.W. (1991). J. Phys. Chem., 95, 8640-8645.
Theoretical study of the bonding of the first- and second-row transition-metal positive ions to acetylene

6088 Sodupe, M., Bauschlicher Jr., C.W. and Partridge, H. (1991a). J. Chem. Phys., 95, 9422-9423.
The metal-ligand binding energies for $Sr(H_2O)_n^+$

6089 Sodupe, M., Bauschlicher Jr., C.W. and Partridge, H. (1992). Chem. Phys. Lett., 192, 185-194.
A theoretical study of $Mg(CO_2)_n^+$ and $Sr(CO_2)_n^+$ for n=1 and 2 and $Mg_2CO_2^+$

6090 Sodupe, M., Lluch, J.M., Oliva, A. and Bertrán, J. (1989a). Organometallics, 8, 1837-1841.
Theoretical study of the conformational preferences in the $Cl_4W{=}CH_2$ complex

6091 Sodupe, M., Lluch, J.M., Oliva, A. and Bertrán, J. (1991b). J. Mol. Struct.(Theochem), 251, 37-47.
Ab initio study of the reaction between $Cl_4W{=}CH_2$ and ethylene

6092 Sodupe, M., Lluch, J.M., Oliva, A., Illas, F. and Rubio, J. (1989b). J. Chem. Phys., 90, 6436-6442.
Ground and low-lying states of FeH^+ as derived from ab initio self-consistent field and configuration interaction calculations

6093 Sodupe, M., Lluch, J.M., Oliva, A., Illas, F. and Rubio, J. (1990). J. Chem. Phys., 92, 2478-2480.
Ab initio study of the ground and low-lying states of FeH

6094 Sodupe, M., Lluch, J.M., Oliva, A., Illas, F. and Rubio, J. (1991c). J. Chem. Phys., 94, 4352-4355.
An ab initio study of the collinear reaction of Fe^+ (^4F) and Fe^+ (^6D) with H_2

6095 Soff, G. (1989). Z. Phys. D, 11, 29-36.
Higher-order vacuum polarization charge densities for spherical symmetric external potentials

6096 Soff, G. (1991). Z. Phys. D, 21, S7-11.
Theory of highly-ionized few-electron systems

6097 Soff, G. and Mohr, P.J. (1988). Phys. Rev. A, 38, 5066-5075.
Vacuum polarization in a strong external field

6098 Soff, G. and Mohr, P.J. (1989). Phys. Rev. A, 40, 2174-2175.
Influence of vacuum-polarization corrections of order $\alpha(Z\alpha)$ and $\alpha(Z\alpha)^3$ in hydrogenlike uranium

6099 Soff, G., Müller, U., Schramm, S., de Reus, T., Mehler, G., Reinhardt, J., Müller, B. and Greiner, W. (1986).
Exotic phenomena in collisions of very heavy ions
In "Atoms in Unusual Situations", Ed. J.P. Briand, Plenum, New York, pp. 343-363.

6100 Soff, G., Schlüter, P., Reinhardt, J., de Reus, T., Wietschorke, K.-H., Schäfer, A., Graf, S., Müller, U., Müller, B. and Greiner, W. (1987). Phys. Scr., 36, 743-755.
Puzzling positron peaks in heavy-ion collisions

6101 Sohn, K.-S., Lee, S., Bylander, D.M. and Kleinman, L. (1989). Phys. Rev. B, 39, 9983-9986.
Nb_2 dimer: Examination of singlet or triplet character

6102 Sokalski, K. (1989). Acta Phys. Pol., A76, 901-904.
On the properties of electron in two dimensions

6103 Sokolov, A.A. and Ternov, I.M. (1986).
Radiation from Relativistic Electrons
AIP, New York, 320 pp.

6104 Solà, M., Balcells, M., Duran, M., Lledós, A. and Bertrán, J. (1992). Int. J. Quantum Chem., 44, 887-895.
Valence-bond calculations on ZnO and HgO using integrals computed throgh the semiempirical AM1 method

6105 Solov'ev, I.V. and Likhtenshtein, A.I. (1989). Fiz. Tver. Tela, 31(11), 253-258; Sov. Phys.-Solid St., 31, 1980-1983.
Discussion of linearized approaches to spin-polarized relativistic calculations for actinides

6106 Solov'ev, I.V., Likhtenshtein, A.I. and Antropov, V.P. (1990). Metallofizika, 12(2), 10-18.
Changes of electron structure and magnetic properties upon alpha - delta transition in plutonium (in Russian)

6107 Solov'ev, I.V., Shik, A.B., Antropov, V.P., Likhtenshtein, A.I., Gubanov, V.A. and Andersen, O.K. (1989). Fiz. Tver. Tela, 31(8), 13-19; Sov. Phys. Solid St., 31, 1285-1289.
New method for calculating the electronic structure of magnetic crystals containing heavy d and f elements

6108 Solovyev, I.V., Liechtenstein, A.I., Gubanov, V.A., Antropov, V.P. and Andersen, O.K. (1991). Phys. Rev. B, 43, 14414-14422.
Spin-polarized relativistic linear-muffin-tin-orbital method: Volume-dependent electronic structure and magnetic moment of plutonium

6109 Solovtsov, I.L. and Chernichenko, Yu.D. (1987). Izv. VUZov SSSR, Fiz., 5, 11-17.

Inverse scattering problem in the relativistic quasiclassical approximation (in Russian)

6110*Sommerfeld, A. (1915). S. Ber. Bayer. Akad. Wiss., 459-500.
Die Feinstruktur der Wasserstoff- und der Wasserstoff-ähnlichen Linien

6111*Sone, J. (1977). J. Phys. Soc. Japan, 42, 1457-1462.
Effect of spin-orbit interaction on the spin susceptibility of small particles

6112 Sonego, S. (1991). Phys. Rev. A, 44, 5369-5382.
Quasiprobabilities and explicitly covariant relativistic quantum theory

6113 Song, J. and Hall, M.B. (1991). Inorg. Chem., 30, 4433-4437.
Bond-stretch isomers of transition-metal complexes. Do they exist?

6114 Spain, E.M. and Morse, M.D. (1992). J. Chem. Phys., 97, 4641-4660.
Ligand-field theory applied to diatomic transition metals. Results for the $d_A{}^9 d_B{}^9 \sigma^2$ states of Ni_2, the $d_{Ni}{}^9 d_{Cu}{}^{10} \sigma^2$ states of NiCu, and the $d_{Ni}{}^8 ({}^3F) d_{Cu}{}^{10} \sigma^2 \sigma^{*1}$ excited states of NiCu

6115 Spence, J.R. and Vary, J.P. (1991a). Phys. Lett. B, 254, 1-6.
Electron-positron scattering resonances from relativistic two-body wave equations

6116 Spence, J.R. and Vary, J.P. (1991b). Phys. Lett. B, 271 27-31.
Electron-proton resonances at low energy from a relativistic two-body wave equation

6117 Spiegelmann, F., Gadea, F.X. and Castex, M.C. (1990). Chem. Phys., 145, 173-189.
Theoretical study of the electronic structure of the ArKr exciplex

6118 Spiegelmann, F., Pavolini, D. and Daudey, J.-P. (1989). J. Phys. B: At. Mol. Opt. Phys., 22, 2465-2484.
Theoretical study of the excited states of the heavier alkali dimers: II. The Rb_2 molecule

6119 Spies, W., Müller, A., Linkemann, J., Frank, A., Wagner, M., Kozhuharov, C., Franzke, B., Beckert, K., Bosch, F., Eickhoff, H., Jung, M., Klepper, O., König, W., Mokler, P.H., Moshammer, R., Nolden, F., Schaaf, U., Spädtke, P., Steck, M., Zimmerer, P., Grün, N., Scheid, W., Pindzola, M.S. and Badnell, N.R. (1992). Phys. Rev. Lett., 69, 2768-2771.
Dielectronic and radiative recombination of lithiumlike gold

6120 Springelkamp, F., de Groot, R.A., Geertsma, W., van der Lugt, W. and Mueller, F.M. (1985). Phys. Rev. B, 32, 2319-2325.
Electronic structure of semiconducting β-NaSn

6121*Spruch, L. (1966). Phys. Rev. Lett., 16, 1137-1139.
Relativistic versus nonrelativistic scattering of slow electrons

6122 Spruch, L. (1986). Phys. Today, 39(11), 37-45.
Retarded, or Casimir, long-range potentials

6123 Srivastava, R., McEachran, R.P. and Stauffer, A.D. (1992c). J. Phys. B: At. Mol. Opt. Phys., 25, 4033-4043.
Excitation of the D states of cadmium and barium by electron impact

6124 Srivastava, R., Zuo, T., McEachran, R.P. and Stauffer, A.D. (1992a). J. Phys. B: At. Mol. Opt. Phys., 25, 1073-1087.
Relativistic distorted-wave calculation of electron excitation of cadmium

6125 Srivastava, R., Zuo, T., McEachran, R.P. and Stauffer, A.D. (1992b). J. Phys. B: At. Mol. Opt. Phys., 25, 2409-2425.
Relativistic distorted-wave calculation of electron excitation of mercury

6126 Srivastava, R., Zuo, T., McEachran, R.P. and Stauffer, A.D. (1992d). J. Phys. B: At. Mol. Opt. Phys., 25, 3709-3720.
Excitation of the $^{1,3}P_1$ states of calcium, strontium and barium in the relativistic distorted-wave approximation

6127 Srivastava, S.K. (1989). J. Math. Phys., 30, 2838-2844.
Solution of the Dirac equation in Kasner's space-time

6128 Srivastava, Y. and Widom, A. (1987). Phys. Reports, 148, 1-65.
Quantum electrodynamic processes in electrical engineering circuits

6129*Stacey, R. (1982). Phys. Rev. D, 26, 468-472.
Eliminating lattice fermion doubling

6130 Staemmler, V. (1990). Z. Phys. D, 16, 167-173.
CEPA calculations on open-shell molecules XI. The two lowest electronic states of $HeAr^+$

6131 Stahlhofen, A. and Biedenharn, L.C. (1988). Lecture Notes Phys. 313, 261-267.
Group-theoretic approach to scattering: The Dirac-Coulomb problem and relativistic supersymmetry

6132 Stauffer, A. (1988).
Relativistic effects in electron-atom scattering
In "Correlation and Polarization in Electronic and Atomic Collisions", Ed. A. Crowe and M. Rudge, World Scientific, Singapore.

6133 Staunton, J., Strange, P., Gyorffy, B.L., Matsumoto, M., Poulter, J., Ebert, H. and Archibald, N.P. (1991).
Theory of magnetocrystalline anisotropy
In "The Effects of Relativity in Atoms, Molecules, and the Solid State", Ed. S. Wilson, I.P. Grant and B.L. Gyorffy, Plenum, New York and London, pp. 295-317.

6134 Steiger, A.D. (1987). J. Quant. Spectrosc. Radiat. Transfer, 38, 81-87.
Relativistic radiative transition rates and energies for the Li-like isoelectronic sequence: Electric and magnetic dipole and quadrupole radiation

6135 Stepanov, S.S. and Tutik, R.S. (1992). Phys. Lett. A, 163, 26-31.
A new approach to the 1/N-expansion for the Dirac equation

6136 Stepanov, S.S. and Tutik, R.S. (1992b). Sov. Phys. JETP, 74, 9-12; Zh. Eksp. Teor. Fiz., 101 18-24.
1/n expansion for the Klein-Gordon equation

6137 Stepanovskij, Yu.P. (1987). Ukr. Fiz. Zh., 32, 1316-1321.
Hydrogen atom in the external field as an anharmonic oscillator

6138 Stern, A. and Entin-Wohlman, O. (1991). Phys. Rev. B, 44, 10976-10979.
Damped conductivity oscillations resulting from the interplay of spin-orbit and Zeeman couplings

6139*Steslicka, M. and Davison, S.G. (1970). Phys. Rev. B, 1, 1858-1860.
Boundary conditions for the relativistic Kronig-Penney model

6140*Steslicka, M., Davison, S.G. and Brown, A.G. (1971a).
One-dimensional relativistic theory of impurity states
In "Electronic Density of States", Ed. L.H. Bennett, NBS Special Publication 323, NBS, Washington DC, pp. 477-481.

6141*Steslicka, M., Davison, S.G. and Srinivasan, U. (1971b).
J. Phys. Chem. Solids, 32, 1917-1923.
Electronic states of heavy diatomic crystals

6142 Stevens, W.J., Krauss, M., Basch, H. and Jasien, P.G. (1992). Can. J. Chem., 70, 612-630.
Relativistic compact effective potentials and efficient, shared-exponent basis sets for the third-, fourth-, and fifth-row atoms

6143 Stewart, J.J.P. (1990). J. Computer-Aided Mol. Design, 4, 1-105.
MOPAC: A semiempirical molecular orbital program

6144 Sticht, J. (1989).
Bandstrukturrechnung für Schwere-Fermionen-Systeme
Dissertation, Darmstadt, 155 p.

6145 Strange, P., Durham, P.J. and Gyorffy, B.L. (1991).
Phys. Rev. Lett., 67, 3590-3593.
Dichroic X-ray fluorescence

6146 Strange, P., Ebert, H., Staunton, J.B. and Gyorffy, B.L. (1989a). J. Phys.: Cond. Matter, 1, 2959-2975.
A relativistic spin-polarised multiple-scattering theory,

with applications to the calculation of the electronic structure of condensed matter

6147 Strange, P., Ebert, H., Staunton, J.B. and Gyorffy, B.L. (1989b). J. Phys.: Cond. Matter, 1, 3947-3955.
A first-principles theory of magnetocrystalline anisotropy in metals

6148 Strange, P., Gyorffy, B.L. and Durham, P.J. (1992). J. Magn. Magn. Materials, 104-107, 755-756.
Relativistic theory of x-ray fluorescence from magnets

6149 Strayer, M.R., Bottcher, C., Oberacker, V.E. and Umar, A.S. (1990). Phys. Rev. A, 41, 1399-1407.
Lattice calculation of muon-pair production with capture in relativistic heavy-ion collisions

6150 Strittmatter, R.J. (1990). Diss. Abstr. Int. B, 51, 4867.
The bonding and electronic structure of tris(cyclopentadienyl) actinide compounds and related transition metal systems

6151 Strittmatter, R.J. and Bursten, B.E. (1991). J. Am. Chem. Soc., 113, 552-559.
Bonding in tris(η^5-cyclopentadienyl) actinide complexes. 5. A comparison of the bonding in Np, Pu, and transplutonium compounds with that in lanthanide compounds and a transition-metal analogue

6152 Strömberg, D., Gropen, O. and Wahlgren, U. (1989). Chem. Phys., 133, 207-219.
Non-relativistic and relativistic calculations on some Zn, Cd and Hg complexes

6153 Strömberg, D., Sandström, M. and Wahlgren, U. (1990). Chem. Phys. Lett., 172, 49-54.
Theoretical calculations on the structure of the hexahydrated divalent zinc, cadmium and mercury ions

6154 Strömberg, D. and Wahlgren, U. (1990). Chem. Phys. Lett., 169, 109-115.
First-order relativistic calculations on Au_2 and Hg_2^{2+}

6155 Stroucken, H.J.A.C. and Dreizler, R.M. (1991). Phys. Rev. A, 43, 3401-3406.
Extended Thomas-Fermi theory for relativistic systems at finite temperatures

6156 Struensee, M.C. and Cohen, J.S. (1988). Phys. Rev. A, 38, 3377-3387.
Transitions between the n=2 fine-structure levels in H+p collisions

6157 Styszynski, J. and Karwowski, J. (1988). J. Phys. B: At. Mol. Opt. Phys., 21, 2389-2397.
Multiconfiguration Dirac-Fock study on the ground-state energies of two-electron atoms

6158 Styszynski, J. and Karwowski, J. (1990). Acta Phys. Pol. A, A78, 855-859.

Screening constants in multi-configuration Hartree-Fock and Dirac-Fock methods

6159 Su, Q.-C. and Eberly, J.H. (1987). Phys. Rev. A, 35, 2962-2964.
Essential states in the relativistic theory of electron-photon interactions

6160 Su, R.-K. and Ma, Z.-Q. (1986). J. Phys. A, 19, 1739-1743.
Confinement properties for the Dirac equation with scalar-like and vector-like potentials

6161 Subramanian, V., Vijayakumar, M. and Ramasami, T. (1991). Chem. Phys. Lett., 182, 232-236.
Ionization energy calculations of MCl_4 (M=Ti, Zr and Hf)-type molecules by the MS Xα method

6162 Sucher, J. (1986).
QED and relativity in atomic physics
In "Atoms in Unusual Situations", Ed. J.P. Briand, Plenum, New York, pp. 251-300.

6163 Sucher, J. (1987). Phys. Scr., 36, 271-281.
Relativistic many-electron Hamiltonians

6164 Sucher, J. (1988). J. Phys. B: At. Mol. Opt. Phys., 21, L585-L591.
On the choice of the electron-electron potential in relativistic atomic physics

6165 Sucher, J. (1989a).
Bound state QED
In "Relativistic, Quantum Electrodynamic and Weak Interaction Effects in Atoms", Ed. W. Johnson, P. Mohr and J. Sucher, AIP Conf. Proc., 189, pp. 28-46.

6166 Sucher, J. (1989b).
Potentials from field theory: Non-uniqueness, gauge dependence and all that
In "Relativistic, Quantum Electrodynamic and Weak Interaction Effects in Atoms", Ed. W. Johnson, P. Mohr and J. Sucher, AIP Conf. Proc., 189, pp. 337-370.

6167 Sud, K.K. and Moattar, S. (1991). Acta Phys. Hung., 69, 19-24.
K-shell ionization of atoms by relativistic protons

6168 Sud, K.K., Sharma, D.K. and Agarwal, V.K. (1987). Indian J. Phys. B, 61B, 317-324.
The L-shell ionization cross section of atoms by relativistic electron

6169 Sugar, J., Kaufman, V. and Rowan, W.L. (1988). J. Opt. Soc. Am. B, 5, 2183-2189.
Aluminumlike spectra of copper through molybdenum

6170 Sugar, J., Kaufman, V. and Rowan, W.L. (1990). J. Opt. Soc. Am. B, 7, 152-158.
Spectra of the Si I isoelectronic sequence from Cu XVI to Mo XXIX

6171 Sugar, J., Kaufman, V. and Rowan, W.L. (1992). J. Opt. Soc. Am. B, 9, 1959-1961.
Rb-like spectra: Pd X to Nd XXIV

6172*Sugiura, Y. and Urey, H.C. (1926). Kgl. Danske Vid. Selskab, Mat.-Fys. Medd., 7(13), 18p.
On the quantum theory explanation of the anomalies in the 6th and 7th periods of the periodic table

6173 Sumathi, K. and Balasubramanian, K. (1990). J. Chem. Phys., 92, 6604-6619.
Electronic states and potential energy surfaces of H_2Te, H_2Po, and their positive ions

6174 Sundholm, D. (1988). Chem. Phys. Lett., 149, 251-255.
Two-dimensional, fully numerical solution of the molecular Dirac equation. Dirac-Slater calculations on LiH, Li_2, BH and CH^+

6175 Sundholm, D. and Olsen, J. (1990a). Chem. Phys. Lett., 171, 53-57.
Large MCHF calculations on the electron affinity of boron

6176 Sundholm, D. and Olsen, J. (1990b). Phys. Rev. A, 42, 1160-1164.
Nuclear quadrupole moments of ^{33}S and ^{35}S

6177 Sundholm, D. and Olsen, J. (1992a). Chem. Phys. Lett., 198, 526-530.
Finite element multiconfiguration Hartree-Fock determination of the atomic quadrupole moment of Ca(3d4s; ^1D)

6178 Sundholm, D., Pyykkö, P. and Laaksonen, L. (1987). Phys. Scr., 36, 400-402.
Two-dimensional, fully numerical solutions of second-order Dirac equations for diatomic molecules. Part 3

6179 Suric, T., Bergstrom Jr., P.M., Pisk, K. and Pratt, R.H. (1991). Phys. Rev. Lett., 67, 189-192.
Compton scattering of photons by inner-shell electrons

6180 Svane, A. (1989). J. Chim. Phys., 86, 911-917.
Theory of ^{119}Sn isomer shifts in solids

6181 Svensson, W.A., Krause, M.O. and Carlson, T.A. (1987). J. Phys. B: At. Mol. Opt. Phys., 20, L271-L275.
Partial cross sections and parameters for the argon spin-orbit components in the $3s3p^6np\ ^1P_{10}$ resonance region

6182 Svensson, W.A., Krause, M.O., Carlson, T.A., Radojevic, V. and Johnson, W.R. (1986). Phys. Rev. A, 33, 1024-1032.
Photoelectron studies of the 4f and 6s subshells in atomic ytterbium

6183 Swainson, R. and Drake, G.W.F. (1986). Phys. Rev. A, 34, 620-623.
Lamb shifts and fine-structure splittings for light muonic

ions: Hyperfine-structure corrections

6184 Swainson, R.A. and Drake, G.W.F. (1991a). J. Phys. A, 24, 79-94.
A unified treatment of the non-relativistic and relativistic hydrogen atom: Pt. 1. The wavefunctions

6185 Swainson, R.A. and Drake, G.W.F. (1991b). J. Phys. A, 24, 95-120.
A unified treatment of the non-relativistic and relativistic hydrogen atom: Pt. 2. The Green functions

6186 Swainson, R.A. and Drake, G.W.F. (1991c). J. Phys. A, 24, 1801-1824.
A unified treatment of the non-relativistic and relativistic hydrogen atom: III. The reduced Green functions

6187 Syromyatnikov, V.G. and Savicius, E. (1986). Liet. Fiz. Rinkinys, 26, 496-499.
Semiempirical method for the study of the energy levels of atoms and ions in the relativistic approximation

6188 Syros, C. (1987). Phys. Scr., 36, 385-392.
Non-relativistic evolution operator in QFT and the relativistic H-atom

6189 Szasz, L. (1992).
The Electronic Structure of Atoms
Wiley-Interscience, New York, 571 pp.

6190 Szmytkowski, R. (1991). J. Phys. B: At. Mol. Opt. Phys., 24, 3895-3904.
The relativistic polarized orbital theory of the elastic electron and positron scattering from closed-shell atoms

6191 Szpunar, B. (1986). Phys. Lett. A, 115, 157-160.
A method for the calculation of the orbital moments for complex crystal structures

6192 Szunyogh, L., Weinberger, P. and Redinger, J. (1992). Phys. Rev. B, 46, 2015-2023.
Relativistic calculation of CCV and CVV Auger-electron spectra: Applications to $Ti_{0.5}Al_{0.5}N$ and Pd

6193 Szwed, J. (1986). Phys. Lett. B, 181, 305-307.
The "square root" of the Dirac equation within supersymmetry

6194 Tadjeddine, M., Flament, J.P. and Teichteil, C. (1988a). J. Mol. Struct.(Theochem), 166, 147-152.
Ionization potentials of CH_3I : Ab-initio calculations including spin-orbit coupling

6195 Tadjeddine, M., Flament, J.P. and Teichteil, C. (1988b). Chem. Phys., 124, 13-28.
Spin-orbit calculations of the molecular states of CH_3I^+, related to photofragmentation experiments

6196 Tahri, Y., Chermette, H., El Khatib, N., Krupa, J.C. and Simoni, E. (1990). J. Less-Common Met., 158, 105-116.
Electronic structures of thorium and protactinium

halide clusters of $[ThX_8]^{4-}$ type

6197 Takahashi, Y. and Ropchan, C. (1986). Progr. Theor. Phys., 76, 1187-1197.
The first order Schrödinger equation, the symmetrization of the stress tensor and the spin-orbit coupling

6198 Takeda, K. and Shiraishi, K. (1992). Chem. Phys. Lett., 195, 121-126.
Electronic structure of chain-like polystannane

6199 Takeuchi, N., Chan, C.T. and Ho, K.M. (1989). Phys. Rev. B, 40, 1565-1570.
First-principles calculations of equilibrium ground-state properties of Au and Ag

6200 Takizawa, S. (1990).
Study on Phase Stability of Noble Metal - Transition Metal Alloys by First Principles Calculation
Thesis, University of Tokyo

6201 Takizawa, S., Blügel, S., Terakura, K. and Oguchi, T. (1991). Phys. Rev. B, 43, 947-955.
Theoretical study of the structural stability of CuPd and CuPt alloys: Pressure-induced phase transition of CuPt alloy

6202 Takizawa, S., Terakura, K. and Mohri, T. (1989). Phys. Rev. B, 39, 5792-5797.
Electronic theory for phase stability of nine AB binary alloys, with A= Ni, Pd, or Pt and B= Cu, Ag, or Au

6203*Taksar, I.M., Kunin, P.E., Zapol', B.P. and Kruglevskii, V.A. (1971).
Relativistic calculation of frequencies and oscillator strengths in spectra of monovalent atoms by the effective potential method

6204 Talman, J.D. (1986). Phys. Rev. Lett., 57, 1091-1094.
Minimax principle for the Dirac equation

6205 Talwar, I. and Wright, L.E. (1989). Comp. Phys. Comm., 55, 367-380.
The radiation tail accompanying elastic electron scattering from the atomic nucleus

6206 Tamura, E. (1992). Phys. Rev. B, 45, 3271-3278.
Relativistic single-site Green function for general potentials

6207 Tamura, E. and Feder, R. (1989). Solid State Comm., 70, 205-209.
Relativistic electronic structure of crystal surfaces

6208 Tamura, E. and Feder, R. (1990). Z. Phys. B, 81, 425-431.
Electronic surface states on Pd(110): theory and comparison with experiment

6209 Tan, Y. and Thouless, D.J. (1992). Phys. Rev. B, 46, 2985-2994.

Total energy for fermions on a two-dimensional lattice in a magnetic field

6210*Tani, S. (1949). Soryushiron Kenkyu(Kyoto), 1, 15-23.
Physical interpretation of the Dirac equation

6211*Tani, S. (1951). Progr. Theor. Phys., 6, 267-285.
Connection between particle models and field theories I. The case spin 1/2

6212 Tanpipat, N. (1988). Diss. Abstr. Int. B, 49, 1199-1200.
Ab initio relativistic effective potential and relativistic configuration interaction calculations on diatomic molecules containing heavy elements

6213 Tatsumi, K., Matsubara, I., Inoue, Y., Nakamura, A., Cramer, R.E., Tagoshi, G.J., Golen, J.A. and Gilje, J.W. (1990). Inorg. Chem., 29, 4928-4938.
A homoleptic uranium thiolate: Synthesis, structure, and fluxional behavior of $[Li(dme)]_4[U(SCH_2CH_2S)_4]$ and reaction with CS_2

6214 Tatsumi, K. and Nakamura, A. (1987a). Organometallics, 6, 427-428.
Unusual Th-C-C angles in bis(cyclopentadienyl)-dialkyl complexes of thorium

6215 Tatsumi, K. and Nakamura, A. (1987b). J. Am. Chem. Soc., 109, 3195-3206.
Actinide-to-carbon bonds in $Cp_2An(alkyl)_2$, -(butadiene), -(metallacyclopentadiene), and -(cyclobutadiene) complexes

6216 Tatsumi, K. and Nakamura, A. (1987c). Inorg. Chim. Acta, 139, 247-251.
An MO study on structures and reactivity of organoactinides based on the extended Hückel method : Actinide bis(cyclopentadienyl) complexes

6217 Tayal, S.S. (1991). Phys. Scr., 43, 270-272.
Energy levels and oscillator strengths for Ga II

6218 Tegze, M. and Hafner, J. (1989). Phys. Rev. B, 39, 8263-8274.
Electronic structure of metallic and semiconducting alkali-metal-lead compounds

6219 Tegze, M. and Hafner, J. (1992). J. Phys.: Condens. Matter, 4, 2449-2474.
Electronic structure of alkali-pnictide compounds

6220 Teitelboim, C. (1988).
Quantum Mechanics of Fundamental Systems I
Plenum, 226 p.

6221 Temmerman, W.M. and Sterne, P.A. (1990). J. Phys.: Cond. Matter, 2, 5529-5538.
The band model for the ground state properties of gadolinium metal

6222*Teo, B.-K. and Lee, P.A. (1979). J. Am. Chem. Soc., 101,

2815-2832.
Ab initio calculations of amplitude and phase functions for extended x-ray absorption fine structure spectroscopy

6223 Terakura, K., Oguchi, T., Mohri, T. and Watanabe, K. (1987). Phys. Rev. B, 35, 2169-2173.
Electronic theory of the alloy phase stability of Cu-Ag, Cu-Au, and Ag-Au systems

6224 Ternov, I.M. (1990). Zh. Eksp. Teor. Fiz., 98, 1169-1172; Sov. Phys. JETP, 71, 654-656.
Evolution equation for the spin of a relativistic electron in the Heisenberg representation

6225 Ternov, I.M., Khapaev, A.M. and Ponomaryov, I.V. (1988). Ann. Physik, 45, 278-284.
On some solutions of relativistic wave equations

6226 Tesevich, B.I. (1991). Izv. Akad. Nauk SSSR, Ser. Fiz., 55, 2284-2286.
Vacuum-polarization potential for a finite nucleus near origin (in Russian)

6227 Thakkar, A.J. (1986). J. Chem. Phys., 85, 4509-4514.
Relativistic kinetic energies and mass-velocity corrections in diatomic molecules

6228 Thakur, J. (1987). Pramana, 28, 1-8.
Semiclassical approximation for relativistic potentials

6229 Thakur, J. (1991). Pramana, 37, 21-38.
Relativistic quantum mechanics of spin-zero and spin-half particles

6230 Thakur, J. (1992). Pramana, 39, 253-264.
Classical mechanics of directly interacting relativistic particles

6231*Thaler, J. (1985). J. Phys. G, 11, 201-208.
Relativistic effects in protonium

6232 Thaller, B. (1986). Lett. Math. Phys., 12, 15-19.
Relativistic scattering theory for long-range potentials of the nonelectrostatic type

6233 Thaller, B. (1988). J. Math. Phys., 29, 249-257.
Normal forms of an abstract Dirac operator and applications to scattering theory

6234 Thaller, B. (1989).
Dirac particles in magnetic fields
In "Recent Developments in Quantum Mechanics. Proceedings of the Brasov Conference", Ed. A. Boutet de Monvel, P. Dita, G. Nenciu and R. Purice, Kluwer, Dordrecht, pp. 357-366.

6235 Thaller, B. (1992).
The Dirac Equation
Springer-Verlag, Berlin, 357 pp.

6236 Thaller, B. and Enss, V. (1986). Ann. Inst. Henri Poincare, 45, 147-171.
Asymptotic observables and Coulomb scattering for the Dirac equation

6237 Thayyullathil, R.B. (1991). Mod. Phys. Lett. A, 6, 1219-1224.
Covariant two-body bound state equation

6238 Theodosiou, C.E. (1989). Phys. Rev. A, 39, 4880-4883.
Accurate calculation of the 4p lifetimes of Ca^+

6239 Theodosiou, C.E. and Curtis, L.J. (1988). Phys. Rev. A, 38, 4435-4445.
Accurate calculations of 3p and 3d lifetimes in the Na sequence

6240 Theodosiou, C.E., Curtis, L.J. and El-Mekki, M. (1991). Phys. Rev. A, 44, 7144-7157.
Calculations of 2p lifetimes in the Li sequence

6241 Thiel, J., Bunker, A., Momberger, K., Grün, N. and Scheid, W. (1992). Phys. Rev. A, 46, 2607-2612.
Electron-positron pair creation with capture and ionization in relativistic heavy-ion collisions by the finite-difference method

6242 Thiel, W. and Voityuk, A.A. (1992). Int. J. Quantum Chem., 44, 807-829.
Extension of MNDO to d orbitals: Parameters and results for the halogens

6243 Thies, B., Fricke, B., Sepp, W.-D. and Hartung, H. (1986). J. Phys. B, 19, L617-L621.
New calculations of P(b) curves for 1s excitation in low-Z (Z 10) ion-atom scattering

6244 Thies, B., Sepp, W.-D. and Fricke, B. (1989a). J. Phys. (Paris), 50, C1-193/C1-198.
Many electron coupled channel calculations for the system F^{8+} - Ne

6245 Thies, B., Sepp, W.-D. and Fricke, B. (1989b). Phys. Lett. A, 139, 161-164.
Many-electron relativistic calculation and interpretation of atomic processes in time dependent heavy-ion scattering

6246 Thies, M. (1988). Nucl. Phys., A478, 571c-577c.
Towards an alternative to the Dirac approach

6247 Thole, B.T. and van der Laan, G. (1987). Europhys. Lett., 4, 1083-1086.
Systematics of the relation between spin-orbit splitting in the valence band and the branching ratio in X-ray absorption spectra

6248 Thole, B.T. and van der Laan, G. (1988). Phys. Rev. A, 38, 1943-1947.
Linear relation between x-ray absorption branching ratio and valence-band spin-orbit expectation value

6249 Thomson, M.J. and McKellar, B.H.J. (1991). Am. J. Phys., 59, 340-346.
The solution of the Dirac equation for a high square barrier

6250 Thumm, U. and Norcross, D.W. (1991). Phys. Rev. Lett., 67, 3495-3498.
Evidence for very narrow shape resonances in low-energy electron-Cs scattering

6251 Thumm, U. and Norcross, D.W. (1992). Phys. Rev. A, 45, 6349-6370.
Relativistic R-matrix calculations for electron-alkali-metal-atom scattering: Cs as a test case

6252 Thümmel, H., Klotz, R. and Peyerimhoff, S.D. (1989). Chem. Phys., 135, 229-245.
Study of the a $^3\Pi$ and X $^1\Sigma^+$ states of the MgO molecule: Spin-forbidden transitions and breakdown of the Born-Oppenheimer approximation

6253 Tisenko, Yu.A. (1989). Izv. VUZov, Fiz., 32(10), 45-50; Sov. Phys. J., 32, 796-800.
Dul'yan-Faustov pseudopotential equation and the fine structure of the helium atom. I

6254 Tiwary, S.N., Kumar, P., Singh, D.D. and Sharma, R.J. (1990). Fizika, 22, 577-581.
Oscillator strengths of the resonance transition in the Rb I, Sr II, Cs I and Ba II systems

6255 Tobisch, S. and Rasch, G. (1990). Chem. Phys. Lett., 166, 311-316.
Ab initio molecular-orbital calculations on Pt, PtH and PtH_2 with a quasirelativistic pseudopotential for Pt

6256*Todorov, I.T. (1973).
Recoil corrections and reduced mass in relativistic calculations
In "Properties of Fundamental Interactions", Part C, Vol. 9, pp. 953.

6257 Todorov, N.S. (1991). Ann. Foundation Louis de Broglie, 16, 73-90.
On position-velocity coexistence in a 'minimal' statistical interpretation of the Dirac electron states

6258 Tománek, D., Sun, Z. and Louie, S.G. (1991). Phys. Rev. B, 43, 4699-4713.
Ab initio calculation of chemisorption system: H on Pd(001) and Pd(110)

6259 Tong, X.-M. and Li, J.-M. (1989). Acta Phys. Sin., 38 1406-1412.
Two photon transitions in atomic inner-shells for Xe-relativistic effect and atomic screening effect (in Chinese)

6260 Tong, X.-M., Li, J.-M., Kissel, L. and Pratt, R.H. (1990).

Phys. Rev. A, 42, 1442-1449.
Two-photon transitions in atomic inner shells: Relativistic and atomic-screening effects

6261 Topol', I.A., Vovna, V.I. and Kazachek, M.V. (1987). Teor. Eksp. Khim., 23, 456-461; Theor. Exp. Chem., 23, 427-432.
Electronic structure and photoelectron spectra of osmium and ruthenium tetraoxides

6262 Tornieporth-Oetting, C., Buzek, P., Schleyer, P.v.R. and Klapötke, T.M. (1992). Angew. Chem., 104, 1391-1392.
Azidoiodoiod(1+)-hexafluoroantimonat: das erste binäre Stickstoff-Iod-Kation

6263 Toshima, N. and Eichler, J. (1988a). Phys. Rev. A, 38, 2305-2316.
Coupled-channel theory of excitation and charge transfer in relativistic atomic collisions

6264 Toshima, N. and Eichler, J. (1988b). Phys. Rev. Lett., 60, 573-576.
Coupled-channels treatment of excitation and charge transfer in $U^{92+}+U^{91+}$ collisions at 1 and 0.5 GeV/u

6265 Toshima, N. and Eichler, J. (1989). Phys. Rev. A, 40, 125-132.
Relativistic coupled-channel calculations including pseudostates

6266 Toshima, N. and Eichler, J. (1990a). Phys. Rev. A, 41, 5221-5224.
Distorted-wave approximations for relativistic atomic collisions

6267 Toshima, N. and Eichler, J. (1990b). Phys. Rev. A, 42, 3896-3900.
Direct reactions in relativistic atomic collisions and the influence of Coulomb boundary conditions

6268 Tóth, A., Földy, L. and Révai, J. (1988). Z. Phys. D, 8, 1-5.
Two-center Dirac equation with separable potentials

6269 Toyama, F.M. (1991). Phys. Rev. C, 44, 2881-2883.
Relation between bound state radius and scattering length: exactly solvable relativistic model

6270 Toyama, F.M., Beachey, D.J., Nogami, Y. and van Dijk, W. (1991). Phys. Rev. C, 44, 67-72.
Relation between the bound state radius and the scattering length: Relativistic effects

6271 Tragin, N., Geindre, J.-P., Chenais-Popovics, C., Gauthier, J.-C., Wyart, J.-F. and Luc-Koenig, E. (1989). Phys. Rev. A, 39, 2085-2089.
Ionization limits in Cu-like and Ni-like high-Z ions from ab initio calculations and wavelength measurements in Rydberg series

6272 Tréglia, G. and Ducastelle, F. (1987). J. Phys. F, 17, 1935-1944.
Is ordering in PtNi alloys induced by spin-orbit interactions?

6273 Trinquier, G. (1990). J. Am. Chem. Soc., 112, 2130-2137.
Double bonds and bridged structures in the heavier analogues of ethylene

6274 Trinquier, G. (1991). J. Am. Chem. Soc., 113, 144-151.
Singly bridged arrangements on Group 14 X_2H_4 potential surfaces

6275 Trinquier, G. (1992). J. Am. Chem. Soc., 114, 6807-6820.
Polymorphism in the heavier analogues of the ethyl cation

6276 Trinquier, G. and Barthelat, J.-C. (1990). J. Am. Chem. Soc., 112, 9121-9130.
Structures of X_2F_4, from carbon to lead. Unsaturation through fluorine bridges in Group 14

6277 Tripathi, G.S., Misra, C.M., Tripathi, P. and Misra, P.K. (1989). Phys. Rev. B, 39, 94-101.
Theory of the Knight shift in solids: Effect of electron-phonon interaction

6278 Tripathi, G.S., Misra, C.M., Tripathi, P. and Misra, P.K. (1990). Hyperfine Interactions, 60, 623-626.
Effect of electron-phonon interaction on the Knight shift

6279 Trubitsin, V.Yu., Trubitsina, N.A. and Shirokovskii, V.P. (1992). J. Phys.: Condens. Matter, 4, 7555-7564.
Relativistic spin-polarized calculation of the electronic structure of antiferromagnetic chromium

6280 Trunschke, A. and Miessner, H. (1986). Z. Chem., 26, 416-417.
Relativistische Modellrechnungen (REX) zur Chemisorption von Kohlenmonoxid an der Platin(111)-Oberfläche

6281 Trusov, V.F. and Eglais, M.O. (1989). Izv. Akad. Nauk SSSR, Ser. Fiz., 53, 907-909; Bull. Acad. Sci. USSR, Phys. Ser., 53(5), 82-84.
Contact charge density calculation in an Fe atom by the Dirac-Fock method

6282*Tsai, W.-Y. (1973). Phys. Rev. D, 7, 1945-1948.
Energy eigenvalues for charged particles in homogeneous magnetic field - an application of the Foldy-Wouthuysen transformation

6283 Tschinke, V. (1989).
The Fermi Hole in the HFS-Method and Its Non-Local Extensions
Thesis, the University of Calgary

6284*Tterlikkis, L., Mahanti, S.D. and Das, T.P. (1969). Phys. Rev., 178, 630-640.
Relativistic effects on the hyperfine interactions in alkali metals

6285 Tulkki, J. (1989). Phys. Rev. Lett., 62, 2817-2820.
Multiple excitation at xenon 5s photoionization threshold

6286 Tulkki, J. and Åberg, T. (1989). J. Phys. B, 22, 3681-3687.
Hydrogenic resonances in photoabsorption of positive ions

6287 Tulkki, J., Åberg, T., Mäntykenttä, A. and Aksela, H. (1992b). Phys. Rev. A, 46, 1357-1366.
Relativistic multichannel calculation of the Ne KLL and Ar $L_2M_{2,3}M_{2,3}$ Auger transition rates

6288 Tulkki, J., Åberg, T., Whitfield, S.B. and Crasemann, B. (1990). Phys. Rev. A, 41, 181-185.
Quantum approach to photoelectron recapture in post-collision interaction

6289 Tulkki, J., Aksela, S., Aksela, H., Shigemasa, E., Yagishita, A. and Furusawa, Y. (1992a). Phys. Rev. A, 45, 4640-4645.
Krypton 4p, 4s, and 3d partial photoionization cross sections below a photon energy of 260 eV

6290 Tulkki, J., Armen, G.B., Åberg, T., Crasemann, B. and Chen, M.H. (1987). Z. Phys. D, 5, 241-252.
Quantum theory of post-collision interaction in inner-shell photoionization

6291 Tupitsyn, I.I. (1991). Radiokhimiya, 33 (5), 63-69.
Calculations of the ionization potentials of uranium by relativistic Hartree-Fock-Dirac method with taking into account interaction of configurations (in Russian)

6292 Turner, D.E. and Harmon, B.N. (1989). Phys. Rev. B, 40, 10516-10522.
Electronic structure of red mercuric iodide

6293*Turro, N.J. and Devaquet, A. (1975). J. Am. Chem. Soc., 97, 3859-3862.
Chemiexcitation mechanisms. The role of symmetry and spin-orbit coupling in diradicals

6294 Tuszynski, J.A. and Dixon, J.M. (1987). Physica A, 143A, 331-345.
A study of effective crystal fields, spin-orbit coupling and residual interactions, due to relativistic effects and configuration mixing, for gadolinium(3+) ion in C_{3h} symmetry

6295 Tuszynski, J.A. and Dixon, J.M. (1989a). J. Less-Common Met., 148, 381-386.
Comments on the role of relativistic crystal field effects with applications to 4f and 3d ion complexes

6296 Tuszynski, J.A. and Dixon, J.M. (1989b). Physica A, 156, 924-935.
Calculations of relativistic crystal-field contributions to the zero-field splittings in 4f and 3d ion complexes

6297 Tutik, R.S. (1992). J. Phys. A, _25_, L413-L417.
Exact solution of the N-dimensional generalized Dirac-Coulomb equation

6298 Tyson, C.C., Bzowski, A., Kristof, P., Kuhn, M., Sammynaiken, R. and Sham, T.K. (1992). Phys. Rev. B, _45_, 8924-8928.
Charge redistribution in Au-Ag alloys from a local perspective

6299 Ulehla, I. and Horejsi, J. (1986). Phys. Lett. A, _113_, 355-358.
Generalized Prüfer transformation and the eigenvalue problem for radial Dirac equations

6300 Ungar, A.A. (1989). Found. Phys., _19_, 1385-1396.
The relativistic velocity composition paradox and the Thomas rotation

6301 Ungar, A.A. (1991). Am. J. Phys., _59_, 824-834.
Thomas precession and its associated grouplike structure

6302 Ungar, A.A. (1992). Am. J. Phys., _60_, 815-828.
The abstract Lorentz transformation group

6303 Uylings, P.H.M. (1988). J. Phys. B: At. Mol. Opt. Phys., _21_, L391-L394.
Energy contributions of relativistic two-body operators

6304 Uylings, P.H.M. (1989). J. Phys. B: At. Mol. Opt. Phys., _22_, 2947-2961.
Extended theory of the spin-orbit interaction

6305 Uylings, P.H.M. (1992). J. Phys. B: At. Mol. Opt. Phys., _25_, 4391-4407.
Applications of second quantization in the coupled form

6306 Vahtras, O., Ågren, H., J rgensen, P., Jensen, H.J.Aa., Helgaker, T. and Olsen, J. (1992a). J. Chem. Phys., _96_, 2118-2126.
Spin-orbit coupling constants in a multiconfiguration linear response approach

6307 Vahtras, O., Ågren, H., J rgensen, P., Jensen, H.J.Aa. and Helgaker, T. (1992b). Int. J. Quantum Chem., _41_, 729-731.
The second-order energy contribution from the spin-orbit interaction operator to the potential energy curve of Cr_2

6308 Vainshtein, L.A. and Safronova, U.I. (1988). Sov. Phys. -Lebedev Inst. Rep., _6_, 37-40.
Transition energies in Ne-like ions. Correlation effects

6309 Vainshtein, L.A., Syunyaev, R.A. and Churazov, E.M. (1986). Kratk. Soobshch. Fiz., _1_, 33-34.
Calculation of the hyperfine-structure parameters for lithium and lithium-type ions. Relativistic correction and taking into account the nuclear structure

6310 Van Alstine, P. and Crater, H. (1986). Phys. Rev. D, _34_, 1932-1935.

Exact parapositroniumlike solution to two-body Dirac equations

6311 van Daelen, M.A., Jansen, A.P.J., Biemolt, W. and van Santen, R.A. (1992). Chem. Phys. Lett., 190, 190-198. Density functional calculations on cation-induced changes in the adsorption of sulphur on Ir_4 clusters

6312 van der Laan, G. (1989). Physica B, 158, 395-397. The X-ray absorption branching ratio of transition metal compounds

6313 van der Laan, G. and Thole, B.T. (1988). Phys. Rev. Lett., 60, 1977-1980. Local prcbe for spin-orbit interaction

6314 van der Laan, G., Thole, B.T., Sawatzky, G.A., Fuggle, J.C., Karnatak, R., Esteva, J.-M. and Lengeler, B. (1986). J. Phys. C, 19, 817-827. Identification of the relative population of spin-orbit split states in the ground state of a solid

6315 van der Waals, J.H. (1986). Int. Rev. Phys. Chem., 5, 219-228. Spin-orbit coupling in the tetroxo anions of the d^0 transition metals and the identification of their lower triplet states

6316 van Dijk, W. (1992). Phys. Scr., 45, 75-78. Relativistic effect on the radius of a model deuteron

6317 Vanquickenborne, L.G., Renders, A., Pierloot, K., Devoghel, D., Görller-Walrand, C. and Goodman, G.L. (1991). Eur. J. Solid State Inorg. Chem., 28, 209-216. A reassignment of the Al-Kα photoelectron spectra of the actinide tetrahalides UF_4, ThF_4, UCl_4 and $ThCl_4$ by relativistic local-density molecular-orbital calculations

6318 van Wezenbeek, E.M. (1992). Relativistic Effects in Atoms and in Uranium Compounds Thesis, VU, Amsterdam.

6319 van Wezenbeek, E.M., Baerends, E.J. and Snijders, J.G. (1991). Theor. Chim. Acta, 81, 139-155. Relativistic bond lengthening of UO_2^{2+} and UO_2

6320 Veltman, H. (1989). Phys. Rev. B, 40, 2810-2819. Radiative corrections to polarized Compton scattering

6321 Verhey, T.R., Das, B.P. and Perger, W.F. (1987). J. Phys. B: At. Mol. Phys., 20, 3639-3644. Multiconfiguration Dirac-Fock calculation of the forbidden $^2P_{1/2} \rightarrow {}^2P_{3/2}$ ground-state M1 transition in the boron isoelectronic sequence

6322 Veseth, L. (1987a). Phys. Rev. A, 35, 1931-1933. Ab initio calculation of the isotope shift in the fine structure of C II and N II

6323 Veseth, L. (1987b). J. Phys. B, 20, L73-L77.

Many-body calculations of g_J factors for second-row atoms

6324 Veseth, L. (1988). J. Phys. B: At. Mol. Opt. Phys., 21, 1719-1738.
Refined many-body calculations of atomic fine structure

6325 Viar, Zh.-F. and Safronova, U.I. (1990). Opt. Spektrosk., 68, 984-988; Opt. Spectrosc., 68, 574-576.
Model-potential method for calculating the atomic characteristics of Ne-like ions. Energy levels

6326 Vidolova-Angelova, E. (1992). J. Phys. B, 25, 3735-3746.
Energies and radiation lifetimes of high lying Rydberg states of lutetium

6327 Vidolova-Angelova, E.P., Angelov, D.A., Krustev, T.B. and Mincheva, S.T. (1989). J. Opt. Soc. Am. B, 6, 2295-2299.
Laser spectroscopy measurement of radiative lifetimes of highly excited thulium Rydberg states

6328 Vidolova-Angelova, E.P. and Ivanov, L.N. (1991). J. Phys. B: At. Mol. Opt. Phys., 24, 4147-4158.
Autoionizing Rydberg states of thulium. Reorientation decay due to monopole interaction

6329 Vidolova-Angelova, E.P., Ivanov, L.N., Ivanova, E.P. and Angelov, D.A. (1986). J. Phys. B: At. Mol. Phys., 19, 2053-2069.
Relativistic perturbation theory method for investigating the radiation decay of highly excited many electron atomic states. Application to the thulium atom

6330 Vijayakumar, M. and Balasubramanian, K. (1992). J. Chem. Phys., 97, 7474-7488.
Electronic states of thallium clusters and their positive ions [Tl_n,Tl_n^+(n=2-5)]

6331 Vijayakumar, M. and Gopinathan, M.S. (1991). Phys. Rev. A, 44, 2850-2859.
Theoretical Auger transition energies for atoms and ions through the relativistic and correlated local-density method

6332 Vijayakumar, M. and Gopinathan, M.S. (1992). J. Chem. Phys., 97, 6639-6643.
Pair-correlation energy and successive ionization potentials of atoms He-Zn

6333 Vijayakumar, M., Vaidehi, N. and Gopinathan, M.S. (1989). Phys. Rev. A, 40, 6834-6851.
Fermi and Coulomb correlated relativistic local-density method for atoms

6334 Villalba, V.M. (1990). J. Math. Phys., 31, 2702-2707.
Separation of variables and exact solution to the Dirac equation in curvilinear orthogonal coordinates with cylindrical symmetry

6335 Villemoes, P., van Leeuwen, R., Arnesen, A., Heijkenskjöld, F., Kastberg, A., Larsson, M.O. and Kotochigova, S.A. (1992). Phys. Rev. A, 45, 6241-6252.
Experimental and theoretical study of the hyperfine structure in the lower configurations in ^{45}Sc II

6336 Visscher, L., Aerts, P.J.C. and Visser, O. (1991a).
General contraction in four-component relativistic Hartree-Fock calculations
In "The Effects of Relativity in Atoms, Molecules, and the Solid State", Ed. S. Wilson, I.P. Grant and B.L. Gyorffy, Plenum, New York and London, pp. 197-205.

6337 Visscher, L., Aerts, P.J.C., Visser, O. and Nieuwpoort, W.C. (1991b). Int. J. Quantum Chem., S25, 131-139.
Kinetic balance in contracted basis sets for relativistic calculations

6338 Visser, O. (1992).
Relativity in Quantum Chemistry with Applications to Group IV Hydrides and EuO_6^{9-}
Thesis, Groningen.

6339 Visser, O., Aerts, P.J.C., Hegarty, D. and Nieuwpoort, W.C. (1987). Chem. Phys. Lett., 134, 34-38.
The use of Gaussian nuclear charge distributions for the calculation of relativistic electronic wavefunctions using basis set expansions

6340 Visser, O., Aerts, P.J.C. and Visscher, L. (1991).
Open shell relativistic molecular Dirac-Hartree-Fock SCF-program
In "The Effects of Relativity in Atoms, Molecules, and the Solid State", Ed. S. Wilson, I.P. Grant and B.L. Gyorffy, Plenum, New York and London, pp. 185-195.

6341 Visser, O., Visscher, L., Aerts, P.J.C. and Nieuwpoort, W.C. (1992a). Theor. Chim. Acta, 81, 405-416.
Relativistic all-electron molecular Hartree-Fock-Dirac-(Breit) calculations on CH_4, SiH_4, GeH_4, SnH_4, PbH_4

6342 Visser, O., Visscher, L., Aerts, P.J.C. and Nieuwpoort, W.C. (1992b). J. Chem. Phys., 96, 2910-2919.
Molecular open shell configuration interaction calculations using the Dirac-Coulomb Hamiltonian: The f^6-manifold of an embedded EuO_6^{9-} cluster

6343 Vitale, M., Palke, W.E. and Ford, P.C. (1992). J. Phys. Chem., 96, 8329-8336.
Origins of the double emission of the tetranuclear copper(I) cluster $Cu_4I_4(pyridine)_4$: An ab initio study

6344 Vokhmin, V.G. and Ionova, G.V. (1988). Zh. Fiz. Khim., 62, 796-797; Russ. J. Phys. Chem., 62, 392-393.
Changes in the electrostatic and spin-orbital interaction parameters of f electrons under the influence of ligands

6345 Volkov, A.B. (1989). Phys. Rev. A, 39, 4406-4413.
Finite-basis-set variational solution of the Dirac equation

6346*Volkov, D.M. (1937). Zh. Eksp. Teor. Fiz., 7, 1286-1289.
Electron in the field of plane nonpolarized electromagnetic waves as viewed upon in light of the Dirac equation

6347 Von-Eiff, D., Haddad, S. and Weigel, M.K. (1992). Phys. Rev. C, 46, 230-237.
Approach to the relativistic extended Thomas-Fermi expansion for Green's function, phase-space densities, and densities

6348 Von-Eiff, D. and Weigel, M.K. (1992). Phys. Rev. C, 46, 1797-1810.
Relativistic Thomas-Fermi calculations of finite nuclei including quantum corrections

6349 Vonsovskii, S.V., Svirskii, M.S. and Svirskaya, L.M. (1990). Teor. Mat. Fiz., 85, 211-221; Theor. Math. Phys., 85, 1159-1167.
Zitterbewegung of a band electron

6350*Voronkov, V.V. and Kolesnikov, N.N. (1960). Zh. Eksp. Teor. Fiz., 39, 189-191.
Electron levels of atoms of super-heavy elements

6351 Vosko, S.H., Chevary, J.A. and Mayer, I.L. (1991a). J. Phys. B: At. Mol. Opt. Phys., 24, L225-L231.
Predictions of stable Yb^- in the $P_{1/2}^{\circ}$ state: The importance of spin-orbit coupling

6352 Vosko, S.H., Lagowski, J.B. and Mayer, I.L. (1989). Phys. Rev. A, 39, 446-449.
Prediction of stable Sr^-, Ba^-, and Ra^- from density-functional theory

6353 Vosko, S.H., Lagowski, J.B., Mayer, I.L. and Chevary, J.A. (1991b). Phys. Rev. A, 43, 6389-6392.
Theoretical study of even- and odd-parity states in La^- and Ac^-: Evidence for the uniqueness of La^-

6354 Vrbik, J., DePasquale, M.F. and Rothstein, S.M. (1988). J. Chem. Phys., 88, 3784-3787.
Estimating the relativistic energy by diffusion quantum Monte Carlo

6355 Vrscay, E.R. and Hamidian, H. (1988). Phys. Lett. A, 130, 141-146.
Rayleigh-Schrödinger perturbation theory at large order for radial relativistic hamiltonians using hypervirial Hellmann-Feynman theorems

6356 Vukajlovic, F.R., Shirley, E.L. and Martin, R.M. (1991). Phys. Rev. B, 43, 3994-4001.
Single-body methods in 3d transition-metal atoms

6357 Wadt, W.R. (1987). J. Chem. Phys., 86, 339-346.
An ab initio study of low-lying 5f-5f excitations in PuF_6

6358 Wague, A. (1989). Z. Phys. D, 13, 123-128.
Intermediate-coupling effects on the n=2 resonant photoionization of multiply-charged helium-like ions

6359 Walch, S.P. (1987). J. Chem. Phys., 87, 6776-6778.
A theoretical study of the excited states of Ag_3

6360 Walch, S.P., Bauschlicher Jr., C.W. and Langhoff, S.R. (1986). J. Chem. Phys., 85, 5900-5907.
Theoretical studies of diatomic and triatomic systems containing the group IB atoms Cu, Ag, and Au

6361 Walker, E.C., Christiansen, P.A. and Fernandez Pacios, L. (1988). Mol. Phys., 63, 139-144.
Orbital partitioning errors in transition metal relativistic effective potentials

6362*Walker, T.E.H. and Richards, W.G. (1970). J. Chem. Phys., 52, 1311-1314.
Molecular spin-orbit coupling constants. The role of core polarization

6363 Wallace, N.M. (1992). Diss. Abstr. Int. B, 53, 2335.
Spin-orbit CI potential curves of coinage metal halides

6364 Wallace, N.M., Blaudeau, J.P. and Pitzer, R.M. (1991). Int. J. Quantum Chem., 40, 789-796.
Optimized Gaussian basis sets for use with relativistic effective (core) potentials: Li-Ar

6365*Waller, I. (1929). Z. Physik, 58, 75-94.
Die streuung kurzwelliger Strahlung durch Atome nach der Diracschen Strahlungstheorie

6366*Waller, I. (1930a). Z. Physik, 61, 837-851.
Die Streuung von Strahlung durch gebundene und freie Elektronen nach der Diracschen relativistischen Mechanik

6367*Waller, I. (1930b). Z. Physik, 62, 673-676.
Bemerkungen über die Rolle der Eigenenergie des Elektrons in der Quantentheorie der Strahlung

6368*Waller, I. (1934). Z. Physik, 88, 436-448.
Über die Rückwirkung der Strahlung bei der Streuung durch freie Elektronen

6369 Walzer, U. (1990). Phys. Stat. Sol. B, 162, 75-88.
Systematization of binary intermetallic compounds

6370 Walzer, U. (1991a). Phys. Stat. Sol. B, 163, 61-74.
Structural classification of AB-type binary compounds

6371 Walzer, U. (1991b). Phys. Stat. Sol. B, 168, 397-412.
Systematics of the binary intermetallic phases

6372 Walzer, U. (1992a). High Temp. High Pressure, 24, 23-34.
The application of a pseudopotential approach to the physics of binary intermetallic compounds

6373 Walzer, U. (1992b). High Temp. High Pressure, 24, 35-44.
Melting temperature systematics of binary intermetallic compounds

6374 Wang, H. and Carter, E.A. (1992). J. Phys. Chem., 96, 1197-1204.
Metal-metal bonding in transition-metal clusters with open d shells: Pt_3

6375 Wang, J.-Z. and Balasubramanian, K. (1989). J. Mol. Spectrosc., 138, 204-221.
Spectroscopic properties and potential energy curves for 27 electronic states of TcH

6376 Wang, J.-Z., Das, K.K. and Balasubramanian, K. (1990). Mol. Phys., 69, 147-159.
Potential energy surfaces for insertion of Zr into H_2

6377 Wang, K.-Y. and Yu, W.-L. (1991). J. Phys.: Cond. Matter, 3, 2529-2533.
Zero-field splitting of a ^6S-state ion in tetragonal symmetry

6378 Wang, L.-S., Lee, Y-T., Shirley, D.A., Balasubramanian, K. and Feng, P. (1990a). J. Chem. Phys., 93, 6310-6317.
Photoelectron spectroscopy and electronic structure of clusters of the group V elements. I. Dimers

6379 Wang, L.-S., Niu, B., Lee, Y.T., Shirley, D.A. and Balasubramanian, K. (1990b). J. Chem. Phys., 92, 899-908.
Photoelectron spectroscopy and electronic structure of heavy group IV-VI diatomics

6380 Wang, M.S. (1986). Phys. Rev. A, 34, 636-637.
Relativistic dispersion relation for x-ray atomic anomalous scattering factor

6381 Wang, M.S. and Chia, S.-H. (1988). Phys. Rev. A, 38, 5639-5641.
Model dependence of the anomalous X-ray atomic scattering factor

6382 Wang, M.S. and Chia, S.-H. (1989). Phys. Rev. A, 40, 5420-5421.
Anomalous x-ray atomic scattering factor for Zr, Nb, and Mo

6383 Wang, P. and Ermler, W.C. (1991). J. Chem. Phys., 94, 7231-7236.
Ab initio study of the electronic spectrum of a $Cs_{11}O_3$ cluster

6384 Wang, S., Fackler Jr, J.P., King, C. and Wang, J.C. (1988). J. Am. Chem. Soc., 110, 3308-3310.
Luminescent organometallic compounds with relativistic metal-metal bonds. Synthesis and characterization of $AuTl[Ph_2P(CH_2)S]_2$, a one-dimensional metal-metal bonded polymer in the solid state

6385 Wang, W.-J. (1992). Acta Phys. Sinica, 41, 726-731.
Relativistic multiconfiguration Dirac-Fock calculation of

fine-structure energy levels and transition wavelengths for N-like K XIII, Ca XIV, Sc XV and Ti XVI

6386 Wang, X.-D., Zhang, X.-G., Butler, W.H., Stocks, G.M. and Harmon, B.N. (1992). Phys. Rev. B, 46, 9352-9358.
Relativistic-multiple-scattering theory for space-filling potentials

6387 Wang, X.W., Louie, S.G. and Cohen, M.L. (1989). Phys. Rev. B, 40, 5822-5825.
Hydrogen interactions in PdH_n (1 n 4)

6388 Wang, Y. and Chou, M.Y. (1991). Phys. Rev. B, 44, 10339-10342.
Pseudopotential plane-wave calculation of the structural properties of yttrium

6389 Wang, Y.J. and Davenport, J.W. (1990). Phys. Rev. A, 41, 4690-4695.
Relativistic effects on the s -> d promotion energy and spin-polarization energy of the transition metals

6390 Wang, Z.-W., Zhu, X.-W. and Chung, K.T. (1992a). J. Phys. B: At. Mol. Opt. Phys., 25, 3915-3927.
Energy and fine structure of $1s^2nd$ and $1s^2nf$ states for the lithium isoelectronic sequence

6391 Wang, Z.-W., Zhu, X.-W. and Chung, K.T. (1992b). Phys. Rev. A, 46, 6914-6919.
Energy of $1s^2ns$(n=3,4,and 5) states for the lithium isoelectronic sequence

6392 Wasson, D.A. (1991). Nucl. Phys. A, 535, 456-478.
Methodology for relativistic Hartree calculations

6393 Watson, R.E., Davenport, J.W. and Weinert, M. (1986). Phys. Rev. B, 34, 8421-8429.
Linear augmented-Slater-orbital method for electronic-structure calculations. IV. 5d-5d alloys

6394 Watson, R.E., Davenport, J.W. and Weinert, M. (1987). Phys. Rev. B, 35, 508-518.
Linear augmented Slater-type-orbital study of Au-5d-transition-metal alloying

6395 Watson, R.E., Fernando, G.W., Weinert, M., Wang, Y.-J. and Davenport, J.W. (1991a). Phys. Rev. B, 43, 1455-1462.
Local-density approximation: Cohesion in the transition metals and s --> d promotion in the transition-metal atoms

6396 Watson, R.E., Fernando, G.W., Weinert, M., Wang, Y.-J. and Davenport, J.W. (1991b). J. Appl. Phys., 69, 5688-5690.
Cohesion and promotion energies in the transition metals: Implications of the local-density approximation

6397 Watson, R.E., Weinert, M., Davenport, J.W. and Fernando, G.W. (1989). Phys. Rev. B, 39, 10761-10769.
Energetics of transition-metal alloy formation: Ti, Zr,

and Hf alloyed with the heavier 4d and 5d elements

6398 Wehman-Ooyevaar, I.C.M., Grove, D.M., de Vaal, P., Dedieu, A. and van Koten, G. (1992). Inorg. Chem., 31, 5484-5493.
A hydrogen atom in an organoplatinum-amine system. 2. The first isolated arylplatinum(IV) hydrides: synthetic and theoretical study (SCF and CAS-SCF calculations) on model complexes with a $N-M(IV)-H/M(II)^{-}...H-N^{+}$ unit (M=Pd, Pt)

6399*Wei, S.-H. and Krakauer, H. (1985). Phys. Rev. Lett., 55, 1200-1203.
Local-density-functional calculation of the pressure-induced metallization of BaSe and BaTe

6400 Wei, S.-H. and Zunger, A. (1989). Phys. Rev. B, 39, 6279-6282.
Negative spin-orbit bowing in semiconductor alloys

6401 Wei, S.-H. and Zunger, A. (1990). Sol. Stat. Comm., 73, 327-330.
Stability of atomic and diatomic hydrogen in fcc palladium

6402 Weigel, M.K., Haddad, S. and Weber, F. (1991). J. Phys. G, 17, 619-644.
Semiclassical expansions of the nuclear relativistic Hartree-Fock theory

6403 Weigold, E. (1992). Nucl. Phys. A, 543, 259c-274c.
Electron momentum spectroscopy of atoms, molecules and solids

6404*Weinberger, P. (1982b). J. Phys. F, 12, 2171-2184.
Electronic structure of Au and Pt and of the $d^{n}s^{1}$ impurities in these metals

6405 Weinberger, P., Banhart, J., Schadler, G.H., Boring, A.M. and Riseborough, P.S. (1990). Phys. Rev. B, 41, 9444-9451.
Calculation of magnetic impurities in a nonmagnetic host: Fe in Au

6406 Weinberger, P., Boring, A.M., Albers, R.C. and Temmerman, W.M. (1988). Phys. Rev. B, 38, 5357-5362.
Theoretical study of the electronic structure of ordered and disordered Cu_3Au

6407 Weinberger, P. and Gonis, A. (1987).
Relativistic studies in actinides
In "Handbook on the Physics and Chemistry of the Actinides", Ed. A.J. Freeman and G.H. Lander, 5, Elsevier, Amsterdam

6408 Weinert, M., Davenport, J.W. and Watson, R.E. (1986). Phys. Rev. B, 34, 2971-2974.
Final-state screening of core holes in metals

6409 Weitsman, J. and Hagelstein, P.L. (1986). J. Phys. B: At. Mol. Phys., 19, L59-L64.
Self-energy in many-electron atoms

6410 Wells, J.C., Oberacker, V.E., Umar, A.S., Bottcher, C., Strayer, M.R., Wu, J.-S. and Plunien, G. (1992). Phys.

Rev. A, 45, 6296-6312.
Nonperturbative electromagnetic lepton-pair production in peripheral relativistic heavy-ion collisions

6411 Wendt, K., Ahmad, S.A., Klempt, W., Neugart, R., Otten, E.W. and Stroke, H.H. (1987). Z. Phys. D, 4, 227-241.
On the hyperfine structure and isotope shift of radium

6412 Werner, H.-J. and Buckingham, A.D. (1986). Chem. Phys. Lett., 125, 433-437.
An accurate ab initio calculation of the dipole moment function of GeH

6413 Wexler, P.A., Wigley, D.E., Koerner, J.B. and Albright, T.A. (1991). Organometallics, 10, 2319-2327.
Preparation, properties, and bonding analysis of tantalum(II) η^6-arene complexes

6414 White, G.B. (1989). Diss. Abstr. Int. B, 49, 3246.
The nonrelativistic limit of the Dirac operator: Splitting of the operator and classification of the discrete spectrum

6415 Whitfield, S.B., Tulkki, J. and Åberg, T. (1991). Phys. Rev. A, 44, R6983-R6986.
Shake-modified resonant autoionization in magnesium

6416 Wiesenfeld, L. (1988). Mol. Phys., 64, 739-745.
Effect of atomic number on parity-violating energy differences between enantiomers

6417 Wiest, R., Leininger, T., Jeung, G.-H. and Bénard, M. (1992). J. Phys. Chem., 96, 10800-10804.
Trioxorhenium as a strong acceptor group: ab initio CI investigations on the Re-L bonding in L-ReO_3 complexes ($L=C_5H_5$, CH_3, Cl, F, CF_3)

6418 Wietschorke, K.-H. (1986).
Continuum of the two-center Dirac equation
Report GSI-86-4, 219 p.

6419 Wietschorke, K.-H., Schlüter, P., Soff, G., Rumrich, K. and Greiner, W. (1987). Phys. Rev. A, 36, 377-380.
Relativistic two-center continuum

6420 Wijesundera, W.P., Grant, I.P. and Norrington, P.H. (1992a). J. Phys. B: At. Mol. Opt. Phys., 25, 2143-2163.
The $6s^2 6p$ and $6s6p^2$ resonances in electron scattering by neutral Hg

6421 Wijesundera, W.P., Grant, I.P. and Norrington, P.H. (1992b). J. Phys. B: At. Mol. Opt. Phys., 25, 2165-2174.
The $6s^2 6p^3$ resonances in electron scattering by neutral Pb

6422 Wijesundera, W.P., Grant, I.P., Norrington, P.H. and Parpia, F.A. (1991a). J. Phys. B: At. Mol. Opt. Phys., 24, 1017-1036.
Electron scattering by Se XXV (neon-like selenium) using the Dirac R-matrix method

6423 Wijesundera, W.P., Parpia, F.A., Grant, I.P. and Norrington, P.H. (1991b). J. Phys. B: At. Mol. Opt. Phys., 24, 1803-1816.
Electron scattering by Kr XXIX (oxygen-like krypton) using the Dirac R-matrix method

6424 Wilkin, A. and Salahub, D.R. (1992). Int. J. Quantum Chem., 42, 977-995.
Relativistic corrections in the LCGTO-local-spin-density method. I. Atomic bases for transition metals

6425 Wills, J.M. and Eriksson, O. (1992). Phys. Rev. B, 45, 13879-13890.
Crystal-structure stabilities and electronic structure for the light actinides Th, Pa, and U

6426 Wills, J.M., Eriksson, O. and Boring, A.M. (1991). Phys. Rev. Lett., 67, 2215-2218.
Theoretical studies of the high pressure phases in cerium

6427 Wills, J.M., Eriksson, O., Söderlind, P. and Boring, A.M. (1992). Phys. Rev. Lett., 68, 2802-2805.
Trends of the elastic constants of cubic transition metals

6428 Wilson, M. (1987). Phys. Lett. A, 125, 205-207.
Pseudo-relativistic screening ratios for Re

6429 Wilson, M. (1990).
Calculation of radiative decay rates and branching ratios in Hg^+
In "Atomic Spectra and Oscillator Strengths for Astrophysics and Fusion Research", Ed. J.E. Hansen, North-Holland, Amsterdam, pp. 144-145.

6430 Wilson, S. (1987).
Basis sets
In "Ab Initio Methods in Quantum Chemistry -I", Ed. K.P. Lawley, John Wiley & Sons Ltd, pp. 439-500.

6431 Wilson, S. (1988).
Relativistic molecular structure calculations
In "Methods in Computational Chemistry", Ed. S. Wilson, 2, pp. 73-108, Plenum, New York and London.

6432 Wilson, S. (1990). Comp. Phys. Comm., 58, 71-81.
Universal basis sets and Cholesky decomposition of the two-electron integral matrix

6433 Wilson, S. (1991).
Relativistic many-body perturbation theory of atomic and molecular electronic structure
In "The Effects of Relativity in Atoms, Molecules, and the Solid State", Ed. S. Wilson, I.P. Grant and B.L. Gyorffy, Plenum, New York and London, pp. 217-251.

6434 Wilson, S. (1992a). J. Mol. Struct.(Theochem), 261, 287-311.
On the diagrammatic analysis of electron correlation

effects in atoms and molecules

6435 Wilson, S. (1992b).
The perturbation theory of electron correlation. III. Relativistic many-body perturbation theory
In "Methods in Computational Molecular Physics", Ed. S. Wilson and G.H.F. Diercksen, Plenum, New York.

6436 Wilson, S., Grant, I.P. and Gyorffy, B.L. (Ed.) (1991).
The Effects of Relativity in Atoms, Molecules, and the Solid State
Plenum, New York and London

6437 Wilson, W.L., Rudolph, R.W., Lohr, L.L., Taylor, R.C. and Pyykkö, P. (1986). Inorg. Chem., 25, 1535-1541.
Multinuclear NMR characterization of anionic clusters of the main-group elements Ge, Sn, Sb, Tl, Pb, and Bi in nonaqueous solution

6438 Winkler, W. and Vetter, R. (1990a). Chem. Phys., 144, 7-28.
Electronic reorganization mechanisms in strongly covalent bonds. I. A Mössbauer isomer shift investigation

6439 Winkler, W. and Vetter, R. (1990b). Chem. Phys., 144, 29-45.
Electronic reorganization mechanisms in strongly covalent bonds. II. Verified by nuclear quadrupole interactions and NMR spin-spin couplings

6440 Winkler, W., Vetter, R. and Hartmann, E. (1987). Chem. Phys., 114, 347-358.
Mössbauer isomer shift calibration of ^{119}Sn using pseudopotential and Xα SW calculations

6441 Wisner, J.M., Bartczak, T.J., Ibers, J.A., Low, J.J. and Goddard III, W.A. (1986). J. Am. Chem. Soc., 108, 347-348.
Geometrical characteristics from experiment and theory of isostructural complexes involving palladium- and platinum-methyl bonds

6442*Wittel, K. (1972). Chem. Phys. Lett., 15, 555-557.
Spin-orbit coupling in I_2^+

6443 Wolniewicz, L. and Poll, J.D. (1986). Mol. Phys., 59, 953-964.
On the higher vibration-rotational levels of HD^+ and H_2^+; Erratum 66, 701.

6444 Wong, C.-Y. (1986). Phys. Rev. Lett., 56, 1047-1050.
Anomalous positron peak in heavy-ion collisions

6445 Wong, C.-Y. (1987). Phys. Rev. Lett., 58, 426.
Reply to Lipkin

6446 Wong, M.K.F. (1986). Phys. Rev. A, 34, 1559-1560.
Infeld-Hull factorization and the simplified solution of the Dirac-Coulomb equation

6447 Wong, M.K.F. (1987). Phys. Rev. A, 36, 2962-2963.

Value of the Dirac-Coulomb wave function at the origin

6448 Wong, M.K.F. (1990). J. Math. Phys., 31, 1677-1680.
Exact solution of the n-dimensional Dirac-Coulomb equation

6449*Wong, M.K.F. and Yeh, E.H.Y. (1985). J. Math. Phys., 26, 1701-1710.
The Dirac Coulomb Green's function and its application to relativistic Rayleigh scattering

6450 Wood, C.P. and Pyper, N.C. (1986). Phil. Trans. R. Soc. Lond. A, 320, 71-105.
Relativistic ab initio calculations of interaction energies: Formulation and application to ionic solids

6451 Wu, C.-M., Chi, H.-C. and Huang, K.-N. (1992). Phys. Rev. A, 46, 1680-1683.
Double-excitation effects on the angular distribution of photoelectrons from strontium

6452 Wu, D.-D. (1992). Phys. Energ. Fortis Phys. Nucl., 16, 183-186.
Ambiguity of perturbative Dirac theory

6453 Wu, G.S., Ma, H. and Lin, S.H. (1992). J. Chem. Phys., 96, 4518-4525.
A modified parametrization scheme for the complete neglect of differential overlap method with relativistic correction of orbital exponents and energy levels with particular applications to rare earth compounds

6454 Wu, J., Bai, J.J. and Cusson, R.Y. (1990). Z. Phys. A, 335, 125-130.
Numerical method for the time evolution of the Dirac equation

6455 Wu, J.-S., Stahlhofen, A., Biedenharn, L.C. and Iachello, F. (1987). J. Phys. A, 20, 4637-4652.
A group theoretical calculation of the S matrix for the Dirac-Coulomb problem

6456 Wu, S.C., Li, H., Li, Y.S., Tian, D. and Jona, F. (1991). Phys. Rev. B, 44, 13308-13314.
Relativistic effects in the electron band structure of Cu{111}

6457 Wu, S.C., Sokolov, J., Lok, C.K.C., Quinn, J., Li, Y.S., Tian, D. and Jona, F. (1989). Phys. Rev. B, 39, 12891-12894.
Relativistic effects in the electronic structure of Cu{001}

6458 Wu, S.C., Wang, Z.Q., Sondericker, D., Sokolov, J., Garrett, R.F. and Jona, F. (1986). Phys. Rev. B, 34, 8418-8420.
Splitting of valence levels in Ag-Au alloys

6459*Wu, T.-Y. and Goudsmit, S. (1933). Phys. Rev., 43, 496.
Low states of the heaviest elements

6460*Wu, Z., Walker, T.G. and Happer, W. (1985). Phys. Rev. Lett., 54, 1921-1924.
Spin-rotation interaction of noble-gas alkali-metal atom pairs

6461 Wulfsberg, G. (1987).
Principles of Descriptive Inorganic Chemistry
Brooks/Cole, Monterey; 2nd Ed., Freeman (1991), 461 p.

6462*Wüst, R. (1973). Math. Z., 131, 339-349.
A convergence theorem for selfadjoint operators applicable to Dirac operators with cutoff potentials

6463*Wüst, R. (1975). Math. Z., 141, 93-98.
Distinguished self-adjoint extensions of Dirac operators constructed by means of cut-off potentials

6464*Wüst, R. (1977). Math. Z., 152, 259-271.
Dirac operations with strongly singular potentials

6465 Wycech, S., Poth, H. and Rook, J.R. (1990). Z. Phys. A, 335, 355-359.
On the fine-structure splitting in p atoms

6466 Xia, J.-B. and Chang, Y.-C. (1992). Phys. Rev. B, 46, 10894-10902.
Magnetic exchange in rare-earth-metal superlattices

6467 Xiao, L.-X., Darewych, J.W. and Horbatsch, M. (1990). Phys. Rev. A, 42, 69-77.
Relativistic wave equations for bound states of two scalar particles from scalar quantum electrodynamics

6468 Xu, J.H., Jarlborg, T. and Freeman, A.J. (1989). Phys. Rev. B, 40, 7939-7947.
Self-consistent band structure of the rutile dioxides NbO_2, RuO_2, and IrO_2

6469 Xu, X.Y., Tang, J.Z., Huang, W., Zhao, W.Z. and Chen, D.Y. (1990). J. Phys. B, 23, 3315-3323.
Rydberg energy levels and their structure in the ruthenium atom

6470 Yamagami, H. and Hasegawa, A. (1990). J. Phys. Soc. Japan, 59, 2426-2442.
Self-consistent symmetrized relativistic augmented plane wave method: Application to -U

6471 Yamagami, H. and Hasegawa, A. (1991). J. Phys. Soc. Japan, 60, 1011-1021.
The Fermi surface of itinerant 4f electrons in CeNi

6472 Yamamoto, M., Mabuchi, M. and Naito, S. (1992). J. Phys. Chem., 96, 10784-10788.
Relativistic self-consistent calculation for hafnium atoms and ions based on a relativistic local density functional theory

6473 Yamasaki, H. (1991). Eur. J. Phys., 12, 218-223.
Geometrical interpretation for relativistic composition

of velocities and the Thomas precession

6474 Yang, D.-S., Bancroft, G.M., Bozek, J.D., Puddephatt, R.J. and Tse, J.S. (1989). Inorg. Chem., 28, 1-2.
Assignment of the valence molecular orbitals of cis-[PtR_2L_2] complexes: UV photoelectron spectra and SCF-MS-X calculations

6475 Yang, D.-S., Bancroft, G.M., Dignard-Bailey, L., Puddephatt, R.J. and Tse, J.S. (1990a). Inorg. Chem., 29, 2487-2495.
Electronic structure of cis-dimethylplatinum(II) complexes from UV photoelectron spectra and SCF-MS-X calculations

6476 Yang, D.-S., Bancroft, G.M. and Puddephatt, R.J. (1990b). Inorg. Chem., 29, 2118-2122.
Electronic structure of (-diketonato)Trimethylplatinum(IV) complexes from UV photoelectron spectra and SCF-MS-X calculations

6477 Yang, D.-S., Bancroft, G.M., Puddephatt, R.J., Tan, K.H., Cutler, J.N. and Bozek, J.D. (1990c). Inorg. Chem., 29, 4956-4960.
Assignment of the valence molecular orbitals of $CpPtMe_3$ and $Me_2Pt(COD)$ using variable-energy photoelectron spectra

6478 Yang, D.-S., Bancroft, G.M., Puddephatt, R.J. and Tse, J.S. (1990d). Inorg. Chem., 29, 2496-2501.
Electronic structure of square-planar cis-bis(trifluoromethyl)platinum(II) complexes from UV photoelectron spectra and SCF-MS-Xα calculations

6479 Yang, H.-Y., Qiu, Y.-H., Li, S.-C., Sun, Y.-S. and Yang, Y.-L. (1992). J. Phys. B: At. Mol. Opt. Phys., 25, 791-798.
Dielectronic recombination for F-like and Ne-like germanium ions

6480 Yang, L., Heinemann, D. and Kolb, D. (1991). Chem. Phys. Lett., 178, 213-215.
An accurate solution of the two-centre Dirac equation for H_2^+ by the finite-element method

6481 Yang, L., Heinemann, D. and Kolb, D. (1992). Chem. Phys. Lett., 192, 499-502.
Relativistic self-consistent calculations for small diatomic molecules by the finite element method

6482 Yang, L. and Li, J.M. (1992). J. Phys. B: At. Mol. Opt. Phys., 25, 649-655.
Relativistic configuration interaction studies of fine structure splittings and transition wavelengths of K-like ions

6483 Yarkony, D.R. (1986). J. Chem. Phys., 84, 2075-2078.
On the use of the Breit-Pauli approximation for evaluating line strengths for spin forbidden transitions. II. The symbolic matrix element method

6484 Yarkony, D.R. (1989). J. Chem. Phys., 90, 7164-7175.

On the quenching of helium 2 ^3S: Potential energy curves for, and nonadiabatic, relativistic, and radiative couplings between, the a $^3\Sigma_u^+$, A $^1\Sigma_u^+$, b $^3\Pi_g$, B $^1\Pi$, c $^3\Sigma_g^+$, and C $^1\Sigma_g^+$ states of He_2

6485 Yarkony, D.R. (1990). J. Chem. Phys., 92, 320-323.
Spin-orbit effects in the decomposition reaction $N_3H(X\ ^1A') \rightarrow N_2(X\ ^1\Sigma_g^+) + NH(X\ ^3\Sigma^-, a\ ^1\Delta)$

6486 Yasui, M. (1988). Physica B+C, 149, 139-142.
Orbital and spin-orbit magnetic susceptibilities in bcc. transition metal: Vanadium, niobium, and tungsten

6487 Yasui, M., Takahashi, T. and Callaway, J. (1991). J. Phys.: Cond. Matter, 3, 2679-2692.
Orbital and spin-orbit magnetic susceptibilities in face-centred cubic transition metals and noble metals

6488 Yelkhovsky, A.S. (1987). J. Phys. B: At. Mol. Phys., 20, L7-L9.
On the quadrupole moment of the hydrogen atom

6489 Yennie, D.R. (1989a).
Introduction to recoil effects in bound state problems
In "Relativistic, Quantum Electrodynamic and Weak Interaction Effects in Atoms", Ed. W. Johnson, P. Mohr and J. Sucher, AIP Conf. Proc., 189, pp. 93-107.

6490 Yennie, D.R. (1989b).
A perturbative approach to many electron energy levels
In "Relativistic, Quantum Electrodynamic and Weak Interaction Effects in Atoms", Ed. W. Johnson, P. Mohr and J. Sucher, AIP Conf. Proc., 189, pp. 393-407.

6491 Yilmazer, A.U. (1986). Fortschr. Phys., 34, 345-359.
Relativistic two-fermion problem with the most general electric and magnetic potentials

6492 Yin, R.Y., Gabriel, O.V. and Pratt, R.H. (1987). Phys. Rev. A, 36, 1207-1219.
High-energy atomic photoelectric effect and bremsstrahlung

6493 Yin, R.Y. and Pratt, R.H. (1987a). Phys. Rev. A, 35, 1149-1153.
Survey of relativistic Cooper minima

6494 Yin, R.Y. and Pratt, R.H. (1987b). Phys. Rev. A, 35, 1154-1158.
Ionic and point-Coulomb relativistic Cooper minima

6495 Yonei, K. (1989). Phys. Lett. A, 138, 193-196.
A relativistic Thomas-Fermi model of compressed atoms

6496 Yoshida, T. (1987). Am. J. Phys., 55, 1133-1136.
Considerations on the precessing orbit via a rotating Laplace-Runge-Lenz vector

6497 Yoshida, T. (1988). Phys. Rev. A, 38, 19-25.
Rotating Laplace-Runge-Lenz vector leading to two relativistic Kepler's equations

6498 Yoshida, T. (1989). Nuovo Cim., 103B, 127-148.
Applications of the rotating Laplace-Runge-Lenz vector to quantum eigenvalue problems

6499 Young, L., Childs, W.J., Dinneen, T., Kurtz, C., Berry, H.G., Engström, L. and Cheng, K.T. (1988). Phys. Rev. A, 37, 4213-4219.
Hyperfine structure of Sc II: Experiment and theory

6500 Yu, W.-L. (1989). Phys. Rev. B, 39, 622-632.
Cubic zero-field splitting of a ^6S-state ion

6501 Yu, W.-L. (1990). Phys. Rev. B, 41, 9415-9423.
Rank-four spin Hamiltonian parameters of a $3d^5$ ion

6502 Yu, W.-L. and Rudowicz, C. (1992). Phys. Rev. B, 45, 9736-9748.
Comprehensive approach to the zero-field splitting of ^6S-state ions: Mn^{2+} and Fe^{3+} in fluoroperovskites

6503 Yuan, J.-M. and Zhang, Z.-J. (1990). Phys. Rev. A, 42, 5363-5373.
Alternative static-exchange formalism: Low-energy electron scattering with heavy alkaline-earth atoms

6504 Yuan, J.-M. and Zhang, Z.-J. (1991a). J. Phys. B: At. Mol. Opt. Phys., 24, 275-285.
Application of the quasirelativistic approach to low-energy electron-atom scattering: Xe

6505 Yuan, J.-M. and Zhang, Z.-J. (1991b). Phys. Lett. A, 160, 81-84.
Relativistic effects and spin polarization on the low-energy shape resonances of electron scattering with magnesium, zinc and cadmium atoms

6506 Yuan, J.-M. and Zhang, Z.-J. (1992a). Phys. Lett. A, 168, 291-295.
Spin polarization of electrons elastically scattered from barium atoms

6507 Yuan, J.-M. and Zhang, Z.-J. (1992b). Chin. Phys. Lett., 9, 359-362.
Spin polarisation of elastically scattered electrons from Hg atom

6508 Yudin, G.L. (1991a). Phys. Scr., T37, 101-108.
The unitary semiclassical theory of Coulomb excitation of atoms

6509 Yudin, G.L. (1991b). Phys. Rev. A, 44, 7355-7360.
Theory of the stopping power of fast multicharged ions

6510 Zamick, L., Zheng, D.C. and Müther, H. (1992). Phys. Rev. C, 45, 2763-2770.
Spin-orbit splitting for single-particle and single-hole energies: interplay of relativity and core polarization

6511 Zangrande, G., Granozzi, G., Casarin, M., Daudey, J.-P.

and Minniti, D. (1986). Inorg. Chem., 25, 2872-2877.
UV PE spectra and pseudopotential ab initio calculations on trans-Pt$(PEt_3)_2X_2$ complexes (X=Cl, Br, I)

6512 Zapryagaev, S.A., Manakov, N.L. and Mogilev, A.V. (1986). Izv. Akad. Nauk. SSSR, Ser. Fiz., 50, 1367-1371.
Relation between the quantum defects, phases and the Green's function in the relativistic quantum defect method (in Russian)

6513 Zapryagaev, S.A., Manakov, N.L. and Mogilev, A.V. (1988). Opt. Spektrosk., 64, 702-704; Opt. Spectrosc., 64, 422-423.
Polarizability of multicharged ions in the quantum-defect relativistic method

6514 Zapryagaev, S.A. and Nefedov, Yu.A. (1991). Opt. Spectrosc., 71, 241-246; Opt. Spektrosk., 71, 417-424.
Photoionization of multiply charged hydrogenlike ions

6515 Zecca, A. (1988). Phys. Lett. B, 213, 210-216.
Fine structure of equal mass q-$\bar{q}$ systems from a relativistic potential

6516 Zel'dovich, Ya.B., Buchachenko, A.L. and Frankevich, E.L. (1988). Usp. Fiz. Nauk, 155, 3-45; Sov. Phys. Usp., 31, 385-408.
Magnetic-spin effects in chemistry and molecular physics

6517 Zeyada, H.M. (1992). Phys. Scr., 46, 538-541.
Influence of spin-orbit coupling and environmental conditions on diamagnetic susceptibility of thin copper film

6518 Zhang, H.-L. and Sampson, D.H. (1989). At. Data Nucl. Data Tables, 43, 1-69.
Relativistic distorted wave collision strengths for excitation to the 88 n=3 and n=4 levels in all 71 neon-like ions with $22 \leq Z \leq 92$

6519 Zhang, H.-L. and Sampson, D.H. (1990). Phys. Rev. A, 42, 5378-5383.
Rapid relativistic distorted-wave approach for calculating cross sections for ionization of highly charged ions

6520 Zhang, H.-L. and Sampson, D.H. (1992). At. Data Nucl. Data Tables, 52, 143-173.
Relativistic distorted-wave collision strengths and oscillator strengths for the 45 $\Delta n=0$ transitions with n=2 in the 85 Be-like ions with $8 \leq Z \leq 92$

6521 Zhang, H.-L., Sampson, D.H. and Clark, R.E.H. (1990a). Phys. Rev. A, 41, 198-206.
Relativistic cross sections for excitation of highly charged ions to specific magnetic sublevels by an electron beam

6522 Zhang, H.-L., Sampson, D.H. and Fontes, C.J. (1990b). At. Data Nucl. Data Tables, 44, 31-77.
Relativistic distorted-wave collision strengths and

oscillator strengths for the 85 Li-like ions with $8 \leq Z \leq 92$

6523 Zhang, H.-L., Sampson, D.H. and Fontes, C.J. (1990c). At. Data Nucl. Data Tables, 44, 273-304.
Relativistic distorted wave collision strengths and oscillator strengths for the 33 Cu-like ions with $60 \leq Z \leq 92$

6524 Zhang, H.-L., Sampson, D.H. and Fontes, C.J. (1991). At. Data Nucl. Data Tables, 48, 91-163.
Relativistic distorted-wave collision strengths and oscillator strengths for the 33 Ni-like ions with $60 \leq Z \leq 92$

6525 Zhang, H.-L., Sampson, D.H. and Mohanty, A.K. (1989). Phys. Rev. A, 40, 616-632.
Fully relativistic and quasirelativistic distorted-wave methods for calculating collision strengths for highly charged ions

6526 Zhang, H.-X. and Balasubramanian, K. (1992a). J. Phys. Chem., 96, 6981-6985.
Potential energy surfaces for insertion of Ni^+, Pd^+, and Pt^+ into H_2

6527 Zhang, H.-X. and Balasubramanian, K, (1992b). J. Chem. Phys., 97, 3437-3444.
Electronic structure of the group V tetramers (P_4-Bi_4)

6528 Zhang, J.-K. and Onley, D.S. (1991). Phys. Rev. C, 44, 1915-1919.
Retardation effect in finite nuclei in relativistic mean field theory

6529 Zhang, S.B., Cohen, M.L. and Phillips, J.C. (1987). Phys. Rev. B, 36, 5861-5867.
Relativistic screened orbital radii

6530 Zhang, T. and Koniuk, R. (1992). Can. J. Phys., 70, 683-686.
Muonium hyperfine and fine splittings in a new bound-state formalism

6531 Zhang, T., Xiao, L.-X. and Koniuk, R. (1992). Can. J. Phys., 70, 670-682.
Bound state QED effects from the Schrödinger equation

6532 Zhang, Y., Zhu, Q.-R. and Pan, S.-F. (1991). Chi. Phys., 11, 910-913.
Radiative transition probabilities for Na-like Fe

6533 Zhang, Y., Zhu, Q.-R. and Pan, S.-F. (1992a). At. Data Nucl. Data Tables, 52, 177-226.
Energy levels, wavelengths, and radiative transition probabilities for the Na-like ions with $38 \leq Z \leq 45$

6534 Zhang, Z.C. and Pyper, N.C. (1988). Mol. Phys., 64, 963-981.
Exact relativistic analogues of the non-relativistic

hyperfine structure operators. II. A numerical study

6535 Zhao, Y.-F. and Pan, S.-F. (1987). J. Natural Sci. Jilin Univ., 3, 59-62.
Relativistic HFS calculations for the first ionization potentials of potassium, copper, rubidium, silver, and cesium (in Chinese)

6536 Zhao, Y.-F. and Pan, S.-F. (1988a). J. At. Mol. Phys., 5, 625-632.
Relativistic energy levels and oscillator strengths in neon-like krypton ion (in Chinese)

6537 Zhao, Y.-F. and Pan, S.-F. (1988b). Acta Opt. Sin., 8, 319-323.
Relativistic multiconfiguration Dirac-Fock calculation of energy levels and oscillator strengths in the iron (Fe XVII) ion (in Chinese)

6538 Zhao, Y.-F. and Wang, Z.-W. (1988). J. At. Mol. Phys., 5, 712-718.
Relativistic multiconfiguration oscillator strengths for $6p^2$-6p7s transition of Pb atom (in Chinese)

6539 Zhao, Y.-F., Wang, Z.-W. and Pan, S.-F. (1986). Acta Opt. Sin., 6, 753-758.
Relativistic calculations of single-electron energy eigenvalues and x-ray energies of actinides actinium, thorium, protactinium and uranium (in Chinese)

6540 Zhao, Y.-F., Wang, J.-X. and Pan, S.-F. (1988). J. At. Mol. Phys., 5, 633-640.
The transition state calculations of the x-ray energies for the iron group atoms (in Chinese)

6541 Zhdanov, V.I. (1991). J. Phys. A, 24, 5011-5027.
Convergence of iteration method in the relativistic two-body problem, taking into account the retardation of interactions

6542 Zhong, X.-F., Xu, Y.-N. and Ching, W.Y. (1990). Phys. Rev. B, 41, 10545-10552.
Orthogonalized linear combinations of atomic orbitals. IV. Inclusion of relativistic corrections

6543 Zhou, B., Kissel, L. and Pratt, R.H. (1992). Phys. Rev. A, 45, 6906-6909.
Many-electron correlation effects on anomalous scattering factors

6544 Zhou, B. and Pratt, R.H. (1992). Phys. Rev. A, 45, 6318-6322.
Mechanism for spurious structures in photoionization calculations in the independent-particle approximation

6545 Zhou, B., Pratt, R.H., Roy, S.C, and Kissel, L. (1990). Phys. Scr., 41, 495-498.
Calculation of anomalous scattering for ions and atoms

6546 Zhou, F. and Rosenberg, L. (1992). Phys. Rev. A, 45,

7818-7830.
Relativistic scattering in a multimode external field

6547 Zhou, H.L. and Norcross, D.W. (1989). Phys. Rev. A, 40, 5048-5051.
Improved calculation of the quadratic Stark effect in the 6 $P_{3/2}$ state of Cs

6548 Zhu, M.J., Bylander, D.M. and Kleinman, L. (1987). Phys. Rev. B, 36, 3182-3185.
Ab initio calculations of the cohesive energy of Mo and W and heat of formation of $MoSi_2$ and WSi_2

6549 Zhu, Q.-R. and Pan. S.-F. (1990). Chi. J. Comp. Phys., 7, 403-406.
MCDF package and relativistic calculations of electric dipole radiative transition probability (in Chinese)

6550 Zhu, Q.-R., Pan, S.-P. and Yang, Y.-F. (1991). Phys. Scr., 43, 170-173.
Calculations of oscillator strengths for n=3 to 2 and 3 transitions in F-like selenium

6551 Zhu, Q.-R., Pan, S.-F. and Zeng, T.-H. (1992). At. Data Nucl. Data Tables, 52, 109-141.
Energy levels, wavelengths, and radiative transition probabilities for the Li-like ions with 20 Z 25

6552 Zhu, X.-Q., Gourishankar, R., Khanna, F.C., Leung, G.Y. and Mobed, N. (1992). Phys. Rev. C, 45, 959-973.
One-time relativistic equation and its application to nucleon-nucleon scattering

6553 Zhuikov, B.L., Glebov, V.A., Nefedov, V.S. and Zvara, I. (1990). J. Radioanal. Nucl. Chem., 143, 103-111.
Relativistic effects in kurchatovium chemistry

6554*Ziegler, T. (1985c). Organometallics, 4, 675-680.
A theoretical study on the proton affinity of some mononuclear d^8 and d^{10} metal carbonyls by the Hartree-Fock-Slater transition-state method

6555 Ziegler, T. (1987).
On the relation between relativity and periodic trends within a triad of transition metals
In "Understanding Molecular Properties", Ed. J. Avery, et al., Reidel, Dordrecht, pp. 521-532.

6556 Ziegler, T., Cheng, W., Baerends, E.J. and Ravenek, W. (1988a). Inorg. Chem., 27, 3458-3464.
Theoretical study on the difference in the relative strengths of metal-hydrogen and metal-methyl bonds in complexes of early transition metals and complexes of middle to late transition metals

6557 Ziegler, T., Nagle, J.K., Snijders, J.G. and Baerends, E.J. (1989a). J. Am. Chem. Soc., 111, 5631-5635.
Theoretical study of the electronic structures and absorption spectra of $Pt(CN)_4^{2-}$ and $Tl_2Pt(CN)_4$ based on density functional theory including relativistic effects

6558 Ziegler, T., Snijders, J.G. and Baerends, E.J. (1989b). Relativistic effects on compounds containing heavy elements In "The Challenge of d and f Electrons", Ed. D.R. Salahub and M.C. Zerner, ACS, pp. 322-338.

6559 Ziegler, T., Tschinke, V., Baerends, E.J., Snijders, J.G. and Ravenek, W. (1989c). J. Phys. Chem., 93, 3050-3056. Calculation of bond energies in compounds of heavy elements by a quasi-relativistic approach

6560 Ziegler, T., Tschinke, V. and Becke, A. (1987a). Polyhedron, 6, 685-693. A theoretical study on the strength of multiple metal-metal bonds in binuclear complexes and transition-metal dimers by a non-local density functional method

6561 Ziegler, T., Tschinke, V. and Becke, A. (1987b). J. Am. Chem. Soc., 109, 1351-1358. A theoretical study on the relative strengths of the metal-hydrogen and metal-methyl bonds in complexes of middle to late transition metals

6562 Ziegler, T., Tschinke, V., Fan, L. and Becke, A.D. (1989d). J. Am. Chem. Soc., 111, 9177-9185. Theoretical study on the electronic and molecular structures of $(C_5H_5)M(L)$ (M=Rh, Ir; L=CO, PH_3) and $M(CO)_4$ (M=Ru, Os) and their ability to activate the C-H bond in methane

6563 Ziegler, T., Tschinke, V. and Ursenbach, C. (1987c). J. Am. Chem. Soc., 109, 4825-4837. Thermal stability and kinetic lability of the metal-carbonyl bond. A theoretical study on $M(CO)_6$ (M=Cr, Mo, W), $M(CO)_5$ (M= Fe, Ru, Os), and $M(CO)_4$ (M=Ni, Pd, Pt)

6564 Ziegler, T., Tschinke, V., Versluis, L., Baerends, E.J. and Ravenek, W. (1988b). Polyhedron, 7, 1625-1637. A theoretical study of metal-ligand bond strengths (M-L: L=OH, OCH_3, SH, NH_2, PH_2, CH_3, SiH_3, CN and H) in the early transition metal systems Cl_3ML (M=Ti, Zr and Hf) and late transition metal systems $LCo(CO)_4$

6565 Zilitis, V.A. (1987). Opt. Spektrosk., 63, 1182-1184; Opt. Spectrosc., 63, 698-699. Values of the energy levels for ions of the potassium isoelectronic series calculated by the Dirac-Fock method

6566 Zilitis, V.A. (1988). Opt. Spektrosk., 65, 999-1002; Opt. Spectrosc., 65, 588-590. Photoionization cross sections for ions of the potassium isoelectronic series calculated by the Dirac-Fock method

6567 Zilitis, V.A. (1989). Opt. Spektrosk., 67, 1009-1012; Opt. Spectrosc., 67, 595-597. Dirac-Fock calculation of oscillator strengths and level lifetimes for ions in the isoelectronic sequence of potassium

6568 Zilitis, V.A. (1990). Opt. Spektrosk., 69, 499-502; Opt. Spectrosc., 69, 298-300.
Energy calculations based on the Dirac-Fock method for Rydberg s and p levels of potassiumlike ions

6569 Zilitis, V.A. (1991). Opt. Spektrosk., 70, 1182-1184; Opt. Spectrosc., 70, 692-694.
Energies of Rydberg d and f levels of potassiumlike ions determined by the Dirac-Fock method

6570 Zilitis, V.A. (1992). Opt. Spectrosc., 72, 427-429; Opt. Spektrosk., 72, 801-804.
Theoretical evaluation of the oscillator strengths for the principal series of potassiumlike ions by the Dirac-Fock method

6571 Zimmerer, P., Grün, N and Scheid, W. (1991). J. Phys. B: At. Mol. Opt. Phys., 24, 2633-2640.
Scaling of relativistic Auger rates with Z for ions with two electrons

6572 Zimmermann, D. (1987). Z. Phys. D, 6, 21-24.
Calculation of hyperfine splitting constants for s-electrons: A clarification

6573 Zou, Y. and Li, J.-M. (1990). Chi. Phys. Lett., 7, 216-218.
Theoretical XUV spectra from carbon-like cobalt ions by relativistic configuration interaction calculation

6574 Zuloaga, F. and Arratia-Pérez, R. (1986). J. Phys. Chem., 90, 4491-4498.
Multiple scattered wave study of the relativistic and nonrelativistic electronic structure and bonding for cis-diamminedichloroplatinum(II)

6575 Zuo, T., McEachran, R.P. and Stauffer, A.D. (1991). J. Phys. B: At. Mol. Opt. Phys., 24, 2853-2870.
Relativistic distorted-wave calculation of electron impact excitation of xenon

6576 Zuo, T., McEachran, R.P. and Stauffer, A.D. (1992). J. Phys. B: At. Mol. Opt. Phys., 25, 3393-3403.
Relativistic distorted-wave calculation of electron excitation of heavy noble gases

6577 Zygelman, B. and Dalgarno, A. (1987). Phys. Rev. A, 35, 4085-4100.
Impact excitation of the n=2 fine-structure levels in hydrogenlike ions by protons and electrons